Optical Switching/Networking and Computing for Multimedia Systems

OPTICAL ENGINEERING

Founding Editor

Brian J. Thompson

Distinguished University Professor
Professor of Optics
Provost Emeritus
University of Rochester
Rochester, New York

Editorial Board

Additional Volumes in Preparation

Optical Switching/Networking and Computing for Multimedia Systems

edited by

Mohsen Guizani
University of West Florida
Pensacola, Florida

Abdella Battou
firstwave Intelligent Optical Networks
Greenbelt, Maryland

MARCEL DEKKER, INC.   NEW YORK • BASEL

ISBN: 0-8247-0707-9

This book is printed on acid-free paper.

Headquarters
Marcel Dekker, Inc.
270 Madison Avenue, New York, NY 10016
tel: 212-696-9000; fax: 212-685-4540

Eastern Hemisphere Distribution
Marcel Dekker AG
Hutgasse 4, Postfach 812, CH-4001 Basel, Switzerland
tel: 41-61-261-8482; fax: 41-61-261-8896

World Wide Web
http://www.dekker.com

The publisher offers discounts on this book when ordered in bulk quantities. For more information, write to Special Sales/Professional Marketing at the headquarters address above.

Current printing (last digit):
10 9 8 7 6 5 4 3 2 1

PRINTED IN THE UNITED STATES OF AMERICA

Foreword

We have witnessed the rapid growth in the popularity of the Internet technology over the past few years. The current network technologies provide ubiquitous access to a vast amount of information on a huge variety of topics, but limited bandwidth has constrained the impact and growth of broadband multimedia applications. To allow new broadband services to reach their potential will require an exponential increase in bandwidth availability to all users. These demanding high data rate applications will require the development of a robust and stable network infrastructure.

At the current time, optical fiber is the medium of choice both to accommodate this bandwidth explosion and to meet the increasing demands for quality-of-service (QoS) required by the emerging multimedia applications. Dense wavelength division multiplexing (DWDM) has provided the technological solution to increased network transport capacity. The next step in this evolutionary process is the implementation of photonic packet switching to offer high-speed data rate/format transparency and configurability. This transformation to a packet-based environment is required to allow all the different forms of data and multimedia content that will need to be supported in tomorrow's networks. While the debate about whether to have an optoelectronic or an all-optical (end-to-end) network rages on, the cost per bit continues to decline, and network operators continue to upgrade and improve their fiber-based networks to try to meet the needs of tomorrow's broadband market.

It is a pleasure to introduce this book, which discusses some of the fundamental concepts necessary for broadband multimedia applications that will need to be supported by the optical network infrastructure. This includes issues relating to optical networking capabilities for the next generation Internet, the design and performance of a lightwave data storage network using computer analysis and simulation, analysis of multimedia access protocols for shared medium networks, scalable electro-optical Clos switch architectures, different routing issues in optical switching and networking, and all-optical computing architectures. I believe this book to be a fine collection of research work written by experts in the field. The presentations stress some fundamental as well as advanced issues in lightwave technology, the future of optics in the multimedia technology, important optical switch and network architectures, routing and control for optical multimedia support, and QoS requirements. The book is intended as a reference for systems designers, hardware and software engineers, R&D managers, and market planners who seek an understanding of optical broadband networks that support multimedia systems.

Scott Hinton
Distinguished Professor and Chair
Electrical Engineering and Computer Science
The University of Kansas
Lawrence, Kansas

Preface

The last few years have witnessed an explosive growth in the development and deployment of networked applications that transmit and receive audio and video content over the Internet. New multimedia networking applications such as entertainment video, IP (Internet Protocol) telephony, Internet radio, teleconferencing, interactive games, and distance education are announced on a daily basis. The service requirements of these applications differ significantly from those of traditional voice-oriented applications. In particular, multimedia applications are sensitive to end-to-end delay and delay variation but can tolerate occasional loss of data. These fundamentally different service requirements suggest that a network architecture that has been designed primarily for data communication may not be well suited for supporting multimedia applications. In this book, we investigate some of the supporting network architectures as well as some multimedia protocols that are supported by the high-speed networks. Our emphasis is on the optical part of the network architecture that includes the hardware as well as the protocol levels.

As telecommunications and computer networking continue to converge, data traffic is gradually exceeding voice traffic. This means that many of the existing connection-oriented or circuit-switched networks will need to be upgraded to support packet-switched data traffic. The concept of wavelength division multiplexing (WDM) has provided us an opportunity to multiply network capacity. Current optical switching technologies allow us to deliver rapidly the enormous bandwidth of WDM networks. Of all the switching schemes, photonic packet switching appears to be a strong candidate because of the high speed, data rate/format transparency, and configurability it offers. Therefore the use of all-optical networking and switching to support the new multimedia applications seems inevitable.

Optical networks are those in which the dominant physical layer technology for transport is optical fiber. They can be opaque or all-optical, and can be single-wavelength or based on dense wavelength division multiplexing (DWDM). In opaque networks, the path between end users is interrupted at intermediate nodes by optical-electronic-optical (OEO) conversion operations. The traditional synchronous optical networks/synchronous digital hierarchy (SONET/SDH) is an opaque single-wavelength system. The intermediate nodes, being either electronic/optical add-drop multiplexes (OADMs) or optical crossconnects (OXCs), actually have electronic bit handling in the signal path. None of these opaque systems has the versatility and power of the all-optical network.

In addition to meeting the newly acquired data transmission demands, an all-optical network has other advantages. Optical fibers offer abundant bandwidth, low attenuation, and high speed. New optical equipment has recently been developed and deployed. Commercial systems with up to 160 OC-48 (10 Gb/s) channels have been announced that can communicate over paths thousands of kilometers long, uninterruped by OEO conversions. Advances in optical materials have contributed to the declining cost of optical elements that has made all-optical networks affordable. Offering service and maintenance is potentially much simpler for all-optical elements than for the traditional OEO systems. An all-optical network also has inherent robustness. The new physical layer can provide not only basic transport but also several other network-wide rerouting functions that allow the optical layer to exhibit extremely high service availability. New protocols at the medium access control (MAC) layer have been developed for new and ever changing environments. Optical networking architects were concerned with MAC protocols for optical packet switching, the combinatorics of wavelength conversion, and optimal forms of network topology. As the field has matured, it has become clear that the more important problems relate to the network "control plane" and to ensuring the service integrity of the optical layer, which increasingly carries large routed chunks of precious traffic on which depends the viable operation of many institutions and businesses, public and private.

One of the exciting things about the two-layer story is that both the optical networking community and the IP community have begun to agree that the way to control both layers is by multiprotocol label switching (MPLS) or, in the case of the optical layer, a slightly modified version called multiprotocol lambda switching (MPλS). Each of these control planes has two phases, the transient one in which paths are set up and the steady-state or traffic phase in which the state information, which has been set up in each node to define the paths, then acts to forward packets in a way that provides much of the long-missing quality-of-service (QoS) capability. The control phases of these two emerging standards provide a unified, agreed-upon way for the nodes in the IP and optical layers to set up and tear down their portions of virtual point-to-point IP packet connections between end users. There are a number of reasons for MPλS to replace the many variants in current practice in the lower communication layers (especially SONET/SDH) and the IP layer. These traditional control software families (IP and SONET/SDH) are very much vendor-dependent, their control structures are totally different from each other, and both are too slow to satisfy anticipated needs, not just for protection but also for restoration and provisioning.

The first of the optical layer integrity processes to receive attention has been *protection switching,* the millisecond-scale substitution of a new lightpath for a failed one. This action usually requires prescanned algorithms akin to the SONET/SDH protection switching algorithms and invokes only a very localized part of the network, usually no more than a single span, string, or ring of notes. Conventional protection switching is often triggered by some bit-level process. In optical protection switching, the trigger can be loss of optical signal-to-noise ratio (OSNR).

The second process is *restoration*, the replacement of the failed optical path within the network, or part of the network, by another—that is, providing a new backup for the former one, now playing the role of the service path. Since one can allow minutes or even longer to do this, it is possible not only to have real-time software do the job but also to involve much larger portions of the network in the process that can be tolerated for protection switching.

Finally, there is *provisioning* or *reconfiguration*, in which one relieves stranded bandwidth conditions, arranges for the brokering of bandwidth between service providers using the optical facility, or even setting up rent-a-wavelength conditions. Provisioning/reconfiguration can involve optimization over the entire optical network portion or even several of them. Today, all these actions are expected to be available in the access, metro, and long-haul environments.

These issues and more are discussed in this fine collection of chapters. This volume brings together work by some of the leading international experts in the field. The first chapter, by Ali and Kurtz, gives an overall picture of optical networking capabilities for the next generation Internet. They address some key issues as to how the legacy SONET-based infrastructure currently in place will make a graceful transition to the next generation datacentric networking paradigm. They carefully address the problems with today's legacy networking infrastructure and identify challenges facing both carriers and service providers. They provide an understanding of the current and future service portfolio and associated requirements.

Next, Wang and Hamdi investigate the analysis of multimedia access protocols for shared medium networks. They propose a methodology for integrating different MAC protocols into a single shared medium network to accommodate efficiently various types of multimedia traffic streams with different characteristics and QoS demands. In particular, they provide an integrated MAC protocol, termed the Multimedia Medium Access Control (Multimedia-MAC) protocol, that can efficiently and simultaneously serve three types of multimedia traffic streams: a constant bit rate (CBR) traffic and two classes of variable bit rate (VBR) traffic. Then a queuing system with *vacation* model is developed to establish a mathematical framework for the analysis and performance evaluation of the proposed Multimedia-MAC protocol. The Multimedia-MAC design approach and performance evaluation schemes are applied to a *wavelength division multiplexing* network, and it is shown that the proposed protocol design approach is general enough to be used in various shared medium networks.

Then, Ramesh and Tanzer present the design and performance of a lightwave data storage network using computer analysis and simulation. This type of network, composed of broadcasting star couplers, cascading optical amplifiers, and intensity-modulated/direct detection transmission at data rates of 1 to 20 Gbps, is simulated using MATLAB and SIMULINK® software to predict network performance factors. This involves both modeling system components and evaluating system performance measures. It is concluded that large network sizes are configurable, and scaling projections at 10 Gb/s show that cascaded optical amplifiers can extend to reach to several thousand end users, while WDM implementation can produce aggregate network capacities of a hundred terabits per second.

The next four chapters discuss different routing techniques in WDM networks. First, Wang and Hamdi present a survey of routed wavelength WDM networks. The survey covers important issues related to their design and operation, including virtual topology mapping, wavelength routing and assignment, dynamic routing, wavelength conversion, QoS provision, and multicasting. Second, Battou, Ben Brahim, and Khan present a thorough study and implementation of an optical network control plane. They describe the design of the optical network control plane, with particular emphasis on the difficult issues in optical network management, and lightpath signaling and routing. Next, Znati and Melhem describe routing and path establishment for point-to-point and point-to-multipoint communications over WDM networks. In order to address the QoS requirements, in terms of bandwidth and end-to-end delay of the underlying applications, a low-cost, conversion-bound heuristic is proposed. The heuristic establishes a tree of semi-lightpaths between a source and a group of destination nodes. The unique feature of this heuristic is that it decouples the cost of establishing the multicast tree from the delay incurred by data transmission due to lightwave conversion and processing at intermediate nodes along the transmission path. Finally, in this group, Shen, Pan, and Horiguchi provide an overview on recent developments on efficient routing in multihop optical networks supported by WDM with limited wavelength conversion. It addresses offline and online routing in both reliable and unreliable networks for point-to-point, multicast, and multiple multicast communication patterns. Group membership updating schemes for the three patterns are also discussed.

The last category is a set of chapters dealing with optical architectures that serve as the backbone for the all-optical network. We provide description of free-space optical switch and network architectures as well as arithmetic, logic, and image processing ones. Guizani gives an overview of different types of photonic ATM architectures based on three different switching techniques. Liotopoulos describes a three-stage Clos switch architecture, based on a two-level hierarchy of optical cross connects, which can be implemented with AWGFs and ODFs. A number of important related research issues are identified. Li and Esener describe a parallel free-space optical interconnection with all design methodologies and requirements. Advances in this area are reviewed. A number of 3D optical components necessary for such architectures are presented. Finally, Cherri and Alam present two chapters that discuss optical architectures that achieve ultrafast arithmetic, logical, and image processing operations using trinary signed-digit arithmetic and polarization-encoded optical shadow-casting.

ACKNOWLEDGMENTS

We would like to take this opportunity to thank the University of West Florida and the NRL for their support while this book was developed. Special thanks to all the authors who contributed to this book—in particular, Mounir Hamdi, Ghassen Ben Brahim, Bilal Khan, and Taieb F. Znati—for their patience and encouragement, without which this work would not have been completed on time. Rita Lazazzaro and Eric Stannard have been very cooperative throughout this project—many thanks to both of them.

We are grateful to our families for their support. Mohsen thanks his parents and family, in particular, his wife, Saida, and all the kids—for their patience and understanding during the many hours consumed in preparing the manuscript. Abdella thanks his family for their love and support during this project.

Further, we acknowledge our teachers, friends, and colleagues, who have provided us with so much intellectual stimulation throughout our careers.

The editors and authors welcome any comments and suggestions for improvements or changes that could be implemented in forthcoming editions of this book. Comments can be sent to mguizani@cs.uwf.edu.

Mohsen Guizani
Abdella Battou

Contents

Contributors

Mohammad S. Alam   *The University of South Alabama, Mobile, Alabama*

Mohamed A. Ali   *City College of the City University of New York, New York, New York*

Abdella Battou   *firstwave Intelligent Optical Networks, Greenbelt, Maryland*

Ghassen Ben Brahim   *Princeton Optical Systems, Princeton, New Jersey*

Abdallah K. Cherri   *Kuwait University, Safat, Kuwait*

Sadik Esener   *University of California, San Diego, La Jolla, California*

Mohsen Guizani   *University of West Florida, Pensacola, Florida*

Mounir Hamdi   *Hong Kong University of Science and Technology, Kowloon, Hong Kong*

Susumu Horiguchi   *Japan Advanced Institute of Science and Technology, Tatsunokuchi, Japan*

Bilal Khan   *Center for Computational Science at the Naval Research Laboratory, Washington, D.C.*

Russ Kurtz   *REALTECH Systems Corporation, Edison, New Jersey*

Guoqiang Li   *University of California, San Diego, La Jolla, California, and Shanghai Institute of Optics and Fine Mechanics, Shanghai, People's Republic of China*

Fotios K. Liotopoulos *Computer Technology Institute, Athens, Greece*

Rami Melhem *University of Pittsburgh, Pittsburgh, Pennsylvania*

Yi Pan *Georgia State University, Atlanta, Georgia*

S. K. Ramesh *California State University, Sacramento, California*

Hong Shen *Japan Advanced Institute of Science and Technology, Tatsuno-kuchi, Japan*

Herbert J. Tanzer *Hewlett Packard Co., Roseville, California*

Lixin Wang *Hong Kong University of Science and Technology, Kowloon, Hong Kong*

Taieb F. Znati *University of Pittsburgh, Pittsburgh, Pennsylvania*

1

Optical Networking Capabilities for the Next-Generation Internet

Mohamed A. Ali
City College of the City University of New York, New York, New York

Russ Kurtz
REALTECH Systems Corporation, Edison, New Jersey

1 INTRODUCTION

Over the past few years, the field of computer and telecommunications networking has experienced tremendous growth. Traffic demand has increased substantially, somewhat unexpectedly, prompting carriers to add capacity quickly and in the most cost-effective way possible. There is no sign that this increase is an anomaly, and both traffic and backbone capacity are likely to continue a rapid increase into the foreseeable future. The leaders of the data communications industry are not the same companies that lead the marketplace today in sales of telecommunications equipment, so one cannot extrapolate to the future by simply assuming the network element types deployed today will increase in speed and number to meet the capacity growth required in the future.

The change in the fundamental character of backbone network traffic, as demonstrated by the current shift in the telecommunications industry from the traditional voice-centric TDM/SONET/SDH (time division multiplexing/synchronous optical networks/synchronous digital hierarchy) optimized circuit-switched networking paradigm to the data-centric packet-optimized networking paradigm, is leading to revolutionary changes in the traditional concepts of how networks are constructed. The primary reason for the paradigm shift in network design is the nature of the traffic crossing today's long-haul backbones. Specifically, internet protocol (IP) applications are the fastest growing segment of a service provider's network traffic. This growth is expected to continue well into this century.

Concurrently with the emerging data-centric networking paradigm, there has been a well-publicized "explosion" of attention centered on the topic of optical networking. With skyrocketing optical networking startups, valuations, mergers, and acquisitions, the rest of the world has turned its attention to optical networking as the vehicle that promises to transform the Internet, blowing away bandwidth bottlenecks and almost eliminating delays. However, in the midst of much debate and excitement over the likely winners in the race to roll out optical networking technologies, no one seems to be certain of what, exactly, our industry is trying to accomplish. How can our industry bring this technology into the network quickly and cost-effectively? Where are we going with technology? Is it simply more bandwidth?

One might argue that the end goal of our industry is the creation of an all-optical Internet, one that is free from all bandwidth and scalability restrictions, as well as distance limitations. While the vision for creating such a network of networks seems to be gaining momentum, the implications of implementing this vision are far-reaching: all conventional aspects of networking functionality have to be reconsidered; the most basic assumptions of how networks are constructed are being called into question. At a minimum, service providers, carriers, and enterprise network managers will be forced to rethink fully routing, switching, and traffic engineering for the emerging data-centric networking paradigm.

The myriad innovations in dense wavelength division multiplexing (DWDM) technology that increase the number of wavelengths, generate more bandwidth, send multigigabit signals faster and longer distances, and those enabling optical crossconnects (OXCs) with the ability to add, drop, and in effect construct wavelength-switched and wavelength-routed networks, are moving us towards this point. High-performance routers (IP layer) plus an intelligent optical transport layer (DWDM layer) equipped with a new breed of photonic networking equipment hold the keys to the two pieces of the puzzle that will comprise the next generation optical Internet. Of course, there are several visions of how best to marry these two technologies. The key question appears to be whether carriers should control their next-generation data-centric networks using only routers, or some combination of routers and OXC equipment.

The realm in which we are likely to see the first real deployments of high-performance routers interconnected by intelligent optical core networks that utilize optical networking technology will likely be within the long-haul backbone. Several carriers have already completed field trials of all-optical systems, and many others have trials in process. So how will the migration to this new data-centric optical networking paradigm evolve?

The key ingredient to the success of any migration strategy is that the new networking paradigm must coexist with the aging TDM circuit switching infrastructure, offer the network reliability inherent in legacy SONET/SDH systems, and at the same time meet the surging demand for packet traffic. In other words, how the legacy SONET-based infrastructure currently in place can make a graceful transition to the next generation data-centric networking paradigm is the real challenge. To assess which architecture will enable the most effective migration strategy, the following two issues must be carefully addressed: (1) problems with today's legacy networking infrastructure as well as challenges facing both carriers and service providers must be fully identified; (2) current and future service portfolio and associated requirements must be thoroughly understood.

1.1   Today's Problems

The fundamental problem facing both carriers and service providers is that Internet traffic is already skyrocketing, and looks certain to accelerate even further as broadband access technologies such as DSL (digital subscriber line) and cable modems are rolled out to the masses. Service providers are faced with daunting challenges that span both the access and backbone transport network requirements. They must offer service breadth—IP, frame relay, ATM, Ethernet, and TDM-based voice—plus the flexibility to add and modify new services as market conditions warrant. While transporting data services, the new data-centric networking paradigm must coexist with the voice infrastructure and offer the network reliability inherent in legacy SONET/SDH systems.

Let us first consider the trouble with today's most widely used standard, SONET. SONET infrastructure has been the standard for wide area networking for the past 15 years, ideally designed to handle circuit-switched voice and point-to-point TDM connections, it began to feel burdened by growing data traffic demands and the need to deliver real-time circuit provisioning. This is because SONET was designed before the explosion of the asynchronous and unpredictable Interne-based traffic profiles.

The asynchronous and unpredictable nature of Internet-based traffic profiles makes forecasting for growth all the more difficult, and the issue of scalability a major concern. Unfortunately, SONET was never designed to deal with these problems. SONET networks are always built under the assumption that transmit/ receive traffic is always balanced and predictable, so that growth would be easy to forecast.

SONET's inability to handle the huge influx of data traffic, and the associated unpredictable traffic patterns, signaled the end of this standard's long reign in the backbone transport network. While long-haul architectures based solely on SONET will be far too expensive and complex to meet the needs of tomorrow's backbones, some of the functionality SONET encompasses will live on in different, more efficient vehicles. The emerging data-centric networking paradigm, a purely optical Internet, by contrast, will bypass SONET altogether, providing for a backbone without capacity restrictions or distance limitations.

SONET provides four primary functions: (1) intermediate multiplexing, (2) network survivability, (3) performance monitoring, and (4) bandwidth provisioning. The incorporation of these into any near-term optical Internet architecture is a prerequisite for any graceful transition. In the near term, optical networking systems and technologies combined with high-performance multigigabit or terabit routers will be able to perform these functions. These systems are already in late-stage development, and we can expect to see true commercial developments beginning by the end of this year. So, while some SONET functionality is still needed, a separate SONET layer is not.

That is not the end of the traditional network's troubles. Another major issue facing today's service providers is the static and cumbersome process of provisioning end-to-end circuits. Provisioning a cross-country OC-48 service is extremely manual, involves complex network planning and roll-out activities, and generally takes several months to accomplish. The roll-out of DWDM has led to another problem; current digital crossconnects simply aren't big enough to handle the number of connections running across backbones. Carriers can foresee needing huge switch fabrics with thousands of ports, and right now, they do not exist.

1.2 Tomorrow's Promises

All of this is about to change. Several dramatic advances in the optical-networking arena in recent years have emerged to set the foundation for the next-generation optical Internet and to challenge both the traditional view of networking (e.g., routing, switching, provisioning, protection, and restoration) and the conventional approaches of networks design that have always addressed technology first and management second. First, the abundance of bandwidth propelled by the explosion of DWDM poses, for the first time, a new challenge to network architects. Whereas in the past, IP, ATM, and related routing protocols have focused on managing the scarcity of bandwidth, the new challenge is managing the abundance of bandwidth.

Second, rapid advances in DWDM technology—dense WDM add/drop multiplexers (ADMs), wideband optical amplifiers, stable single frequency lasers, novel optical crossconnects (OXCs)—with the ability to add, drop, and in effect construct wavelength-switched and wavelength-routed networks, are now beginning to shift the focus more toward optical networking and network-level issues. Thus, there is an attractive opportunity to evolve DWDM technology toward an optical networking infrastructure with transport, multiplexing, switching, routing, survivability, bandwidth provisioning, and performance monitoring, supported at the optical layer.

Harnessing the newly available bandwidth is difficult because today's core network architecture, and the networking equipment it is built on, lack both the scale and functionality necessary to deliver the large amounts of bandwidth required in the form of ubiquitous, rapidly scalable, multigigabit services. Today's core network architecture model has four layers: IP and other content-bearing traffic, over ATM for traffic engineering, over SONET for transport, and over WDM for fiber capacity. This approach has functional overlap among its layers, contains outdated functionality, and will not be able to scale to meet the exploding volume of future data traffic, which makes it ineffective as the envisioned architecture for a next-generation data-centric networking paradigm.

A simplified two-tiered architecture that requires two types of subsystems will set the stage for a truly optical Internet: service delivery platforms that enforce service policies, and transport platforms that intelligently deliver the necessary bandwidth to these service platforms. If IP can be mapped directly onto the WDM layer, some of the unnecessary network layers can be eliminated, opening up new possibilities for the potential of collapsing today's vertically layered network architecture into a horizontal model where all network elements work as peers dynamically to establish optical paths through the network.

High-performance routers (IP layer) plus an intelligent optical transport layer (DWDM layer) equipped with a new breed of photonic networking equipment hold the keys to the two pieces of the puzzle that will compose the optical Internet. Closer and efficient interworking of these two pieces is the key to solving the puzzle. Specifically, linking the routing decision at the IP layer with the point-and-click provisioning capabilities of optical switches at the transport layer, in our opinion, is the cornerstone to the puzzle's solution. This will allow routers and ATM switches to request bandwidth where and when they need it.

Now that the basic building blocks are available for building such a network of networks, the key innovations will come from adding intelligence that enables the interworking of all the network elements (routers, ATM switches, DWDM transmission systems and OXCs). The IETF has already addressed the interworking of routers and optical switches through the multiprotocol lambda switching (MPλS) initiative [1]. The main goal of this initiative is to provide a framework for real-time provisioning of optical channels, through combining recent advances in multiprotocol label switching (MPLS) traffic-engineering control plane with emerging optical switching technology in a hybrid IP-centric optical network. However, only a high-level discussion has been presented.

This work presents a balanced view of the vision of the next generation optical Internet and describe, two practical near-term optical Internet candidate model architectures along with a high-level discussion of their relative advantages and disadvantages: (1) a router-centric architecture deployed on a thin optical layer, and (2) a hybrid router/OXC-centric architecture deployed on a rich and intelligent optical transport layer. A router-centric model assumes that IP-based traffic is driving all the growth in the network, and the most cost-effective solution is to terminate all optical links directly on the router. The granularity at which data can be switched is at the packet level. In a hybrid router/OXC model, all optical links coming into a core node are terminated on an OXC. The cross-connect switches transit traffic (data destined for another core node) directly to outbound connections, while switching access traffic (data destined for this core node) to ports connected to an access router. The granularity at which data can be switched is at the entire wavelength level, e.g., OC-48 or OC-192.

Both models are based on the two-layer model, IP directly over WDM, and therefore DWDM equipment is available to provide the underlying bandwidth. The main issue is really about the efficiency of a pure packet-switched network versus a hybrid, which packet switches only at the access point and circuit switches through the network. The distinction between these two models is based on the way that the routing and signaling protocols are run over the IP and the optical subnetwork layers.

A special consideration will be given to the hybrid router/OXC-centric model, referred to as the integrated model in a recent IETF draft [2], since it is our opinion that this model represents a credible starting point toward achieving an all-optical Internet. This model aims at a tighter and more coordinated integration between the IP layer and the underlying DWDM-based optical transport layer.

The work presented here builds on the recent IETF initiatives [1,2] and other related work on the (MPλS) [3,4] and addresses the implementation issues of the path selection component of the traffic engineering problem in a hybrid IP-centric DWDM-based optical network. An overview of the methodologies and associated algorithms for dynamic lightpath computation, that is based on a fully distributed implementation, is presented. Specifically, we show how the complex problem of real-time provisioning of optical channels can be simplified by using a simple dynamic constraint-based routing and wavelength assignment (RWA) algorithm that computes solutions to three subproblems: the routing problem, the constrained-based shortest route selection problem, and the wavelength assignment problem.

We present two different schemes for dynamic provisioning of the optical channels. The first scheme is conceptually simple, but the overhead involved is excessive. The second scheme is not as overhead consuming as the first one but has the drawback that the chosen wavelength is only locally optimized along the selected route; consequently the utilization of network resources is not globally optimized. Two different scenarios are then considered: provisioning lightpaths in a network with full wavelength conversion capability, and provisioning lightpaths in a network without wavelength conversion capability

The remainder of this chapter is organized into four sections. Section 2 presents the case for building an optical Internet. Section 3 describes two near-term optical Internet candidate models. In Section 4, we present an overview of real-time provisioning at the optical layer and describe two different schemes for dynamic provisioning of the optical channels. Finally, section 5 offers summary and conclusion.

2 THE CASE FOR BUILDING AN OPTICAL INTERNET

2.1 Overview

There is considerable debate in the telecommunications industry about the best technology for transporting IP services. Carriers are generally more committed to ATM as the common network technology. One of the underlying assumptions of the carriers is that data services would be one of many competing services in a portfolio of service types. As such, ATM networks are optimized to carry a mix of different service types rather than being optimized for one specific service type. The ATM virtual paths make it easier to construct router networks by giving routers the appearance of being only one hop apart, whereas in reality the virtual path may traverse many intermediate ATM switches. ATM does also provide a powerful set of capabilities in terms of traffic engineering.

On the other hand, there are many in the Internet community who believe that soon, just about everything will ride over IP, so that a network optimized to carry IP is the most appropriate. While there are compelling arguments for both sides of the debate, the most likely near-term outcome is that IP over ATM and IP/MPLS directly over WDM services will exist in parallel to meet the spectrum of customer requirements for IP networking. This is particularly true with the advent of DWDM-based optical transport systems as the main (and sole) underlying transport infrastructure that can support a multitude of transport service delivery mechanisms from traditional SONET/SDH services to the new optical ATM and IP architectures.

The case therefore for an optical Internet essentially rests on the predicted volumes for Internet traffic growth and the expected predominant types of Internet applications. If Internet traffic continues to grow exponentially, and if for the bulk of traffic all that is required is a "best efforts" or an "ensured" delivery service, then high-volume IP pipes would seem to be the most appropriate technology. In that scenario, it seems appropriate to build a network, first and foremost, that is optimized for the delivery of that type of service. The remaining services can then be delivered on top of an IP network (which may or may not be less than optimal) or continue to be delivered over a parallel ATM network. We will refer to this data-centric networking solution as a router-centric architecture. The details of this architecture will be considered in the next section.

On the other hand, if data services are treated as one of many competing services in a portfolio of service types, then it seems appropriate to build a network that is optimized to carry a mix of different service types rather than being optimized for one specific service type. We will refer to this networking solution as a hybrid router/OXC-centric architecture, the details of which will also be considered in the next section.

2.2   The Envisioned Data-Centric Networking Paradigm: Optical Internet

Recently there have been several initiatives in building an optical Internet where IP routes are interconnected directly with WDM links [5]. A simplified two-tiered architecture that require two types of subsystems will set the stage for a truly optical Internet: service delivery platforms that enforce service policies and transport platforms that intelligently deliver the necessary bandwidth to these service platforms. High-performance routers plus an intelligent optical transport layer equipped with a new breed of photonic networking equipment hold the keys to the two pieces of the puzzle that will compose the optical Internet.

However, closer and efficient interworking of these two pieces is the key to solving the puzzle. Specifically, linking the routing decision at the IP layer with the point-and-click provisioning capabilities of optical switches at the transport layer, in our opinion, is the cornerstone to the puzzle's solution. This will allow routers and ATM switches to request bandwidth where and when they need it.

As high-performance routers emerge with high bit rate optical interfaces (like OC-48c and OC-192c), the high bit rate multiplexing traditionally performed by a SONET terminal is no longer necessary. These routers, which represent the first piece of the two-tiered networking model, can also handle equipment protection, further diminishing the need for SONET. On the other hand, in order for this new two-tiered model to be truly viable, optical networking technologies and systems must support additional functionality, including multiplexing, routing, switching, survivability, bandwidth provisioning, and performance monitoring. These emerging sets of optical networking functionality and equipment is the second piece of the two-tiered puzzle that will compose the optical Internet.

The following are some of the salient features and associated dramatic technological changes that should take place en route towards implementing the envisioned optical Internet: (1) SONET transport should give way to optical transport; fast restoration (50 ms) is necessary; without it, a significant availability and reliability benchmark set by SONET would be lost; (2) bandwidth is provisioned, not at TDM granularities, but rather at wavelength granularity; to meet exponential growth rates, rapid provisioning must be an integral part of the new architecture; (3) as future data network cores continue to scale—as traffic between pairs of backbone routers reaches just a single OC-48c/OC-192c—ATM's virtual path becomes equivalent to a wavelength; (4) ATM cell granularity and traffic engineering are made obsolete by the use of MPLS at the IP level and wavelength traffic engineering at the optical transport layer; (5) restoration will take place at the optical transport layer in a fast, scalable manner. The service layer will be informed of the event only if the transport layer cannot restore due to lack of transport resources. Then the service layer will be advised to perform its restoration functions. The two will perform in a nonoverlapping, predictable, and scalable manner.

This simplified two-tiered architecture optical Internet is defined here as mapping IP traffic directly over the optical layer, as there is no underlying SONET/SDH and ATM transport layer/protocol. This architecture is commonly referred to as the IP-over-WDM problem. The IP-over-WDM problem has two fundamental sub-problems: the user-plane protocol stack problem and the routing problem. The user-plane protocol stack problem consists of defining the protocol stack for carrying IP packets over multiwavelength optical links, and the corresponding network architecture.

Candidate protocol stacks include (1) IP over AAL/ATM over SONET/SDH over WDM, (2) IP over PPP/HDLC over SONET/SDH over WDM, (3) IP over PPP/HDLC over WDM, and (4) IP directly over WDM. In accordance with the two-tiered architecture definition of an optical Internet described above, this work considers the two-layer protocol stack (IP directly over WDM) as the only viable candidate model for building the next generation optical Internet.

3 POSSIBLE NEAR-TERM GRACEFUL OPTICAL INTERNET TRANSITIONAL ARCHITECTURAL OPTIONS

3.1 Overview

Clearly, there are several feasible optical Internet architectures that will emerge to challenge embedded infrastructure. Near-term transition strategies for wider deployment need to look at service requirements and product availability. High-performance routers (IP layer) plus an intelligent optical transport layer (DWDM layer) equipped with a new breed of photonic networking equipment hold the keys to the two pieces of the puzzle that will compose the next generation optical Internet. Of course, there are several visions of how best to marry these two technologies. The key question appears to be whether carriers should control their next generation data-centric networks using only routers, or some combination of routers and OXC equipment. The debate is really about the efficiency of a pure packet-switched network versus a hybrid, which packet switches only at the access point and circuit switches through the network.

The realm in which we are likely to see the first real deployments of high-performance routers interconnected by intelligent optical core networks that utilize optical networking technology will likely be within the long-haul backbone. Several carriers have already completed field trials of all-optical systems, and many others have trials in process. The key ingredient to the success of any migration strategy is that the new networking paradigm must coexist with the aging TDM circuit switching infrastructure, offer the network reliability inherent in legacy SONET/SDH systems, and at the same time meet the surging demand for packet traffic.

3.2 The Router-Centric Model

The notion of this model is to build a dedicated optical Internet network for large-volume backbone pipes that does not require an underlying multiservice SONET/SDH and ATM transport protocols. In this model, IP routers are interconnected directly with point-to-point physical links on which multiple WDM channels are carried. A router-centric model assumes that IP-based traffic is driving all the growth in the network, and the most cost-effective solution is to terminate all optical links directly on the router. The router handles all traffic in a PoP (point of presence). All packets go through the router whether they are accessing the network through that PoP or entering and exiting via backbone connections. The granularity at which data can be switched is at the packet level.

The case for building such a router-centric model is based on the following facts/predictions: (1) if networks of the future are predominantly "datagram based," then there may not be a need for complex optical switching technology; (2) as interface rates approach line rates, aggregation is pushed to the edge of the network and handled locally by routers, thus making the multiplexing functionality in the backbone nodes redundant.

The router at the edge, rather than an optical switching device at the core, becomes the prime intelligent device for routing and switching packets between various optical links. Traffic engineering of this network can now be accomplished only at the IP layer as there is no underlying transport layer, and so protocols like MPLS becomes essential for the fast restoring, protection, effective management, and engineering of the network.

The optical layer is completely separate from the IP layer, and optical transport offers only higher capacity and higher reliability. In this scenario, the optical network provides point-to-point optical links for the transport of IP packets through the optical domain. IP is more or less independent of the optical subnetwork. The IP/MPLS routing, topology distribution, and signaling protocols are independent of the routing, topology distribution, and signaling protocols at the optical layer. This model is conceptually similar to the conventional IP over ATM or MPOA models, but it is applied to an optical subnetwork directly.

If full layer-3-based restoration speed is acceptable for IP applications, then significant bandwidth efficiency can be gained at the transport layer by the elimination of the protection bandwidth. It is anticipated that layer 3 restoration capabilities will reach Carrier class performance using MPLS (this is not yet quite clear). But if bandwidth is cheap (WDM has significantly reduced the cost/bit), protection bandwidth is a small cost, so additional motivations are required to build the case for full layer 3 restoration.

Router-only advocates stress that as the majority of traffic becomes IP (predicted to be more than 95% by early next decade), it is more efficient to build a network strictly to optimize for IP. This minimizes the cost by requiring fewer boxes in the PoP and decreases the network management load. The downside to the router approach is that carriers need to map multiple protocols to IP, it requires the creation and management of IP routing adjacencies over the optical network, and routed networks are not considered as reliable or as capable of fast restoration.

3.3 The Hybrid Router/OXC-Centric Architecture Integrated Model

As architectures and needs evolve, we see that the functionality within IP and the DWDM-based optical networking technologies and components converge. While these components overlap in some capabilities, they complement each other's strength. High-performance routers plus a smart optical transport layer equipped with a new breed of photonic networking components and subsystems together are setting the foundation for the next generation networking paradigm. These are the elements that will form the essence of the hybrid router/OXC-centric architecture or the integrated model.

In a hybrid router/OXC model, all optical links coming into a PoP are terminated on an OXC. The crossconnect switches transit traffic (data destined for another PoP) directly to outbound connections while switching access traffic (data destined for this PoP) to ports connected to an access router. The router handles only the packets accessing the network through that PoP. Packets entering and exiting via backbone connections are handled by a crossconnection, either at a TDM level or as an entire wavelength. All connections will need to terminate on the crossconnect to provide reconfigurability of the network. Under this hybrid model, the IP and optical networks are treated together as a single integrated network managed and traffic engineered in a unified manner. In this regard, the OXCs are treated just like any other router as far as the control plane is concerned.

One of the main goals of the integration architecture is to make optical channel provisioning driven by IP data paths and traffic engineering mechanisms. This will require a tight cooperation of routing and resource management protocols at the two layers. Multiprotocol label switching (MPLS) for IP packets is believed to be the best integrating structure between IP and WDM. MPLS brings two main advantages. First, it can be used as a powerful instrument for traffic engineering. Second, it fits naturally to WDM when wavelengths are used as labels. This extension of the MPLS is called multiprotocol lambda switching [1–4]. The main characteristics of this model is that the IP/MPLS layers act as peers of the optical transport network, so that a single routing protocol instance runs over both the IP/MPLS and optical domains.

Hybrid advocates argue that routing is expensive and should be used only for access, because circuit switching is better for asynchronous traffic, and restoring traffic at the optical layer is faster and well understood. In addition, this model allows seamless interconnection of IP and optical networks. The downside to this approach is that hybrid networks have the disadvantages inherent to a circuit switch model—they waste bandwidth and introduce an N^2 problem, e.g., to connect 30 routers in the east with 30 routers in the west requires more than 900 connections. Another disadvantage of the hybrid approach is that it requires routing information specific to optical networks to be known to routers. Finally, as it is assumed they will need a router, hybrid solutions include another device to manage.

In the following subsections, we start off with an overview of MPLS. The architecture of the hybrid router/OXC-centric model considered here is then presented. Finally, we consider the implementation issues of the dynamic path selection component of the traffic engineering problem in the hybrid router/OXC-centric model.

3.3.1 Overview of MPLS

MPLS introduces a new forwarding paradigm for IP networks. The idea is similar to that in ATM and frame relay networks. A path is first established using a signaling protocol; then a label in the packet header, rather than the IP destination address, is used for making forwarding decisions in the network. In this way, MPLS introduces the notion of connection-oriented forwarding in an IP network. In the absence of MPLS, providing even the simplest traffic engineering functions (e.g., explicit routing) in an IP network is very cumbersome.

Two signaling protocols may be used for path setup in MPLS: the Constraint-Based Routed Label Distribution Protocol (CR-LDP) and extensions to Resource Reservation Protocol (RSVP) [6]. The path set up by the signaling protocol is called a label-switched path (LSP). Routers that support MPLS are called label-switched routers (LSRs). An LSP typically originates at an edge LSR, traverses one or more core LSRs, and then terminates at another edge LSR. The ingress edge LSR maps the incoming traffic onto LSPs using the notion of a forwarding equivalence class (FEC). An FEC is described by a set of attributes such as the destination IP address prefix, type of service (TOS) fields, or IP protocol. All packets that match a given FEC will be sent on the LSP corresponding to that FEC. This is done by prepending the appropriate label to the IP packet. The core LSRs forward labeled packets using only information contained in the label; the rest of the IP header is not consulted. When an LSR receives a packet, it looks up the entry in its label information base (LIB), and determines the output interface and new outgoing label for the packet. Finally, the egress edge LSR will remove the label from the packet and forward it as a regular IP packet.

3.3.2 The Hybrid Network Model

The network model considered here consists of IP/LSR routers attached to an optical core network. As shown in Fig. 1, the optical network consists of multiple optical crossconnects (OXCs) interconnected via WDM links in a general mesh topology. Each OXC can switch high-speed optical signals (e.g., OC-48,

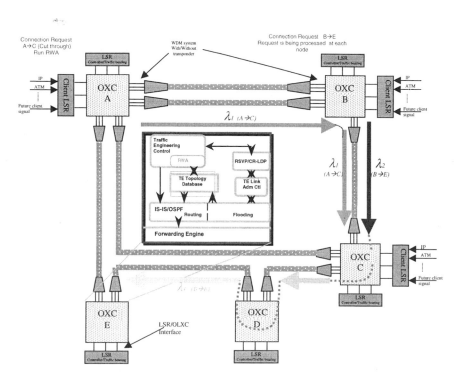

Figure 1 A hybrid router/OXC-centric architecture.

OC-192) from input ports to output ports. The switching fabric may be purely optical or electrical or a combination. The LSRs are clients of the optical network and are connected to their peers over switched optical paths (lightpaths) spanning potentially multiple OXCs. LSRs process traffic in the electrical domain on a packet-by-packet basis. The OXC processing unit is one wavelength.

A lightpath is a fixed bandwidth connection between two network elements such as IP/LSR routers established via the OXCs. Two IP/LSR routers are logically connected to each other by a single-hop channel. This logical channel is the so-called lightpath. A continuous lightpath is a path that uses the same wavelength on all links along the whole route from source to destination.

Each node in this network consists of an IP router and a dynamically reconfigurable OXC. The IP/LSR router includes software for monitoring packet flows. The router is responsible for all nonlocal management functions, including the management of optical resources, configuration and capacity management, addressing, routing, topology discovery, traffic engineering, and restoration. In general, the router may be traffic bearing, or it may function purely as a controller for the optical layer and carry no IP data traffic [2]. In this work, it is assumed to be a combination of both. The node may be implemented using a stand-alone router interfacing with the OXC through a defined interface, or it may be an integrated system, in which the router is part of the OXC system.

These OXCs may be equipped with full wavelength conversion capability, limited wavelength conversion capability, or no wavelength conversion capability at all. If the WDM systems contain transponders or if electronic OXCs are used, then it is implied that a channel associated with a specific wavelength in the WDM input can be converted to an output channel associated with a different wavelength in the WDM output (i.e., wavelength conversion is inherent). However, if the switching fabric is optical and there is no transponder function in the WDM system, then wavelength conversion is only implemented if optical-to-electronic conversion is performed at the input or output ports, or if optical wavelength converters are introduced to the OXC.

Some of the processing capabilities and functionality of the hybrid nodes (wavelength conversion capable or incapable) comprising the core network are shown in Figs. 2, 3, and 4, respectively. These include, but are not limited to, wavelength merging, flow aggregation (grooming) into a continuous lightpath, flow switching to a different lightpath, and flow processing/termination at an intermediate node along a continuous lightpath. Wavelength merging is a strategy for reusing precious wavelengths that allows tributary flows to be aggregated by merging packets from several streams. The OXC requires enhanced capabilities to perform this merging function [7]. The device would be able to route the same wavelength from different incoming fibers into a single outgoing fiber. The key property of this device is that contention between bits on the wavelength must be resolved before they are multiplexed into the common outgoing fiber.

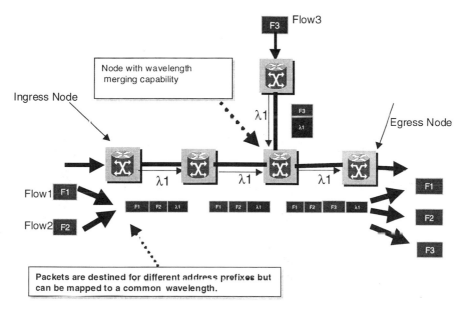

Figure 2 Backbone nodes without wavelength conversion capability.

We define a flow as a sequence of packets that travel together along a subset of the same route in the network before exiting the network. This definition is a generalization of the more common, fine-granularity definition, which identifies a flow as sequence of packets with the same source and destination IP addresses and transport port numbers.

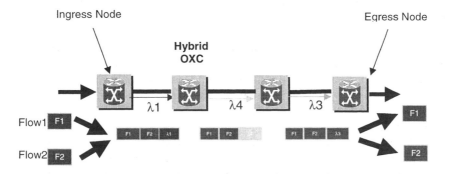

Figure 3 Backbone nodes with full wavelength conversion capability.

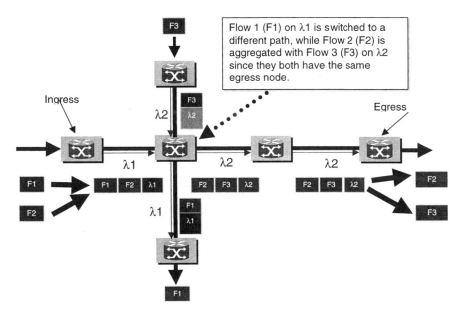

Figure 4 Backbone nodes with partial wavelength conversion and O-E-O processing capabilities.

The MPLS label switching concept is extended to include wavelength-switched lightpaths. Optical nodes are treated as IP-MPLS devices, termed hereafter optical lambda switch router (O-LSR) nodes, where each O-LSR node is assigned a unique IP address [3]. We assume that the OXCs are within one administrative domain and run interdomain routing protocols, such as OSPF and IS-IS, for the purpose of dynamic topology discovery. For optimal use of network resources, the routing protocols have been enhanced to disseminate additional link-state information over that currently done by standard OSPF and IS-IS protocols.

The request to set up a path can come either from a router or from an ATM switch connected to the OXCs (O-LSRs). The request must identify the ingress and the egress OXCs between whom the lightpath has to be set up. Additionally, it may also include traffic-related parameters such as bandwidth (OC-48/Oc-192), reliability parameters and restoration options, and setup and holding priorities for the path. The lightpaths are provisioned by choosing a route through the network with sufficient available capacity. The lightpath is established by allocating capacity on each link along the chosen route, and appropriately configuring the OXCs. Protection is provided by reserving capacity on routes that are physically diverse from the primary lightpath.

There should be a mechanism to exchange control information between OXCs, and between OXCs and other LSRs. This can be accomplished by using either an in-band or an out-of-band signaling channel (i.e., a 1510 nm supervisory channel), using the same links that are used to carry data-plane traffic. An OXC must be able to provide the MPLS traffic engineering control plane with pertinent information regarding the state of individual fibers attached to that OXC, as well the state of individual lightpaths or optical channel trails within each fiber.

4 REAL-TIME PROVISIONING AT THE OPTICAL LAYER

4.1 Overview

Provisioning end-to-end circuits is an endless source of struggle for service providers and a frustration for end-users. While a technology like multiprotocol lambda switching (MPλS) promises to ease and expedite provisioning, its effectiveness is yet to be tested and/or demonstrated. In addition, it is difficult to speculate whether the deployment of highly dynamic data-driven cut-through lightpaths within core networks will be widely adopted by carriers and ISPs. The reasons are numerous: possible CPU overhead in core network elements, complexity of proposed solutions, stability concerns, lack of true economic drivers for this type of service, and above all, the argument is put forth, Do we really need real-time provisioning? And if so, what is the target interval of provisioning? Is it a day, a week, a month, or perhaps much less or much more? In any case, we emphasize that although the scenario for the dynamic provisioning scheme proposed here needs more optimization (work currently going on), it is definitely a step forward in the right direction that will ultimately lead to much more advanced real-time provisioning schemes in the near future.

Dynamic computation of a lightpath involves the implementation of two traffic-engineering components: an information distribution mechanism that provides knowledge of the relevant attributes of available network resources, and a path selection process that uses the information distributed by the dynamic link-state advertisement algorithm to select a path that meets the specific requirements of the traffic flow.

Note that while MPLS provides a method of setting up explicit paths and forwarding traffic on them, it does not address the issue of how to find paths with constraints. To simplify the complex problem of real-time provisioning of optical channels, we develop a simple dynamic constraint-based routing and wavelength assignment (RWΛ) algorithm that computes solutions to three subproblems: the routing problem, the constrained-based shortest route selection problem, and the wavelength assignment problem.

Two comments are in order here. First, note that the dynamic RWA algorithm is based on a fully distributed model in which all nodes maintain a synchronized and identical topology and link-state information (traffic-engineering database, TED). Second, the simplicity of the wavelength assignment algorithm is a function of whether the network nodes have full wavelength conversion capability, partial wavelength conversion capability, or no wavelength conversion capability.

4.2 Dynamic Lightpath Computation

The two traffic-engineering components that involve the dynamic computation of a lightpath (in a fully distributed implementation) in a hybrid IP-centric DWDM-based optical network are

1. An information distribution mechanism that provides knowledge of available network resources. This component is implemented by defining relatively simple extensions to the interior gateway protocol (IGP), e.g., open shortest path first (OSPF) so that link attributes are included as part of each router's link-state advertisement. Some of the traffic-engineering extensions that need to be added to the IGP link-state advertisement include maximum link bandwidth, maximum reservable link bandwidth, current bandwidth reservation, current bandwidth usage, and link coloring. These extensions capture optical link parameters and any constraints specific to optical networks. The topology and link-state information maintained by all nodes are identical. Another important component is to define naming and addressing conventions for different elements of the physical plant hierarchy. For example, primary and backup paths between ingress OXC and an egress OXC should follow routes that are physically separate. Control architecture and signaling protocols used in MPLS, such as CR-LDP and RSVP, are not cognizant of the details of the physical plant connecting the network elements (LSRs). In fact, no standard terminology to describe the physical plant hierarchy exists. Therefore we have defined and assigned the naming and addressing convention for different elements of the physical plant hierarchy. We have implemented our own dynamic simple link-state advertisement algorithm (a simple version of an extended OSPF). This algorithm is capable of periodically updating and advertising, to maintain and refresh a synchronized data base for all nodes, all of the above link attributes. The link-state updates can be triggered, for instance, based on a given threshold of the number of available wavelengths per fiber, below which the updates can be triggered.

2. A path selection process that uses the information distributed by the dynamic link-state advertisement algorithm to select an explicit route that meets the specific requirements of the traffic flow. This process can be performed either off-line or on-line using a constraint-based routing calculation. The source router is basically responsible for computing the complete path all the way to the destination through the optical domain, and then initiating path setup using the signal-

ing protocol (e.g., CR-LDP or RSVP). The route may be specified either as a series of nodes (routers/OXCs) or in terms of the specific links used (as long as IP addresses are associated with these links). Once each node has a representation of the full physical network topology and the available resources on each link (these are obtained and updated via the link-state advertisement protocol), we need a path selection algorithm. To simplify the real-time provisioning problem, the path selection process, that is, the dynamic constraint-based RWA algorithm, computes solutions to three subproblems: the routing problem, the constrained-based shortest route selection problem, and the wavelength assignment problem.

4.2.1   The Routing Problem

We emphasize that it is necessary to use explicit routing for constructing lightpaths. This is valuable for traffic engineering and load optimizations in the network. A shortest path algorithm is run off-line to calculate the shortest path between every source–destination node pair throughout the nodes composing the simulated arbitrary network topology. This procedure is repeated k times for every source–destination pair in order to select the other k-1 alternative shortest routes in an orderly fashion. The off-line calculated k shortest routes for every source–destination pair is then stored in the data base of each CSR.

The routing scheme used here is a fixed alternate routing algorithm. This routing scheme is a constraint path selection in which a path is selected from a predetermined set of candidate paths. These candidate paths may include the shortest path route, the second shortest path route, the third shortest path route, etc.

We have examined the performance of two fixed alternate routing selection algorithms, the k link-disjoint shortest routes, and the k best routes.

The k link-disjoint routes. In this routing scheme, the alternate routes should be link-disjoint. Two alternate routes between a source S and a destination D cannot share any links. This routing approach provides a straightforward approach to handling link failure. The set of shortest k-disjoint paths between all node pairs is computed in advance and stored in routing tables at each node.

The k best-routes. In this approach, two alternate routes can share any number of common links but should differ in at least one link. The set of shortest k-best paths between all node pairs is computed in advance and stored in routing tables at each node.

4.2.2   The Constraint-Based Shortest Route Selection Problem

To optimize the utilization of the network resources, the route with the least average number of used wavelengths (least-congested path) per the entire route is then selected out of the k routes. This is achieved by applying a simple constrained shortest path selection algorithm on the k routes stored off-line at the ingress node. This algorithm uses the information stored and updated, via the link-state advertisement protocol, at the first-hop router's TED. Specifically, these are the link attributes that are included as part of each router's link-state advertisement.

As an illustrative example, consider a lightpath request from A to C as shown in Fig. 1. There are two alternative disjoint routes ($k = 2$ in this example) that should be stored at node A. The first route (first shortest path route) is A–B–C, and the second route is A–E–D–C. The first route has two links, AB and BC. Assume that the number of used wavelengths on link AB = 4, and that the number of used wavelengths on link BC = 12. Then the average number of used wavelengths on route A–B–C = 8 wavelengths.

The second disjoint alternative route (second shortest path route) has three links, AE, ED, and DC. Assume that the number of used wavelengths on link AE = 5, the number of used wavelengths on link ED = 2, and that the number of used wavelengths on link DC = 8, Then the average number of used wavelengths on this route = 5 wavelengths. Because the average number of used wavelengths [5] on the second route (A–E–D–C) is less than those [8] on the first route (A–B–C), the second route is selected. Note that route A–B–C was not selected, even though it is the shortest path. This is a constraint-based shortest route selection problem.

4.2.3 The Wavelength Assignment Problem

For a given request, once a constrained shortest path (the path with the least average number of used wavelengths per the entire route) is selected out of the k routes, a wavelength assignment algorithm is invoked (on-line) to assign the appropriate wavelength(s) across the entire route. A wavelength assignment algorithm must thus be used to select the appropriate wavelength (S).

Several heuristics wavelength assignment algorithms can be used, such as the Random wavelength assignment (R), least used (LU), most used (MU), least-loaded (LL), and max-sum algorithms [8–12]. The implementation of these algorithms requires the propagation of information throughout the network about the state of every wavelength on every link in the network. As a result, the state required and the overhead involved in maintaining this information would be excessive. However, the advantage gained is that the chosen wavelength is globally optimized, and consequently the utilization of network resources is globally optimized, since the wavelength assignment algorithm has global information about the state of every wavelength on every link in the network.

Note that, as mentioned above, the complexity of the wavelength assignment algorithm depends on whether the network nodes are wavelength conversion capable or incapable. In general, the complexity of the proposed approach and the overhead involved in maintaining and updating the TED depends on the value of k. The smaller the value of k, the lower the overhead involved and the simpler the algorithm. A reasonable value of k may range from 4 to 6.

4.3 Provisioning Lightpaths in a Network with Full Wavelength Conversion Capabilities

In an optical network with full wavelength conversion capability, channel allocation (wavelength assignment) is a straightforward process and can be performed independently on different links along a route. On the other hand, if the utilization of the network resources is to be globally optimized, the wavelength assignment algorithm can be very complicated. The route on which a new lightpath is to be established is specified in the lightpath setup message.

Once a lightpath request from a source is received by the first-hop router (ingress node), it computes the complete path all the way to the destination through the optical domain using the above approach. The output of this calculation is an explicit route consisting of a sequence of hops that provides the shortest path through the network that meets the constraints. This explicit route is then passed to the signaling component that initiates path setup (to reserve resources) using the signaling protocol (e.g., CR-LDP or RSVP).

The first-hop router creates a lightpath setup message and sends it toward the destination of the lightpath where it is received by the last-hop router (egress node). The lightpath request can be sent from the first-hop router as the payload of a normal IP/MPLS packet with router alert. A router alert ensures that the packet is processed by every router in the path. A channel is allocated for the lightpath on the downstream link at every node traversed by the setup. The identifier of the allocated channel is written to the setup message. After a channel has been allocated at a node, the router communicates with the OXC to reconfigure the OXC to provide the desired connectivity.

If the setup fails, the first-hop router issues a release message to release resources allocated for the partially constructed lightpath. Upon failure, the first-hop router may attempt to establish the lightpath over an alternate route, before giving up on satisfying the original user request. This alternate route is selected out of the remaining k-1 routes, subject to the same constraints, i.e., the route with the least average number of used wavelengths per the entire route.

After processing the setup, the destination (or the last-hop router) returns an acknowledgment to the source. The acknowledgment indicates that a channel has been allocated on each hop of the lightpath. It does not, however, confirm that the lightpath has been successfully implemented (i.e., the OXCs have been reconfigured). It may be desirable to have the acknowledgment confirm that every hop has completed the OXC configuration. Either way, the channel becomes available immediately after the request is sent, at the discretion of the user. Once established, the lightpath may carry arbitrary traffic, such as ATM, frame relay, or TDM circuit.

If the user requests a restored lightpath, then capacity must be reserved within the network. This is performed independently, although potentially overlapping in time, with the setup of the primary lightpath; but it may take a significantly longer time. The first-hop router is responsible for ensuring that restoration capacity is reserved for all restorable failures. The first-hop router informs the source once this is completed. The establishment of a restored lightpath is completed when the primary capacity is allocated and the restoration capacity is reserved [2].

4.4   Provisioning Lightpaths in a Network Without Wavelength Conversion Capabilities

A network or a subnetwork that does not have wavelength converters will be referred to as being wavelength continuous. In the case where wavelength converters are not available, the procedures used to provision a lightpath in the above section, where wavelength converters are available, are identical, except that the wavelength assignment algorithm is now different.

In this case, where wavelength converters are not available, a common wavelength must be located on each link along the entire route. Whatever wavelength is chosen on the first link defines the wavelength allocation along the rest of the section. A wavelength assignment algorithm must thus be used to choose this wavelength.

Finally, in the case where the lightpath is wavelength continuous (cut-through), optical nonlinearities, chromatic dispersion, amplifier spontaneous emission, and other factors together limit the scalability of an all-optical network. Routing in such networks will then have to take into account noise accumulation and dispersion to ensure that lightpaths are established with adequate signal qualities. This work assumes that the all-optical (sub-) network considered is geographically constrained so that all routes will have adequate signal quality, and physical layer attributes can be ignored during routing and wavelength assignment. However, the policies and mechanisms proposed here can be extended to account for physical layer characteristics.

It is important to emphasize that while the proposed scheme is conceptually simple, to provision the network one needs to propagate information throughout the network about the state of every wavelength on every link in the network. As a result, the states required and the overhead involved in maintaining this information would be excessive.

To relieve the excessive overhead problem, another approach is first to select the shortest path between source and destination based only on the conventional OSPF algorithm; except that the overhead consuming heuristics wavelength assignment algorithms used above should now be replaced with a simple wavelength assignment signaling protocol (SWASP) that requires wavelength usage information from only the links along the chosen route [2,7].

In this case, we must select a route and wavelength upon which to establish a new lightpath, without detailed knowledge of wavelength availability. Constructing the continuous lightpath requires SWASP to pick a common free wavelength along the flow path; therefore the algorithm must collect the list of free wavelengths for each node. If there is one free wavelength common to all the nodes, it will be picked; if not, the algorithm returns a message to the first-hop router informing it that the lightpath cannot be established. The signaling can be initiated either at the, end of the path (last-hop router) or at the beginning of the path (first hop router) [7].

If the signaling is initiated from the first-hop router, the algorithm picks the set of free wavelengths along the first link of the path to be established. When the next hop receives that set it intersects it with its own free wavelength set and forwards the result to the next hop. If the final set is not empty, the last-hop router picks one free wavelength from the resulting set, configures its local node, and sends an acknowledgment back to the previous hop with the chosen wavelength. Upon receiving an acknowledgment, the previous hop configures its local node and passes the acknowledgment to its previous hop, until the packet is received by the first-hop router (ingress node) [7]. The signaling can also be initiated from the last-hop router, but it will take more time.

Although this second scheme is not as overhead consuming as the first one, the drawback is that the chosen wavelength is now only locally optimized, and consequently the utilization of network resources is not globally optimized.

5 CONCLUSION

An overview of the vision of the next generation optical Internet has been presented. We have described two practical near-term optical Internet candidate model architectures along with a high-level discussion of their relative advantages and disadvantages: a router-centric architecture deployed on a thin optical layer, and a hybrid router/OXC-centric architecture deployed on a rich and intelligent optical transport layer. The fundamental distinction between these two models is based on the way that the routing and signaling protocols are run over the IP and the optical subnetwork layers.

This work has addressed the implementation issues of the path selection component of the traffic-engineering problem in a hybrid IP-centric DWDM-based optical network. An overview of the methodologies and associated algorithms for dynamic lightpath computation has also been presented. Specifically, we have shown how the complex problem of real-time provisioning of optical channels can be simplified by using a simple dynamic constraint-based routing and wavelength assignment (RWA) algorithm that computes solutions to three subproblems: the routing problem, the constraint-based shortest route selection problem, and the wavelength assignment problem.

We have also presented two different schemes for dynamic provisioning of the optical channels. The first scheme is conceptually simple, but the overhead involved is excessive. The second scheme is not as overhead consuming as the first one, but, its drawback is that the chosen wavelength is only locally optimized along the selected route, and consequently the utilization of network resources is not globally optimized. Two different scenarios are then considered: provisioning lightpaths in a network with full wavelength conversion capability, and provisioning lightpaths in a network without wavelength conversion capability

Finally, it is important to emphasize that although the scenarios for the dynamic provisioning schemes proposed here need more optimization (work currently going on), it is definitely a step in the right direction that will ultimately lead to much more advanced real-time provisioning schemes in the near future.

REFERENCES

1. D. Awduche et al. Multiprotocol lambda switching. Internet draft, work in progress, Nov. 1999.
2. N. Chandhok et al. IP over Optical Networks: A Summary of Issues. Internet draft, work in progress, July 2000.
3. N. Ghani. Lambda-labeling: a framework for IP-over-WDM using MPLS. Optical Networks, Vol. 1, pp. 45–58, April 2000.
4. Yinghua Ye, S. Dixit, and M. A. Ali. On joint protection/restoration in IP-centric DWDM-based optical transport networks. IEEE Communication, Vol. 38, No. 6, pp. 174–183, June 2000.
5. B. Arnaud. Architectural and engineering issues for building an optical Internet. SPIE Proceedings, Vol. 3531, pp. 358–377, Nov. 1998.
6. D. Awduche, J. Malcolm, J. Agogbua, M. O'Dell, and J. MeManus. Requirements for traffic engineering over MPLS. Internet draft, RFC-2702, Sept. 1999.
7. J. Bannister, J. Touch, A. Willner, and S. Suryaputra. How many wavelengths do we really need? A study of the performance limits of packet over wavelengths. Optica Networks, Vol. 1, No. 2, pp. 17–28, April 2000.
8. O. Gerstel and S. Kutten. Dynamic wavelength allocation in all-optical ring Networks, Proc. IEEE ICC 97, Montreal, Quebec, Canada, Vol. 1, pp. 432–436, June 1997.
9. S. Subramnian and R. A. Barry. Wavelength assignment in fixed routing WDM networks. Proc. ICC '97, Montreal, Canada, Vol. 1, pp. 406–410, June 1997.
10. R. A. Barry and S. Subramanian. The MAX-SUM wavelength assignment algorithm for WDM ring networks. Proc. OFC 97, Feb. 1997.
11. G. Jeong and E. Ayanoglu. Comparison of wavelength-interchanging and wavelength-selective cross-connects in multiwavelength all-optical networks. Proc. IEEE INFOCOM '96, San Francisco, CA, Vol. 1, pp. 156–163, March 1996.
12. E. Karasan and E. Ayanoglu. Effects of wavelength routing and selection algorithms on wavelength conversion gain in WDM optical networks. IEEE/ACM Transactions on Networking, Vol. 6, No. 2, pp. 186–196, April 1998.

2
Analysis of Multimedia Access Protocols for Shared Medium Networks

Lixin Wang and Mounir Hamdi
Hong Kong University of Science and Technology, Kowloon, Hong Kong

1 INTRODUCTION

Future local and metropolitan area networks (LANs/MANs) will be required to provide a wide variety of services, as shown in Fig. 1. The low-speed and non-quality-of-service-oriented services of Fig. 1 could be handled by evolutionary versions of the presently available networks. However, the high-speed and QoS-oriented services require a new generation of LANs and MANs. The target is therefore integrated services (multimedia) LANs and MANs that support the whole spectrum of the traffic shown in Fig. 1. However, with QoS-oriented multimedia LANs and MANs come various challenges. The most important challenge is the design of *Medium Access Control* or MAC protocols for these networks, since their topology is mostly based on a shared medium technology. One reason for this challenge is that all higher layer services (e.g., the ones shown in Fig. 1) are built on the fundamental packet transfer service provided by the MAC sub-layer, and it is the MAC protocol that determines the characteristics of this fundamental service. Hence improvements to MAC services result in improved system performance, while the provision of new MAC services means that new applications and services can be developed.

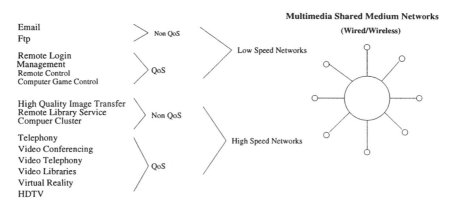

Figure 1 Various applications depend on the MAC protocols used in shared medium networks.

A plethora of MAC protocols have been proposed for future-generation LANs/MANs [4]. Unfortunately, most of these MAC protocols are not suitable for multimedia applications because they have been designed with one *generic* traffic type in mind. As a result, they perform quite well for the traffic types they have been designed for but poorly for other traffic streams with different characteristics. This observation is true for MAC protocols for wireline networks as well as wireless networks.

The objective of this paper is to propose a methodology and framework for integrating different MAC protocols into a single shared medium network efficiently to accommodate various types of multimedia traffic streams with different characteristics and QoS demands. The proposed integrated MAC protocol is termed *Multimedia Medium Access Control* (*Multimedia*-MAC) protocol. We also develop efficient analytical methodologies for the performance evaluation of our MAC protocols using different traffic and networking environments.

This paper is organized as follows. Section 2 gives an overview of MAC protocols. Section 3 introduces our *Multimedia*-MAC protocol. In Sec. 4, we derive an analytical model for the performance evaluation of our protocol. We present an example application of our Multimedia-MAC using wavelength division multiplexing networks in Sec. 5. Finally, Sec. 6 concludes the paper.

2 OVERVIEW OF MAC PROTOCOLS

MAC protocols have been the subject of a vast amount of research over the past two decades. Without going into detail about any specific MAC protocol, we can classify MAC protocols into three categories [4]: preallocation access protocols, reservation protocols, and random access protocols.

> *Preallocation-based protocols.* The nodes access the shared medium in a predetermined way. That is, each active node is allocated one or more slots within a frame, and these slots cannot be accessed by other nodes.
>
> *Reservation-based protocols.* A node that is preparing a packet transfer has to reserve one or more time slots within a frame before the actual packet transmission can take place. The reservation is typically done using dedicated control slots within a frame or a separate reservation channel. As a result, access to the data slots is achieved without the possibility of collision.
>
> *Random access protocols.* The nodes access the shared medium with no coordination between them. Thus when more than one packet is transmitted at the same time, a collision occurs, and the information contained in all the transmitted packets is lost. As a result, a collision resolution mechanism must be devised to avoid high packet delays or even unstable operation under high loads.

To serve various multimedia applications, the traffic loaded on the network is no longer of a single type but is a mixture of different types of traffic streams to be served simultaneously. Different traffics have different characteristics and thus have different transmission requirements. We can classify multimedia traffic streams as a function of their data burstiness and delay requirements as shown in Fig. 2. For example, video/audio streams and plain old telephone service (POTS) have small data burstiness but require almost constant transmission delay and almost fixed bandwidth in order to guarantee their QoS. On the other hand, applications such as image networking and distance learning are less stringent in terms of their delay requirements, but their traffic streams are very bursty. Finally, there are other applications that require a very low delay while their traffic streams are bursty. Examples of this type of applications include control messages for video-on-demand systems or interactive games, and network control and management.

These different traffic streams are better served by different MAC protocols. Video/audio data streams and other constant-bit-rate (CBR) traffic streams benefit best from allocation-based MAC protocols, since they can guarantee that each node has a cyclic and fixed available bandwidth. The best MAC protocols for this purpose would be a simple round-robin *time division multiplexing access* (TDMA) scheme. On the other hand, reservation-based MAC protocols are very well suited for applications where the traffic streams are bursty (i.e., VBR Traffic) or the traffic load of the nodes is unbalanced, since reservation-based MAC protocols schedule the transmission according to a particular transmission request.

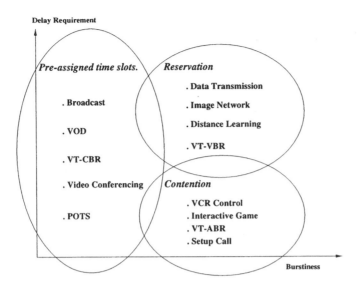

Figure 2   Multiple traffic stream features and their appropriate MAC protocols.

Finally, random access (contention) MAC protocols have the potential of meeting the delay requirements of very urgent messages, since their access delay is the lowest when collisions are avoided (or are infrequent). Some examples of these urgent messages (e.g., call setup) are listed in Fig. 2. These applications may not generate a lot of traffic data when compared to the other applications. But, once a certain traffic (message) is generated, they require a very low delay. Random access protocols perform very well, especially with respect to packet delays when the traffic load is low. However, if the traffic load gets high, then packet collisions start to occur more frequently. As a result, the performance can be quite poor.

As we can see, none of these MAC protocols serves all types of traffic well, although each one of them is ideal for certain types of traffic streams. When these types of traffic are simultaneously loaded onto a single network, no single protocol can efficiently deal with all traffic types.

With this observation, we propose an efficient access scheme for shared medium networks that

1. Integrates different types of MAC protocols into a single MAC protocol.
2. Efficiently supports different types of traffic.
3. Can be widely applied into various kinds of shared medium networks.

We term this protocol the *multimedia medium access control* (*Multimedia-MAC*) protocol for shared medium networks. More specifically, a single *Multimedia*-MAC network can simultaneously support a wide range of multimedia applications. The functions required by these applications are supported by the subprotocols that are integrated into the *Multimedia*-MAC.

3 DESIGN OF *MULTIMEDIA*-MAC PROTOCOLS

In this section we will introduce the principles behind the design of the *Multimedia*-MAC. This design is general enough to be applied to a wide variety of shared medium networks. A *Multimedia*-MAC consists of multiple *subprotocols*, each of which serves a certain type of traffic. As shown in Fig. 3, the three types of protocols, preallocation, reservation, and contention protocols, which are denoted by TDM, RSV and CNT, respectively, are the most typical protocols. Therefore integration of these three protocols is expected to cover most multimedia traffic streams.

To accommodate the three different control strategies into a single network, a time division multiplexing method is a natural choice. The idea is that each protocol in turn controls the shared medium. During that time, a protocol uses the medium, and the medium access is controlled according to the discipline of

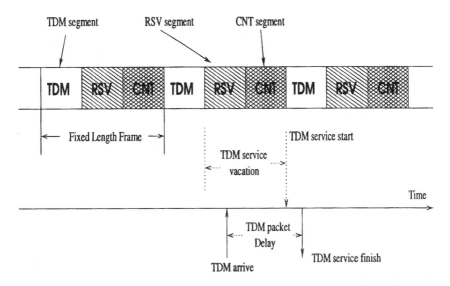

Figure 3 *Multimedia*-MAC protocol construction.

this protocol. Thus it can be thought of as if each protocol controls a subchannel of the overall channel, as shown in Fig. 3.

We can properly arrange the priority of the three subprotocols in terms of occupying the medium. In this paper, a TDM subprotocol is considered the most time-restricted protocol because it follows a predefined order. Any delay in one cycle may affect the consecutive cycles, and consequently affect the QoS. Thus TDM is designated to have the highest priority to start its service.[1] To ensure the fairness and the cyclic timing of the TDM sub-protocol, all three subprotocols should fill in the TDM framework. We can think of a cycle as a *time frame*, which is identified by a TDM cycle *id*. Because the TDM cycle is usually of constant time period, the frame is of fixed length. The frame is further divided into three parts, each of which is called a *segment*, corresponding to the three subprotocols. That is, each subprotocol takes a segment in a frame. The segment is identified by the subprotocol names. For instance, the segment occupied by a TDM subprotocol is called a TDM segment. The other two segments are named an *RSV segment* and a *CNT segment*, respectively.

The RSV subprotocol is based on reservation, so before the data transmission, the required bandwidth has to be determined. Therefore the RSV subprotocol should be allocated after the TDM subprotocol. Because the RSV type traffic is bursty traffic, the RSV subprotocol bandwidth requirement is variable.

Since the frame size is fixed, the CNT segment can only be the remaining part of the frame after the TDM and the RSV segment allocation. The CNT is a random access protocol. Both the available bandwidth and the traffic load determine its performance. When the TDM and the RSV traffic is high, the performance of the CNT subprotocol can be poor because the CNT protocol has the lowest priority to get the bandwidth. However, due to the burstiness of the RSV traffic, a large amount of bandwidth would be available after a burst has been served. The utilization of this bandwidth can achieve much better performance in this period (RSV idle period) because the CNT traffic load is relatively low, while the available bandwidth is almost the summation of the CNT and the RSV segments. It is feasible for the CNT subprotocol to fill in the idle period since the CNT subprotocol is a kind of immediate access protocol in which there is no extra signaling before the transmission. The channel can be accessed as soon as it is available.

4 ANALYTICAL MODELING OF THE *MULTIMEDIA*-MAC PROTOCOLS

In our analysis of *Multimedia*-MAC, we are particularly interested in its performance with respect to QoS metrics. For example, we are interested in finding if

---

[1] We can easily assign other priorities if it is more appropriate.

a packet has missed its transmission deadline or not. We propose a general approach to find the *deadline missing rate* (DMR), which is defined as the probability that a packet transmission exceeds an arbitrary given deadline. This model can also be used in finding traditional metrics, such as mean packet delay and throughput. We use a queueing system with a vacation model to model the waiting time distribution of the subprotocols. Then we employ a numerical method to evaluate the distribution and obtain its moments. Based on the moments, we approximately determine the DMR of these subprotocols (Fig. 4).

In a *Multimedia*-MAC protocol, the traffic streams are transmitted under the control of different subprotocols, and are multiplexed into the shared channel(s). Thus the services given to each of the data streams are not continuous, i.e., the service may be unavailable sometimes. As a result, we employ a *queueing with vacation model* to model the *Multimedia*-MAC protocol. Corresponding to different features of the different traffics and their services, we use a different queueing model to analyze their performance. More precisely, we use a $D/D/1$ queue with vacation system to model a TDM subprotocol operation; we use a $M^{(x)}/M/1$ queue with *Erlang* vacation to model the reservation type protocol; and we use a $M/G/1$ queue with deterministic vacation to model the random access protocol. To determine the DMR of a packet for a given traffic load and deadline, the packet waiting time distribution should be derived. However, it is a difficult problem to get the distribution of *non-M/M/1* queues with general vacation [3].

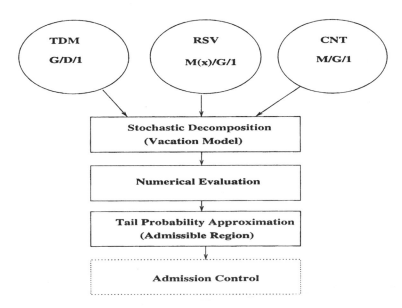

Figure 4 Our approach to modeling and analyzing the *Multimedia*-MAC protocol.

We successfully apply an approach that combines the decomposition properties of general queueing vacation systems, numerically calculating the moments of the waiting time distribution from their Laplace–Stieltjes Transform (LST) forms, and approximately computing the tail probability from the moments.

4.1 General Queueing Model Without Vacation

Let us start from a simple case: a general arrival and a general service with a single server model, denoted a $GI/G/1$ queue.

4.1.1 G/G/1 Queue

Let w_n denote the waiting time of the nth packet in the queue. Let u_n denote the interarrival time between the nth packet and the $n + 1$th packet, and v_n denote the service time of the nth packet. We choose the packet departure time as the *embedded point*, denoted as D_n. Because we assume there is no server vacation at the moment, the transmission is continuous if the packet buffer is nonempty, i.e., the waiting time of a packet is equal to the previous packet waiting time plus its service time minus the interarrival time between them. Hence we have

$$w_{n+1} = w_n + v_n - u_n \tag{1}$$

$$= w_n + x_n \tag{2}$$

where $x_n = v_n - u_n$. If the buffer is empty when the $n + 1$th packet arrives, the packet is transmitted immediately without queueing, so $w_{n+1} = 0$ in this case. Combining these two cases, we get

$$w_{n+1} = \text{MAX}(0, w_n + v_n - u_n) \tag{3}$$

From the above equation, we can get the *idle period* in which the server (the channel) is idle. The idle period is denoted by $I_n = -\text{MIN}(0, w_n + v_n - u_n)$. Then

$$w_{n+1} - I_n = w_n + x_n \tag{4}$$

$$E[e^{-s(w_{n+1} - I_n)}] = E[e^{-sw_n}]E[e^{-sx_n}] \tag{5}$$

Because we have $w_{n+1} = 0$, when $I_n > 0$, and $w_n \neq 0$ when $I_n = 0$, we get

$$E[e^{-s(w_{n+1} - I_n)}] = E[e^{-sw_n}|I_n > 0]\Pr(I_n > 0) $$
$$+ E[e^{-sw_{n+1}}|I_n = 0]\Pr(I_n = 0) \tag{6}$$

Because $w_{n+1} = 0$ when $I_n > 0$, then $E[e^{-sw_{n+1}}|I_n > 0] = 1$. Let $\Pr(I_n > 0) = a_0$ which is the probability of a new packet arrival seeing an empty buffer. Using Eq. (7), we get

$$E[e^{-sw_n}]E[e^{-sx_n}] = E[e^{-s(-I_n)}|I_n > 0]\alpha_0 + E[e^{-sW_{n+1}}] - a_0 \tag{7}$$

In a steady state, we get

$$\lim_{n\to\infty} E[e^{-sw_n}] = \lim_{n\to\infty} E[e^{-sw_{n+1}}] = E[e^{-sW}]$$

Using the LST definition, the LST of the waiting time in the queue is W^* $(s) = E[e^{-sW}]$, and the LST of the idle period I is I^* $(s) = E[e^{-sI}]$. We define the LST of the accessory random variable $x = \lim_{n\to\infty} x_n$ to be K^* $(s) = E[e^{-sX}]$. Then by Eq. (7), we can get the packet waiting time in a $G/G/1$ queue as

$$W^* (s) = \frac{\alpha_0[1 - I^* (-s)]}{1 - K^* (s)} \tag{8}$$

From equation (8), we can derive the LST forms of different queueing models for different arrivals and service disciplines.

4.1.2   D/D/1 Queue

In a $D/D/1$ queue, the packet arrival rate and service rate are deterministic. Let λ denote the arrival rate and μ represent the service rate. Then $u_n = 1/\lambda$, and $v_n = 1/\mu$. Hence

$$I^* (s) = e^{-s(1/\mu - 1/\lambda)}$$

$$K^* (s) = E[e^{-s(v_n - u_n)}] = e^{-s(1/\lambda - 1/\mu)}$$

Thus the LST form of a $D/D/1$ queue is $W^* (s) = 1$.

4.1.3   M/G/1 Queue

For a Poisson arrival system, the idle time distribution is the same as the inter-arrival-time distribution because of the property of *Poisson Arrival See Time Averages* (PASTA) which implies the memoryless property of the Poisson process (17). Then

$$A^* (s) = I^* (s) = \frac{\lambda}{(\lambda + s)} \tag{9}$$

where A is the random variable of packet arrivals and $A^* (s)$ is its LST. We denote the packet service time distribution by $B(t)$ and its LST by $B^* (s)$. Its mean is $\mu = B^{*\prime} (0)$.

Recall that

$$K(s) = E[e^{-s (v_n - u_n)}] = E[e^{-sv_n}]E[e^{-s(-u_n)}] = B^* (s) \frac{\lambda}{\lambda - s} \tag{10}$$

Using Eq. (8), we can get the LST of an $M/G/1$ queueing system as

$$W^* (s) = \frac{(1 - \rho) [-s/\lambda - s]}{1 - [\lambda/\lambda - s]} \tag{11}$$

$$= \frac{s(1 - \rho)}{s - \lambda + \lambda B^* (s)} \tag{12}$$

4.1.4 $M^{(x)}/G/1$ Queue

Let us consider the case of the bulk arrival where more than one packet arrives at a time instant. We assume the number of arrivals at each instance to be given by a random variable G, which is of the *probability generating function* (PGF) $G(z) = \sum_j g_j z^j$. Its mean is $g = E[G] = \sum_j j g_j$. We arbitrarily choose a packet in a bulk arrival, termed *tagged packet*. In a bulk arrival queueing system, the packet delay can be thought of as having two components: (1) the waiting time from the bulk arrival time to the time when the first packet in the bulk gets transmitted, denoted by $W_1^* (s)$ (LST form), and (2) the waiting time from when the first packet in the bulk gets served to the time when the tagged packet gets transmitted, denoted by $W_2^* (s)$. We will discuss them separately before we combine them.

To obtain $W_1^* (s)$, we can treat the whole bulk as a *message* which arrives in Poisson distribution. Then the service time distribution of the whole bulk is

$$B_b^* (s) = \sum_{k=1}^{\infty} a_k [B^* (s)]^k = G[B^* (s)] \tag{13}$$

Then for the *message*, we have

$$K^* (s) = E[e^{-s(v_n - u_n)}] \tag{14}$$

$$= E[e^{-s v_n}] E[e^{-s(-u_n)}] \tag{15}$$

$$= G[B^* (s)] \frac{\lambda}{(\lambda + s)} \tag{16}$$

Using Eq. (8), we can get the LST of this queueing system as

$$W_1^* (s) = \frac{a_0 s}{s - \lambda[1 - G[B^* (s)]]} \tag{17}$$

From the property that $W_1^* (0) = 1$ we can get $a_0 = 1 - \rho$ where $\rho = \lambda g/\mu$. Thus we have

$$W_1^* (s) = \frac{s(1 - \rho)}{s - \lambda[1 - G[B^* (s)]]} \tag{18}$$

The waiting time of the tagged packet in a bulk is determined by the bulk size and the position of the tagged packet in the bulk. Suppose g_i is the probability that a bulk is of size i. The probability that the tagged packet in the bulk is of exactly size i is p_i. Then we get $p_i = ig_i/g$. Assume there is no difference among the packets in a bulk. Then the probability that the tagged packet appears at position j in the bulk with size i is $1/i$. Then the waiting time for the tagged packet to be transmitted in the bulk is the time that all the packets ahead of it in the queue are transmitted. Let R be a random variable of the number of packets before the tagged packet in a bulk. The probability that there are k packets placed before the tagged packet for any number of packets in the bulk is

$$\Pr[R = k] = \sum_{j=k+1}^{\infty} \frac{g_i}{g} \tag{19}$$

$$= \frac{1}{g} \sum_{j=k+1}^{\infty} \tag{20}$$

$$= \frac{\Pr[G > k]}{g} \tag{21}$$

The PGF of R can be obtained by

$$R(z) = \sum_{k=0}^{\infty} z^k \Pr[R = k] \tag{22}$$

$$= \frac{1}{g} \sum_{k=0}^{\infty} z^k \sum_{j=k+1}^{\infty} g_i \tag{23}$$

$$= \frac{1}{g} \sum_{j=1}^{\infty} \sum_{k=0}^{(} j - 1)z^k \tag{24}$$

$$= \frac{1}{g(1-z)} \sum_{j=1}^{\infty} (1 - z^j)g_i \tag{25}$$

$$= \frac{1 - G(z)}{g(1 - z)} \tag{26}$$

The time the tagged packet has to wait in the bulk is the total transmission time of the k packets that are placed before the tagged packet. Thus the LST of the tagged packet waiting time in a bulk is

$$W_2^* (s) = \sum_{k=0}^{\infty} (B^* (s))^k \Pr[R = k] = R[B^* (s)] \tag{27}$$

$$= \frac{1 - G[B^*(s)]}{g[1 - B^*(s)]} \tag{28}$$

Consequently, we obtain the LST of the queueing time for a $M^{(x)}/G/1$ system as

$$W^*(s) = \frac{(1 - \rho)}{s - \lambda + \lambda G[B^*(s)]} \frac{1 - G[B^*(s)]}{g[1 - B^*(s)]} \tag{29}$$

4.2 Vacation Model for a *Multimedia*-MAC Protocol

The *Multimedia*-MAC protocol integrates different protocols in a single shared medium network. These protocols are active in an alternative manner to utilize the medium using different control strategies. When one of the protocols is active, the others are in an idle state even though there are packets pending in their corresponding transmission buffers. This idle period is referred to as *vacation* period. Therefore a packet waiting time from its arrival to its departure consists of three components: the queueing time, the service time, and the vacation time.

According to the decomposition property [11,14,17], the waiting time distribution (in the form of LST) can be decomposed into two components: the waiting time in the queueing system without vacation and the *forward recurrent time* in the vacation. Let V be a random variable of the time since the start of the vacation and $V^*(s)$ be its LST form. Then we can represent the LST of the waiting time distribution in the queueing vacation as

$$W^*(s) = W^*(s) \, B^*(s) V/^*(s) \tag{30}$$

Note that $V/^*(s)$ denotes a function of vacation length. For instance, it can be in a form of PGF that gives the probability of the number of packets in the system when the system is observed at an arbitrary time during the vacation. By this relation, we can further derive the LST of the waiting time distribution of the *Multimedia*-MAC subprotocols. As we discussed earlier, the transmission using a *Multimedia*-MAC protocol is organized in time frames, which are of fixed size. Let T_{frame} denote the frame time and L_{frame} the frame length in bits. Given the channel transmission rate $\mathcal{R}$ and neglecting the gaps between frames, we have $\mathcal{R} = L_{frame}/T_{frame}$. The frame is further divided into several segments corresponding to different subprotocols. The segment lengths of the the TDM subprotocol, the RSV subprotocol, and the CNT subprotocols are denoted by L_{TDM}, L_{RSV}, and L_{CNT}, respectively. The segments are further divided into fixed-length *slots* which are of length l_{slot} bits. For simplicity, we will measure the L_{frame}, L_{TDM}, L_{RSV}, and L_{CNT} in units of l_{slot}.

4.2.1 TDM Subprotocol Model

The transmission using a TDM subprotocol is controlled by a preallocated scheduling table. Each node in turn takes a TDM segment to transmit its packets.

Hence, from the point of view of the queue on each node, the transmission can be permitted only in its allocated TDM segment. In other times, the service is in vacation. Because we assume the transmission is slotted using fixed-length slots, which are of the same size as the packet size, the service time can be considered as a *deterministic service*. Thus

$$B^*(s) = e^{-s/\mu} \tag{31}$$

where μ is the service rate.

Because of the fixed transmission cycle of the TDM subprotocol, the vacation period is also fixed in length. The traffic load is deterministic with mean λ. Hence, during the vacation, the number of packets arriving at an arbitrary time is uniformly distributed, thus

$$V^*(s) - \frac{1 - e^{-\lambda L_v s}}{s\lambda L_v} \tag{32}$$

where $L_v = T_{\text{frame}} - T_{\text{TDM}}$ is the vacation time. As a result, the waiting time distribution of the packets for the TDM subprotocol can be expressed as

$$W^*_{\text{TDM}}(s) = \frac{(1 - e^{-\lambda L_v s})e^{-s/\mu}}{s - L_v\lambda} \tag{33}$$

4.2.2 RSV Subprotocol Model

Using the RSV subprotocol model, the transmission is no longer in fixed cycles but determined by the reservation and the scheduling algorithm that is affected by the traffic load and the transmission requests from all nodes, instead from only the local traffic load as in the case of TDM subprotocol. In other words, at the end of the service time, the start of the next service time is determined by the sum of the transmission time received by all the other nodes in the network. In this case, the distribution of the vacation time is quite complicated to obtain [17]. However, we can assume that the transmission time at each node is independent and has an exponential distribution. Then the summation of the transmission time can be modeled as an *Erlan-k* distribution [17], where k is the number of transmitting nodes. Assume a symmetric system where the traffic load is identical on each node, and a fair system where each node gets the same portion of service time. Then the probability that a node gets a transmission chance in a network with N nodes is $p = 1/N$. Suppose the mean time of the service time obtained by each node is T_r, then the LST of the vacation time for each of the nodes is

$$V^*(s) = \left(\frac{N}{N + sTr}\right)^N \tag{34}$$

The RSV subprotocol is designed for burst traffic transmission. In particular, the packet arrival is modeled as a bulk arrival. Applying the results for a $M^{(x)}/G/1$ system as given by Eq. (29), we can get the LST of the packet delay of the RSV subprotocol as

$$W^*_{\text{RSV}}(s) = \frac{(1 - \rho)}{s - \lambda_{\text{RSV}} + \lambda_{\text{RSV}} G[B^*(s)]} \frac{1 - G[B^*(s)]}{g[1 - B^*(s)]} \left(\frac{N}{N + G[B^*(s)]Tr} \right)^N$$

(35)

Where $G(\ )$ is the arrival distribution of the bulk length with mean g which has a geometric distribution.

4.2.3 CNT Subprotocol Model

The CNT subprotocol is a random access protocol. It serves applications that produce small amounts of data but require quick responses. We assume the packet arrival to have a Poisson distribution with mean λ_{CNT}.

Although it is still a slotted protocol, the service time of a packet is no longer deterministic because the packets may collide with each other and should be retransmitted. Before a packet is successfully transmitted out, the packet is considered to be in service (server is busy). Because the trial procedure of the CNT subprotocol can be thought of as a Bernoulli sequence of trials, the elapsed time that a packet stays in the service state has a geometric distribution [6] with a parameter p, the probability that a packet gets successfully transmitted.

More precisely, any two or more packets that are transmitted at the same time in the same channel may collide. Thus the probability that a packet gets successfully transmitted is the probability that there is only one packet being transmitted at a time in a single channel. Let n be the number of backlogged nodes at the beginning of a given slot. Each of these nodes will transmit a packet in the given slot, independently of each other, with probability q_r. Each of the $m - n$ other nodes will transmit a packet in the given slot if one (or more) such packets arrived during the previous slot. Since such arrivals are Poisson distributed with mean λ_{CNT}/m, the probability of no arrivals is $e^{-\lambda/m}$. Thus the probability that an unbacklogged node transmits a packet in the given slot is $q_a = 1 - e^{-\lambda/m}$. Let $Q_a(i,n)$ be the probability that i unbacklogged nodes transmit packets in a given slot, and let $Q_r(i,n)$ be the probability that i backlogged nodes transmit, then we have

$$Q_a(i,n) = \binom{m - n}{i} (1 - q_a)^{m-n-i} q_a^i$$

(36)

$$Q_r(i,n) = \binom{n}{i} (1 - q_r)^{n-i} q_r^i$$

(37)

Let p_i denote the probability that the number of backlogged nodes is i. The state increases when there are new arrivals on the backlogged nodes, and decrease by one if a packet is transmitted successfully. It can be shown that the state transitions form a Markov chain [7]. Then the transition probability matrix can be formed by

$$P_{n,n+1} = \begin{cases} Q_a(i,n), & 2 \leq i \leq (m\ n) \\ Q_a(1,n)[1 - Q_r(0,n)], & i = 1 \\ Q_a(1,n)Q_r(0,n) + Q_a(0,n)[1 - Q_r(1,n)] & i = 0 \\ Q_a(0,n)Q_r(1,n), & i = -1 \end{cases} \tag{38}$$

$$p_n = \sum_{i=0}^{n+1} p_i P_{i,n} \tag{39}$$

$$\sum_{n=0}^{m} p_n = 1 \tag{40}$$

By solving Eq. (39) and Eq. (40), we can get $p = p_1$, the probability that there is only one packet to be transmitted, which leads to the successful transmission, that is,

$$p = \left[\frac{(m-n)q_a}{1-q_a} + \frac{nq_r}{1-q_r} \right] (1-q_a)^{m-n}(1-q_r)^n \tag{41}$$

By knowing p, we can express the service time of a packet using a CNT subprotocol as

$$B^*_{CNT}(s) = \frac{p}{(1-p)e^{-s} - 1} \tag{42}$$

Using a *Multimedia*-MAC protocol, the CNT subprotocol takes a fixed portion of the bandwidth, so the vacation period is deterministic. Using an *M/G/1* queue with deterministic vacation, we can get the LST of the CNT subprotocol waiting time distribution as

$$W^*_{CNT}(s) = \frac{s(1-\rho)}{s - \lambda + \lambda B^*_{CNT}(s)} B^*_{CNT}(s) \frac{1}{s - \lambda L_v} \tag{43}$$

4.3 Numerical Evaluation of the Waiting Time Distribution

We have got the waiting time distribution (LST form) of the different subprotocols within the framework of the *Multimedia*-MAC. However, directly inverting from the LST form to their PGF is very difficult [3]. Fortunately, a Fourier-series method that can numerically invert the Laplace transforms and generating functions by a numerical method has been proposed [1,2,3,9].

For a nonnegative random variable X, let $p_k = \Pr\{X = k\}$, $k = 0, 1, \ldots$ Then its probability generating function is defined by

$$G(z) = E[z^x] = \sum_{k=0}^{\infty} pk^{z^k} \tag{44}$$

When X is a continuous variable, the cumulative density function of X is denoted by $F(x)$. Then the LST of $F(x)$ is given by

$$F^*(s) = E[e^{-sX}] = \int_0^{\infty} -e^{-sx}\,dF(x) \qquad \text{for } \operatorname{Re}(s) \geq 0 \tag{45}$$

Let μ_n represent the nth moment of X. Then the moment generating function (MGF) is defined as

$$M(z) = E[e^{zX}] \tag{46}$$

then

$$\mu_n = M^{(n)}(0) \tag{47}$$

The relation between the probability generating function (PGF), the moment generating function (MGF), and the Laplace–Stieltjes Transform (LST) is

$$M(z) = F^*(-z) = G(e^z) \tag{48}$$

To avoid confusion, we define a z-transform formula by

$$W(z) = \lim_{n=0}^{\infty} \omega_n z^n \tag{49}$$

where ω_n is the factor sequence with the z-transform. Let Γ denote the close contour (counterwise direction) in the complex z-plane and $W(z)$ be *analytic* within and on the contour Γ. Then by Cauchy's integral formula [3]:

$$\omega_n = \frac{1}{2\pi i} \int_{\Gamma} \frac{W(z)}{z^{n+1}}\,dz \tag{50}$$

Choose a radius r_n of the contour, i.e., $z = r_n e^{i\theta}$, and apply the *m-point trapezoidal rule* of the numerical integral computing, we get

$$\omega = \frac{1}{mr_n^n} \sum_{j=1}^{m} W(r_n e^{2\pi ij/m}) e^{-2\pi ij/m} - \bar{e} \tag{51}$$

where $\bar{e}$ is the discretion error. It can be shown that [3]

$$\bar{e} = \sum_{j=1}^{\infty} \omega_{n+jm} r_n^{jm} \tag{52}$$

In particular, let $m = 2nl$, where l is some integer factor to control the round-off error. The accuracy of the formula can be expressed as

$$r_n = 10^{-\gamma/m} = 10^{\gamma/2nl} \tag{53}$$

Then Eq. (51) becomes

$$\omega_n = \frac{1}{2nlr_n^n} \{ W(r_n) + (-1)^n W(-r_n) \tag{54}$$

$$+ 2 \sum_{j=1}^{nl-1} \text{Re}(W(rne^{\pi ij/nl} e^{-\pi ij/l}) \} - \bar{e}$$

$$\bar{e} = \sum_{j=1}^{m} \omega_{n+2ljn} 10^{-\gamma j} \tag{55}$$

Let $W(z)$ be the moment generating function (MGF) of $M(z)$. The calculation of ω_n can be used to compute the moments of X. We can expand Eq. (48) as

$$M(x) = E[e^{zX}] = \sum_{n=0}^{\infty} \frac{\mu_n}{n} z^n \tag{56}$$

so that

$$\omega_n = \frac{\mu_n}{n!} \tag{57}$$

from which we can obtain the moments of X as $\mu_n = \omega_n n!$.

During the calculation of ω_n, it may grow too fast as n increases, which causes large discretion errors. On the other hand, it may get too small, causing round-off errors. To solve the problem, an *adaptive decay rate* α_n is added in as follows:

$$\alpha_n = \frac{\mu_{n-2}/(n-2)!}{\mu_{n-1}/(n-1)!} = (n-1)\frac{\mu_{n-2}}{\mu_{n-1}} \tag{58}$$

Reconstructing the factor sequence $\omega_{nk} = \alpha_n^k \mu_k/k!$, we get

$$W_n(z = \sum_{k=0}^{\infty} \alpha_n^k \frac{\mu_k}{k!} z^k = M(\alpha_n z) \tag{59}$$

$$= F^*(-\alpha_n z) \tag{60}$$

$$= G(e\alpha_n z) \tag{61}$$

Consequently,

$$\mu_n = \frac{n!\,\omega_{nn}}{\alpha_n^n}$$

$$= \frac{n!}{2nlr_n^n\alpha_n^n}\{W_n(r_n) + (-1^n W_n(-r_n)$$

$$+ 2\sum_{j=1}^{nl-1} \mathrm{Re}(W_n(r_n e^{\pi ij/nl})e^{-\pi ij/l})\} - \bar{e} \tag{62}$$

$$= \bar{e} = \sum_{j=1}^{\infty} \alpha_n^{2ljn} \frac{n!}{(n + 2ljn)} \mu_{n+2ljn} 10^{-\gamma j} \tag{63}$$

That is, by applying the above method, we can obtain the moments of a distribution from its PGF, MGF, or LST. Many examples show that this method is quite accurate [1].

4.4 Deadline Missing Rate: The Tail Probability

The objective of analyzing the distribution of the waiting time is to obtain the distribution of the probability that the waiting time is over a given deadline D, i.e., $\mathrm{Pr}\{w > D\}$. By knowing the *cumulative density function* (CDF) of distribution of the waiting time, $W_{cdf}(x) = \mathrm{Pr}\{w \le x\}$, the probability distribution that the waiting time exceeds a given deadline D is $F_{DMR} = \mathrm{Pr}\{d = D\} = \mathrm{Pr}\{w > D\}$ $= 1 - W_{cdf}(x)$. It is equivalent to computing the tail probability of $W(t)$, which is the waiting time distribution. However, an arbitrarily given deadline makes computing of the W_{cdf} quite difficult because the number of moments required to produce adequately accurate results is unknown, although we can get quite a number of moments with high accuracy by the method we discussed in the previous subsection. Therefore, we need to find an approximation to obtain the tail probability. Chudhury proposed an approximation for this purpose [9], which is in the form

$$F_{DMR}(x) \approx Ae^{-\eta x} \tag{64}$$

where $\eta = \lim_{n\to\infty} \eta_n$ and $\eta_n = n\mu_{n-1}/\mu_n$, whereas $A = \lim_{n\to\infty} A_n$ and $A_n = \eta_n^n \mu_n/n!$.

With this calculation, we can obtain the deadline-missing rate of an arbitrarily given deadline.

4.5 Admissible Region

As we can see from above discussion, the DMR depends on a given deadline and the packet elapsed time that a packet should be waiting in the system before it is transmitted out. The waiting time is affected by the network traffic load (e.g., the distribution of the arrival and its mean λ), service discipline (e.g., B^* (s)), and the vacation regulation (e.g., V^* (s)). Thus the DMR not only reflects the requirement of the traffic to be transmitted (i.e., the deadline D) but also the current network status and operation situation. The significance of evaluating the DMR is that it can *estimate* the QoS of the network. That is, by knowing the $F_{DRM}^t(t)(r)$, which is the DMR at time t, we can estimate the DMR $F_{DMR}^{t+1}(x+)$, when a new traffic arrives, which is associated with the arrival distribution, amount of load, deadline requirement $(x+)$, and expected DMR, Ψ. If $F_{DMR}^{t+}(x+)$ $> \Psi$, this traffic QoS requirement cannot be satisfied; otherwise, the traffic can be admitted safely. Note that after a new traffic is admitted, the network state is changed. If a new traffic arrives, the estimation should be performed based on the latest network states, i.e., the $F_{DMR}(x)$ should be reevaluated.

Using our *Multimedia*-MAC protocol, we consider only three types of sub-protocols, i.e., the three sets of arrivals, service, and vacation disciplines. Any new traffic will be admitted in one of the sets. Because the traffic streams within one of the sets are transmitted using the same subprotocol, the service discipline is deterministic. Therefore many of the parameters to evaluate the DMR become deterministic, which results in simplification of the admission complexity.

More precisely, within each of the sets, the parameters are only the traffic load λ and a given deadline D for the evaluation of the $F_{DMR}(x,d)$. Therefore the relation between the DMR and the mean traffic load λ and the given deadline D can be illustrated in a three-dimensional graph as shown in Fig. 5. As we can see, the space is divided by a surface. The points on the surface indicate the critical state at which, for the corresponding traffic load and deadline, the *minimum deadline missing rate*, denoted by $\mathscr{F}$, can be achieved. So at this point, if the required DMR is above $\mathscr{F}$, this traffic can be safely admitted. Therefore we call the region that is above the surface an *admissible region*.

Walking on the surface, we can see that when the deadline requirement is not strict, then more traffic can be loaded. Also, when the traffic load is low, more tight deadline requirement traffic can be admitted.

As shown in Fig. 5, the admissible region can be used to estimate the QoS of the network before the traffic is really loaded in. Suppose the current network state is at $\mathscr{F}(x, d)$, a new traffic with load $x+$ and deadline requirement $d+$ comes in. Along the surface, we can find the *minimal deadline missing rate* as shown at $\mathscr{F}(x+, d+)$. By comparing the $\mathscr{F}(x+, d+)$ with the deadline missing rate requirement of the newcomer, we can decide if the new traffic should be admitted or not.

For a given traffic distribution, service distribution, and vacation discipline, the admissible region is relatively stable. Thus it is significant in practice to estimate quickly the QoS by possibly constructing a lookup table, or by approximately computing the surface.

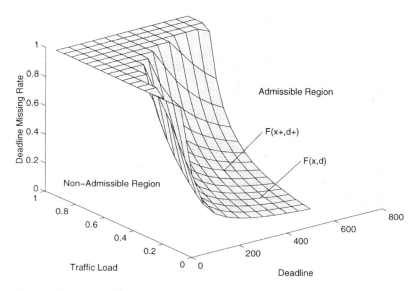

Figure 5   Admissible region as a function of DMR, deadline, and traffic load.

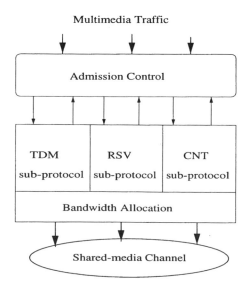

Figure 6   Admission control and *Multimedia*-MAC.

We should note that, although our admissible region is obtained by approximation, the information that it intends to supply to the admission control is instantaneous, rather than based on long-term statistics as in many existing admission control schemes. Thus the admission control strategy is expected to be much more accurate. As a result, both channel utilization and QoS can be improved. This idea is shown in Fig. 6.

5 THE APPLICATION OF *MULTIMEDIA*-MAC ON WDM NETWORKS

The *Multimedia*-MAC design and the analytical model that have been presented in the previous sections can be applied to a wide variety of shared medium networks. For example, we have already applied the design principles of *Multimedia*-MAC to the design of a multimedia wireless access protocol [19,20]. We have shown (through extensive simulations) that our multimedia wireless access scheme was able to provide QoS guarantees under stringent conditions and outperforms state-of-the-art wireless access schemes. For more details the reader is referred to Ref. 19.

In this section, we consider the application of the *Multimedia*-MAC and its analytical model on *wavelngth division multiplexing* (WDM) networks. WDM networks use fiber-optic links that can potentially deliver terabits/s. WDM is an effective technique for utilizing the large bandwidth of an optical fiber. It partitions the optical fiber into a number of noninterfering wavelengths (channels). Then, by allowing multiple messages to be simultaneously transmitted on a number of channels, WDM has the potential to improve significantly the performance of optical networks. The nodes in such a network can transmit and receive messages on any of the available channels by using and tuning one or more tunable transmitter(s) and/or tunable receiver(s). Several topologies have been proposed for WDM networks [15,16]. Of particular interest to us in this paper is the single-hop topology where a WDM optical network is configured as a broadcast-and-select network in which all the inputs from the various nodes are combined in a passive star coupler, and the mixed optical information is broadcast to all destinations.

To unleash the potential of single-hop WDM passive star networks, efficient multiple-access protocols are needed to allocate efficiently and coordinate the system resources while satisfying the traffic QoS constraints. Based on the idea of our proposed *Multimedia*-MAC protocol, together with the consideration of the physical transmission characteristics of WDM networks, we propose a MAC protocol for single-hop WDM networks, which combines the advantages of three types of MAC protocols within a single framework to serve well a wide

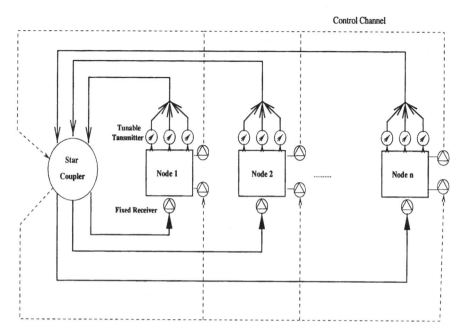

Figure 7   An M-WDMA network architecture.

range of traffic streams in multimedia application. This MAC protocol is termed *Multimedia-WDMA*, M-WDMA, for short.

5.1   The M-WDMA Architecture

Consider an M-WDMA network with N nodes connected by a star coupler and having C channels as shown in Fig. 7. Each node has a fixed channel (*home channel*) with wavelength λ_i ($i = 1, 2, \ldots, C$). The home channels are intended for the destination nodes to receive packets. In case the number of available channels $C < NT$ several nodes ($[N/C]$) may share one single home channel. The destination nodes can then accept or discard the packets by checking the addresses associated with these packets.

   The M-WDMA MAC protocol has three integrated *MAC subprotocols*, namely, a TDMA protocol, a reservation-based protocol, and a random-access protocol to serve all types of traffic streams. In our M-WDMA protocol, each node has three tunable transmitters.[2] The three tunable transmitters are used to

---

[2] For detailed justification and thorough analysis on the cost-effectiveness of using three transmitters, the reader is referred to Ref. 21. Because of lack of space and to avoid putting the chapter out of focus, we left that material out.

serve three different classes of traffic streams according to the corresponding *subprotocols*. The transmitters are named TDM transmitters, RSV transmitters, and CNT transmitters. The TDM stands for time division multiplexing, meaning that this transmitter is driven by a TDM-like subprotocol. Similarly, the RSV and the CNT transmitters are used for the reservation-based subprotocol and the contention-based sub-protocol, respectively.

To complete a packet transmission, wavelength coordination has to be established in a WDM network. In a M-WDMA network, the transmitters have to tune to the home channel of the destination node before the transmission takes place. However, since there is only one receiver at each node, if more than one transmitter sends data to the same receiver, that would produce what is known as *receiver collision*. When two or more receivers share the same wavelength (the case when $N > C$), only one of them can be active during the communication, otherwise it may cause *channel collision*. Finally, if at a pre-scheduled transmission time, the transmitter is occupied by another transmission at the same node, this may result in a *transmitter collision*. Consequently, with the occurrence of any of the above collisions, the transmission would be unsuccessful. The major function of the MAC protocol is properly to avoid these collisions and efficiently transmit the packets.

Different types of WDM MAC protocols use different schemes to avoid or resolve the above collisions. In a TDMA protocol, packet transmissions avoid collision by properly arranging the schedule maps. In a reservation-based protocol, packet collisions are avoided by actively arranging the packet transmission coordination according to the transmission requests. In a contention-based MAC protocol, when packets collide, they are required to be retransmitted again using, for example, a binary back-off algorithm.

The three transmitters of a M-WDMA network can operate in a *pipeline* fashion. That is, when one transmitter is transmitting a packet, the other transmitters can start their tuning process. As a result, the three types of transmissions (TDM, CNT, and RSV) from a node cannot take place at the same time. This has lead us to organize the three types of transmissions into a time *frame*. A frame consists of three *segments*: a TDM segment with length L_{TDM} slots, an RSV segment with length L_{RSV} slots, and a CNT segments with length L_{CNT} slots. The TDM transmissions can only be started in a TDM segment, and so with the CNT and RSV transmissions, respectively. To make sure that the TDM segments appear in periodic constant intervals, the frame length, L_{frame}, is fixed (i.e., $L_{frames} = L_{TDM} + L_{RSV} + L_{CNT}$ is set as a fixed parameter). A transmission example of our M-WDMA MAC protocol is illustrated in Fig. 8.

In this figure, the horizontal direction denotes time and the vertical direction denotes spatial location of the backlogged nodes (the number in the vertical axis represents the location of the nodes, 1 means the location of node 1). The white segments denote the TDM segments, the light-shaded segments are the RSV segments, and the dark segments are the CNT segments. The numbers in the TDM

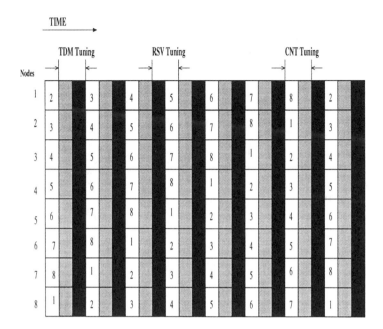

Figure 8 An M-WDMA transmission schedule example.

segments identify the home channel number that the underlying TDM segment ties to, according to the TDM protocol. For example, in the second row and fourth frame, the frame *id* is 6, which means that at that time the current TDM segment is transmitted on channel 6. Next, the fifth frame will be transmitted on channel number 7. All these transmissions happen at node 2, i.e., the TDM transmitter on node 2 is tuned to channel 7 at the time the fifth frame comes. From the figure, we can see, ideally, that all channels can be utilized and all nodes transmitting even though some of their transmitters may be in a nontransmitting mode (tuning to a certain channel).

The frame length can have an implication on the operation and performance of an M-WDMA network. Hence it should be properly chosen so that the tuning time can be efficiently masked. That is,

$$L_{TDM} + L_{RSV} > \Gamma$$
$$L_{TDM} + L_{CNT} > \Gamma$$
$$L_{CNT} + L_{RSV} > \Gamma$$

Since $L_{frame} = L_{TDM} + L_{RSV} + L_{CNT}$, then $L_{frame} > 3\Gamma/2$.

In particular, the length of the frame directly affects the bandwidth allocated to TDM segments. Hence, the total TDM bandwidth is given by

$$B_{\text{TDM}} = \frac{L_{\text{TDM}}B}{L_{\text{frame}}}/(N - 1)$$

When L_{frame} increases, B_{TDM} decreases.

In a M-WDMA MAC protocol, a frame segment is further divided into slots (with length L bits) as shown in Fig. 9, so that multiple nodes can share the same segments. Since the TDM segments and the RSV segments are used in a predetermined manner, then at any one time, only one source node will be using a specific segment, while the CNT segments of a frame can be used by multiple nodes.

The control channel is a shared-access channel that is used by all nodes, which is controlled by a pure TDMA scheme. Its cycle time is T_{ctrl}. A control channel cycle, T_{ctrl}, consists of N minislots, each of which is assigned to a node. Every node can put on its reserved minislot any necessary information that need be known by all other nodes. At the end of a cycle, all nodes get global information about the status of the whole M-WDMA network. Based on this information, all control procedures can be done by the nodes locally and network-wide synchronously. The control procedures include (1) reservation requests, which are used by the reservation subprotocol, and (2) collision acknowledgements.

5.2 The M-WDMA Protocol

The M-WDMA MAC protocol is an integrated protocol. It includes three subprotocols: a TDM sub-protocol, an RSV subprotocol, and a CNT subprotocol. Under

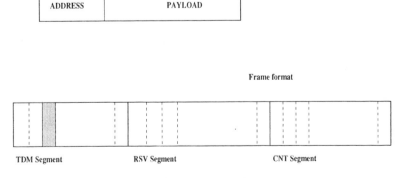

Figure 9 An M-WDMA frame and slot format.

the regulation of a M-WDMA frame format, these three subprotocols operate independently.

5.2.1 The TDM Subprotocol

The operation of the TDM subprotocol within our M-WDMA network is basically an *interleaved TDMA* MAC protocol [8]. The only difference between our TDM subprotocol and ITDMA is that in a M-WDMA network we take tuning time into consideration. Using the M-WDMA protocol, at the border between a TDM segment and an RSV segment, the TDM transmitter starts to tune to the next channel. Since the tuning order is prefixed, no extra information exchange is needed in the control channel.

5.2.2 The RSV Subprotocol

In a M-WDMA MAC protocol, the RSV packet transmission is controlled using a *multiple token* method [22]. Each channel is associated with a "token." A node can send its packets onto the destination channel only if it holds the corresponding token. An obvious advantage of using a token-based scheme in our RSV subprotocol is that it can efficiently support very bursty traffic streams, and its implementation can be simple. The disadvantage of this scheme is that the channel efficiency may not be high when compared to a perfect scheduling scheme, since in a frame only one single node is allowed to access the RSV segment at a time.

In our implementation of the multiple token mechanism, the tokens are not explicitly issued. They are in fact *virtual* tokens. In each cycle of the control channel, the M-WDMA nodes broadcast their transmission requests to all nodes. At the end of the cycle, all nodes synchronously execute a *token rotation algorithm* to determine the token distribution in the coming frame. The *token rotation algorithm* is depicted in Fig. 10.

Here a prefixed order can adopt any order. We usually use an order of node *id* so that all nodes could get identical results when the algorithm is executed in a distributed manner. If an M-WDMA network has an unbalanced load (e.g., client–server traffic), we can change the token order to favor the loaded nodes. In this way, more transmission chances are allocated to those nodes. Note that there are no transmission conflicts among nodes when using the token rotation algorithm, since each node can only send one transmission request, and there is only one token available for any particular channel. The last part of the token rotation algorithm is to allocate arbitrarily the spare token to some idle nodes. To ensure that the algorithm gets identical results, the allocation is also carried out in a prefixed order.

The token rotation is executed according to the control channel timing. Thus the rotation results have to be synchronized with the data transmission (i.e.,

[Token Rotation Algorithm]

Put all the tokens with their holder into a *token list*;
Put all the nodes with a request into *checking list*;
For every token do {
Check if the token holder still needs the token
     If (it is true and does not exceed the token holding time)
     then {
          allocate the token to the current holder;
          remove the node from the checking list;
          remove the token from the token list;
          continue to the next token;
     };
     Find another node waiting for the token in a pre-fixed order;
       if (Found) {
          Allocate the token to that node.
          remove the node from the checking list;
          remove the token from the token list;
       }
}

For each of the token remaining in the token list, do {
Find a node in the checking list, which dose not have any token;
Allocate the token to the node;
remove the node from the checking list;
remove the token from the token list;
}

Figure 10 The token rotation algorithm.

the frame timing). In particular, any rotation results are effective for the first coming RSV segment after each execution of the token rotation algorithm.

5.2.3 The CNT Subprotocol

The CNT subprotocol of our M-WDMA MAC protocol is similar to the *interleaved slotted ALOHA* [8]. The active nodes compete for the slots in the current CNT segment. In case there is a collision, retransmission is scheduled according to a binary back-off algorithm. In the context of the M-WDMA, two problems have to be solved. The first one is how to handle the acknowledgments. In M-WDMA, there are two receivers at each node. One is the home channel receiver and the other is the control channel receiver. Although we can detect the collision at the receiver side, the notification of collision can consume a portion of the control channel bandwidth. In our design, the collision notification is issued in the early part of control channel cycle. Considering the round trip propagation delay, the total acknowledgment delay is the propagation time plus the control channel cycle time. In case the propagation delay is large, we can adopt a larger *sliding window* scheme to reduce its effect.

The second problem that needs to be addressed is the retransmission of collided packets. When a collision occurs, retransmission is inevitable. But the scheduled retransmission should not cross the border of a CNT segment, otherwise it will collide with other types of transmissions. However, if the back-off is limited to within one CNT segment, we may cause even more collisions. Thus a CNT segment counter is set at each node to resolve this problem. When a node encounters the last slot of a CNT segment, its CNT segment counter stops counting. When the first slot of a CNT segment arrives, the counter starts to tick again. In this way, the retransmission time is calculated in terms of the counter rather than the real slot numbers in a frame. As a result, retransmission can be carried out across the frames.

5.3 Control Channel Configuration

The control channel operates in a TDMA manner independent of the data transmission. One cycle consists of N *minislots*, each of which is designated to a node. The cycle length is determined by the amount of information required and the number of nodes involved. Since the TDM subprotocol is a preallocated protocol, there is no need to exchange control information before transmission through the control channel. The RSV subprotocol is a multiple token passing protocol, so the information needed to process the token rotation is only the states of the queue. We can use a bit map to represent the reservation request, each node taking a bit. If the bit is "1", that means the queue is nonempty and there is at least one packet waiting for transmission. Otherwise, when it is "0", that means the queue is empty and there are no transmission requests. Therefore for each node one bit is sufficient. If there are N nodes in the network, each minislot of the control channel needs N bits to indicate the request information (including itself for simplicity). For the CNT transmission acknowledgement, we can use the bit map, again, to indicate the success or failure during the contention. Since there are L_{cnt} slots in a frame for contention, the L_{cnt} bits are enough for a CNT transmission acknowledgment. If a bit is "1", it implies the corresponding slot is successful. Since this information is broadcast to all nodes through the control channel, the source node can obtain this information and start to prepare the next transmission (or tuning). If a bit is "0", then there is a collision in the corresponding slot; thus the source nodes involved in this collision have to start the retransmission process. Consequently, the length of a minislot is determined by $N + L_{cnt}$. As a result,

$$T_{ctrl} = \frac{N(N + L_{cnt})}{R}$$

where R is the channel bit rate. To realize the frame-by-frame reservation and collision detection, the control cycle has to complete within a frame time, that is,

$$\frac{N(N + L_{cnt})}{l_{slot}} < L_{frame}$$

where l_{slot} is the slot length in bits. Solving the inequality, we get

$$N < \frac{L_{cnt}}{2}\left(\sqrt{1 + \frac{4l_{slot}L_{frame}}{L_{cnt}^2}} - 1\right)$$

This leads to a network scale limitation. For example, if we choose $L_{cnt} = 10$ slots, $L_{frame} = 30$ slots, $l_{slot} = 424$ bits, then the number of nodes should be $N < 107$.

By adding another receiver to do the collision detection, the L_{cnt} can be eliminated. Then the scale limitation becomes $\sqrt{l_{slot} L_{frame}}$ (which is 113 for the example above). Alternatively, by releasing the requirement of frame-by-frame reservation, the square-root item can increase considerably, so that the restriction become more relaxed.

5.4 Modeling of the M-WDMA Network

This section investigates the performance of the M-WDMA MAC protocol analytically and through simulations. In an M-WDMA MAC protocol, three subprotocols coexist, but they are not active simultaneously. When one of the subprotocols is in use, the other subprotocols are idle from the viewpoint of channels and nodes. In particular, the three subprotocols can be thought of as if they operate in three different networks. These three virtual networks have a bandwidth equal to the bandwidth portion that is allocated to that subprotocol in an M-WDMA network. This is reasonable under the assumption that the three classes of traffic streams are independent from each other.[3] We term this assumption the *protocol*

---

[3] Under a very dynamic environment, where traffic loads and types change frequently, extensive discrete-event simulations have already been conducted on an M-WDMA using *dynamic bandwidth allocation algorithms* [18,21]. We leave those results for a companion paper. In particular, analytically modeling such a system is an extremely difficult problem that is one research direction being taken by the authors.

independent assumption. With this assumption, the three subprotocols can be modeled separately

Given a certain packet, no matter which subprotocol would serve it, it has to wait in its queue for some time until it is its turn to be served. That is, the server is not always available. It simply alternates between the states of *idle* and *busy.* In other words, a server (transmitter) would poll the queues containing the packets each time it wants to serve a packet. As we discussed earlier, this system can be modeled using a queueing with vacation system [5,10,12].

We assume that all the nodes in the M-WDMA network have equal probability to generate packets and also have equal probability to be the destination of a packet. Let λ be the network normalized traffic load. Then λ_{TDM}, λ_{RSV}, and λ_{CNT} are the mean traffic loads for the individual segments of the subprotocols, respectively. The relationships between these values are

$$\lambda_{TDM} = \frac{L_{TDM}}{L_{frame}} \lambda \tag{65}$$

$$\lambda_{RSV} = \frac{L_{RSV}}{L_{frame}} \lambda \tag{66}$$

$$\lambda_{CNT} = \frac{L_{CNT}}{L_{frame}} \lambda \tag{67}$$

We consider a system of N nodes and C channels. Each node has $3C$ queues, which correspond to C channels and to one of the three types of traffic. We assume that each of the queues has infinite capacity and uses an FCFS discipline. Packets arrive with a mean rate of λ for each station. Let $B(x)$ be the distribution function for the service time, with $1/\mu$ being its mean and $B^*(s)$ being its Laplace–Stieltjes transform (LST). At the end of each segment, a different subprotocol is started, while the former subprotocol comes into *vacation.* We denote the RV of vacation length (in terms of slot time) as v, its LST as $V^*(s)$, and its mean as $E[V]$. We assume all stations are statistically identical in terms of all the parameters and states, i.e., are considering a symmetric system. By applying the analytical results introduced in Sec. 4, we can get the queueing model with vacation for these three subprotocols.

5.4.1   The TDM Subprotocol Model

According to a TDM sub-protocol, node i gets a change to transmit L_{tdm} packets to node j in every $N-1$ frame, and this frequency is a constant. So from queue q_j's point of view, the vacation length of the queue is $v_i = (N-1) L_{frame} T_{slot}$. Due to the symmetric properties of the system, we can simplify the vacation period of all TDM queues on every node as

$$v = (N - 2)L_{\text{frame}} T_{\text{slot}} + L_{\text{TDM}} T_{\text{slot}} \tag{68}$$

Because the slot is of fixed size, we have

$$L_v = (N - 2)L_{\text{frame}} + L_{\text{TDM}} \tag{69}$$

According to [33] and the numerical evaluation method depicted in Section 4, we can get the necessary moments of the waiting time distribution. In particular, the mean waiting time is given by

$$E[W] = W^{*(1)}(0) = \mu_1 \tag{70}$$

Figure 11 compares the results obtained by the analytical model and discrete event simulation in terms of mean delay. As we can see, the two sets of results are quite close to each other and confirm the accuracy of our analytical model.

With these moments, we can get the deadline-missing rate (DMR) according to the approximation [64] derived in Sec. 4. Figure 12 shows the TDM model validation in terms of DMR when compared to discrete-event simulation. Although they go through a quite complicated computation process, the results are quite close to each other, which implies a good accuracy of our TDM model.

5.4.2 The RSV Subprotocol Model

In an M-WDMA, the RSV subprotocol is a token-based protocol. Once a node i gets a token corresponding to the destination (or channel) j, when queue j becomes empty or the number of packets transmitted exceeds a quota (length of

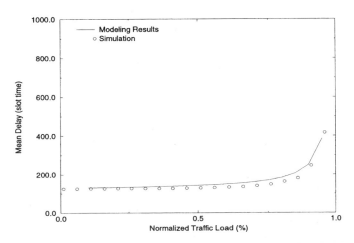

Figure 11 TDM model validation in terms of mean delay.

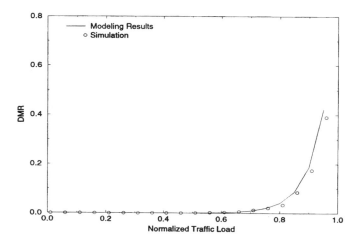

Figure 12   TDM model validation in terms of DMR.

RSV segment), the service is stopped and switched to the next node. During the service time, new arrivals are automatically involved into the transmission in case the slots of the RSV segment are not yet used.

For simplicity, we assume all the token rotations to be mutually independent, and the rotation direction is of round-robin fashion. Thus, from the point of view of each RSV queue on each node, the vacation time is the time that the corresponding token is rotated through all other nodes, if the nodes have packets to transmit using this token. This service discipline is identical to what we discussed in Sec. 4, so the analysis result can be applied directly.

Note that here we assume a bulk arrival with a mean λ_{RSV} in which the bulk has an exponential distribution with mean g. That is the PGF of G:

$$G(z) = \frac{g}{g - \ln z} \tag{71}$$

As a collision free and slotted transmission system, the service time distribution is $B^*(s) = e^{-s/\mu}$ with mean $E[B] = 1/\mu$.

The time waiting on each node is expected to be $T_r = L_{frame}$. There are totally $k = N - 1$ nodes, so they are considered as $N - 1$ stages in an Erlang service model. Therefore we can model the token rotation as an *Erlang-k* distribution [17]. In particular, we can apply the results [35] in Sec. 4 to obtain the LST of the waiting time $W^*_{RSV}(s)$, and then get the necessary moments for calculating the DMR.

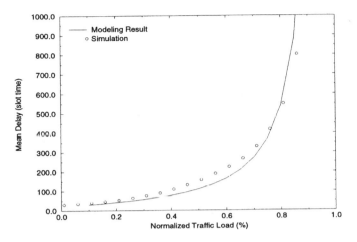

Figure 13   RSV model validation in terms of mean delay.

Figure 13 shows the validation of the model in terms of mean delay. The result shows that the model is reasonably accurate.

As with the TDM model, we can calculate the DMR according to [64]. Figure 14 compares the analytical results with the simulation results. The two results are close to each other, especially when the traffic load is light.

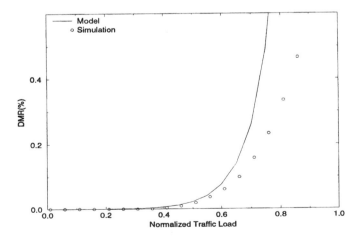

Figure 14   RSV model validation in terms of DMR.

5.4.3 CNT Subprotocol Model

The CNT subprotocol in an M-WDMA network is a random access protocol. Every node can access the CNT segment of each frame. But due to collisions, a packet may not be successfully transmitted out. In this case, the packet is kept at the front of the queue until it is successfully transmitted. Thus from the point of view of a CNT queue, the service time is no longer deterministic but in a geometric distribution. In addition, because all the slots in the CNT segments can be used by any node, the vacation is the only time when TDM and RSV are active, i.e., $v = (L_{TDM} + L_{RSV})T_{slot}$. Again, the model detailed in Sec. 4 can be used here as well. By choosing the exact same parameters for both modeling and simulation, we validate the CNT model in Fig. 15 in terms of mean delay and Fig. 16 for DMR comparison. As we can see, again the results are reasonably close to each other.

5.4.4 Admissible Region Comparison

As illustrated previously, the admissible region can be effectively used by an admission control policy to decide whether to admit a traffic stream or not. By calculating the DMR of TDM, RSV, and CNT subprotocols for normalized traffic load and a range of required deadlines, we can obtain the corresponding *admissible region* as shown in Figs. 17, 18, and 19.

In Figs. 17 and 18, we show three types of surfaces. The lowest surface is generated from our analytical modeling results. The top layer surfaces are upper-bound computed according to *Chebyshev's inequality* [13] and are used

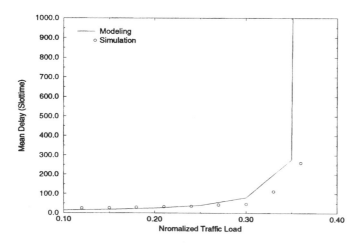

Figure 15 CNT model validation in terms of mean delay.

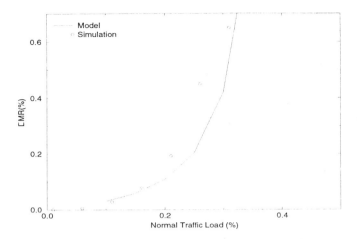

Figure 16 CNT model validation in terms of DMR.

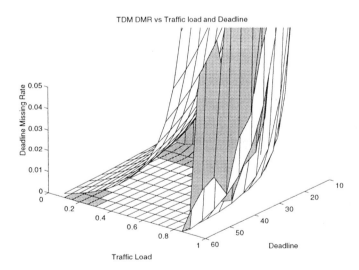

Figure 17 Admissible region for CBR traffic.

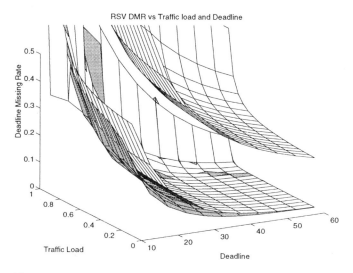

Figure 18   Admissible region for RSV traffic.

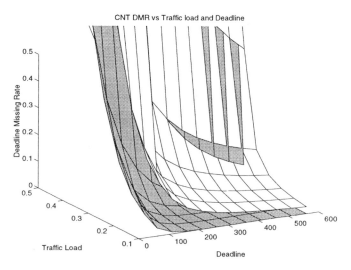

Figure 19   Admissible region for CNT traffic.

as modeling references (in fact,/a simple upper bound). The middle layer surface is obtained through intensive simulation. Similar comparison on CNT traffic is shown in Fig. 19.

In general, the simulation results are much closer to the analytical modeling results than that of the upper bound.

By comparing these three figures, we can see that the admissible region of a TDM subprotocol is quite flat. This is because in a M WDMA, the TDM sub-protocol has the highest priority to take the bandwidth. Hence the bandwidth is guaranteed and the DMR can be kept low even when the traffic load is high.

The RSV admissible region can extend to very high traffic loads when the deadline is not very strict. This is the advantage of the RSV subprotocol because the RSV subprotocol is primarily used for bursty traffic with relatively loose delay requirements. However, when the deadline becomes strict, the QoS of an RSV subprotocol is eventually worse than that in TDM—as is expected.

The CNT admissible region is valid only in very low traffic areas, that is, under light traffic conditions—which is again expected from contention-based protocols. However, a very low DMR can be achieved even when the deadline is very strict. Thus some urgent packets (e.g., for network control, for example) can exploit this feature when the traffic load is low.

6 CONCLUSION

This paper introduces a new methodology that combines different types of MAC protocols into a single shared medium network the better to serve a wide variety of multimedia applications. Some of the goals of this approach are (1) to keep the advantages of the individual MAC protocols with respect to specific types of traffic streams, (2) efficiently to support a large range of traffic streams with different characteristics and QoS requirements in a single shared medium net-work, and (3) to be applied to a wide variety of shared medium networks. We have also derived a detailed analytical model that can be used to determine the admissible region given various practical parameters (e.g., traffic load, type of traffic, deadline missing rate). This admissible region can readily be used by an admission control policy to decide whether to admit an arriving stream while satisfying its QoS and at the same time not altering the QoS of the already admit-ted streams.

We have illustrated the usage of our MAC protocol design and analytical model through a wavelength division multiplexing network. We have shown that our general framework can readily be used for such networks. If addition, we have shown that our analytical results are reasonably accurate when compared to simulation. In particular, our analytical model can effectively be used with admission control algorithms for providing QoS guarantees to multimedia applications.

We believe our framework can be applied to many other types of shared medium networks, especially in the next generation of networks, where the integrated or multimedia services are supplied in a single physical network.

REFERENCES

1. J. Abate, G. Choudhury, and W. Whitt. Exponential approximations for tail probabilities in queues: I: Waiting times. *Operations Research*, 43(5):885–901, Sept.–Oct. 1995.

2. J. Abate, G. Choudhury, and W. Whitt. Exponential approximations for tail probabilities in queues II: Sojourn time and workload. *Operations Research*, 44(5):758–763, Sept.–Oct. 1995.

3. J. Abate and W. Whitt. The Fourier-series method for inverting transforms of probability distributions. *Queueing Systems*, 10:5–88, 1992.

4. H. R. V. As. Media access techniques: the evolution towards terabit/s LANs and MANs. *Computer Networks* and *ISDN Systems*, 7(5):603–656, Sept. 1994.

5. R. Al-Naami. Queueing analysis of slotted ALOHA with finite buffer capacity. *IEEE GLOBCOME '93*, 2:1139–43, 1993.

6. A. O. Allen. *Probability, Statistics, and Queueing Theory with Computer Science Application*. Academic Press, INC., 1990.

7. G. Bersekas. *Network and Communication Systems*. Princeton, 1992.

8. K. Bogineni, K. M. Sivalingam, and P. W. Dowd. Low-complexity multiple access protocols for wave-length-division multiplexed photonic networks. *IEEE Journal on Selected Areas in Communications*, 11(4):590–604, May 1993.

9. G. Choudhury and D. M. Lucantoni. Numerical computation of the moments of a probability distribution from is transform. *Operations Research*, 44(2):368–381, Mar.–Apr. 1996.

10. E. de Souza e Silva, H. R. Gail, and R. R. Muntz. Polling systems with server timeouts and their application to token passing networks. *IEEE/ACM Transactions on Networking*, 3(5):560–575, Oct. 1995.

11. M. Doshi. G/G/1 queue with general vocation—a survey. *Queueing System*, November 1989.

12. W. P. G. and H. Levy. Performance analysis of transaction driven computer systems via queueing analysis of polling models. *IEEE Transactions on Computers*, 41(4): 455–465, April 1992.

13. G. R. Grimmett and D. R. Stirzaker. *Probability and Random Processes*. Oxford Science Publications, 1992.

14. J. Medhi. *Stochastic Models in Queueing Theory*. Academic Press, 1991.

15. B. Mukherjee. WDM-based local lightwave networks Part I: Single-hop systems. *IEEE Network*, 6(3):12–27, May 1992.

16. B. Mukherjee. WDM-based local lightwave networks Part II: Multi-hop systems. *IEEE Network*, 6(4):20–32, July 1992.

17. H. Takagi. *Queueing Analysis: A Function of Performance Evaluation.* North-Holland, 1991.

18. L. Wang and M. Hamdi. Efficient protocols for multimedia streams on WDMA networks. *Proceedings of the Twelfth International Conference on Information Networking (ICOIN-12)*, 1998.

19. L. Wang and M. Hamdi. M-WMAC: an adaptive channel access protocol for multimedia wireless networks. *Seventh International Conference on Computer Communications and Networks (ICCN'98)*, October 1998.

20. L. Wang and M. Hamdi. Multimedia wireless channel access protocol for personal communication system. *Wireless Personal Communication Systems*, 1999 (to appear).

21. L. Wang. *Multimedia access protocols for shared medium networks.* Ph.D. dissertation, Computer Science Department, Hong Kong University of Science and Technology, 1999.

22. A. Yan, A. Ganz, and C. M. Krishna. A distributed adaptive protocol providing real-time services on WDM-based LANs. *Journal of Light-Wave Technology*, 14(6): 1245-1254, June 1996.

3

The Design and Performance of a Lightwave Data Storage Network Using Computer Analysis and Simulation

S. K. Ramesh
California State University, Sacramento, California

Herbert J. Tanzer
Hewlett Packard Co., Roseville, California

1 INTRODUCTION

Advanced optical gigabit network technology, protocols, and standards are being developed at a rapid pace. However, while there have been significant advances and system implementations in the long-distance networks used for telecommunications, the implementation of optics into the short-haul connection of computer networks has begun more recently. This work is an attempt to apply appropriate hardware technology and selection factors into the design and modeling of an all-optical and multichannel storage area network called a SAN. At the physical layer, a SAN is very similar to a LAN, with the major distinction stemming from the higher degree of data-intensive and data-secure architecture appropriate for an enterprise's data warehousing and processing needs.

The objective of this work is to utilize computer analysis and simulation models to predict system performance, and to aid in the design of a next-generation SAN operating at 1 to 20 Gbps electronic data rates. Wave-division multiplexing is expected to enhance greatly network throughput through multichannels. It is desired to develop an engineered system that is realizable for testbed evaluation in the near term. However, development of gigabit lightwave LANs and SANs is inherently an involved systems issue and relates a large number of technology factors.

It is first necessary to define network topologies, and then to select the appropriate optical components and transmission techniques. The physical architecture chosen utilizes a broadcast-and-select, single-hop type of network protocol that relies on passive star couplers as the basic hub/switch interconnect devices, optical amplifiers to overcome power merge and splitting losses, and wavelength tunable transmitters and receivers. The widely used on-off, intensity modulate/direct detect (IMDD) transmission scheme is selected. The modeling process is started with the calculation of power budgets for the SAN that grows in size and complexity, thus requiring addition of cascaded optical amplifiers. A prototypical SAN topology that was modeled is shown in Fig. 1.

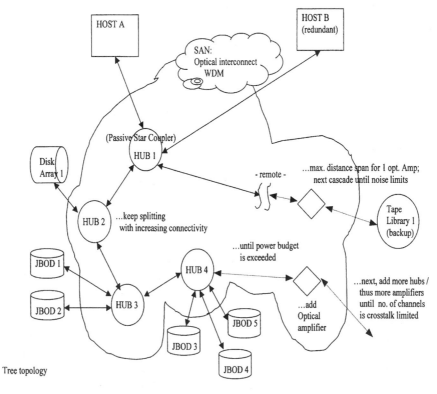

Figure 1   Tree topology.

Characteristics of the optical components within the transmitter, network link, and receiver were built into a software simulation model of the complete optical system. Configuration, noise, and other key parameters are made selectable via lookup tables. Signal power budgets and noise impairments to receiver detectability are identified in terms of signal-to-noise ratio (SNR) and bit error rates (BER). Noise sources at the laser transmitter, within the optical amplifiers in the network, and within the receiver are included. Details of system operation such as electrical and optical waveforms, eye diagrams, and noise levels are displayed. The system model allows prediction of configurations that reach to maximum end node connectivity and capacity.

Time scaling was required to achieve data input and sampling rates at 1 to 20 Gbps due to MATLAB modeling limitations associated with run times. The fundamental relationship in communication theory between SNR and BER was used to correlate simulation error rate results, and to permit projecting to very low error rates (10^{-9} to 10^{-12}).

An example eye diagram for meeting system BER requirements of 10^{-12} and having several hundred in-line optical amplifiers is shown in Fig. 2. Optical amplifier placement in the network topology involves tradeoffs between gain and noise factors. Scaling to maximum network systems at 10 Gbps shows that cascaded optical amplifiers can extend the reach to several thousand end users, while WDM implementation can produce aggregate network capacities of a hundred terabits per second.

1.1   Lightwave System Design and Selection

There are technical and operational requirements for lightwave network systems. Advanced optical gigabit network technology, protocols, and standards are being developed, and a major challenge exists to marry short distance networks used to connect computer networks with long distance networks used for telecommunications [1]. The short-haul networks are commonly known as LANs and MANs and are classified as multipoint-to-multipoint (many sources, many destinations). Specifically characterized in this work is a data-intensive and data-secure architecture type of local-area networks known as storage area networks (SAN) using fiber-channel (FC) protocols [2].

1.2   Summary Description of the Selected Lightwave Data Storage Network

For clarity, the physical architecture and characteristics of the selected SAN system are summarized. In general, the physical hardware constraints as dictated by the present-day FC standards will be used only as a baseline. They or next-generation standards will be amended as the need grows for higher end node reach and

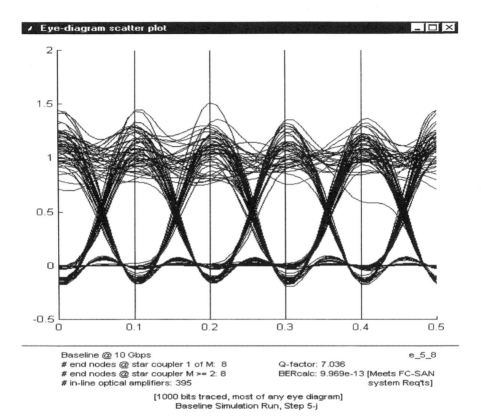

| Eye-diagram scatter plot |

Baseline @ 10 Gbps   e_5_8
end nodes @ star coupler 1 of M: 8   Q-factor: 7.036
end nodes @ star coupler M >= 2: 8   BERcalc: 9.969e-13 [Meets FC-SAN
in-line optical amplifiers: 395   system Req'ts]
[1000 bits traced, most of any eye diagram]
Baseline Simulation Run, Step 5-j

Figure 2   Eye diagram with inline optical amplifiers.

system capacity. Work on this project will address only the physical layer aspects of the WDM optical network, which mainly includes the device and transmission technologies. Eventual implementation of WDM systems and networks such as the SAN also involve significant challenges in the software control and management of data.

Operating wavelengths are in the third transmission wavelength window that is centered at 1550 nm.

Link distances between any nodes within the SAN fabric will be kept within 10 km, except for the possible extension beyond this limit to a remote backup storage node. Modeling is done with much shorter distances, and focus is put on broadcast merging and splitting losses in star couplers.

The key interconnect devices within the network fabric will be lightwave star couplers, which act as broadcast hubs. In this topology, active fiber amplifiers are essential to overcome power merging and splitting losses, as the network becomes large in node count.

Tunable laser transmitters and tunable optical filters at the receivers will be used in association with the broadcasting couplers to form a single-hop network protocol (optical/electrical conversions only at end nodes). Tunable transceivers can also be considered to function as wavelength multiplexers and demultiplexers. With full-duplex transmission, data can flow in both directions simultaneously; but this requires two independent channels, one for data in each direction.

Tuning speeds of selected transmitters and receivers will be on the order of a few milliseconds, which enables network circuit switching to allow relatively slow reconfiguration of end-node setups.

Wavelength division multiplexing (WDM) will be used to introduce multichannel signals at different wavelengths, each operating at peak electronic data rates.

The physical size of the SAN will be on the order of several tens to several thousand nodes, with up to several hundreds of channels. These limits will be investigated by using analytical models for power budgets and simulation models for noise impact on BER and number of channels. The SAN topology is illustrated in Fig. 1.

1.3   Description of Components and Transmission Technique to be Modeled

A precise summary of the link components and transmission techniques selected for this investigation is provided. The link schematic corresponding to these selections is shown by Fig. 3.

The transmitter end shall consist of a tunable laser diode operating at 1550 nm center frequency and utilizing ASK optical modulation. Circuit-switching tunability will be assumed by use of a mechanically (external cavity) Fabry–Perot filter laser. The effective linewidth reduction associated with tuning to achieve ''single-frequency'' laser diodes will be simulated with Gaussian noise sources that vary the single-frequency signal wavelength. The reduction of chirp will be evaluated in the simulation in relation to the effect that nonzero extinction has on reduced received signal power. Note that chirp can be eliminated from an engineered system by introducing an external laser modulator, but the available bandwidth is reduced from the full useful 200 nm (26,800 GHz) spectrum of the 1550 nm transmission window to a mere 18 GHz. This would allow closer spacing of individual channels because of reduced crosstalk, but the total number of channels in the bandwidth space will be limited.

The transmission line will consist of single-mode fiber, with attenuation, insertion, coupling and splicing losses. Dispersion-induced mode partition noise

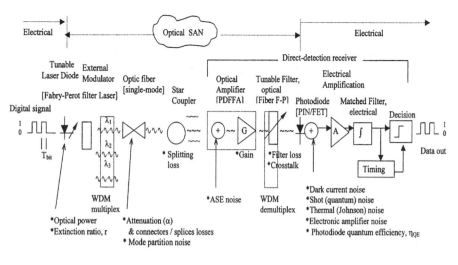

Note:

1) The active-fiber optical amplifier (PDFFA, for operation in the 1300-nm window) is used as an in-line amplifier as well as a preamplifier for the direct detection receiver.

Figure 3   Block diagram of selected optical link for simulation.

in the fiber is avoided by use of narrowband sources. Passive star couplers are used to broadcast WDM optical signals, and in-line active-fiber amplifiers are used to regain the splitting losses associated with the couplers. The appropriate 3rd transmission window amplifiers are the EDFAs, whose optical bandwidth is about 35 to 40 nm. Power budget analytical computations will be used to establish network link limitations.

The receiver end shall be direct-detection with PIN photodiodes. To enhance detectability, the optical EDFFA will be used as a preamplifier, in addition to polarization filters to reduce ASE noise from these amplifiers. To enable circuit switching and WDM, the tunable fiber Fabry–Perot optical filter is used.

1.4 Optical System Performance Factors

At the direct-detect digital receiver, the amplified and filtered electrical signal emerging from the equalizer is compared with a threshold level once per time slot to determine whether a pulse is present at the photodetector in that time slot. Ideally, this output signal would always exceed the threshold voltage when a ''1'' is present and would be less than the threshold when no pulse ''0'' was sent. In actual systems, statistical deviations from the average value of the output signal are caused by various noises, interference from spreading of adjacent pulses (or intersymbol interference ISI), interference from adjacent frequency channels (or crosstalk) in WDM systems, and conditions of the light source not being completely extinguished during a zero pulse. The standard method for measuring the rate of error occurrences in a digital data stream is called the probability of error P_e in the decoding of any bit, more commonly known as the bit error rate or BER. The FC SAN standard BER is 10^{-12}; no more than one error can occur in every 10^{12} bits sent. There is a fundamental relationship between BER and signal-to-noise ratio, SNR. This is discussed in detail in the next section.

Link Signal Power Budget. Among the variables that determine the received signal power at the point where the decision is made are

1. Laser output power—watts, actually delivered into the fiber pigtail. This number is dictated by the choice of components available.
2. Fiber attenuation—dB/km
3. Connector and splice losses—dB
4. Combining and splitting losses—dB, in star couplers, also refection and insertion losses
5. Extinction ratio r_e—for reducing chirp in ASK modulation. This r is the ratio of optical power for binary "0" to that for binary "1" ($r_e = P_0/P_1 \leq 1$). The received power SNR is multiplied by ($1 - r_e$), thus reduced.
6. Active-fiber optical amplifier gain—G
7. Quantum efficiency—with which the PIN photodiode converts photons to electron carriers, generally less than 100%

Link Signal Noise Items. There are two types of noise: additive and multiplicative. Additive noise remains when the signal disappears, and multiplicative noise either is inherent randomness within the signal itself or is produced in a device only when the signal is present.

The sources of multiplicative noise that must be considered are

1. Mode partition noise—in the case of nonmonochromatic laser sources driving single-mode fiber. By utilizing "single-frequency" laser diodes as part of the engineered system these effects are considered to be minor.
2. Modal noise—present in multimode fibers only, which are not used in this investigation.
3. Laser phase noise—the process that causes a laser to exhibit most of its nonzero spectral linewidth. For direct-detection systems, is usually not an issue; therefore not considered.
4. Laser amplitude noise—expressed by the RIN factor. Not usually an issue in direct-detection systems, but is modeled in this investigation due to reflections in short links.

5. Gain-factor noise—present only in APD photodiodes, not in the PIN diodes selected.
6. Quantum (shot) noise—results from the random multiplied Poisson nature of the photocurrent produced by the photodiode used.

Sources of additive noise include:

7. Amplified spontaneous emission (ASE) noise—generated in any optical amplifier that appears in the path between laser and photodiode. This noise is broadband, occurring over the entire gain bandwidth.
8. Dark current noise—within the photodiode, caused by thermal processes independent of illumination.
9. Thermal noise or Johnson noise—arises from the detector load resistor of the electronic amplifier that follows the photodiode. Thermal noise is Gaussian in nature.
10. Electronic amplifier noise—arise-within the amplification stages following the photodiode due to the amplifier current and voltage current noise sources. These are statistically independent and can be represented by Gaussian statistics.
11. Crosstalk—in multichannel systems, from adjacent channels at the receiver or from intermodulation distortion and gain saturation-induced gain variations in laser-diode amplifiers. Since this investigation uses active-fiber amplifiers having extremely long gain response times, these are considered to be largely immune from interference phenomena between optical channels simultaneously injected into the amplifier.

Number of Channels in Multichannel Systems. The maximum number of resolvable channels is limited by optical device technology; primary considerations are the total available bandwidth or spectral range of the components and the channel spacing. Although the bandwidths of the fiber medium in the low attenuation regions around 1300 nm and 1550 nm are each approximately 200 nm, the optical networks cannot necessarily take advantage of this entire range due to the bandwidth limitations of the optical components. Optical amplifiers have a bandwidth of around 30 to 40 nm. Choosing very fast (nanoseconds) packet-switching devices such as DFB lasers and electro-optic filters at the receivers would provide tuning ranges of only about 10 nm, thus severely restricting the number of channels. Assuming instead the use of mechanically tuned Fabry–Perot filter lasers and fiber Fabry–Perot filters at the receivers provides tuning ranges that encompass the entire 200 nm. These tuning devices are commercially available and provide slower switching speeds (in milliseconds) adequate for circuit switching.

2 POWER BUDGET MODELING AND ANALYSIS

The power budget analysis for the lightwave SAN was performed as an adjunct to the Simulink model. The calculated loss/power budgets are used within Simulink in the form of lookup tables for parameter input to the appropriate modeling blocks.

2.1 Network Power Calculations

The star coupler and end node connectivity scheme of the network is shown by Fig. 4.

It includes the numbering scheme of 1 through M star couplers, 1 through N end nodes connected to the first star coupler, and 1 through N_r end nodes connected to all stars after the first one. This dual end node numbering scheme offers some flexibility in modeling different network configurations during the simulation modeling. There are also 1 through $(M-1)$ optical amplifiers in the network. The first star coupler is logically associated with the predetect optical amplifier in the receiver node, while within the network each star coupler is associated with an adjoining optical amplifier as a "set." Since each star coupler is expected to merge and broadcast all signals, there are both combining and splitting losses that must be accounted for. For example, the loss for an $N \times N = 8 \times 8$ star coupler that attaches to 16 end nodes is

$$N \times N \text{ coupler loss} = 2 * \log_{10}N)_{\text{merge}} + 2 * \log_{10}N)_{\text{split}} \tag{1}$$
$$= 4 * \log_{10}N = 18.06 \text{ dB}$$

The power budget is set up in Table 1 for each of the four data rates being modeled: 1, 5, 10, and 20 Gbps. The optical power source is rated at 0 dBm (1 mW). Minimum received powers based on the quantum limit for ASK with preamplifier and at a BER of 10^{-12} [1] are used in the simulation. A system margin of 3 dB is added to these quantum power levels to obtain the system power budgets.

The detailed loss calculations that meet the power budget constraints of both the overall system and within a star coupler–optical amplifier set are done in spreadsheet format in Table 2.

Parameters varied are the data rate and the number of end nodes attached to the star couplers.

2.2 Allowable Distances to Remote Location

Typically, a SAN includes a single link to a remotely located backup tape site in order to provide data integrity due to disastrous losses at the main enterprise

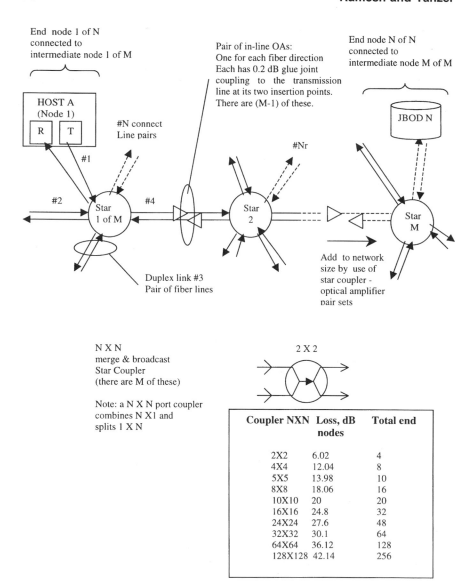

Figure 4 Star coupler and end node connectivity scheme.

Table 1 Analysis of Network Signal Power Budget

$P_T = P_S - P_R$

where P_T = total optical power loss allowed in the system
   P_S = optical source power
      use   0 dBm (1 mW) max. ave. for DFB laser, also linewidth $\sigma_\lambda = 3$ nm
   P_R = receiver sensitivity
      use   ASK w/optical preamp quantum limit, using InGaAs pin diode at $\lambda = 1550$ nm, at BER of 10^{-12} has a minimum received power sensitivity vs. data rate
         $P_R = -49$ dBm @ 1 Gbps/-45.5 dBm @ 2 Gbps/-42 dBm @ 5 Gbps-39 dBm @ 10 Gbps/-36 dBm @ 20 Gbps

This total optical power loss is allocated to the network link elements:

$P_T = l_c + \alpha_f L + l_{coup} + M_s$

where l_c = connector loss; use 0.3 dB per
      use   duplex link, i.e., 2 cables w/ 1 signal direction in each, thus 2 connectors associated with each end node
   $\alpha_f L = \alpha_f$, fiber attenuation (dB/km) over the transmission distance, L
      use   0.22 dB/kM for high-grade 1550 nm fiber
      Assume $L = 20$ m between all end nodes, except to remote backup tape (which is separately characterized as a long-distance link path)
   l_{coup} = star coupler combining and splitting losses
      use   power loss of $10\log_{10}N$ for every N-fold combining operation as well as every splitting operation
      e.g.,   A $4 \times 1 = N \times 1$ combining operation followed by a $1 \times 4 = 1 \times N$ splitting operation is the same as a $4 \times 4 = N \times N$ merge and broadcast operation, or $2 \times 10\log_{10}N = 12.04$ dB
      Include   insertion loss of 0.2 dB per facet or each N-way for each of M couplers
      Include   fiber lengths losses between couplers (assume $L = 20$ m)
   M_s = system margin, typically is 3 to 6 dBm; use 3 dBm

Therefore the operating regime based on purely power budget limitations is

$P_S - P_R \geqq = l_c + \alpha_f L + l_{coup} + M_s$

1 Gbps: 0-(-49) dBm $\geqq$ Losses + 3 dBm, or losses $\leqq -46$ dBm

2 Gbps: 0-(-45.5) dBm $\geqq$ losses + 3 dBm, or losses $\leqq -41.5$ dBm

5 Gbps: 0-(-42) dBm $\geqq$ losses + 3 dBm, or losses $\leqq -39$ dBm

10 Gbps: 0-(-39) dBm $\geqq$ losses + 3 dBm, or losses $\leqq -36$ dBm

20 Gbps: 0-(-36) dBm $\geqq$ losses + 3 dBm, or losses $\leqq -33$ dBm

Table 2 Spreadsheet Computation of Network Losses

Allocate losses per user or end node (N):
  Connector: 0.3 dB/connector × (2 conn./end node) = 0.6 dB * N
  Fiber length: 0.22 dB/kM × 20kM = 0.0044 dB * N
  _____
    or end node losses = 0.6044*N, dB

Allocate losses per coupler or intermediate node (M):
  Each combine operation: 10 log$_{10}$(N/2)
  Each split operation: 10 log$_{10}$(N/2)
  Coupler insertion loss: 0.2 dB * (N)
  M node interconnect fiber loss: (1) × 022 dB/kM × 50 m = 0.011 dB
  _____
    or coupler losses = 2 * (10 log$_{10}$(N/2)) + 0.2(N) + 0.011 dB

Allocate gains per in-line optical amplifier (M − 1)
  Max. gain = 30 dB (includes 2 × 0.2 dB glue coupling to line)

Data rate B, Gbps	Star M	Node N	Losses end node, dB	Losses coupler, dB	Losses total, dB	Power budget, dB	Notes/clarification
1	1	8	4.835	13.652	18.487	46	ASK w/ preamplifier at PIN
1	1	18	10.879	22.696	33.575	46	
1	1	26	15.714	27.490	43.204	46	
1	1	28	16.923	28.534	45.457	46	Max. nodes @ M = 1 is 28

1	2	6	3.626	10.753	14.380	30	Insert an in-line OFA (pair) add a star coupler 2 of M
1	2	10	6.044	15.990	22.034	30	Max: 15 nodes @ $M = 2$
1	2	15	9.066	20.512	29.578	30	
1	3	8	4.835	13.652	18.487	30	Add a 3-d coupler ($M = 3$), with another OFA pair
1	3	15	9.066	20.512	29.578	30 ↗	
							Repeats, eventually limited by the cascaded ASE noise from the optical amplifiers
2	1	10	6.044	15.990	22.034	41.5	ASK w/preamplifier at PIN
2	1	24	14.506	26.395	40.900	41.5	Max. nodes @ $M = 1$ is 24
5	1	12	7.253	17.974	25.227	39	ASK w/preamplifier at PIN
5	1	22	13.297	25.239	38.536	39	Max. # nodes @ $M = 1$ is 22
10	1	14	8.462	19.713	28.175	36	ASK w/preamplifier at PIN
10	1	19	11.484	23.365	34.849	36	
10	1	20	12.088	24.011	36.099	36	Max. # nodes @ $M = 1$ is 20
20	1	16	9.670	21.273	30.943	33	ASK w/preamplifier at PIN
20	1	17	10.275	21.999	32.274	33	Max. # nodes at $M = 1$ is 17

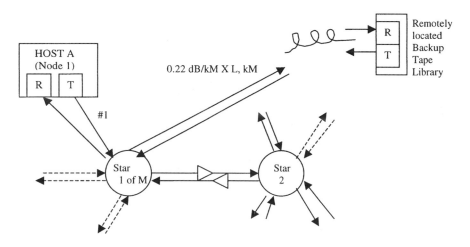

Figure 5 Link to remote end node.

facilities. It is desired to know the allowable link distances possible for connecting to this remote site. The physical link detail of this special case connect segment is shown in Fig. 5, with star coupler 1 of *M* being arbitrarily chosen.

One possible attachment scenario is to keep the full complement of end nodes on all the star couplers as allowed by the power budget, and then add an OA as a power booster immediately following the coupler on its path to the remote end node. Example allowable distances can be calculated:

30 dB OA gain $= 0.22$ dB/kM $\times$ L, kM

$L = 30/0.22 = 136$ kM, the maximum

allowable link distance to this remote node

Another scenario is one in which a star coupler–optical amplifier node set only has, say, eight end nodes attached for local storage (a 12 dB loss). This leaves 18 dB available for the remote node, and for this case the maximum allowable link distance becomes 82 kM.

3   DIGITAL LIGHTWAVE SYSTEM SIMULATION MODEL

3.1   Matlab® Simulink® System Modeling Description

SIMULINK® is a MATLAB® software package for modeling, simulating, and analyzing dynamical systems that was used to build and model the optical network system. For brevity, the simulation model that was built using this software package is referred to as OpSAN. The MATLAB application libraries used with OpSAN included the Communications Toolbox, the Signal Processing Toolbox, and the DSP Blockset. For modeling, SIMULINK provides a graphical user interface for building modeling blocks as block diagrams, using click-and-drag mouse operations. The simulation model is built in a hierarchical fashion, with high level blocks that can be mouse double-clicked on to go down through the layers to see increasing levels of detail. Parameters for the simulation model can be entered either from the SIMULINK menus or from MATLAB's command window using "M-files."

3.2 Modeling the Intensity-Modulated Direct-Detection (IM-DD) Optical Digital Communication System

OpSAN is built and modeled based on the on–off modulation of the light source and direct detection of the received lightwave impinging on the photodetector. The use of an optical amplifier (OA) as a preamplifier in each receiver node has allowed the IM-DD to perform close to the quantum limit. The basic optical system that is modeled is shown in Fig. 6. All input parameters, both fixed and selectable input parameters, are included as part of the graphical interface that was developed using SIMULINK. M-files were used in the early stages of model building, but the direct interaction with the block diagrams was deemed more

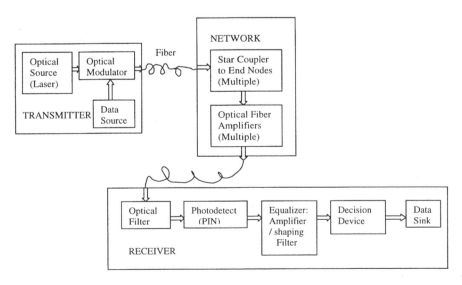

Figure 6 Basic optical IM-DD network.

useful. OpSAN provides lockup tables that provide choices for inputting of selectable parameters.

3.3 Intensity Modulated (IM) Transmitter Model

The direct imposition of a square-wave ASK signal on the bias current of a laser diode has not been modeled. Instead, only the end result of this operation is applied as the source using a random number generator, a square-wave shape of 1s and 0s. NRZ coding is used, with two bits of data transmitted for every clock cycle. Source signal parameters for data rates modeled are shown in Table 3.

If the physical implementation were actually ASK modulation, a nonzero extinction ratio (r_e) is usually applied to avoid chirp phenomena by keeping the operating regime in the linear region of the laser diode P–I curve. The extinction ratio is defined as the ratio between the power, P_0, associated with the binary 0 and P_1 associated with the binary 1:

$$r_e = \frac{P_0}{P_1} \qquad 0 \leq r_e \leq 1 \tag{2}$$

The average power falling on the photodetector is thus reduced and has the effect that the received power SNR is reduced by the factor $(1 - r_e)$. A resulting power penalty curve as a result of extinction ratio can be evaluated in terms of the Q factor [3], and by using the assumption that noise dependency on received power is minimal (i.e., thermal noise is dominant over shot), the penalty in is given by [4]

$$\Delta_e \triangleq \frac{1 + r_e}{1 - r_e} \tag{3}$$

In the range of r_e of 0–0.5, the maximum power penalty obtained using the above formulation is 9.54 dB. The effect of having an extinction ratio greater than zero is investigated during the detectability sensitivity analysis. An alternative method is to utilize external modulation in order totally to avoid chirping and other laser

Table 3 Source Signal Parameters

Data rate, R_d, Gbps	Symbol time period, $T_d = 1/R_d$, ns	Baud rate or clock speed $f_d = 1/(2 * T_d) = R_d/2$, GHz
1	1	0.5
5	0.2	2.5
10	0.1	5
20	0.05	10

Table 4 RIN Noise Values

Reflected power, dB	RIN, power spectral density, dB/Hz	Noise level, r_i^2, mW	Δ_i [dB], $Q = 6$ (BER $= 10^{-9}$)
-20	-110	0.05	0.819
-30	-120	0.005	0.031
-40	-130	0.0005	negligible
-50	-140	5×10^{-5}	negligible
-60	-150	5×10^{-6}	negligible

transients. This latter method completely separates the light generation and modulation processes, and the laser can be allowed to run continuously at almost constant power level. OpSAN allows either method to be modeled: the r_e parameter can be turned off or a value can be selected.

The amplitude fluctuation is also modeled at the laser. The quantity relative intensity noise (RIN) is a form of noise density–to–signal power ratio. It is power spectral density (expressed in dB/Hz), or the mean square intensity fluctuation of the laser output divided by the average laser light intensity. Based on reported experiments [6], RIN may be expected as low as -150 dB/Hz if reflections from splices and connectors are kept below -60 dB, but as bad as -110 dB/Hz for reflected power around -20 dB. With the relatively short distance runs of a SAN, reflections can be a factor. RIN is modeled as a Gaussian noise source and adds an extra noise to the receiver. An example of this calculation [4] at -110 dB/Hz and an input data rate of 10 Gbps (having system bandwidth $B = 5$ GHz) follows, and results are tabulated in Table 4.

$$\text{RIN noise level (Gaussian)} \triangleq r_i^2 = 10^{\text{RIN}/10} \times B$$
$$= 10^{-11} \times 5 \text{ GHz} = 0.05 \text{ mW} \tag{4}$$

The resulting penalty, from the definition of the Q factor, can be obtained from [4]

$$\Delta_i \triangleq \frac{1}{1 - r_i^2 Q^2} \tag{5}$$

For RIN below -120 dB/Hz, corresponding to r_i^2, the penalty is negligible when designing for system BER of 10^{-9}.

3.4 Baseband Modulation and Demodulation

In the simulation of modulation techniques, there are two options for representing the modulation/demodulation process: passband and baseband. In passband simulation, the carrier signal is induced on the transmission model. The frequency of the optical carrier at the wavelength of 1550 nm ($f_0 = 1.935 \times 10^{15}$ Hz) is much greater than the highest frequency of the message input signal (with f_d on the order of 1 to 10 GHz). For the simple IM-DD transmission scheme, baseband simulation can readily be used and thus avoid the very slow and inefficient simulation of the high frequency signal. In OpSAN's baseband scheme, large increases in run times were used to obtain total bit throughput. By Nyquist sampling theory, the simulation sampling frequency f_s must be at least twice the bandwidth of the input signal being modeled. Thus

$$BW_{\text{input signal}} = f_d \tag{6}$$

$$f_s = \frac{1}{\text{sampling time } t_s} \geq 2 * VW_{\text{input signal}} = 2 * f_d \tag{7}$$

The ratio f_s/f_d is selected as a positive integer of 2 or more. Therefore a data rate of 10 Gbps (via time scaling) having an input digital transfer frequency of 5 GHz requires a computational sampling frequency of at least 10 GHz. This is important in the selection of lowpass filters in demodulation, as described in the Receiver Modeling section.

Although attempted, it was not possible to use the Digital Mo/Dem from the Communications Baseband Library. This Matlab method maps an input signal and then modulates the mapped signal with a complex envelope (bipolar $+1 = 1$, and $-1 = 0$). This approach is interrupted by the squaring function of the photodiode, which turns all waveforms into positive magnitudes. The Demod portion of this scheme will not function when only positive waveforms are presented.

3.5 Model of Network Intermediate Nodes with Passive Star Couplers and Optical Amplifiers

The intermediate nodes of the network consist of multiple passive optical star couplers, each with its own multiple end nodes, and optical amplifiers to overcome the dominant power splitting losses at these star couplers. All distances between nodes are assumed to be short (15 m) and of constant length, thus minimizing attenuation and dispersion effects in the fiber as compared to star coupler splitting and optical amplifier noise. Minor effects only are present in the fiber when implemented in a short-haul network configuration and are thus neglected, and dispersion with its resultant intersymbol interference (ISI) has not been modeled.

Power budget calculations have been provided in the previous section. Inclusion into the OpSAN model is done via lookup tables and manual parameter selection provided in the SIMULINK graphical layout. It is not done via line coding and calculation within MATLAB itself. It is important to note the distinction in end node connections to the first star coupler (1 of M) compared to all of the rest ($M \geq 2$). It is allowable for the first star coupler to have as many end nodes (N) as does not exceed the minimum required incident power at the optical receiver. This power level must be above the ASK preamp quantum limit, and accordingly, the system power budget calculations have included a margin of 3 dB. Minimum received optical signal power is obtained by selecting the largest allowable number of end nodes that is within the system power budget. Thus, in OpSAN, the selectable number of end nodes at the first star coupler varies from 17 at 1 Gbps to 10 at 20 Gbps (refer to Table 2). Subsequent star couplers have each only the power budget that has been gained by the preceeding optical amplifier, and the two together constitute an intermediate network node pair, or set. OpSAN allows a selectable number (N_r) of 14 to four end nodes attached to each star coupler beyond the first. The case of $N_r - 14$ represents the maximum allowable mount of power loss at the star coupler, or 28.2 dB, which is within the 30 dB gain capability used for the optical amplifiers (refer to Table 2). OpSAN is set up to model the network configuration this way, with N (the number of end nodes on the first star coupler), N_r (the number of end nodes on star couplers $M \geq 2$), and M (the number of star coupler/in-line optical amplifier node sets) all being selectable parameters.

Each optical fiber amplifier is assumed to provide a constant gain, as selected, over its gain bandwidth of 30 nm. As a result of the repeating chain of equal gain and loss, the signal power is the same at the output of each amplifier. The incoherent ASE noise component, however, builds up linearly as the cascade of star coupler–optical amplifier node set increases. The modeling approach taken is to artificially separate the cascaded ASE noise stream (noise power spectral densities) from the multiple in-line optical amplifiers from the main signal stream, and then recombine after the optical filter operation as total noise power in the predetector of the receiver. This is more fully described in the Receiver section.

3.6   Direct Detect (DD) Receiver Model

The first element encountered in the receiver is the optical amplifier used as a preamplifier for the photodiode detector. An optical filter is placed between this amplifier and the detector in order to attenuate ASE significantly noise, if applied in a multichannel WDM system, it will demultiplex many optically amplified channels by passing a selected channel and blocking all the other channels. If the network system requires wavelength-tunable receivers, then the optical filters must be tunable. The ASE-generated noise is very broadband and occurs over the entire gain bandwidth of the optical amplifier. It can be considered as Gaussian white noise [3,4]. ASE noise is difficult to model in SIMULINK. A high-frequency bandpass optical filter, having a center frequency $f_0 = 193.5$ THz at the nominal wavelength of $\lambda = 1550$ nm, and a bandwidth B_0 of at least three times [4] the modulation bandwidth, cannot be modeled within the much-lower-frequency baseband signal stream that is on order of several GHz. MATLAB is incapable of sampling at such extreme rates in any reasonable computational time.

Figure 7 is a block diagram of an optically amplified signal impinging on a detector. The noise component due to ASE impinging on the detector is critically dependent on both optical and electrical bandwidths of the corresponding filters. The spectral shape of these filters can take many forms. Since the optical and electrical filters must be at least as wide as the modulated signal bandwidth for undistorted transmission of the signal, some ASE must leak through the filter within the same bandwidth as the signal. In other words, ASE beat noise can be reduced by reducing filter bandwidths but cannot be eliminated. The characterizing ASE noise equations from Kazovsky et al. [4] are listed below.

The spectral density of the ASE noise power at the amplifier output can be approximated by

$$\text{ASE}_{\text{spectral}} = \eta_{\text{sp}}(G - 1)h\nu, \text{ W/Hz} \tag{8}$$

where η_{sp} = the spontaneous emission factor = 1.4, $h\nu$ = photon energy = 1.282×10^{-19} J/cycle, and G = gain of optical amplifier = 1000 (30 dB).

The total ASE optical noise power over the optical bandwidth of the active medium ($\Delta\nu$) in one polarization is

$$P_{\text{sp}} = \eta_{\text{sp}}(G - 1)h\nu(\Delta\nu) \qquad \text{W} \tag{9}$$

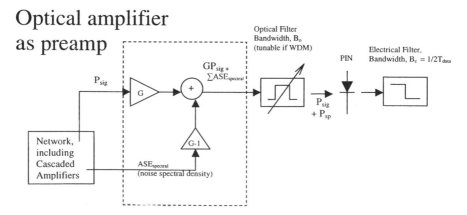

Figure 7 Optical amplifier at the direct-detect receiver.

ASE power contributes to beat noise terms as the result of two incident signals (P_{sig} and P_{sp}) or field waves each at specific frequencies responding to the square-law photodetector (i.e., the two waveforms get multiplied together). The filter shapes can be approximated to be uniform within a certain bandwidth, resembling a rectangular form. The resulting signal-spontaneous beat noise variance, $\sigma^2_{sig\text{-}sp}$, that falls within the optical filter and electrical detector bandwidths is

$$\sigma^2_{sig\text{-}sp} = 4q\left(\frac{q}{h\nu}\right)P_{sig}G(G-1)\eta_{sp}B_e \qquad amp^2 \qquad (10)$$

where q = electron charge = 1.6×10^{-19} Coulomb, and $q/h\nu$ = conversion factor between photons and electrons = 1.28.

After integration and convolution and only considering the noise passing through the two filters, the spontaneous-spontaneous electrical beat noise variance, $\sigma^2_{sp\text{-}sp}$, is

$$\sigma^2_{sp\text{-}sp} = 4q^2(G-1)^2\eta^2_{sp}B_eB_o \qquad amp^2 \qquad (11)$$

When using a narrow passband optical filter at the receiver, the signal-spontaneous beat noise dominates over the signal-signal beat noise; this is the optimum regime in which to operate an optically amplified system. All other noise sources in the receiver, e.g., thermal and shot noise, are negligible. In our model we separated the Gaussian-driven ASE noise stream (including all of the cascaded optical amplifiers within the network) from the signal stream. The optical noise power, P_{sp} (but not the true noise waveform nor the proper attenuation of the noise), is obtained by passing the spectral density of the ASE noise stream through a multiplier representing an optical filter. Some modeling error is expected here. This ''filtered'' optical noise stream is then recombined with the signal stream before passing through the PIN photodetector and electrical amplifier. Table 5 shows the large dependence of noise power on the optical filter bandwidth $\Delta\nu = B_o$.

Modeling of optical demultiplexing is complicated by additional effects. The ASE passing through the filter attains a spectral shape similar to that of the filter passband itself; thus using common Lorentzian-lineshaped FP optical filters will alter the equations above that assumed an approximate rectangular shape. In addition, each of the rejected channels will itself beat with the ASE that passes through the filter and is spectrally colocated with any given rejected channel. No attempt was made to model these added filter complexities; therefore the simulation is not expected to predict system power penalty as a function of channel separation. As a design guideline, the channel separation should be more than 1.5 times the optical filter bandwidth [4].

Table 5   ASE Noise Parameters as Functions of Optical Bandwidth

$\Delta\lambda$, nm	$\Delta\nu = (\nu/\lambda)\Delta\lambda = 1.2487 \times 10^{20}\Delta\lambda$ $= 3 \times 10^8/(1.55 \times 10^{-6})^2\Delta\lambda$, GHz	P_{sp}, mW
0.1	12.487	0.0022389
0.2	24.97	0.004478
0.3	37.46	0.006717
1	124.87	0.022389
2	249.7	0.04478
3	374.6	0.06717
30 = full bandwidth of opt. amp.	3.746×10^{12} Hz	0.6717

The power spectrum of the electrical current $(I_{sig})^2$ out of the PIN photo-diode is the result of the squaring operation upon the incident optical power. The performance of a photodiode is characterized [5] by its responsivity, R:

$$R = \frac{I_{sig}}{P_{sig}} = \eta\left(\frac{q}{h\nu}\right)M = 0.63 * 1.28 * 1 = 0.8 \ \mu A/\mu W \tag{12}$$

where η = quantum efficiency (for InGaAs at 1550 nm it is 63%), $(q/h\nu)$ = 1.28 A/W at 1550 nm, and M = multiplication factor (for = PIN it is 1.0).

There are various noises that accompany and contaminate the conversion of the optical signal into the electrical signal. These noise terms and their characterizing equations are summarized in Table 6. Also included in this table are the ASE-noise terms as well as the source (laser) RIN noise. Note that all noise terms are linear functions of electrical bandwidth, so that B_e becomes an important variable parameter within the OpSAN noise calculations. The primary photocurrent generated in the photodiode is a Poisson random arrival process, and the noise currents generated are shot, surface leakage, and dark current. For a process following Poisson statistics, the probability density function (PDF) includes the parameter λ, in which the variance is equal to the mean. Of these, shot noise is typically the most significant. The Gaussian noise sources are assumed flat in spectrum (characterizing "white" noise containing all frequency components) and are statistically independent. The Gaussian PDF is described by two parameters: the mean (μ) and the standard deviation (σ) or its square (σ^2), also called the variance. With a PIN diode, the thermal and amplifier (Gaussian) noise source typically dominate the Poisson noises. However, when using optical amplifiers, all noises are typically negligible compared to the ASE noises.

Detection of the weakest optical signals requires that the photodiode and the following electrical amplifier circuitry be optimized to maintain a specific

Table 6 Complete List of Noise Sources

Noise component	Expression	Noise type	Statistical nature
Laser RIN, $I_{RIN}^2 = \sigma_i^2$	$RIN(I_{sig})^2 B_e$	Multiplicative	Gaussian
ASE signal-spontaneous beat, $\sigma_{sig\text{-}sp}^2$	$4q(q/h\nu)P_{sig}G(G\text{-}1)\eta_{sp}B_e$	Multiplicative	Gaussian
ASE signal-signal beat, $\sigma_{sp\text{-}sp}^2$	$4q^2(G\text{-}1)^2\eta_{sp}^2 D_e D_0$	Additive	Gaussian
Receiver shot (quantum), $I_{shot}^2 = \sigma_{shot}^2$	$2qI_{sig}B_e$	Multiplicative	Poisson
Receiver amplifier thermal (Johnson), $I_{th}^2 = \sigma_{th}^2$	$4k_{Bc}TR_c/R_L$	Additive	Gaussian
Receiver surface leakage current, $I_{leak}^2 = \sigma_{leak}^2$	$2qI_{leak}B_e$	Additive	Poisson
Receiver dark current, $I_{dark}^2 = \sigma_{dark}^2$	$2qI_{dark}B_e$	Additive	Poisson
Receiver elect. amplifier, $I_{amp}^2 = \sigma_{amp}^2$	$2qI_{amp}B_e$	Additive	Gaussian

signal-to-noise ratio (SNR). The detected electrical power SNR at the output of an optical receiver is the signal mean squared divided by the sum of all the noise mean currents squared (i.e., noise variances),

$$\text{SNR}_{power} = \frac{(RP_{sig})^2}{\sum I_{noise}^2} = \frac{I_{sig}^2}{\sum \sigma_{noise}^2} \quad \text{dimensionless ratio} \quad (13)$$

or, in dB,

$$[\text{SNR}_{power}]_{dB} = 10 \log_{10}\left[\frac{I_{sig}^2}{\sum \sigma_{noise}^2}\right]$$

The electrical amplifier that follows the photodiode typically has an input impedance represented by the parallel combination of a resistance and a shunt capacitance. Within OpSAN, this amplifier is modeled only to overcome all preceding system power losses and thus provide the gain necessary to bring the signal power back to a mean value equal to 1 mW for a digital ''1''. A lookup table is provided within OpSAN to select the appropriate gain parameter, which is dependent on network configuration and received optical power at the photodetector.

The DD receiver uses an electrical filter matched to the square signal waveform. The matched-filter condition is between the incoming waveform (square pulse) and the effective receiver impulse response (an identical square pulse). The Analog Filter Design from the Matlab DSP Blockset is used to implement a low-pass configuration. The best resulting filtered waveforms were obtained by the Butterworth filter, followed by the Chebyshev Type I. The filter order was kept at the default value of 8. The bandwidth B_e of the low-pass filter equals its cutoff frequency ω_c, that is, $B_e = \omega_c$. The cutoff frequencies for the range of simulation rate rates are delineated in Table 7.

The low-pass filter is followed a by a decision process that involves several SIMULINK blocks to derive a threshold decision, a retiming of the bit clock to account for SIMULINK system delays, and comparison with the reference input signal before making a "right" or "wrong" (error rate) decision on the output data. A threshold decision is made by use of the Relational Operator block to determine if the filter output is above or below the threshold line. The threshold level for the on–off binary signal is set to 0.5, but it is selectable, e.g., to account for a laser extinction ratio that is above 0 during the sending of a 0 bit. Sampling time for this block is derived from the driving block, i.e., the filter (refer to Table 7).

Thus the output of this operation is a square-wave set of 1s and 0s, with time duration of each output sample the same as the input data rate. Some (slight) inconsistency of this operation was noticed when viewing a scope of these waveforms; a "marginal" sample sometimes resulted in an opposite spike for a fraction of the decision waveform. However, the decision output from the Relational Operator block was based on the full sample time duration and was not influenced by the opposite spike. SIMULINK delay times were determined by use of the Eye Diagram block to determine visually about when the height of the eye opening was largest, and by using a visual display which superimposed on the same screen scope (the Mux block) the reference signal, the filtered output, and the threshold decision output. Determining the delay time requires multiple iterations to achieve the "best results" or lowest error rates. Better accuracy was actually achieved by relying on the Mux scope. The Error Meter block was then used to make the final decision based on comparing the reference input waveform to the appropriately delayed output waveform. Again, as an aid to simulation runs, a lookup table is included in OpSAN in order to select the appropriate delay times based on data rate.

Table 7   Bandwidth of Low-Pass Electrical Filter

Input data rate, R_d, Gbps	$f_d = BW_{\text{message signal}}$, GHz	$f_s = (2 * f_d) = B_e$, GHz or (ω_c in Grad/s)	
1	0.5	1	(6.283)
2	1	2	(12.5664)
5	2.5	5	(31.415)
10	5	10	(62.83)
20	10	20	(125.664)

4 ANALYSIS METHODS

4.1 Data Bit Rates and Run Times

The intent of the OpSAN simulation is to model high data rates of 1, 5, 10, and 20 Gbps. However, the direct setting of simulation bit times to 1 to 0.05 nanoseconds is not possible. SIMULINK computational times are too large, and display devices will not record any visible signal trace in any reasonable amount of time. Instead, an indirect method of inputting and sampling of data rates at 1, 5, 10, and 20 bps and then greatly increasing the run times to get as many bits through the system as possible is used. This can certainly affect delay times for the various modeling blocks but should not introduce inaccuracies or error in the system. There are a number of modeling factors that influence the amount of time it takes to get one bit pulse through the Simulink system, including filter parameters, tolerance accuracies, and types of displays used. Based on actual timed simulation runs, the average run rates turned were in the range of 10 to 15×10^3 bits per hour. This run rate appeared to be fundamental for SIMULINK, since an increase in bit rates by factors of ten and a hundred did not speed up the throughput rate. Therefore, to get a throughput of $1-2 \times 10^5$ bits typically took an overnight run, as shown by Table 8. The largest throughput actually run in this work amounted to 4.2×10^5 bits and took about 26 hours. In light of the numerous simulation runs actually executed to develop, test, and see the effects of varying model parameters, a validation method became necessary to extrapolate error rate results in OPNet from practical bit total levels to standard data communication network error rates of one in 10^9 to one in 10^{12}. This method involves the Q function and will be explained in the next section.

4.2 Signal-to-Noise Ratio (SNR) and Bit Error Probability (BER) Calculations

Determining the average rate of error occurrences is the most common criterion for performance in a communication system [3,4,5]. In a digital receiver, the amplified and filtered output signal is compared with a threshold level once per time slot to determine whether a pulse is (a ''1'' bit) or is not (a ''0'' bit) present in that time slot. Deviations from the actual value of the output signal are caused by various random noise sources and can cause error in pulse detection. By divid-

Table 8 Bit Throughput and Durations at Rate of 15,000 bits/h

1.5×10^5	3.6×10^5	2.52×10^6	1×10^9	1×10^{12}
10 h	1 day	1 week	397 weeks	39,700 weeks

ing the errors occurring over a particular time interval by the total number of pulses (1s or 0s) transmitted during this time interval, one obtains the bit-error-rate (BER), also known as the probability of error (P_e). Typical error rates for telecommunication systems are 1 in 10^9 (a BER of 10^{-9}). For the Fibre Channel data network modeled in this simulation, the standard is a BER of 10^{-12}. The BER calculation is based on the assumption that there is no statistical correlation between successive bits at the transmitter, there is no correlation between the noise waveforms in successive bit periods, and that "0" and "1" are equally likely. In practice, and as was done for this simulation, the eye diagram technique is used to account for system delay time. The best time to sample the received waveform and to derive the bit-clock signal is when the height of the eye opening is the largest, or at the midpoint of the eye.

The error rate depends on the received signal-to-noise ratio (SNR), defined as the calculated ratio of signal power to noise power at the threshold point. The system error rate requirements and the receiver noise levels thus set a lower limit on the optical power signal that is required at the photodetector. In order to achieve a given bit error rate, the minimum acceptable received optical SNR at the receiver input will be somewhat different for different modulation formats and forms of receiver (due to differences in the electrical SNR as seen at the point where the threshold decision is made). Curves for required minimum received peak optical power for various optical receivers, various bit rates and for several values of bit error rates are described in [1]. For example, the quantum limit for the ideal IM-DD receiver (using ASK with optical preamplification) at 10 Gbps data rate and at a BER of 10^{-12} is −39 dBm or 103 photons/bit. The quantum limit is a bound on the performance that is ultimately achievable; stated in terms of the minimum number of photons per bit that will allow one to achieve a given bit error rate with a specific modulation format and type of receiver.

To compute the bit error rate at the receiver, i.e., the likelihood that the sampled value of the received pulse will detect a "0" when a "1" was sent, P(error/1), or detect a "1" when a "0" was sent, P(error/0), it is necessary to know the relative distribution of amplitudes of the random noise at the equalizer/filter output. Such distribution is specified by the PDF of the noise and is assumed to be Gaussian with zero mean at each signal level. Although the shot and several others noise currents have Poisson distributions, the inaccuracy resulting from the Gaussian approximation is small [3,5]. This PDF is described by an integral that cannot be obtained in closed form and has instead been computed numerically and tabulated in standard mathematical tables. The integral is a function of peak signal amplitude to root-mean-square noise (V_p/σ_n) and is denoted as the Q function for convenience, or $Q(V_p/\sigma_n)$. For the on−off binary pulse train case considered in this simulation, the separation between amplitudes to be distinguished is $(1/2) * V_p$, and consequently the error probability is

$$P(\text{error}/0) = P(\text{error}/1) = Q\left(\frac{V_p}{2\sigma_n}\right) \tag{14}$$

$$\text{With } \text{SNR}_{\text{power}} = (\text{SNR}_{\text{amplitude}})^{1/2} = \left(\frac{V_p}{2\sigma_n}\right)^{1/2} \tag{15}$$

This relationship is a fundamental one in communication theory, since it relates the BER to the electrical SNR. It can be used to validate OpSAN error rate results with calculated SNRs in the model. Good correlation of OpSAN results with Q function theory at low SNRs and thus high error rates (on the order of 10^{-2} to 10^{-4}) gives credibility to projecting to the low error rates (10^{-9} to 10^{-12}) required for data communication networks. This approach is a direct necessity of MAT-LAB modeling limitations associated with run times of Gigabit data rates, as explained in the previous section. A good approximation to the Q function is from Lathi [6]:

$$Q(\gamma) = \left\{\frac{1 - 0.7}{\gamma^2}\right\} * \exp\left(-\frac{\gamma^2}{2}\right) * \frac{1}{\gamma(2\pi)^{0.5}} \tag{16}$$

Within OpSAN, a separate Calculation block attached to the Receiver SIM-ULINK chart is used to calculate SNRs and BERs based on the Q function curve. The next section provides some example results showing the tracking between the simulation error rates and the predicted BER using calculated SNRs and the Q function relationship. Error floors can sometimes occur in which the normal Q function behavior breaks down as the argument increases and error becomes greater than it should be. In other words, simply cranking up the SNR to achieve higher BER rates may not work because of the effects of a noise such as nonzero linewidth or laser phase noise. However, these noise effects are detrimental in coherent systems and not usually an issue in direct detection systems; accordingly this simulation does not model these noise sources.

5 SIMULATION EXAMPLES AND RESULTS

5.1 Baseline Input Parameter Set for the OpSAN Simulation Model

An input set of baseline parameters for 10 Gbps that represents good engineering design and produces reasonable results is identified in Table 9. Many of the parameters, or their ranges, were selected after judicious "what-if" attempts at seeing the responses of numerous simulation runs. They can be used to establish

Table 9 Baseline Input Parameter Set

Parameter type	Parameter value	Comment
Data rate R_b/Time period T_d	10 bps/0.1 s	Scale to 10 Gbps/0.1 ns
Laser output power	1 mW (0 dBm)	Assume constant level
Laser extinction ratio r_e	0	Assumes external modulator
Laser relative intensity noise RIN	0.05 mW (high)	Represents −20 dB of reflected power, resulting from splices/connectors in short runs
Power budget @ 10 Gbps	36 dB	Includes 3 dB margin above quantum limit
Incident minimum optical power @ desired BER of 10^{-12}	0.2512 μW (2.512e-04mW)	Represents a power loss of 36.099 dB, or a power level at receiver of 0.0002455 mW, which is within the quantum limit of 39 dBm
End nodes, N, of star coupler 1 of M	20 maximum, reduce to improve on SNR	
Number of star couplers, $M \geq 2$	Start with 0, determine maximum via simulation	Equals number of OAs (as a set)
Loss and gain of each Star Coupler/OA set	Start with gain limit of OA (30 dB max) or 14 end nodes (28.2 dB), then reduce to optimize system reach	Assume equal gain and loss at each star/optical amplifier set, therefore net power effect is zero

Number of in-line optical amplifiers, M-1	Start with 0, determine maximum via simulation	Dominant noise source in system; drives SNR into Q value regime that allows reasonable run times (short)
Gaussian noise for optical amplifiers (OA's), also for RIN, thermal, ampl.	Mean, μ = 0 variance, σ = 1 to 2, sample times, t_s = 0.01 to 0.1 s	Optical amplifier noise parameters are the major contributor for obtaining Q function correlation (in particular, sample times)
Poisson noise for Shot, Surface Leakage, Dark Current in Rec'r	Lambda = 0.5 sample time = 0.01 s	Mean and variance are equal; not significant in IM-DD system
Optical filter bandwidth, $B_0 = \Delta v$	0.2 nm = 24.97 GHz	Design guide: should be (2-3) × B_e
Electrical amplifier gain	Selectable based on network configuration	Accounts for all power losses/gains and brings amplitude backing to starting level of 1 (mW)
Electrical filter type, electrical filter bandwidth, B_e	Butterworth low-pass, 10 Hz = 62.83 rad/s	Achieved best results (lowest P_e) per Nyquist: 2 × f_d
Sampling threshold level (ampl.)	0.5	(1/2) * V_p
Delay time at Error meter (reclock)	0.07 s	Based on eye diagram and Mux displays; iterative until lowest error achieved

Table 10   Baseline Simulation Run Results Tabulation

Step no.	No. of end nodes (N) @ star coupler 1 of M	Gain = loss @ each star/OA node set, from end nodes (N_t)	No. of in-line OAs @ $M \geq 2$	Calculated Q factor	Calculated BER from Q factor	Simulation error meter @ 5000 bits [gaussian $\sigma = 1.0$, $\mu = 0$, $t_s = 0.05$]	Comments
1a "noisiest"	20	28.2 dB 14	11	1.644	.04656	.0586	Fig. 8
1b			8	1.949	.02497	.0238	Fig 9.a,b,c
1c			6	2.25	.01215	.009	
1d			4	2.709	.00339	.0027	Fig. 10
2a "fewer OAs"	20	14	2	3.547	.000197	large # of bits req'd	
2b		0	0	6.063	6.743e-10	—	
3a "Increase incident P_0"	18	0	0	7.872	1.754e-15	—	Fig. 11
3b		14	2	4.752	1.015e-6	—	
3c			4	3.681	0.0001176	—	
3d			14	2.031	0.02074	.0234	
4a "More P_0"	14	14	11	4.433	4.697e-6	—	
4b			20	3.335	.0004313	—	
4c			30	2.711	.003374	—	
4d			40	2.325	.01002	.011	Fig. 12
5a "optimize for reach; system BER"	14	14	6	5.732	4.448e-9	—	
5b	14	14	5	6.18	3.222e-10	—	Meets 10^{-9}
5c	14	10	14	7.738	5.152e-15	—	
5d	14	10	18	6.974	1.546e-12	—	
5e	12	10	30	7.582	1.706e-14	—	
5f	12	10	40	6.669	1.058e-11	—	
5g	10	10	60	7.777	3.74e-15	—	
5h	10	10	80	6.865	3.339e-12	—	
5i	8	8	300	5.929	1.537e-9	—	
5j	8	8	395	7.036	9.696e-13	—	Meets 10^{-12} Fig. 13
5k	8	8	560	6.003	9.76e-10	—	Meets 10^{-9}

a baseline from which to vary specific parameters during subsequent sensitivity evaluation.

5.2 Baseline Simulation Run Results and Displays

After many simulation trials, a systematic approach for test run characterization turned out to be one that used the following steps. Table 10 lists the baseline simulation run results. Displayed output from the OpSAN simulation are shown in Figs. 8 through 13 for some of the key results.

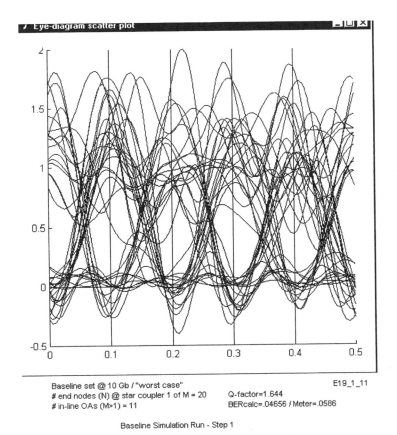

Baseline set @ 10 Gb / "worst case"   E19_1_11
\# end nodes (N) @ star coupler 1 of M = 20   Q-factor=1.644
\# in-line OAs (M>1) = 11   BERcalc=.04656 / Meter=.0586

Baseline Simulation Run - Step 1

Figure 8 Eye diagram for baseline simulation run—lowest ''noisiest'' SNR. Maximum end node connect; lowest incident signal power. 11 in-line optical amplifiers. Showing correlation of calculated BER with simulation error meter.

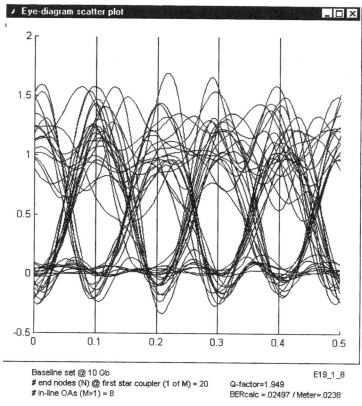

Baseline set @ 10 Gb   E19_1_8
end nodes (N) @ first star coupler (1 of M) = 20   Q-factor=1.949
in-line OAs (M>1) = 8   BERcalc =.02497 / Meter=.0238

Baseline Simulation Run, Step 1b

(a)

Figure 9   (a) Eye diagram for baseline simulation run—low SNR. Reduce number of star coupler/optical amplifier pair set to eight. Showing correlation of calculated BER with simulation error meter. (b) Error rate meter for baseline simulation run—low SNR. Example count for 5000 bits through. (c) MUX waveforms display for baseline simulation run—low SNR. Example display comparing signal in vs. signal out.

Step 1   Go straight to the noisiest configuration in order to validate simulation meter error rates with the BER from Q function (calculated SNR). This consists of attaching the maximum number of end nodes to the first star coupler relative to minimum allowable incident power at the photodetector. Add the maximum number of in-line OAs (with their maximum complement of attached end nodes, Nr, equal to 14, or nearly 30 dB loss) to incur large cascaded ASE noise. Try several variations to get multiple data points for error rate comparison. Corre-

Error rate	_ □ ×

File Edit Tools Window Help

Sender	Receiver
0	0
0	0
1	1
0	0
1	1
1	1
1	1
1	1
0	0
1	1
Symbol Transferred	5000
Error Number	119
Error Rate	0.0238
Reset error count	Close

BERcalc = 0.02497   M19_1_8

Baseline Simulation Run, Step 1-b

(b)

late the low SNR (Q factor in the range of 1 to 3) with the error rates (in the range of 10^{-1} to 10^{-3}). Figures 8 through 10 show example SIMULINK eye diagrams, error meter tabulation, and waveforms display. Note that as the number of star coupler/optical amplifier pair sets are reduced, the cascaded ASE noise decreases and leads to a larger SNR.

Step 2   Reduce the number of intermediate node sets (thus further reducing cascaded ASE noise) to a point where SNR is higher (Q factor in the range of

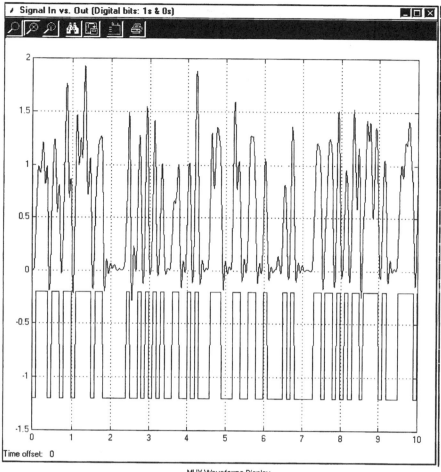

MUX Waveforms Display
Reference Signal - bottom | Receiver Output - top
(100 bits at data rate of 10 bps)
Baseline Simulation Run, Step 1-b

(c)

Figure 9 Continued

4 to 6) to obtain lower BER (in range of 10^{-5} to 10^{-9}). Although it is not possible to run the simulation long enough to obtain error meter numbers in this range, if the previous runs are correlated between Q function calculated BER and simulated BER, then there is some confidence in the projected high-performance system. Figure 11 shows a representative ''clean'' eye diagram at a predicted BER

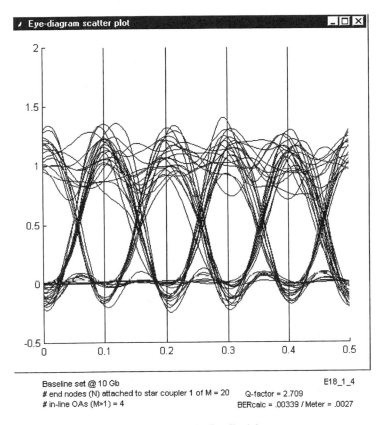

Baseline set @ 10 Gb   E18_1_4
end nodes (N) attached to star coupler 1 of M = 20   Q-factor = 2.709
in-line OAs (M>1) = 4   BERcalc = .00339 / Meter = .0027

Baseline Simulation Run, Step 1-d

Figure 10 Eye diagram for baseline simulation run—improving SNR. Reduce number of star coupler/optical amplifier pair set to four. Showing correlation of calculated BER with simulation error meter.

of 6.743×10^{-10}, thus meeting system requirements. In the following steps, as BER continued to get smaller, the eye diagrams did not visibly improve.

Step 3 Start to reduce the number of end nodes attached to the first star, which begins to increase the incident power on the photodetector. When the number of intermediate network node sets is reduced to 0, points are reached where SNR is very high (Q factor in the range of 7 to 8) to obtain very low system BER (in the range of 10^{-12} to 10^{-15}). Again, the validity of this approach is based on error rate correlation.

Step 4 Reduce the amount of end nodes at the first star, equal to the maximum amount at each intermediate coupler/amplifier node set [14] and thus

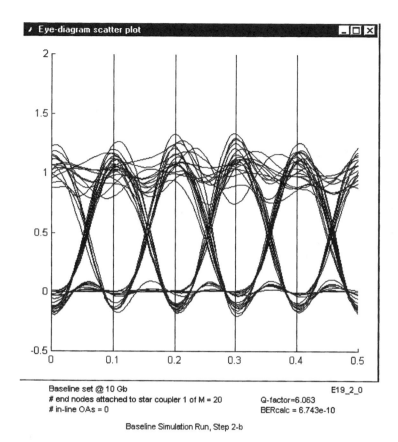

Baseline set @ 10 Gb   E19_2_0
end nodes attached to star coupler 1 of M = 20   Q-factor=6.063
in-line OAs = 0   BERcalc = 6.743e-10

Baseline Simulation Run, Step 2-b

Figure 11   Eye diagram for baseline simulation run—meeting system BER. No in-line optical amplifiers. Representative "clean eye" meeting BER of 10^{-9}.

create a relatively high incident signal power on the photodetector. A correlated eye diagram for a noisy system with 40 OAs is shown in Fig. 12.

Step 5   A level of system optimization for farthest reach at required system error rates is started by reducing the number of attached nodes at both the first and all following star couplers. For example, if $N = N_r$ is reduced to 8, several hundred intermediate nodes are possible, thus many user end nodes. Fig. 13 is the representative eye diagram for meeting FC-SAN system BER requirements of 10^{-12}. The vertical eye opening and the attendant noise margin is somewhat reduced, and indications are that the threshold level should be lowered from 0.5 to about 0.4 for ultimate better error performance.

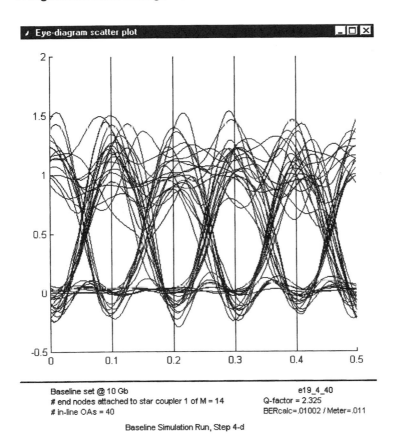

Baseline set @ 10 Gb
end nodes attached to star coupler 1 of M = 14
in-line OAs = 40

e19_4_40
Q-factor = 2.325
BERcalc=.01002 / Meter=.011

Baseline Simulation Run, Step 4-d

Figure 12 Eye diagram for baseline simulation run—increase incident power. Reduce end node attach to star couplers; increase incident optical power. Allows further end node reach (higher system end node count). Showing correlation of calculated BER and simulation error meter.

6 SIMULATION SENSITIVITY ANALYSIS

6.1 Optical Amplifier Gaussian Noise Parameters

As discussed in the receiver modeling section, trying to represent ASE noise and passing it through an optical filter was troublesome. During the simulation trial runs, it was impossible to achieve matching of the Q function calculated BER with simulation error meter rates when using higher frequencies for one of the selectable parameters in the Gaussian Noise Generator modeling block (or t_{sample}

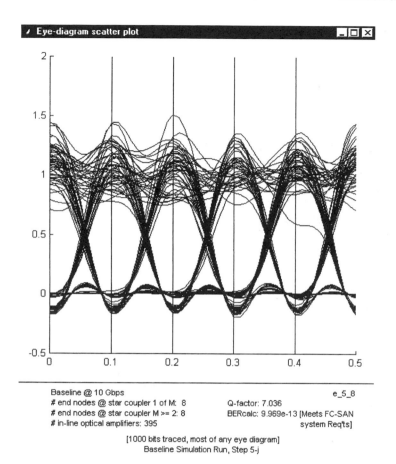

Baseline @ 10 Gbps   e_5_8
end nodes @ star coupler 1 of M: 8
end nodes @ star coupler M >= 2: 8   Q-factor: 7.036
in-line optical amplifiers: 395   BERcalc: 9.969e-13 [Meets FC-SAN
                                                              system Req'ts]
[1000 bits traced, most of any eye diagram]
Baseline Simulation Run, Step 5-j

Figure 13   Eye diagram for baseline simulation run—high reach/meet system BER. Reduce end node count at all star couplers to eight. Allows far reach or large system end node count (3160 users). Meets FC-SAN system BER requirements of 10e-12. Jitter ≈ 20%.

of say, one-tenth the baseband symbol time period) typical of white noise. The simulation error rates were much less than the Q function predictions, indicating that the ASE noise impact was not properly accounted for. Only when using noise sample times very near the baseband message sample time did error rates align. In effect, the best ASE model turned out to be a "binary noise" sent through the system. Perhaps this binary noise model more accurately matches the approximate rectangular shape of the ASE spectrum when passing through an actual Lorentian-shaped Fabry–Perot optical filter [4]. The results of the sensitivity simulation runs are shown in Fig. 14. In addition to sample time, also varied

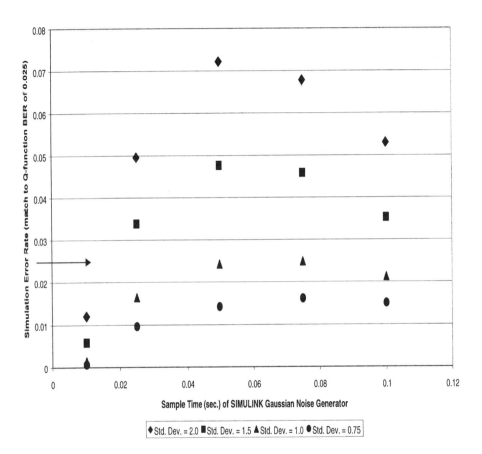

Figure 14 Optical amplifier Gaussian noise parameters. Parameters for the simulation run and BER calculation are: data rate = 10 Gbps; N = 20 end nodes at star coupler 1 of M; N_r = 14 end nodes for each star coupler M ≥ 2; number of star couplers = 8; Q-factor = 1.949 → BER = 0.02497 "noisy"; and simulation runs = 5000 bits through.

was the standard deviation (σ) which is a measure of the spread of the standard deviations, but the mean (μ) was kept at zero. Note that the Gaussian noise generator sample times that produce simulation error rates matching the specifically chosen BER are close to the message time period (T_d). For example, for a data rate of 10 bps, the best Gaussian sample times are 0.025, 0.05, 0.075, and 0.1 s.

6.2 Tuning the Model for 1, 5, 10, and 20 Gbps Data Rates

The error rate matching procedure just described was also undertaken for the several other data rates. The simulation runs for each data rate were done for a configuration that produced a Q factor near two (at the maximum allowable number of end nodes for star couplers, and about 8 or 9 optical amplifiers). In addition, the iterative process of using the Eye and Mux displays to determine the optimum Error Meter block delay times for each data rate was performed. Table 11 summarizes these findings.

6.3 Laser Amplitude Intensity Noise (RIN) Effects

For the baseline runs, it was assumed that relections (20 dB) were present from connections and couplings of the short network link lengths, and a relatively high value of RIN was used (0.05 mW) (Table 12). It is desired to know the effect that laser RIN levels have on system performance. Starting with a 10 Gbps baseline system configured to achieve error rates of better than 10^{-9} ($Q = 6$), simulations were run at decreasing levels of RIN. Beyond a RIN of 0.0005 (representing -40 dB of relections), no improvement was visible. Power penalty results are tabulated below, and Fig. 15 shows the improved eye diagram of a system show-

Table 11 Gaussian Noise Parameters and System Delay Times for Multiple Data Rates

	1 Gbps	5 Gbps	10 Gbps	20 Gbps
Q factor	1.993	2.102	1.949	2.172
BER from Q factor	0.02262	0.01752	0.02497	0.0148
Select Gaussian $T_{sample} = \frac{1}{2} * (T_{data})$	0.5 s	0.1 s	0.05 s	0.025 s
Choose $\sigma_{Gaussian}$ (producing error meter rate @ 5000 bits)	1.0 (0.0224)	1.1 (0.0172)	1.0 (0.0242)	1.15 (0.0148)
Error meter delay time (s)	0.85	0.17	0.07	0.042

Table 12 RIN Sensitivities

RIN, r^2	RIN noise variance, σ_i^2	Total noise, $\Sigma \sigma_{noise}^2$	Q factor	BER from Q factor	Power penalty, $\Delta_i \triangleq 1/(1 - r_i^2 Q^2)$
0.05 mW (−20 dB refl.)	9.64e-11	9.015e-10	6.063	6.743e-10	$1/0.9081 = 1.101$ (0.837 dB) This $= 1/[36.756/40.475]$
0.0158 (−25)	9.629e-12	9.627e-10	6.33	1.233e-10	$1/0.9899 = 1.01$ (.088 dB)
0.005 (−30)	9.643e-13	9.54e-10	6.359	1.024e-10	$1/0.9989 = 1.001$ (.0096 dB)
0.0005 (−40)	9.643e-15	9.531e-10	6.362	1.003e-10	1 (assumed no penalty at this level)
0.00005 (−50)	9.64e-17	9.531e-10	6.362	1.003e-10	1

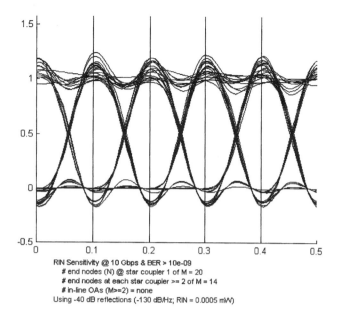

RIN Sensitivity @ 10 Gbps & BER > 10e-09
end nodes (N) @ star coupler 1 of M = 20
end nodes at each star coupler >= 2 of M = 14
in-line OAs (M>=2) = none
Using -40 dB reflections (-130 dB/Hz; RIN = 0.0005 mW)

Figure 15   Eye diagram void of any RIN effects.

ing no RIN effects (compare to the diagram in Fig. 11, having lower eye opening or noise guard).

6.4   Laser Extinction Ratio Effects

The effect of having a laser extinction ratio (r_e) present in the transmitting laser (i.e., greater than zero) has been investigated. Starting with the 10 Gbps system evaluated above in an essentially nonexistent RIN state (RIN = 0.0005 mW, corresponding to -40 dB reflections), simulations were run as extinction ratios and were increased from 0 to 0.6. The decreasing signal level power effects are shown in Table 13, and eye closure at the large extinction ratio of 0.6 is readily apparent in Fig. 16. Note that the noise variances are also affected by r_e because of the shot (σ_{shot}^2) and in particular the ASE ($\sigma_{sig\text{-}sp}^2$) dependencies on the received power.

6.5   System Noise Dependence on Optical Filter Bandwidth

Both the optical and electrical filter bandwidths have a significant effect on the system SNR. The relative dependencies of the noise for varying optical filter

Table 13 Extinction Ratio Effects

Ext. ratio	Psignal @ PIN output	Total noise, $\sum \sigma^2_{noise}$	Q factor	BER from Q factor	Power penalty (dB) $\Delta_e \triangleq (1 + r_e)/(1 - r_e)$ (approx.), dB
$r_e = 0$	3.587e-08	9.531e-10	6.362	1.003e-10	1 (no penalty at $r_e = 0$)
0.1	3.124e-8	8.606e-10	6.026	8.482e-10	1/0.897 = 0.944
0.2	2.469e-8	7.681e-10	5.669	7.217e-9	1/.794 = 2.002
0.3	1.89e-8	6.756e-10	5.289	6.183e-8	1/691 = 3.206
0.4	1.389e-8	5.831e-10	4.88	5.345e-7	1/.5886 − 4.604
0.5	9.643e-9	4.906e-10	4.434	4.676e-6	1/.486 = 1/(19.66/40.47) = 6.27 this ≈ $\Delta_e = 20 \log_{10}(3) = 9.54$
0.6	6.172e-9	3.981e-10	3.937	4.159e-5	1/.383 = 8.336

bandwidths (B_o) are investigated here. Starting with the identical configuration as in the baseline simulation run per Fig. 13 (or Step No. 5j), the optical filter bandwidth was selected as 24.97 GHz (*). This is based on the design rule by Kazovsky [4], which states that the optical filter bandwidth should be about three times the channel bandwidth, which at 10 Gbps is 10 GHz. Table 14 shows the

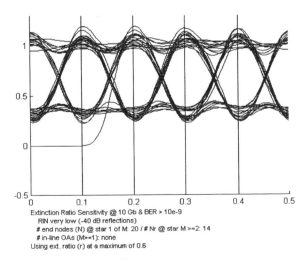

Extinction Ratio Sensitivity @ 10 Gb & BER > 10e-9
RIN very low (-40 dB reflections)
end nodes (N) @ star 1 of M: 20 / # Nr @ star M >=2: 14
in-line OAs (M>=1): none
Using ext. ratio (r) at a maximum of 0.6

Figure 16 Eye diagram showing closure at large extinction ratio.

Table 14 Optical Filter Bandwidth Effect on System Noise

Δλ (nm)	B_0 (Hz)	Q factor	Calculated BER	ASE $\sigma^2_{\text{sig-sp}}$	ASE $\sigma^2_{\text{sp-sp}}$	ASE Total	All other noise
0.1	1.2487e10	7.061	8.277e-13	2.37e-06	1.899e-08	2.256e-06	3.214e-07
0.2*	2.497e-10	7.036	9.69e-13	"	3.798e-08	2.275e-06	"
0.3	3.746e10	7.01	1.197e-12	"	5.697e-08	2.294e-06	"
1.0	1.2487e11	6.838	4.026e-12	"	1.899e-07	2.427e-06	"
2.0	2.497e11	6.614	1.885e-11	"	3.798e-07	2.616e-06	"
3.0	3.746e11	6.41	7.329e-11	"	5.697e-07	2.806e-06	"
10	1.2487e12	5.37	3.976e-08	"	1.899e-06	4.136e-06	"
20	2.497e12	4.496	3.487e-06	"	3.798e-06	6.034e-06	"
30	3.746e12	3.946	4.022e-05	"	5.697e-06	7.933e-06	"

* Per Fig. 13.

system noise sensitivities as B_0 is increased. Once the selected channel power has completely passed through the filter, any increase in B_0 will only increase the ASE spontaneous-spontaneous noise term. A filter bandwidth of less than about 3 nm is necessary to keep the error rate below system requirements, and for multichannel WDM, smaller bandwidths are desired for closer spacing.

7 SCALING TO THE NETWORK SYSTEM

7.1 Optical Amplifier Placement in the Network

Calculations and simulations have shown that the bit error in a system decreases as the ratio of received signal to noise increases. Optical amplifier placement is critical to the IM-DD system that has been modeled. In order to include optical amplifiers in a network in the correct way topologically, it is necessary to examine tradeoffs between a number of factors that revolve around gain and noise. In the absence of photonic amplification, the dominant form of receiver noise is typically thermal noise appearing at the input of the electrical amplifier. OAs can increase signal power to overcome thermal noise, but they add their own form of ASE noise. Another factor at work that has not been modeled is the phenomenon of amplifier gain saturation: the amount of gain produced by the OA decreases with the applied total applied input signal power. Based on typical gain saturation curves for the EDFA operating at 1550 nm [7,8], a small signal level of P_{in} of -60 dBm to -40 dBm that produces a near-constant gain G_0 of 30 dB is reduced in gain to about 24 dB when the input signal power is raised to -27 dBm, to 20 dB gain with input power of -16 dBm input power, and to 17 dB gain with input power of -12 dBm.

The three basic roles for optical amplifiers are as transmitter power amplifiers, receiver front-end preamplifiers, and in-line amplifiers. Placing the OA immediately following the laser transmitter in the power mode will incur large input signal power (0.1–1 mW) and will put a premium on maximizing saturation output power characteristics. As mentioned, this application has not been modeled or investigated, but the latter two have been incorporated into this modeling effort. By placing an OA directly before the receiver, it functions as a preamplifier. When using only a predetection OA at the receiver, it has been shown that a high system BER of 10^{-9} or better is achieved by keeping the splitting loss associated with the number of end nodes at the single star coupler to within detectable limits. Referring to the baseline simulation runs at 10 Gbps of Table 10 and Fig. 8, attachment to the star coupler was 20 end nodes (power loss of 36 dB, which matches the system power budget minimum) and the result was BER of 6.743×10^{-10}. The use of an OA at the threshold signal levels occurring at PIN photodiodes (IM-DD) allows one to approach receiver detectability limits very close to those achievable with more exotic forms of receivers (e.g., coherent heterodyne ASK). However, the use of just a preamplifier OA is uninteresting for network application; numerous OA's are also placed in-line along the transmission path to overcome distribution splitting losses and thereby greatly extend the reach of the optical network. The in-line OA exists in cascade form; a given amplifier's output signal and noise feeding into a subsequent amplifier with attenuation between them. At the receiver detector, high gain and low OA noise is the main figure of merit. Optical filtering and optical isolation must be considered. The receiver is thus limited by ASE noise and not by the receiver thermal noise, as all of the noise calculations within the simulation runs have shown.

Each OA can produce a gain maximum of about 30 dB. As modeled, this is associated with splitting losses of a star coupler to no more than 14 end nodes (actually calculated as a loss of 28.2 dB). The variability afforded by the node attachment to the first star has been an interesting complication during the numerous trial simulation runs. It can be increased in node count until limited by detectability at the receiver (and this varies with data rate, e.g., -46 dBm at 1 Gbps to -33 dBm at 20 Gbps), or it can be decreased in node count, thus reducing loss until all the cascaded OAs see high input signal power and are in essence operating in saturation power mode. An optimum placement of this amplifier exists, as described by Green [3]. A simplifying assumption is made at this point for the following network scaling investigations, namely, that all star couplers in the cascaded chain will have the same number of attached end nodes.

A representation of this power level dynamic associated with node attach count is represented in the Fig. 17.

7.2   End Node Reach Configurations

Figure 18 of the predicted maximum reach of the network at 10 Gbps shows that it is desirable to have a larger number of cascaded low-gain OAs than a smaller number. Further extensions are deemed unrealistic from the standpoints of not having modeled in longer fiber line lengths to accommodate physically such large node quantities, bumping into saturation gains, and the impractical cost of numerous OAs.

Shown in Fig. 19 is the predicted maximum reach of the network as a function of the data rate. The assumption is that $N = 12$ end nodes at each star coupler safely stay away from any amplifier saturation effects, and that fewer numbers of OAs are desired from a system economic standpoint. The decrease in connect count to total number of end nodes goes down very near a logarithmic rate as data rate increases.

7.3   Allowable WDM Channels

In order to exploit the fiber's THz bandwidth, WDM can be used to transmit simultaneously a number of baseband-modulated channels across a single fiber

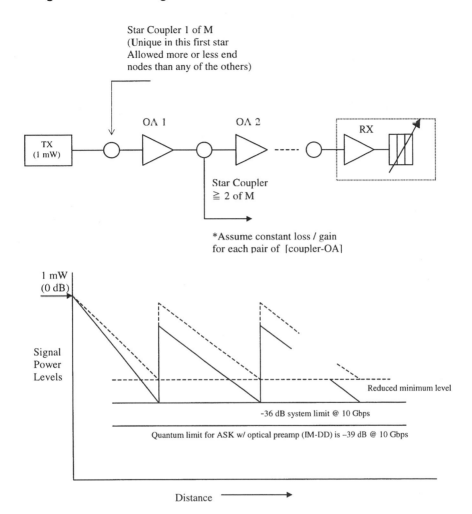

Figure 17   Cascaded amplifier and signal power representation.

path, but with each channel located at a different wavelength. The aggregate system is transmitting at *K* times the individual laser speed, thus providing significant capacity enhancement. Various device, system, and network issues will limit the WDM utilization of the full fiber bandwidth. Among the critical component technologies required for a WDM system are multiple-wavelength transmitters, multiport star couplers, optical amplifiers, and tunable optical filters. These components are involved in a number of critical issues that need addressing in a WDM system; channel broadening, crosstalk, nonlinearities, and power penalties.

As was previously discussed, both wavelength tunable lasers and multiple-

# End Nodes at each Coupler; $N = N_r$	Max. allowable # of OAs	Calculated Q- factor	BER from Q-factor	System Total # of end Nodes = (#OAs x #N)
14 (28.175 dB gain:loss)	5	6.18	3.22e-10	5 x 14 = 70
	3	7.414	6.008e-14	42
12 (25.23 dB)	24	5.989	1.06e-09	288
	17	6.959	1.72e-12	204
10 (22.03 dB)	107	6.109	8.831e-10	1070
	76	7.022	1.096e-12	760
8 (18.49 dB)	560	6.003	9.76e-10	4480
	395	7.036	9.969e-13	395 x 8 = 3160

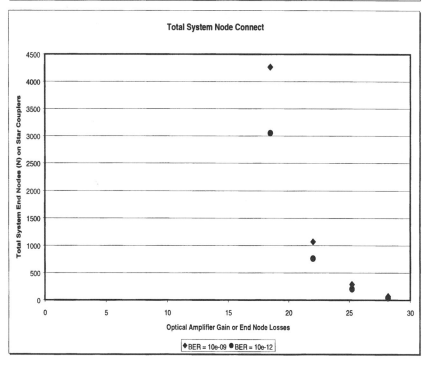

Figure 18 Network node reach limits at 10 Gbps.

wavelength laser arrays are viable at the transmitter (using one tunable device or an array of fixed devices). External-cavity tunable lasers such as the Fabry–Perot have the ability to reduce significantly laser linewidth. When considering the ability to utilize the maximum number of WDM channels, a trade exists between the limiting physical size of a laser array versus the need to limit chirp in a tunable laser by using an external optical modulator. There is significant packaging challenge when coupling a large array of vertical-cavity lasers into a single output fiber [4]. If chirp is not minimized in a tunable laser, energy is lost from the desired channel, and it appears as undesired crosstalk on adjacent chan-

Data Rate	Max. allowable # of OAs	Calculated Q-factor	BER from Q-factor	System Total # of end nodes = (#OAs x 12 nodes per star)
1 Gbps	248	6.001	9.855e-10	248 x 12 = 2976
	175	7.03	1.04e-12	2100
5 Gbps	49	5.99	1.058e-09	588
	34	7.047	9.181e-13	408
10 Gbps	24	5.989	1.06e-09	288
	17	6.959	1.72e-12	204
20 Gbps	12	5.88	2.064e-9	144
	8	6.957	1.741e-12	96

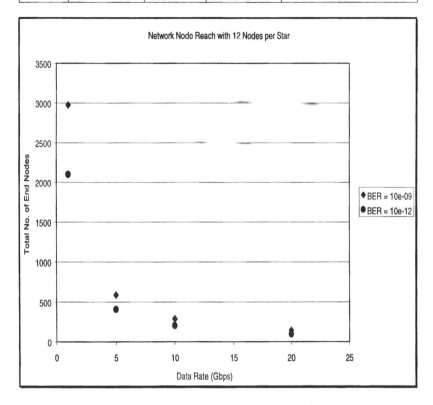

Figure 19 Network node reach as function of data rate.

nels. Using external amplitude modulators such as the Mach–Zehnder interferometer does not produce any chirp, which should allow closer spacing of individual channels. However, they have limited bandwidth of about 18 GHz (which can severely limit the total number of WDM channels that can be crowded together) in addition to an insertion loss of about 6 dB [3]. A significant number of tradeoff analysis and development work is necessary here to optimize for a particular system configuration.

The EDFA is an almost ideal optical amplifier in WDM systems, except for one major flaw; a nonuniform gain spectrum. When passing several different channels at different wavelengths through an EDFA, each channel will experience a different gain, causing a differential in signal power and SNR. In a cascaded system, the gain does not accumulate linearly from stage to stage, and the resultant wavelength-dependent gain dramatically changes shape in a cascade. Methods exist and have been reported [4] to equalize the nonuniform EDFA gain; this assumption is used for network scaling in this work.

The most basic function of an optical filter is to demultiplex many WDM channels so that only one channel is received and all the other channels are blocked. Interchannel crosstalk will act to reduce the power contrast ratio of the selected recovered signal [4]:

$$S = P_S - P_{cr} \tag{16}$$

where P_s = power in the selected wavelength that passes through the filter and P_{cr} = total crosstalk power from rejected channels that is not blocked.

The crosstalk due to power leakage from the other channels is deterministic and affects the signal mean (signal power) and not the signal variance (noise power). The effective recovered signal power signal (S) is thus reduced and will impact the SNR of the recovered data in the selected channel. The crosstalk due to power leakage from the other channels should not exceed a few percent of the selected channel power to maintain good system performance [4]. When demultiplexing many optically amplified WDM channels with an optical filter, there exists in addition wideband ASE noise that will leak through the filter passband. The ASE noise will produce a signal-spontaneous beat noise term (N_{s-sp}) between the selected channel and the ASE, and now also there exists a beat noise term between the rejected channels (N_{r-sp}) and the ASE passing through the filter [4].

By using design guidelines [4,9], an approximation can be made of the allowable number of WDM channels for the optically amplified network:

> The optical filter bandwidth should be three or more times the channel bandwidth (f_d).
> The channel separation should be more than 1.5 times the optical filter bandwidth (B_0).

The effect of SNR of the rejected channels can be described as a system power penalty as a function of channel spacing given the number of channels. An optical channel separation of 1.5 times the optical filter bandwidth is incorporated in the calculations of Table 15, and for 100 rejected channels, a system power penalty of 2 dB has been predicted [10]. The OPSAN simulation model has included a system margin of 3 dB in its power budget, which should be sufficient to account for this power penalty.

It may also be possible to infer crosstalk effects in the optical filters from

Table 15 Approximation of allowable WDM channels

Input data rate, R_d, Gbps	For margin, use, $B_e = 2 \times f_d$ GHz	$B_0 = 3 \times B_e$, GHz	Channel separation = $1.5 \times B_0$, GHz	Total BW per channel = $7.5 \times B_e$, GHz	Channels per OA bandwidth where $B_{OA} = 30$ nm or 3.746×10^{12} Hz
1	1	3	4.5	7.5	$3.746 \times 10^{12}/7.5 \times 10^9 = 500$
5	5	15	22.5	37.5	100
10	10	30	45	75	50
20	20	60	90	150	25

the simulated eye diagrams of the electrical signal after photodetection and matched filtering. The sampling instant is chosen to minimize the intersymbol interference (ISI) or, equivalently, to maximize the eye diagram opening. ISI is an error source that comes from pulse spreading in the optical fiber, i.e., some of the transmitted energy will spread into neighboring time slots as the pulse propagates along the fiber. The presence of energy in adjacent time slots results in an interfering symbol. It can narrow the horizontal eye width, thus reducing the possible range of sampling instants and increasing the sensitivity to their fluctuations, known as timing jitter. Since OpSAN did not model any fiber dispersion linearities due to short network link lengths, any jitter present in the eye diagrams must be due to ASE noise effects in the receiver. Timing jitter is given by

$$\text{Timing jitter (percent)} = \frac{\Delta T}{T_b} \times 100\% \tag{17}$$

where ΔT = measured amount of distortion from the eye diagram and T_b is one bit interval.

Using the eye diagram of Fig. 13, which represents meeting the system BER requirement of of 10^{-12}, the jitter is approximated as 20% within the single channel modeled in OpSAN. If there is this much crosstalk in a single channel, then multichannels in a WDM system should be separated by at least this amount, or 1.2 times the electrical filter bandwidth. Using the above design guideline that channel separation should be of 1.5 times the optical filter bandwidth ($1.5 \times B_0 = 1.5 \times 3B_e = 4.5B_e$) provides ample margin.

7.4 Prediction of Aggregate Network Capacity

Starting with the predicted maximum reach of the network as a function of data rate, and using the allowable WDM channels from Table 15, the aggregate multichannel SAN system capacity can now be scaled (Table 16).

This ultimate predicted network capacity must, of course, be adjusted for the complexities of layered software management protocols, consider nonsimulta-

Table 16 Predicted Aggregate Network Capacity

Input data rate, R_d, Gbps	End users = total # end nodes @ 12 nodes per star coupler	Total # of channels	Ultimate network bandwidth = total # of users × total # of channels × data rate of each channel
1	2100 at BER of 10^{-12}	500	$2100 * 500 * \text{Gbps} = 1.015 \times 10^{15}$ bps
5	408	100	2.04×10^{14}
10	204	50	1.02×10^{14}
20	96	25	4.8×10^{13}

neous usage patterns, and accommodate error rates of some computer data links going as low as 10^{-15}. However, the technical feasibility exists for fiber-optic networks with terabit/s capacities at the physical layer and has been reported elsewhere in the literature [11–14].

8   SUMMARY AND CONCLUSIONS

The implementation of short-haul all-optical networks will likely start on a small scale and rapidly increase toward larger complexity and capacity. The progression and performance of the following technology enablers can be predicted based on results of this optical network system investigation, modeling, and simulation.

The IM-DD transmission scheme with the broadcast-and-select implementation of a passive star coupler is proven and simple. External modulation of the laser can avoid the relatively serious extinction ratio effect. The incorporation of a predetection optical amplifier at the receiver front end can enable a small network of 20 user end nodes (at 10 Gbps and BER of 10^{-10}).

By overcoming power splitting losses at multiple star couplers, the optical amplifier is *the* major enabler for dramatically increasing the physical size or reach of the network. In-line optical amplifier placement configurations can increase end node reach to be in the range of several hundred to several thousand (with tens to hundreds of cascaded amplifiers at 10 Gbps and BER of 10^{-9} to 10^{-12}). The industry need to incorporate optical amplifiers will accelerate making the hard choices associated with 1300 nm versus 1550 nm transmission bands. The 1550 nm EDFA is effective and field proven. The 1330 nm PDFFA has high gain similar to the EDFA but has the drawback of requiring high pump power. Semiconductor optical amplifiers can amplify well at 1300 nm but exhibit considerable more crosstalk than does the active fiber type; this is a negative for WDM.

The next major enabler is WDM, which from this simulation has been projected to increase aggregate network capacity that can approach several hundred THz (at 10 Gbps, 50 channels, BER of 10^{-12}). Multiple wavelength laser transmitters and tunable optical filters are an essential part of making WDM work, from the standpoint of tuning range and maximizing channel selectivity within the optical amplifier bandwidth of 30 nm, and from the standpoint of tuning speeds for switching and network routing protocols. Significant development and optimization of both tunable lasers and tunable filters is required to minimize crosstalk, thus enabling closer channel spacing.

REFERENCES

1. D. E. Tolmie. Gigabit Networking. IEEE LTS, pp. 28–36, May 1992.
2. *www.fibrechannel.com.*

3. Paul E. Green, Jr. Fiber Optic Networks. Prentice-Hall, Englewood Cliffs, NJ, 1993.
4. L. C. Kazovsky, S. Benedetto, and Alan Willner. Optical Fiber Communication Systems. Artech House, Norwood, MA, 1996.
5. Gerd Keiser. Optical Fiber Communications. McGraw-Hill, New York, 2nd. ed., 1991.
6. B. P. Lathi. Modern Digital and Analog Communication Systems. Saunders College Publishing of Holt, Rinehart and Winston, Philadelphia, 2nd ed., 1989.
7. G. R. Walker, N. G. Walker, R. C. Steele, M. J. Creaner, and M. C. Brain. Erbium-doped fiber amplifier cascade for multichannel coherent optical transmission. IEEE Journal of Lightwave Technology, Vol. 9, No. 2, February 1991.
8. C. R. Giles and E. Desurvire. Modeling erbium-doped fiber amplifiers. IEEE Journal of Lightwave Technology, Vol. 9, No. 2, Feb. 1991.
9. C. Brackett. Dense wavelength division multiplexing networks: principles and applications. IEEE Journal on Selected Areas in Communications, Vol. 8, pp. 948–964, August 1990.
10. A. Hill, R. Wyatt, J. Massicott, K. Blyth, D. Forrester, R. Lobbett, P. Smith, and D. Payne. 40-million-way WDM broadcast network employing two stages of erbium-doped fibre amplifiers. Electronics Letters, Vol. 26, pp. 1882–1884, 1990.
11. N. A. Olsson. Lightwave systems with optical amplifiers. IEEE Journal of Lightwave Technology, Vol. 7, No. 7, July 1989.
12. R. Ramaswami. Multiwavelength lightwave networks for computer communication. IEEE Communications, pp. 78–88, Feb. 1993.
13. H. T. Kung. Gigabit local area networks: a systems perspective. IEEE Communications, pp. 79–89, April 1992.
14. Paul E. Green, Jr. Optical networking update. IEEE Journal on Selected Areas in Communications, Vol. 14, No. 5, pp. 764–779, June 1996.

4

Routed Wavelength WDM Networks

Lixin Wang and Mounir Hamdi
Hong Kong University of Science and Technology, Kowloon, Hong Kong

1   INTRODUCTION

An optical fiber can potentially have more than 30 THz of bandwidth that can be exploited by future generation broadband ISDN networks. However, because of the mismatch between optical transmission and electronic processing, the network transmission rate cannot exceed few Gbps. One popular solution to this problem is to employ *wavelength division multiplexing* (WDM) by fragmenting the whole bandwidth of the optical fiber into a number of narrow band (say, 1 Gbps) channels. The source party can communicate with its destination party through these channels under the control of a given protocol.

In WDM networks, there are essentially two types of architectures, *passive switch networks* and *active switch networks*. A passive switch network uses a passive star coupler to split the inlet lightwave to all other outlets. Because the passive star coupler can be considered as a broadcasting device, this single-hop system is known as a broadcast and select WDM network [7]. In this type of system, network-wide status and control information can be easily obtained. As a result, through proper scheduling, the channel efficiency can be considerably high. In addition, due to the accuracy of the control information, a given quality-of-service (QoS) can be controlled and guaranteed relatively easy. However, to make a complete connection, the transceivers on each node have to frequently tune among the channels. Therefore the performance of this type of network is very sensitive to the network parameters, such as the number of channels, the number of nodes in the network, and the transmission overhead (e.g., transceiver tuning time). These restrictions can hamper the scalability of the network [7,31,32]. These problems can be overcome by using *strong multihop* systems [10]. Note that we can have a *weak multihop system*, which uses a *broadcasting and selection* scheme for its nodes to communicate but the network is organized in a multihop way. This approach can be used to reduce the tuning effect and adapt to the case where the transceiver tuning range is small [7]. However, this solution does not result in a scalable network system. The details of weak multihop WDM networks are out of the scope of this paper, and we will exclude this type of network when we introduce *multihop* networks in the rest of the paper.

In a multihop WDM system, the nodes are connected by relatively static channels. A pair of source and destination nodes may have to transmit packets through some intermediate nodes. Instead of working in a *broadcast and select* manner as in single-hop WDM systems, the nodes in a multihop system operate using a *store-and-forward* scheme. Upon arrival to an intermediate node, packets from an inlet link are buffered and then forwarded out according to a given *routing strategy*. As a result, most tuning operations are avoided (tuning may still be necessary when the network needs to be reconfigured). Because the connections between any two nodes can span through different paths, potentially the network capacity increases as the number of nodes increases. Hence the network can very scalable. However, an irregular network topology, an asymmetric traffic load, or a nonadaptable routing algorithm may cause congestion on some intermediate nodes [32]. The congestion causes considerable uncertainty in terms of packet delay, packet loss, inefficiency of network utilization, and complexity of network control. More importantly in these WDM networks, the buffering operations, which are currently implemented using electronic devices, strictly limit the optical bandwidth utilization. This results in an opto-electronic bottleneck problem.

The combination of a single-hop system and a multihop system, by employing a *wavelength routing switch* (WRS), results in a more efficient architecture, known as *wavelength routed optical network* (WRON), which is the main theme of this paper. This optical network employs wavelength multiplexers and optical switches in the routing nodes, so that any arbitrary topology can be accomplished, and very large areas can be covered. Nodes can obtain ''single-hop'' access in a WRON by setting up an all-optical *lightpath* through one or more WRS. Along the lightpath, packets are transmitted using identical wavelengths, without any optical-electronic conversion. Certain store-and-forward nodes can be used to conjunct two or more different lightpaths. The wavelength of these lightpaths are not necessarily the same. Thus if we think of the group of nodes that are connected by a lightpath as a *supernode*, the WRON can be treated as a multihop network of the supernodes. Theoretically, every node in the network can be involved in one or more lightpaths, at the same time. By properly designing *wavelength routing and assignment* (WRA) schemes, a WRON can be very scalable and flexible. On the other hand, efficient routing and network control and management that are needed in these networks are very challenging and present interesting issues that need to be solved.

In a WRON, a lightpath exclusively takes one of the wavelengths that are possibly used by all nodes connected in the path. Otherwise, different lightpaths may use the same wavelength at the same routing node, which leads to transmission conflicts. This is known as a *wavelength-continuity constraint*. By routing through a store-and-forward node, we can "change" the wavelength. However, a buffering operation has to be performed that may cause additional delay. The solution to this problem is to use a *wavelength conversion*. By adding a wavelength conversion device into a WRS, the switch becomes a *wavelength convertible routing switch* (WCRS). By applying this technique, several lightpaths with different wavelengths can be chained together, forming an all-optical path termed a *semilightpath*. With a semilightpath, a wavelength in the network can be reused, and the network utilization can be increased.

Because of the combined advantages and flexible features as well as the great potential connecting capability of the WRON, more and more researchers are interested in this subject, and many excellent works are published in the literature. This was the driving force behind writing this paper. In this paper, we survey the recent research related to wavelength routed optical networks that may serve as a good starting point for researchers starting to explore this interesting area.

The rest of the paper is organized as follows. Section 2 surveys the essential issues related to the design and operation of wavelength routed optical networks without wavelength conversion. Section 3 focuses on the special issues raised when employing wavelength convertible routing optical networks. Section 4 discusses QoS issues related to wavelength routed optical networks. Section 5 introduces multicasting on wavelength routed optical networks, and Sec. 6 concludes the paper.

2 WAVELENGTH ROUTING WDM NETWORKS

This section discusses the major issues that need to be considered when designing wavelength routing WDM networks without wavelength conversion.

2.1 Virtual Topology Mapping

Regular topology networks are well studied, and the routing on these networks is relatively simple and efficient. Therefore mapping a regular topology network onto the WRON can simplify the process of routing. Because the regular topology network is not physically implemented, we refer to *virtual topology*, while the WRON onto which the virtual topology is mapped is called a *physical network*. For example, hypercube networks can be a mapped onto a wavelength routing network such as NSFNET [32] to minimize the lightpath in terms of number of hops. A torus network can be mapped onto a WRON and the corresponding protocols of the torus can be easily applied, e.g., deflecting algorithm, slotted token grid protocol [39].

Suppose we are given a virtual topology $G_p = (V, E_p)$, where V is the set of network nodes, and E_p is the set of links connecting the nodes. Assume node i is equipped with a $D_p(i) \times D_p(i)$ wavelength-routing switch. For a network with M available channels and N nodes, the mapping of graph G onto the network can be described as (a) mapping each of the V nodes in G to the network ($V \leq N$) and (b) for each of the links in G, find a lightpath in the network so that the lightpath connects the two nodes and the lightpath takes the unique wavelength which is in the range of M. Performance of these mapped networks is evaluated in [22,26].

2.2   Routing and Wavelength Assignment (RWA)

We can address the routing problem in a more general sense. Given a *traffic matrix* that represents the source–destination pair of a connection and its required bandwidth, we construct sets of lightpaths so that the requirement of the traffic matrix can be satisfied. In fact there is usually more than one solution to this problem. This problem is known is the *routing and wavelength assignment* (RWA): to find an *optimal* solution from all possible solutions [30,29,11]. The *optimal* solution here can explained and applied in different aspects. The methods that find a minimum number of *hops* (the number of WRSs in a lightpath) are called *delay-oriented* optimization. Some algorithms attempt to find the maximum number of lightpaths that can be accommodated in a given network. This type of algorithm is known as a *utilization-oriented* optimization algorithm. In other algorithms, the cost, which includes the switches, amount of bandwidth, connection durations, etc., is taken into account for certain applications. This type of algorithm is known as a *cost-oriented* optimization algorithm. The first two types can be treated as special cases of the third one. But because the first two criteria are of special importance and application independence, they are typically considered separately. We can extend the concept to *QoS oriented* optimization algorithms where the cost is some function of QoS requirements and multimedia transmission overhead.

The RWA problem can be divided into two subproblems: the *routing problem* and the *wavelength assignment* problem. The routing problem is to construct a *virtual topology* that can optimally meet the requirement of the traffic. This problem can be formally described as follows [29]. Let λ_{sd} denotes the traffic (in terms of a lightpath) [23] from any source s to any destination d. Let F_{ij}^{sd} denotes the traffic that is flowing from source s to destination d on link ij. It can also be used to represent the *cost* of the traffic flow. The RWA problem can be described so as to *minimize F_{max}* where

$$F_{max} \geq \sum_{s,d} F_{ij}^{sd} \quad \text{for all } i,j$$

given the constraint

$$\sum_{i} F_{ij}^{sd} - \sum_{k} F_{jk}^{sd} = \begin{cases} -\lambda_{sd} & \text{if } s = j \\ \lambda_{sd} & \text{if } d = j \\ 0 & \text{otherwise} \end{cases}$$

where λ_{sd} and F_{ij}^{sd} are measured in terms of number of lightpaths.

This graph construction algorithm can be thought of as an integer linear programming (ILP) with the object function being to minimize the flow in each link. It is shown to be an NP-complete problem [6,23]. Suboptimal results can be obtained using *genetic, heuristic,* or hybrid methods [4,11].

Once the lightpaths are chosen, the wavelength assignment can be done by employing *graph coloring* algorithms, which are, again, NP-complete problems. Ramaswami and Sivarajan and Chen [18] and Banerjee and Chen [11] studied the upper and lower bounds of the connections that can be accommodated.

2.3 Dynamic Wavelength Routing

Given that the traffic matrix can be changed when the traffic patterns change, the RWA should generate different results that will lead to different wavelength assignments. Furthermore, if the physical network size and topology are unknown (this is a more realistic case in wide-area networks), an RWA may not be able to work properly due to the lack of global information. These cases need a *dynamic lightpath establishment* (DLE) ability. An early work on the topic is known as *Least Congested Path* (LCP) by Chan and Yum [38]. The idea is to keep the *spare route* as large as possible for a lightpath. When a new connection comes, LCP finds a path that least reduces the *spare route set*. In this way, the least congestion can be expected to be produced. More recent work is proposed by Harai et al. [26]. They consider a routing method with limited trunk reservation in which connections with more hops are prepared for more alternate routes. In addition, the performance improvement is investigated by introducing a wavelength assignment policy and a dynamic routing method. The effectiveness of the proposed method is investigated through simulation. Struyve and Demeester discussed different dynamic routing methods on various lightpaths [2]. They defined *nonprotection, dedicated protection,* and *shared protection* modes to deal with the coming calls according to different usage strategies of spare routes. Three routing strategies are analyzed which adopt the shortest path routing for *fixed minimum hop, dynamic minimum hop,* and *dynamic competitive* routing.

3 WAVELENGTH CONVERTIBLE ROUTED OPTICAL NETWORKS (WCRON)

A wavelength converter is a device that can *convert* the wavelength inlet from an input into another wavelength in the output. With this function of wavelength conversion, the lightpaths in different wavelengths can be chained, without an electronic store-and-forward process, into one optically connected route which is referred to as a semilightpath. This function solves the *wavelength continuity constraint* [28] and results in the capability of wavelength reuse, more flexibility, and higher utilization of network bandwidth. However, there are impacts that have to be considered in a WCRON.

The benefits obtained by using the wavelength conversion are referred to as *wavelength conversion gain.* Suppose a network has W wavelengths per link. Let ρ be the probability that a wavelength is used in any fiber link. For a lightpath in the network, there are H links. With wavelength conversion, a connection cannot be allocated only if all the W wavelengths in one of the H links are occupied, i.e., the blocking probability of the lightpath is

$$P_c = 1 - (1 - \rho^W)^H$$

Define q to be the utilization corresponding to the blocking probability P_c. Then

$$q = [1 - (1 - P_c)^{1/H}]^{1/W} \approx \left(\frac{P_c}{H}\right)^{1/W}$$

While in the wavelength routing networks without wavelength conversion, the connection is blocked when all the wavelengths are used, at least, in one of the H links. The blocking probability P_s is

$$P_s = [1 - (1 - \rho)^H]^W$$

and the corresponding utilization, p, is given by

$$p = \left[1 - (1 - P_s\left(\frac{1}{W}\right))\right]^{1/H} \approx -\frac{1}{H}\ln(1 - P_s^{1/W})$$

Thus the wavelength conversion gain G for the same blocking probability P is

$$G = \frac{q}{p} \approx H^{1-(1/W)} \frac{P^{1/W}}{-\ln(1 - P^{1/W})}$$

As can be seen, as W increases, also G increases until it gets to the peak around $W = 10$, where a maximum gain of $H/2$ is achieved. Generally, the larger the number of wavelengths (W) and the longer the lightpaths (H), the higher the wavelength conversion gain (G) would be.

3.1 Converter-Based Wavelength Routing Assignment

Under the assumption that any of the wavelengths can be converted to any other wavelength used in the network, Ramaswami and Sivarajan show that in case all the WRS are equipped with a wavelength converter, the WCRON is equivalent to a circuit-switching telephone network [18]. As a result, the wavelength routing problem becomes equivalent to a circuit-switching network routing problem. Thus all circuit-switching routing algorithms can be applied. However, not every routing node (WRS) necessarily has wavelength conversion ability due to the existence of the lightpath. To optimize the number of converters (WCRS) needed in a WCRON for-given number of wavelengths and number of nodes can be obtained by various types of algorithms [6,24].

Ramaswami and Sasaki study the case where there is a limited number of wavelengths that can be converted [45]. They investigate the ring, star, tree, mesh, and hub-based networks with fixed wavelength conversion capability in the nodes. They show that with the limited number of wavelengths that can be converted, the connection can still be efficiently routed.

3.2 Dynamic Wavelength Routing with Wavelength Conversion

There are a number of ways to implement wavelength converters, such as opto-electronic conversion, coherent effects, and cross-modulation [23,28]. Different implementation methods lead to different costs. Moreover, reconfiguration of the WCRS results in an additive delay and extra overhead. In a static wavelength routing network, these costs just occur once at the system initialization time, so they can be ignored. But in dynamic wavelength routing networks, the reconfiguration happens frequently, so the costs associated have to be taken into account. Chlamtac et. al. proposed a distributed *shortest path algorithm* to find the cost-effective path in a given routing network [30], which was later improved by Liang et al. [3] by separately considering the lightpath cost and the semilightpath cost. Karasan and Ayanoglu proposed a least-loaded routing algorithm that jointly selects the least-loaded routed-wavelength pair [1]. This algorithm produces a large wavelength conversion gain.

The dynamic routing approach can also be used to improve the reliability of the network services. When a routing node or a fiber link is broken, all related lightpaths or semilightpaths are in failure. By dynamic routing with wavelength conversion, locally reconfiguring some of the neighbor routing nodes can be achieved by passing the fault nodes or links without affecting other nodes.

4 QoS ISSUES ON WAVELENGTH ROUTING NETWORKS

The quality-of-service (QoS) provisioning is the idea that the transmission rates, error rates, and other characteristics can be measured, improved, and to some extent, guaranteed in advance. Different types of applications have different QoS requirements. Accordingly, a network should supply multiple levels of transmission services to meet the needs of various traffic streams. In the context of wavelength routing networks, the QoS can be measured in terms of *connection blocking probability*, i.e., the probability that a requested connection cannot find a route that satisfies the QoS requirement of the connection. However, the implication behind the probability can be varied as a function of the different traffic characteristics. For example, a video stream can be admitted if we found a lightpath that satisfied the bandwidth and delay jitter requirements for it. A file downloading request does not necessarily result in the establishment of a dedicated lightpath; a certain feasible route may be a multihop route, because usually the file transmission does not have very strict delay and bandwidth requirements. Although the QoS performance of these two types of traffic can be evaluated in terms of blocking probability, their resource occupation are obviously different. As a result, wavelength assignment strategies are different.

The QoS oriented wavelength routing problem can be formally described as follows [37]. Given a network $G(V,E)$, with a maximal rate R_l and a link delay d_l for each $l \in E$ and for each of the connections i, $1 \le i \le I$, where I is the connection set, and given the source s^i, destination t^i, burst δ^i, packet size c^i, delay constraint D^i, and bandwidth requirement b^i, then a *feasible path* p^i should satisfy

$$\frac{\delta^i + \mathrm{hops}(p^i)c^i}{r^i} + \sum_{l \in p^i} d_l \le D^i \qquad \text{for } \forall i$$

$$\sum_{i, l \in p^i} r^i \le R_l \qquad \text{for } \forall l$$

$$r^i \ge b^i \qquad \text{for } \forall i$$

It is shown that for $I > 1$, this is an NP-hard problem.

As networks grow in size and complexity, full knowledge of the network parameters is typically unavailable. Hence routing must rely on partial or approximate information and still meet the QoS demands. Lorenz and Orda study the end-to-end delay guarantees for networks with uncertain parameters. They formulated two generic routing problems within the framework where the bandwidth can be reserved and guaranteed. They treat the problem as a maximum flow algorithm. They show that with a delay jitter constraint, the problem is NP-complete. A polynomial-time approximation algorithm is proposed [36].

Jukan and van As study the effects of the *quality attributes* on the routing approach and present simulation results [15]. Huang et al. study the isochronous path selection problem [14]. They show that given a set of established isochronous connections and a set of new isochronous requests, the problem of using a minimal amount of an isochronous bandwidth to serve this isochronous traffic, including the established connections and new requests, is NP hard. They also propose an isochronous path selection algorithm based on paths merging and splitting.

5 MULTICASTING IN WAVELENGTH ROUTING NETWORKS

Various applications demand *group* communication, i.e., more than two parties are involved in an instance of communication. Since using point-to-point methods to implement this function may result in longer delays and extra network resources consumption, a more natural approach is usually considered in this context. That is to make the group of parties share the same communication channel so that one transmission operation can produce the packets for the rest of the members in the communication group, which is referred to as *multicasting*. In wavelength routing networks, the issues involved for multicasting are no longer to find a lightpath, which is usually represented by a point-to-point communication, but to find a subgraph in the network so that the shared communication channel can be established. To avoid multiple paths between any pair of parties in the group, the subgraph must be a tree. The minimization of the tree cost has traditionally been formulated as a *Steiner Minimal Tree* (SMT) [46,41], and the MST has been shown to be NP-complete.

Ramanathan presents a polynomial-time algorithm [41] that provides for tradeoff selection using a single parameter k between the tree cost (Steiner cost) and the run time efficiency. He involves a *directed Steiner Tree* (DST) to describe the wavelength routing network. Accordingly, the *Directed Steiner Minimal Tree* is obtained by the proposed algorithm, *Selective Closest Terminal First* (SCTF), which selects a set of vertices from a partially grown tree and adds a path to a terminal closest to this set.

Tridandapani and Mukherjee study the star-coupler-based multicasting problem [42] by developing a general analytical method for modeling such a system in the context of multicasting traffic. It appears that there is an optimal number of channels that balances the tradeoff of queuing delay and hop distance.

Rouskas and Baldine discuss the multicasting problem in the presence of delay constraints [43]. They show that the problem of finding a multicasting tree with delay constraints among end-to-end users is an NP-complete problem. They propose a heuristic method to solve the problem.

Lin and Lai [44] introduce a dynamic multicasting routing problem in which nodes are allowed to leave dynamically and join the communication group. Hence the multicasting becomes tailorable dynamically. In this algorithm, a virtual trunk (VT), which is a tree of the underlying graph, is introduced and is used as a template for constructing multicasting trees.

6 CONCLUSION

This chapter presents a survey about wavelength routing WDM networks. The topics covered are virtual topology mapping and static and dynamic wavelength routing and assignments, with and without wavelength conversion. The quality of service provision and multicasting in wavelength routing networks are also discussed. Although it is widely believed that wavelength routing networks will assume a very important role in the next generation of WAN/MAN systems, there are very few tutorial papers on this area. We hope that this chapter can be useful to researchers wanting to get a brief introduction into this area.

REFERENCES

1. E. Karasan and E. Ayanoglu. Effects of wavelength routing and selection algorithms on wavelength conversion gain in WDM optical networks. *IEEE/ACM Transactions on Networking*, Vol. 6, No. 2, pp. 186–196, April 1998.
2. K. Struyve and P. Demeester. Dynamic routing of protected optical paths in wavelength routed and wavelength translated networks. *ECOC 97*, 22–25, September 1997.
3. W. Liang, G. Havas, and X. Shen. Improved lightpath (wavelength) routing in large WDM networks. *Proceedings 18th International Conference on Distributed Computing Systems*, pp. 516–523, 1998.
4. M. C. Sinclair. Minimum cost routing and wavelength allocation using a genetic algorithm. Heuristic hybrid approach. *Sixth IEEE Conference on Telecommunications*, pp. 67–71, 1998.
5. O. Gerstel. On the future of wavelength routing networks. *IEEE Network*, Vol. 10, No. 6, pp. 14–20, Nov.-Dec. 1996.

6. T. K. Tan and J. K. Pollard. Determination of minimum number of wavelengths required for all-optical WDM networks using graph coloring. *Electronics Letters*, Vol. 31, No. 22, pp. 1895–1897, Oct. 1995.

7. R. Ramaswami. Multi-wavelength light-wave networks for computer communication. *IEEE Communication Magazine*, Vol. 31, No. 2, pp. 78–88, Feb. 1993.

8. R. J. et al. Cascading of a non-blocking WDM cross-connect based on all-optical wavelength converters for routing and wavelength slot interchanging. *Electronics Letters*, Vol. 33, No. 19, pp. 1647–1648, Sept. 1997.

9. E. A. Medova. Network flow algorithms for routing in networks with wavelength division multiplexing. *IEEE Proceedings on Communication*, Vol. 142, No. 4, pp. 238–242, Aug. 1995.

10. A. Bononi. Weakly versus strongly multihop space-division optical networks. *Journal of Lightwave Technology*, Vol. 16, No. 4, pp. 490–500, April 1998.

11. S. Banerjee and C. Chen. Design of wavelength-routed optical networks for packet switched traffic. *1995 IEEE International Conference on Communications*, pp. 444–448, 1996.

12. C. Chen and S. Banerjee. A new model for optimal routing and wavelength assignment in wavelength division multiplexed optical networks. *Proceedings IEEE INFOCOM '96*, pp. 164–171, 1996.

13. S. L. Danielsen, C. Joergensen, B. Mikkelsen, and K. E. Stubkjaer. Analysis of a WDM packet switch with improved performance under bursty traffic conditions due to tunable wavelength converters. *Journal of Lightwave Technology*, Vol. 16, No. 5, pp. 729–735, May 1998.

14. N. F. Huang, G. H. Liaw, and C. C. Chiou. On the isochronous paths selection problem in an interconnected WDM network. *Journal of Lightwave Technology*, Vol. 14, No. 3, pp. 304–314, March 1996.

15. A. Jukan and H. R. van As. Quality-of-service routing in optical networks. *11th International Conference on Integrated Optics and Optical Fiber Communications*, pp. 160–163, 1997.

16. E. D. Lowe and D. K. Hunter. Performance of dynamic path optical networks. *IEEE Proceedings: Optoelectronics*, Vol. 144, No. 4, pp. 235–239, Aug. 1997.

17. G. Venkatesan, G. Mohan, and C. S. R. Murthy. Probabilistic routing in wavelength-routed multistage, hypercube, and Debruijn networks. *Proceedings Fourth International Conference on High-Performance Computing*, pp. 310–315, 1997.

18. R. Ramaswami and K. N. Sivarajan. Routing and wavelength assignment in all-optical networks. *IEEE/ACM Transactions on Networking*, Vol. 3, No. 5, pp. 489–500, Oct. 1995.

19. R. Ramaswami and K. N. Sivarajan. Design of logical topologies for wavelength-routed optical networks. *IEEE Journal on Selected Areas in Communications*, Vol. 4, No. 5, pp. 840–851, June 1996.

20. Z. Zhang and A. S. Acampora. A heuristic wavelength assignment algorithm for multihop WDM networks with wavelength routing and wavelength reuse. *IEEE/ACM Transactions on Networking*, Vol. 3, No. 3, pp. 281–288, June 1995.

21. Z. Zhang, D. Guo, and A. Acampora. Logarithmically scalable routing algorithms in large optical networks. *Proceedings IEEE INFOCOM '95*, Vol. 3, pp. 1290–1299, 1995.

22. D. Guo, A. S. Acampora, and Z. Zhang. Hyper-cluster: a scalable and reconfigurable wide-area lightwave network architecture. *IEEE Global Telecommunications Conference*, pp. 863–867, 1997.

23. B. Mukherjee. *Optical Communication Networks*. McGraw-Hill Series on Computer Communications, McGraw-Hill, 1997.

24. K. C. Lee and V. O. K. Li. A wavelength-convertible optical network. *Journal of Lightwave Technology*, Vol. 11, No. 5–6, pp. 962–970, May-June 1993.

25. S. Subramaniam and R. A. Barry. Wavelength assignment in fixed routing WDM networks. *1997 IEEE International Conference on Communications*, pp. 406–410, 1997.

26. H. Harai, M. Murata, and H. Miyahara. Performance evaluation of routing methods in all-optical switching networks. *Trans. Inst. Electron. Inf. Commun. Eng.*, Vol. J80B-I, No. 2, pp. 74–86, Feb. 1997.

27. A. Birman. Computing approximate blocking probabilites for a class of all-optical networks. *IEEE Journal of Selected Areas in Communications*, Vol. 14, No. 5, pp. 852–857, June 1996.

28. R. A. Barry and P. A. Humblet. Models of blocking probability in all-optical networks with and without wavelength changers. *IEEE Journal of Selected Areas in Communications*, Vol. 14, No. 5, pp. 858–867, June 1996.

29. D. Banerjee and B. Mukherjee. A practical approach for routing and wavelength assignment in large wavelength-routed optical networks. *IEEE Journal of Selected Areas in Communications*, Vol. 14, No. 5, pp. 903–908, June 1996.

30. I. Chlamtac, A. Farago, and T. Zhang. Lightpath (wavelength) routing in large WDM networks. *IEEE Journal of Selected Areas in Communications*, Vol. 14, No. 5, pp. 909–913, June 1996.

31. B. Mukherjee. WDM-based local lightwave networks, Part I. Single-hop system. *IEEE Network*, pp. 12–27, May 1992.

32. B. Mukherjee. WDM-based local lightwave networks, Part II. Multihop system. *IEEE Network*, pp. 20–32, July 1992.

33. A. Mokhtar and M. Azizoglu. In: G. Cameron, M. Hassoun, A. Jerdee, and C. Melvin. Adaptive techniques for routing and wavelength assignment in all-optical WANs. *Proceedings of the 39th Midwest Symposium on Circuits and Systems*, pp. 1195–1198, 1996.

34. L. Kleinrock, M. Gerla, N. Bambos, J. Cong, E. Gafni, L. Bergman, and J. Bannister. The supercomputer supernet: a scalable distributed terabit network. *Journal of High Speed Networks*, Vol. 4, No. 4, pp. 407–424, 1995.

35. E. Basturk, et al. Design and implementation of a QoS capable switch-router. *Proceedings Sixth International Conference on Computer Communications and Networks*, pp. 276–284, 1997.

36. D. H. Lorenz and A. Orda. QoS routing in networks with uncertain parameters. *Proceedings of INFOCOM '98*, Vol. 1, pp. 3–10, 1998.

37. A. Orda. Routing with end to end QoS guarantees in broadband networks. *Proceedings of INFOCOM '98*, Vol. 1, pp. 27–34, 1998.

38. K. M. Chan and T. P. Yum. Analysis of least congested path routing in WDM lightwave networks. *Proceedings IEEE INFOCOM '94*, pp. 962–969, 1994.

39. M. Hamdi and L. X. Wang. All-optical MAC protocol for Gbits/sec fiber optic LANs/MANs. *Proc. SPIE*, Vol. 2919, pp. 297–305, 1996.

40. M. S. Borella and B. Mukherjee. Limits of multicasting in a packet-switched WDM single-hop local lightwave network. *Journal of High Speed Networks*, Vol. 4, No. 2, pp. 155–167, 1995.

41. S. Ramanathan. An algorithm for multicast tree generation in networks with asymmetric links. *Proceedings IEEE INFOCOM '96*, 1996, pp. 337–344.

42. S. B. Tridandapani and B. Mukherjee. Multicast traffic in multi-hop lightwave networks: performance analysis and an argument for channel sharing. *Proceedings IEEE INFOCOM '96*, pp. 345–352, 1996.

43. G. N. Rouskas and I. Baldine. Multicast routing with end-to-end delay and delay variation constraints. *IEEE Journal on Selected Areas in Communications*, Vol. 15, No. 3, pp. 346–356, April 1997.

44. H. C. Liu and S. C. Lai, VTDM—a dynamic multicast routing algorithm. *Proceedings IEEE INFOCOM '98*, pp. 426–432, 1998.

45. R. Ramaswami and G. G. Sasaki. Multiwavelength optical networks with limited wavelength conversion. *Proceedings IEEE INFOCOM '97*, pp. 489–498, 1997.

46. F. K. Hwang and D. S. Richards. Steiner tree problems. *Networks*, Vol. 22, pp. 55–89, 1992.

5

The Optical Network Control Plane

Abdella Battou
firstwave Intelligent Optical Networks, Greenbelt, Maryland

Ghassen Ben Brahim
Princeton Optical Systems, Princeton, New Jersey

Bilal Khan
*Center for Computational Science at the Naval Research Laboratory,
Washington, D.C.*

1 NETWORK MANAGEMENT

Network management functions can be broadly partitioned into the two classes of monitoring and control.

1. *Network Monitoring.* Network monitoring is the term used to describe passive management functions that cause minimal interference in the network's state. The main goal of such is to collect data and analyze it to extract useful information about the status of the network. Typical data collected includes static information regarding network configuration and dynamic information about network events as they occur. An example of network monitoring management function is the collection of amortized performance statistics about a network element or subnetwork.

2. *Network Control.* Network control is the term used to describe active network functions that change the underlying state of the network. An example of this is the provisioning of connections.

A network management system is composed principally of two types of processes: agents running on managed nodes and network managers collecting and processing information collected by agents. In large networks, management processes may be further stratified into a hierarchy. The managers use predefined management protocols such as SNMP to talk to the agents. The information that is exchanged is organized in a tree-like address space referred to as a management information base (MIB).

Network management system architectures fall into two general categories.

1. *Centralized network management.* In this architecture, all manager processes live on a single machine, perhaps even in the same process. This type of management architecture has been implemented and used extensively, predominantly because of its simplicity.

2. *Distributed network management.* Managers reside on several machines distributed throughout the network, perhaps replicated and/or mobile. Although distributed management architectures have not been used extensively in practice because of the necessary complexity of their software implementations, they offer superior features like scalability, fault tolerance, and availability. We discuss these advantages briefly in what follows.

Scalability. Many NMS functions require considerable processing power. An example of such a function is the aggregation required during upward propagation of information through hierarchies of network managers. This aggregation requires computation of compact representations of the composite state of lower-level managers. In addition, distributing network management across multiple machines permits event filtering and processing to be moved closer to the agents generating these events. This not only reduces the latency of the management system but also reduces the communication costs attributable to control traffic. A centralized NMS approach implicitly forces aggregation schemes and even reporting mechanisms to be simplistic in order to avoid choking the system.

Fault Tolerance. An NMS must be tolerant to failures, both of network components that it is managing *and* of individual managers/agents. Distributing the NMS across many machines minimizes the impact of network hardware failures and software instabilities in any particular set of agents/managers, by isolating the side-effects of such failures to some region of the distributed management system. A sophisticated distributed NMS must be able to survive network partitions: it must permit independent control of the pieces of a partitioned network, while supporting reintegration of NMS functions once the partition heals.

Availability. Distributing the NMS increases its availability, thereby permitting its services to be efficiently utilized by administrators and users located at any network node. At the same time, the latency of issuing such requests is reduced, while the load of handling the requests is favorably distributed across many machines.

Currently, networks maintain their management information in a management information base (MIB) maintained by each of the agents. Typically these MIBs are passive collections of data without smarts and possess very unfriendly interfaces. A better design for an information base would use an object-oriented model where object attributes are recursively defined as the result of effective computations of several simple or composite attributes. Such an architecture facilitates network managers to define dynamically new composite templates customized to their individual needs, which can then be ''filled in'' and maintained by the NMS. Such composite templates might include aggregations that summarize detailed low-level attributes, or refinements of such data obtained by detailed computation, analysis, and annotation. Composed objects should

1. Be dynamically specifiable and be defined both spatially or temporally. By spatial attributes we mean quantities derived from collections of managed network objects, e.g., from all the nodes in a subnetwork, or all the subnetworks in a network. The NMS should permit these spatial groupings to be made dynamically. In contrast, temporal attributes refer to quantities derived from collections of values of low-level attributes over a specified time interval.
2. Be persistent to avoid recomputation of their state.
3. Provide secure access to their internal attributes.
4. Provide an event filter installation interface (e.g., via Java MBean registration).
5. Be able to handle event subscription and notification.
6. Provide visualization through HTML and Java interfaces.

Lastly, they should

7. Use a standard naming convention to allow efficient distribution and discovery of management information using distributed directory or Jini services.

1.1 Distributed Network Management Architecture

Recently, distributed architectures for network management have left the realm of research laboratories and made their debut in commercial environments. This debut was largely facilitated by the impressive progress in the Java development community. Exemplary of this progress are the Entreprise JavaBeans (EJBs), which make it possible to use Java for building distributed three-tier enterprise applications. Other examples include the Java Management Extensions (JMX) and JIRO, which are new Java technologies for network and service management [1,2]. Here we shall describe a distributed network management architecture design based on these technologies; such an architecture can be implemented easily in the Java language.

The agents are developed using the Java Management Extensions (JMX). An agent's intelligence is implemented in MBeans (managed beans) that are registered dynamically in an MBean server. These MBeans can generate alarms and notifications and take actions for active management. An MBean represents one or more composed objects and may implement some preprocessing/filtering that normally is done by a manager. Performing this preprocessing/filtering nearer the source reduces the amount of management traffic. To be accessible, these

MBeans have to be registered with the MBean server. This registration happens dynamically and can be done by the agent itself or by a manager. The MBean server and the MBeans constitute the agent. MBeans can be registered as either volatile or persistent within the MBean server.

To interface to these MBeans remotely, clients need a proxy object. These proxies or client Beans are generated using a special compiler mogen. Any operation performed locally on a proxy is propagated to the remote MBean. MBeans resolve concurrent access by multiple clients based on their own concurrency model, the simplest being strong sequential consistency via mutual exclusion of client operations.

Adapters are used by proxies when talking to MBeans. Adapters implement protocols to talk to the agent, and they are designed as MBeans so they can be dynamically downloaded. In this way, new protocols can be instantiated dynamically and registered on the agent on demand. Currently, the following adapters are available: RMI, HTTP, HTTP over SSL, IIOP, SNMP, and HTML. For example, once the HTML adapter is installed on the agent, any application can point its web browser to the agent to view all MBeans on that agent. Here the HTML adapter generates automatically the HTML pages which are in turn rendered on the client's browser. A collection of adapters together with the proxies using them together from a manager. A proxy may use multiple adapters simultaneously.

Interestingly, an agent can register a client Bean and become a manager to another agent that is implementing the corresponding remote MBean from which the client Bean is generated. This lifts the classical strict separation between agents and managers and gives us a more symmetric, aesthetically pleasing model. The benefit of such an architecture is that there is no need to change the agent to allow a new application to talk to it. All that is needed is to register a new adapter for that application; the remaining agent code stays unchanged.

Serveral services are supported to simplify agent and manager development. Here is a partial list of the services provided and a brief description of each.

1. *Repository service*. Allows MBeans to be registered in an MBean server and permits them to be specified as either volatile or persistent.
2. *MetaData service*. Allows determination of MBean properties and methods, by using the Java Reflection API.
3. *Filtering service*. Allows attaching filters to specific subsets of MBeans in the MBean server. The sets of MBeans could be all registered MBeans, for example, or only those that satisfy certain rules specified in terms of methods, attributes, and object names.
4. *Discovery service*. Allows discovery of other MBean servers. This is achieved by ensuring that all active MBean servers that have registered

the Discovery Responder will acknowledge an IP broadcast of a special "Who is out there?" message.

5. *Management Applet (MLet) service*. This is the service that downloads and configures MBeans dynamically. The Applet loads an HTML page that specifies all the necessary information about the MBean. It then downloads the implementation of the MBean (Java classes) and finally registers the MBean in the MBean server.

1.2 Directory Service

The directory service is a lightweight database to facilitate the management and configuration of agents and managers. This directory could be designed as a branch in the organization LDAP server or built dynamically as a distributed directory service by agents and managers.

1.2.1 LDAP Server Approach

This approach is standard and very easy to implement. It requires laying the configuration data using the LDAP Data Interchange Format (LDIF) in the LDAP directory server. LDIF is a textbased format to display entry information using mnemonic named attributes and text values. A text-based format makes writing tools to process or create entries straightforward. Commercial as well as public domain (openLDAP) LDAP servers are available [14].

1.2.2 Distributed Approach

The LDAP approach has the advantage of being simple but has the drawback of being a single point of failure. A distributed approach where local directories are dynamically combined to form a global directory is a more robust alternative. Each Agent and Manager maintains a local directory tree that is part of the distributed directory. It contains the following information:

Agent/Manager configuration and the location of the current distributed directory root node
Capabilities of the Agent/Manager
MBeans installed at this Agent/Manager and their capabilities

The local directory can be implemented either as an MBean or as a Jini service. We describe the case where it is implemented as an MBean.

Local directories combine to form the distributed directory. The manager holding the root node of the tree is called the Root Manager. The tree is created by an algorithm that is capable of handling network partitioning and failures. An election algorithm based on priorities is started to elect the root node.

Each manager is initialy configured with a list of managers that can become Root Managers, each with a priority. On startup a root node is created in the local manager, and the local directory is registered with it. This way, MBeans are instantaneously provided with a working distributed directory service allowing locally scoped MBeans to start executing successfully.

As each manager polls its root manager at regular intervals to check for its availability, any changes to the root of the distributed directory are then passed down to the managers. If the root directory fails or a network failure causes the network to partition, the root election algorithm is run again. Each manager that lost contact with the root elects itself as root and later merges with the other managers creating a distributed directory in the partition.

1.3 Managers

Managers form the middle tier of the distributed network management architecture, agents being the lower and network management client applications the upper. This discussion has excluded the service management layer as well as the business layer. Managers present two facades, the functional and the management. These are sometimes called the data path and the control path, or service functionality and service administration. The manager architecture should offer basic services in both facades. These basic services facilitate and speed the development of new managers. Synchronization and transaction support, security, and persistence should be part of the basic services, as well as deployment tools. JIRO for example provides additional basic services such as a distributed publish/ subscribe event service, a logging service, and a task scheduling service. The hierarchy of managers is dynamically built from configuration information in the LDAP server. We describe two types of managers, Topology and Connection, in detail and only mention the Fault Manager.

1.3.1 Topology Manager

The Topology Manager defines a set of topology objects, and the operations clients use them to build network topologies. It also defines some semantic rules for understanding network object behaviors such as containment rules. The model used should be simple enough to allow easy integration with the other managers and generic enough to allow cross-domain integration. Examples of such models are ITU-T G.805 and ITU-T M.3100. Here we will borrow from the ITU-T G.805 and build a simpler model. The model uses four types of objects to represent the physical and logical structures in a network and build a hierarchical view of a network (hypergraph).

Network
Subnetwork

Link
Link termination point

Subnetworks are recursive, and termination points are not shared between networks. Each subnetwork/network has two views: an internal view (shown as a dotted oval in Fig. 1) and an external view. The external view publishes the subnetwork access points to its supernetwork. Figure 1 shows network1 with six access points for setting up end-to-end optical trails. Internally, Network1 contains two linked subnetworks, Network2 and Network3. So the external terminations of Network1 must be directly mapped to the external terminations of Network2 and Network3. Both Network2 and Network3 have four external terminations advertised to Network1. One of them is used as a link termination between them by Network1. The network is built recursively and stored in the configuration file at the root manager in the distributed directory. Using a discovery protocol, managers register for topology events, which are generated by agents. Managers can aggregate these events into a subnet graph, which they send to their parents. Finally, at the root manager we get the full view of the network. To get detailed views of the network, a client can navigate the directory service or can request a flat view of a subnetwork or the full network. Topology managers at different levels are either elected or configured through the LDAP server. Having them all run on the same switch makes the data exchange fast but leads to a load

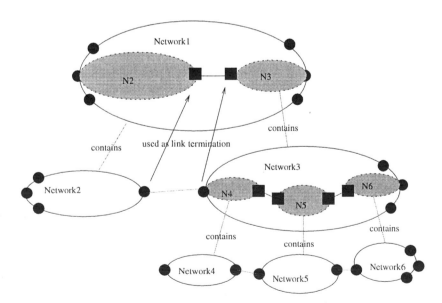

Figure 1 Network hierarchy.

balancing problem. The usefulness of this approach is its ability to model any technology as a subnetwork whose services are made available through its termination points. Mapping the subnetwork termination points at the network level provides subnetwork/technology abstractions and a software approach to integrate different technologies and offer services seamlessly on top of all them.

1.3.2 Connection Manager

The goal of the connection manager is to manage connections across multiple domains. As in the topology manager, the connection manager defines objects to model connection services and semantic rules for the behavior of these objects. Some of these rules include

> *Partition.* As the network layer is partitioned into a set of subnetworks.
> *Contain.* As a network contains many trails or a link termination point contains many network connection termination points (NCTP), one per wavelength for a DWDM network.
> *Delimit.* As a subnetwork is delimited by a set of network connection termination points.
> *Bind.* As a link connection is bound to two network connection termination points (NCTP).

The goal, as in the topology manager, is for these objects to be generic enough to offer integration across multiple technologies. Connections have to have service information such as protocol (WDM optical path, ATM SVC/PVC, Frame Relay, IP, etc.), quality of service (QoS), and bandwidth. The connection model is built on top of the topology model; see Fig. 2 for some of the relationships between topology manager objects and connection manager objects.

The key design criterion here is to ascertain that both the connection model and the topology model are generic enough to support the service management functions. So here the concept of connection is abstract and supports the end-to-end view of the network topology irrespective of the individual network technologies. This concept is valid in ATM networks, WDM networks, and Ethernet networks. The major benefit from such a design is that it provides a simple view of the network functions to integrate services in a global and scalable network. Services do not require detailed knowledge of the underlying technology they ride. In addition, it provides the integration point for multiple technologies.

The connection manager defines the following objects:

> *Link Connection (LC).* A resource that transfers data across a link between two subnets—a wavelength channel in WDM and an SVC in ATM.
> *Connection Termination Point (CTP).* End point of a subnet connection or link connection.

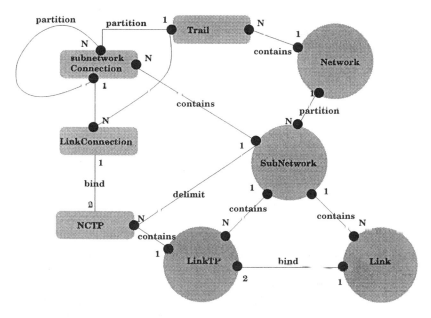

Figure 2 Connection manager objects and relationships with topology manager.

Subnetwork Connection (SC). Resource that transfers data between two CTPs located at the edge of a subnet. The degenerate case is a cross-connect in the switch fabric.

Trail Termination Point (TTP). Access point for clients.

Trail. Resource that transfers client information between two TTPs.

These store the state of the network from connections point of view. This state is shared by all network management components including applications and must be kept consistent. The connection manager objects are dynamic; they are created and deleted by users dynamically. Figure 3 shows a point-to-point connection and a point-to-multipoint connection at the network level, while Fig. 4 shows an example of an end-to-end trail across two subnets and showing all the objects involved.

A connection/trail object represents the service provided by the connection manager. A network management station is used to add, delete, find, and navigate through these trails. There are two ways to set up a trail:

A smart setup uses the signaling protocol. Here the connection manager uses the signaling API to set up a trail between source A and destination B.

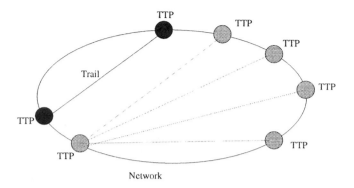

Figure 3 Network, trails, and trail termination points.

A manual setup uses the network management. The connection manager
either interfaces with the routing protocol to get a route from source A
to destination B, or it uses the topology manager to get the route. Once
a route is found, it recursively sends setups through the subnets until it
reaches the network elements. The connection is set up as a transaction.

Routing protocols used to select routes are discussed below.

1.3.3 Fault Manager

The goals of the fault manager are to identify precisely the fault and automatically
to restore services. Faults include fiber cuts, hardware failures, software bugs,
procedural errors, and security breaks. Fault management is distributed across
the network to bring its intelligence closer to the network element to get real-

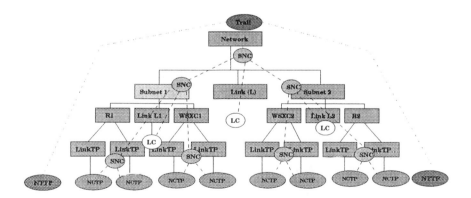

Figure 4 Trail setup between routers across two WSXCs.

time restoration. A subnetwork fault manager can collect all alarms, do some processing/filtering, and only report higher level errors to network managers. One of the hardest parts is the expert engine/tool for alarm correlation and fault identification. This engine takes the state of the network as input and diagnoses problems and provides corrective actions. Its actions come from inference rules on the knowledge base composed of human expertise and domain knowledge based largely on previous experiences. Real-time expert systems are still an active area of research in the distributed artificial intelligence field. Their goal is to provide a framework for autonomous expert systems cooperatively to work to allow expertise sharing and load balancing. This framework includes a language for communication, a negotiation strategy, and efficient ways to represent and exchange domain expertise. Such a framework should guarantee the convergence of the negotiations algorithm with the ultimate goal of designing autonomous networks. The fault manager has to work in concert with the other managers. Topology related failures are forwarded to the topology managers, which translate them into graphical events in the network maps at different levels. The difficulty is to correlate same-layer and inter-layer faults. In addition, one needs to translate topology, connection, and performance faults into service faults that are meaningful to the user.

2 SIGNALING IN AN OPTICAL NETWORK USING THE OVERLAY MODEL

The goal of a signaling protocol is dynamically to set up unidirectional or bidirectional optical paths with minimum latency to transport data across an all-optical network. These paths remain active for an arbitrary amount of time and are not reestablished after a network failure. On the other side, the permanent optical paths, which are provisioned paths, are destined to remain established for long periods of time. In addition, provisioned paths are persistent, so they are reestablished after network failures. This section describes dynamic optical path setup, while the network management section describes the provisioning of permanent optical paths. These optical paths terminate on optical add/drop multiplexers (OADMS), to deliver an optical signal, or terminate on access line interfaces (ALIs), where the optical signal is converted to electronic signal (O/E), as illustrated in Figure 5.

Several signaling protocols are currently under study: ODSI signaling, OIF signaling, and GMPLS.

2.1 Addressing

In the overlay model, clients and optical networks use different addressing spaces. Clients can be IP routers or ATM switches. Clients are attached to the optical

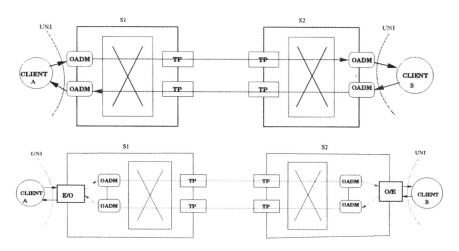

Figure 5 UNI in an all-optical switch with optical (OADM) and electronic interfaces (E/O).

network through physical interfaces on the optical network elements. These physical interfaces are assigned an optical network address (ONA), which is either an IPv4 address or an IPv4 address along with an interface index. The second approach is more economical; the IPv4 address corresponds to the optical network element (ONE) IP address, while the interface index is used as a descriminator. These ONAs are used to identify lightpath termination points. In other words, the signaling protocol uses the network address space to identify lightpath endpoints. The drawback of this approach is that a mechanism to map client addresses to ONAs is needed. A connection termination point that is a client side interface can now be identified as an ONA if there is no multiplexing (ONA, channel) or (ONA, channel, sub-channel). This approach allows for two levels of multiplexing if desired. These connection termination points are configured locally either manually through the network management or dynamically through neighbor discovery at the source and destination of the lightpath between the corresponding ONE and its attached client. There is also provision in the signaling protocol to carry this information from user-side UNI to network-side UNI at the source and from network-side UNI to user-side UNI at the destination.

2.2 UNI Neighbor Discovery and Address Resolution Services

The Neighbor discovery service at the UNI interface permits automatic configuration of the different links between a client and an ONE. This includes the configuration of the control channel as well as the ports on each side of every link. The multiple links connecting a client to an ONE are called *bundles*. To refer to a link in a bundle, the word *component* is used. Thus there is a single bundle between a client and an ONE.

UNI Neighbor Discovery consists of the following steps:

Bring up a control channel between the client and the ONE.
Configure this control channel.
Start a Hello Protocol to maintain this channel.
Determine port connectivity on each side.

The UNI neighbor discovery varies depending on whether the control channel is

InFiber, InBand: Each data channel carries control information.
InFiber, OutOfBand: On each component a separate wavelength such as 1510 is used to carry control information.
OutOfFiber, OutOfBand: The control channel is independent of the data channels. An example will be to have an ethernet connection between the client and the ONE.

The link management protocol (LMP) is used to handle the configuration.

The address resolution service allows a client to register several addresses with an attachment point to the optical network. Addresses such as IPv4, IPv6, and ATM NSAPs are all allowed with the restriction that all addresses registered at an attachment point to the optical network should be of the same type. The address registration (type, length, value) TLV is carried in the LMP config message. Its format is shown in Fig. 6.

The question is how a client sitting at one edge of the optical network learns of other clients, addresses registered at other points.

In a static configuration, every client is made aware of these addresses through a manual configuration either in a file/etc/hosts or in a table populated using a command line interface (CLI) or a network management interface. A better way of course is to run a dynamic routing protocol such as OSPF that will advertise these addresses to all optical switches in the Autonomous System (AS). There is still another solution not quite as powerful as the previous one; it consists in using an Address Resolution Server. Every ONE is made to point to this server and runs an Address Resolution Client responsible to upload to the Address Resolution Server all addresses discovered through the neighbor discovery protocol. The Address Resolution Server timeouts its entries to avoid stale mappings and adapt to new clients being attached at the optical network. These clients are also responsible to maintain a cache of mappings between client addresses and ONAs. Address Resolution Clients talk to Address Resolution Server using an address resolution protocol. ONEs could point at multiple Address Resolution Servers

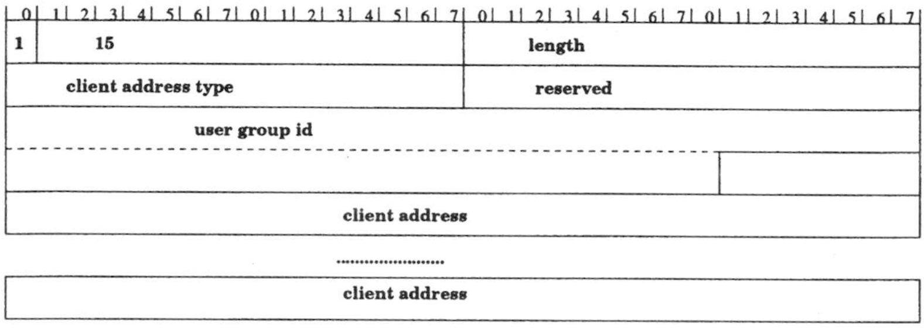

| 0 1 2 3 4 5 6 7 0 1 2 3 4 5 6 7 0 1 2 3 4 5 6 7 0 1 2 3 4 5 6 7 |

| 1 | 15 | length |

| client address type | reserved |

| user group id |

| client address |

....

| client address |

client address type		
Type	**Value**	
0x960	IPv4	
0x961	IPv6	
0x962	ATM NSAP	
0x963	E.164	
0x964	ICD	

UserGroup id: is a 7 bytes VPN ID based on RFC 2685

client address: variable field based on the address type.

Figure 6 Address registration TLV.

for redundancy. A typical example is the RFC1577 protocol used to resolve IP addresses to ATM addresses (NSAPs) before a switched virtual circuit (SVC) is established. Here the situation is a little different: a client **A** gives the address of client **B**, to which it desires to establish a lightpath. The ONE to which client **A** is attached needs to figure out the ONA of the ONE to which client **B** is attached. This is again because the optical network uses ONAs to set up lightpaths. Now every time a lightpath is requested to a destination client **X**, the ONA of the ONE to which **X** is attached is dynamically resolved and the lightpath is established. The ONA either comes from the cache of the source ONE or dynamically resolved through the resolution protocol which issues a GET packet to the Address Resolution Server to obtain the mapping. The Address Resolution Server typically runs in an ONE in the optical network.

2.3 Signaling Transport

It is assumed that the optical path characterized by the dedicated wavelength 1510 is used for all signaling. It is available on every fiber and is extracted on the input port at every switch and converted to electronics so that the switch controller can act on the signaling messages, and reinserted back on the output port after it is converted to optical, as shown in Fig. 7.

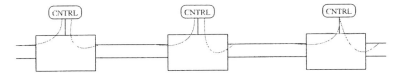

Figure 7   Signaling transport on wavelength 1510

2.4   Description of the UNI Signaling Attributes

The OIF has identified the required attributes and classified them into identification, service, routing, policy, and default classes. Depending on the UNI signaling implementation, each attribute is encoded either as a TLV (UDP implementation), shown in Fig. 8, or as an RSVP object (RSVP implementation), shown in Fig. 9.

1. Identification
    Destination address
    Destination port, channel, and subchannel
    Connection ID
    Source address
    Source port, channel, and subchannel
    User group ID

2. Service
    Service type
    Framing type
    Overhead termination type
    Bandwidth
    Directionality

0	1	2	3	4	5	6	7	0	1	2	3	4	5	6	7	0	1	2	3	4	5	6	7	0	1	2	3	4	5	6	7
U	F	TLV type														TLV length															

Figure 8   TLV format.

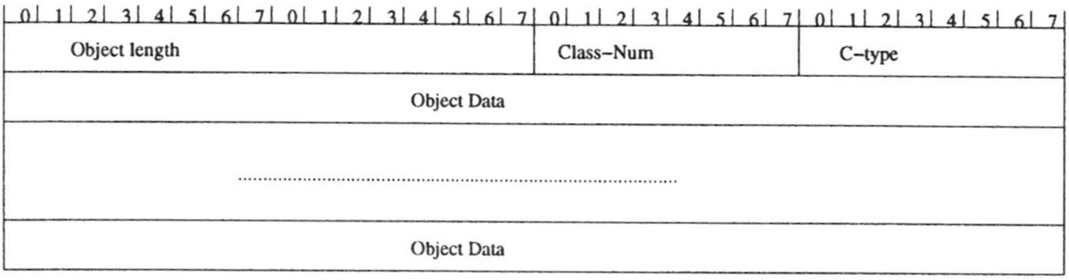

0	1	2	3	4	5	6	7	0	1	2	3	4	5	6	7	0	1	2	3	4	5	6	7	0	1	2	3	4	5	6	7	
Object length	Class–Num	C–type
Object Data		
....		
Object Data		

length includes the header size and is expressed in bytes.

Figure 9 RSVP object format.

3. Routing
   Diversity

4. Policy
   Contract ID

5. Default
   Error code

The TLV header is 4 bytes long, the leftmost 2 bits (U,F) of a TLV being used for behavior when a TLV type is unknown. The next 14 bits specify the TLV type, and the next 16 bits the TLV length. When a TLV type is not recognized, if the U bit is set to 0 (U = 0) a notification must be returned to the message originator and the entire message is ignored. However, if the U bit is set to 1 (U = 1), the TLV is ignored and the rest of the message is processed. The F bit applies when the U bit is set. If the F bit is set to 0 (F = 0), the unknown TLV does not make it into the forwarded message; otherwise (F = 1) it does. The TLV body is of variable length.

The RSVP object header consists of a 2-byte length that includes the header size expressed in bytes, followed by a 1-byte Class Number, and a 1-byte Class Type. It is illustrated in Fig. 9.

For example, the bandwidth, the framing type, and the overhead termination type are part of the LDP Generalized Label Request Object shown in Fig. 10.

While in RSVP, For example, the bandwidth attribute is expressed as a Sender TSpec. This RSVP object is used by the source to specify the traffic characteristics of its data. It is a mandatory object in the Path message. It uses the IntServ (RFC 2210) parameters for token buckets. The object specifies token rate, token size, peak rate, minimum policy unit, and maximum packet size and is illustrated in Fig. 11. Currently, only the peak rate is examined. The hexadecimal value defined for OC48 is 0 × 4D9450C0.

0 1 2 3 4 5 6 7 0 1 2 3 4 5 6 7	0 1 2 3 4 5 6 7	0 1 2 3 4 5 6 7
length	Class–Num = 19	C–type = 5

LSP encoding Type	Link Protection Flags	GPID	
RGT	RT	reserved	RNC

LSP encoding type = 6 for SONET and 5 for SDH
Link Protection Flag: unprotected = 0x01 is the only one specified
Generalized PID: identifies the payload
Request Grouping Type (RGT): 0 is default meaning no concatenation or bundling at the SDH/SONET
Requested Transparency (RT): type of SDH/SONET transparency such as Regenerator Section, Multiplex Section,
　　　　　　　　Virtual Concatenation.
Requested Number of Components(RNC): indicates the number of signals concatenated in the lightpath.
　　　　　　　　zero means no concatenation.

Figure 10　Generalized label request object format.

2.5 RSVP Extensions for UNI Signaling

This section describes the extensions to the Resource reSerVation Protocol (RSVP), an IETF protocol defined for establishing resources for IP flows (RFC 2205), RSVP-TE as a UNI signaling between a client and an ONE.

The mapping of UNI signaling attributes to RSVP objects is as follows:

1. Bandwidth—sender Tspec
2. Contract ID—policy data
3. Destination address—session
4. Destination port, channel, and subchannel—Explicit Route Object, Explicit Label Control
5. Directionality—Upstream Label

0 1 2 3 4 5 6 7 0 1 2 3 4 5 6 7	0 1 2 3 4 5 6 7	0 1 2 3 4 5 6 7
length = 36	Class Num = 12	Class Type = 2
version number	reserved	IS length	
service number	reserved	Service Data length	
parameter ID = 127	parameter flags	Parameter data length	
token rate			
token bucket size			
peak data rate			
minimum policy unit			
maximum packet size			

Figure 11　Sender TSpec format.

6. Diversity—Diversity
7. Framing type—Generalized Label Request
8. Connection ID—connection ID
9. Error code—error spec
10. Service type—session attribute
11. Source address—sender Template
12. Source port, channel, and subchannel—Label Set or Suggest label
13. Overhead termination type—Generalized Label Request
14. User group ID—Address Resolution Request/Response (new)

Some of these objects are new to the original RSVP protocol (RFC2205) and are defined in the Internet Draft *RSVP Extensions in Support of OIF Optical UNI Signaling draft-yu-mpls-rsvp-oif-uni-00.txt.*

The new objects are

LABEL_REQUEST
LABEL
EXPLICIT_ROUTE
RECORD_ROUTE
SESSION_ATTRIBUTE
SENDER_TEMPLATE
FILTER_SPEC
FLOWSPEC

and are optional with respect to RSVP (RFC2205).

2.5.1 RSVP Messages

The signaling protocol uses the RSVP messages defined in RFC 2205. The following list gives the required RSVP messages and their object contents:

1. **Path** Message (type = 1): in the original RSVP this message is used to install a *Path State* along the route from source to destination. The *Path State* also stores the traffic characteristics as specified by the source. Simply stated, this message carries downstream what the source wants. Notice that no resources are reserved. In the case of a UNI signaling only, this message must be delivered to the destination by the internal NNI.

message_id
session
rsvp_hop
time_value
generalized_label_request
sender_template
sender_tspec
upstream_label

2. **Resv** Message (type = 2): In the original RSVP this message is sent by the destination upstream hop by hop following the path established by the **Path** message. Each switch along the path decides either to commit the resources required for this lightpath or refuse the request. Each switch has the right to update the parameters before sending this message upstream. Again for UNI signaling, this message must be delivered by the internal NNI, which also is responsible for committing the resources.

        message_id
        session
        rsvp_hop
        time_value
        resv_confirm
        ff_style
        flowspec
        filter_spec
        generalized_label

3. PathErr Message (type = 3) is used to report any errors that may occur during the path binding. This message travels upstream back through each hop already bound. For a UNI signaling only, this message is generated by the network side of the UNI interface at the source in response to a path failure from the internal NNI.

        message_id
        session
        error_spec
        sender_template
        sender_spec

4. ResvErr Message (type = 4) is used to report any errors that may occur during reservation of resources due to admission control or policy control for example or an error. This message travels downstream back through each hop toward the destination that sent the offending reservation. Similarly, this message is generated by the network side of the UNI interface at the destination in response to a reservation failure from the internal NNI.

        message_id
        session
        rsvp_hop
        error_spec
        ff_style
        flowspec
        filter_spec

5. PathTear Message (type = 5) is used to remove explicitly the path bindings established by the **Path** message.

    message_id
    session
    rsvp_hop
    sender_template
    sender_spec

6. ResvTear Message (type = 6) is used to free explicitly the resources committed by the **Resv** message.

    message_id
    session
    rsvp_hop
    style
    flow_spec
    filter_spec

7. ResvConf Message (type = 7) is used to inform the application that the reservation is indeed accomplished, in response to successful lightpath establishment through the internal NNI signaling.

    message_id
    session
    error_spec
    resv_confirm
    style
    flow_spec
    filter_spec

8. Ack message (type = 13) is sent to the destination to inform it that the reservation is indeed accomplished again in response to a positive acknowledgement from the internal NNI.

    message_id_ack

RSVP runs on top of **IP** and as such is not reliable. It needs to ascertain that lost packets are handled gracefully, and that switches along a path are all keeping state in lockstep with one another. To achieve this, RSVP uses a *Soft State* approach, that is, state must be continuously refreshed or it will be automatically removed. In other words, **Path** and **Resv** messages must be periodically retransmitted to keep a reservation along a path. This technique solves the problem of packet losses and route changes. The timeout interval is specified in the **Path** message. Each time the path is refreshed a new timeout is armed. This timeout is typically three times the interval between RSVP refreshes. **PathErr** are **ResvErr** are not refreshed. The **PathTear** and **ResvTear** messages are used explicitly to remove/free the path bindings/resources, respectively. This is needed if the timeout values are too long, especially if the user is getting billed.

For a UNI signaling interface only, the reliability inside the optical network is handled by the internal NNI.

2.5.2 Lightpath Establishment at the Originating Interface

A new lightpath setup request is initiated by transferring an RSVP Path message with session type set to **LSP_TUNNEL_IPv4** and a Generalized Label Request or an Upstream Label object, a Sender Template object and a Sender TSpec object across the **UNI** interface. Upon sending the **Path** message, the client starts a timer and goes to the lightpath-initiated state *LP _Initiated*. If no response is received before the timer expires, the client either retransmits the **Path** message or clears the lightpath internally. If the optical network can admit the path request, it sends back an **Ack** message and instructs its internal NNI signaling to establish the lightpath; otherwise it sends a **PathErr** back with appropriate error code. When the client gets the **Ack** message, it moves to the state *oLP_proceeding*. If instead it gets **PathErr**, then it clears the lightpath internally. Upon a successful internal lightpath setup through the NNI signaling, the network side of the UNI sends a **Resv** message to the client. Upon receiving the **Resv** message, the client moves to the state *LP_active* and sends a **ResvConf** message. At this time, the source client can start sending data. The optical network will respond with an **Ack** message to the **ResvConf** message.

2.5.3 Lightpath Establishment at the Destination Interface

The optical network indicates the arrival of a lightpath request by transferring the **Path** message across the UNI interface. The UNI-N then goes into the state *LP_present*. Upon receiving the **Path** message, the client goes into the state *LP_present* and upcalls the call control. The call control carries the compatibility check, and if everything is good it accepts the lightpath and sends a reservation request to UNI-C. UNI-C transfers a **Resv** message and goes to the state *LP_waitConf*. Upon getting the **Resv** messages, the UNI-N goes into the state *iLP_proceeding* and instructs its internal NNI to commit the resources along the path from destination to source hop by hop, and it sends an **Ack** message back to UNI-C. Upon receiving a confirmation from its internal NNI it transfers a **ResvConf** message through the UNI interface and goes into the state *LP_active*. Upon receiving the **ResvConf** message, the UNI-C goes into the state *LP_active* and sends back an **Ack** message.

Table 1   Source Network Side UNI Finite State Machine

Current state	Event	Next state	Action
Null	PathInd (UNI-C)	LP-initiated	1. Create Pathstate
			2. Send PathInd to CC
			3. Start a refresh timer
LP-initiated	AckReq (CC)	oLP-proceeding	1. Send AckInd to UNI-C
	PathErrReq (CC)	LP-pathTearReq	1. Send PathErrInd to UNI-C
oLP-proceeding	ResvReq (CC)	LP-active	1. Send ResvInd to UNI-C
	PathErrReq (CC)	LP-pathTearReq	1. Send PathErrInd to UNI-C
LP-active	ResvConfInd (UNI-C)	LP-active	1. Send ResvConfInd to CC
			2. Send an AckInd to UNI-C
	PathTearInd (UNI-C)	LP-pathtearInd	1. Send PathTearInd to CC
	ResvTearReq (CC)	LP-pathtearReq	1. Send ResvTearInd to UNI-C
	PathErrReq (CC)	LP-pathtearReq	1. Send PathTearInd to UNI-C

2.5.4   UNI-N FSM

A simplified (no error handling) version of the source and destination network side finite state machine (FSM) and the states are described below.

1.   LP-null: no lightpath exists.
2.   LP-initiated: this state exists for an outgoing lightpath when the network has received a PathInd but has not yet responded.

Table 2   Destination Network Side UNI Finite State Machine

Current state	Event	Next state	Action
Null	PathReq (CC)	LP-present	1. Create PathState
			2. Send PathInd to UNI-C
			3. Start a refresh timer
LP-present	ResvInd (UNI-C)	iLP-proceeding	1. Send AckInd to UNI-C
			2. Send ResvInd to CC
	PathErrInd (UNI-C)	LP-pathtearInd	1. Send AckInd to UNI-C
iLP-proceeding	ResvConfReq (CC)	LP-active	1. Send ResvConfInd to UNI-C
LP-active	PathTearReq (CC)	LP-pathtearReq	1. Send PathTearInd to UNI-C
	PathTearInd (UNI-C)	LP-pathtearInd	1. Send PathTearInd to CC

Table 3 Network Side UNI Lightpath Clearing Finite State Machine

Current state	Event	Next state	Action
LP-pathtearReq	PathTearResp (CC)	LP-null	1. Clear LP internally
			2. Send PathTearConf to CC
			3. Send AckInd to UNI-C
	AckInd (UNI-C)	LP-null	1. Send PathTearConf to CC
LP-pathtearInd	AckReq (CC)	LP-null	1. Clear LP internally

3. oLP-proceeding: this state exists for an outgoing lightpath when the network has sent an Ack that it can proceed the lightpath.
4. LP-present: this state exists for an incoming lightpath when the network has sent a PathInd to UNI-C but has not received a response yet.
5. iLP-proceeding: this state exists for an incoming lightpath when the network has received an ResInd from UNI-C but is still waiting for the ResvConfInd.
6. LP-active: this state exists for an established incoming or outgoing lightpath.
7. LP-pathtear-req: this state exists when the network has received a PathTearInd from the user.
8. LP-pathtear-ind: this state exists when the network has released the lightpath and sent a PathTearInd to UNI-C.

3 ROUTING IN OPTICAL NETWORKS

The central algorithmic problem in WDM routing is this: Given a WDM network's physical topology (including the characteristics of its links) and a set of source–destination pairs, (1) (*route assignment*) compute a route for a lightpath between each source–destination pair, and (2) (*wavelength assignment*) for each link traversed by this lightpath, determine the wavelength to be allocated for the lightpath on the given link. Together, these two assignment problems are often referred to as the *routing and wavelength assignment (RWA)* problem. Typically, the RWA problem is solved by considering each of its two constituent subproblems in turn [25].

The RWA problem is complicated by the fact that an OXC switch may be (optionally) equipped with *wavelength conversion* hardware that permits lightpaths transient through the switch to enter and leave the switch on different wavelengths. Consider Fig. 12, which illustrates a WDM network consisting of five OXC switches interconnected by optical fiber links. If no switches are

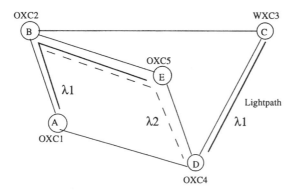

Figure 12 A WDM routing network with five OXCs.

equipped with wavelength converters, then a lightpath in this network must always occupy the same wavelength on every fiber link it traverses. This restriction is commonly known as the *wavelength continuity constraint* for nonwavelength converting switches. An OXC switch that is equipped with wavelength conversion hardware is exempt from this constraint. The reader is referred to the book by Ramaswami and Sivarajan [23] for a comprehensive introduction to WDM network technologies.

A solution to an instance of the RWA problem must necessarily [3,4] respect the following constraints: First, two lightpaths traveling on a given link must be assigned different wavelengths. Second, any lightpath that transits through an OXC switch that is not wavelength conversion capable must use the same ingress and egress wavelengths. The central responsibility of the WDM routing protocol is to ensure that the information required for route assignment and wavelength assignment is maintained by the network in a scalable manner.

1. **Routing Assignment.** There are three broad classes of strategies being used presently to address the routing assignment problem [25]. These strategies are referred to as *fixed routing, fixed alternate routing, and adaptive routing.*

Fixed routing is a simple technique that involves maintaining a fixed routing table at each candidate source node. The routing table consists of one entry for each candidate destination node, where the entry specifies the path from the source to the destination. Fixed routing is simple to implement but is subject to unacceptably high blocking probabilities as wavelength availability on links becomes scarce, and when link failures occur.

In contrast, *fixed alternate routing* attempts to address the shortcomings of fixed routing by augmenting each entry in the routing table to be a prioritized set of paths from source to destination, rather than just a single one. Fixed alternate routing is less sensitive to link failures and wavelength and thus offers lower connection blocking probabilities.

Finally, *adaptive routing* schemes attempt to select a path between a source and destination pair based on dynamically collected information concerning the network's state. This technique is much more resilient to link failures and less sensitive to wavelength scarcity Thus it offers lower connection blocking probability than fixed adaptive routing approaches.

2. **Wavelength Assignment.** The second half of the RWA problem, wavelength assignment, is addressed algorithmically using strategies based either on heuristics or on *graph coloring* algorithms. Presently considered heuristics include Random Wavelength Assignment, First-Fit, Least-Used (SPREAD), Most-Used (PACK). Min-product, Least Loaded, MAX-SUM, Relative Capacity Loss, Distributed Relative Capacity Loss, Wavelength Reservation, and Protecting Threshold. For a detailed comparison between the performance of these heuristics, the reader is referred to the review by Zang et al. [25].

3. **RWA and Traffic Engineering.** One popular extension of adaptive routing schemes is the so-called *least congested path* (LCP) scheme, which attempts to choose the path between source–destination pairs based on current traffic patterns in the network. By considering each wavelength on a link to be a fraction of the link's available bandwidth, LCP schemes are easily adapted to operate in a WDM environment. Unfortunately, LCP schemes introduce significant complexity, both in terms of the protocols that must operate to maintain information about network state, and the algorithms responsible for carrying out the route computation. Together, these concerns and their resolution are referred to as *traffic engineering (TE)*.

TE is responsible for mapping traffic flows to the managed physical network resources. Specifically, TE provides the ability to move traffic flows away from the shortest path and into less congested paths in order to satisfy the *quality of service (QoS)* requirements of higher layers. The main purpose of TE is to load balance the network traffic across various links, routers, and switches, thereby minimizing congestion and ensuring that the network resources are not over- or underutilized. Regrettably, many existing routing protocols do provide QoS and TE adequately to model WDM networks, and so require augmenting the protocol and packet specifications.

Major classes of TE attributes are described by Awduche et al. [8]. These include (1) *traffic parameter attributes* used to capture the characteristics of the traffic streams, (2) *generic path selection and maintenance attributes* specifying rules for selecting the route taken by an incoming traffic as well as the rules for maintenance of paths that have already been established, (3) *priority attributes* specifying relative importance of different traffic flows, (4) *preemption attributes* stating rules for preemption of resources, (5) *resilience attributes* specifying the behavior of certain traffic under fault conditions, and (6) *policing attributes* defining the actions that should be taken by the underlying protocols when incoming traffic becomes noncompliant. Fedyk and Ghanwani [13] present additional metrics and resource information for links tailored specifically for traffic engineering, including hop count metrics, bandwidth-based metrics, delay- based metrics, economic-cost- or expense-based metrics, and administrative weight.

To summarize, the responsibility of an optical routing protocol is to facilitate lightpath establishment between OXC switches, i.e., it must provide access to sufficient information to solve the problems of routing assignment and wavelength assignment while permitting TE information concerning the current network traffic and the QoS requirements of higher layers to be taken into consideration.

3.1 Architectural Models for Optical Routing Information Exchange

In this section, we examine how to carry IP over optical networks. We assume that IP routers are attached to an optical core network and are connected to other IP routers over dynamically established optical paths. In order to establish these optical paths, routers and optical switches need to exchange routing information. Two architectural models are currently under consideration, the overlay model and the peer model.

1. **Overlay Model**. In this model, the optical network is considered quasistatic, and optical paths are viewed as fibers between routers. This approach hides all of the optical network complexities from routers. The optical network runs a separate routing entity, and one ends up with two routing systems. This is the model chosen for IP over ATM, where PNNI is the routing system in the ATM network. Standard interfaces are defined between the routers and the optical switches, *Optical User Network Interface (OUNI)*, and between the optical switches, *Optical Network to Network Interface (ONNI)*. Both the *Optical Domain Service Interconnect (ODSI)* and the *Optical Internetworking Forum (OIF)* have chosen this model and are in the early stages of standard development.

> *Manual configuration*: Each router attached to the optical network is manually configured with optical endpoint addresses corresponding to IP destinations behind it.
> *Reachability protocol across OUNI*: Each router attached to the optical network runs a reachability protocol across the OUNI with its corresponding optical switch to obtain IP addresses of all routers attached to the optical network. This information is then used to build the initial set of IP routing adjacencies, which run an IP routing protocol to acquire all IP destination addresses reachable over the optical network.

2. **Peer Model**. In this model, the IP router is entirely aware of the details of the underlying optical network and can compute end-to-end paths across the optical network. The optical switch to which the IP router is attached acts just like any other router. Thus the usage of the term peer. Optical switch controllers will run an updated version of OSPF, which is the standard routing protocol in the Internet and will exchange reachability information with routers. IETF is presently considering this model, because of the issues encountered in the IP over ATM experience.

> *Routing protocol across OUNI*: Each router and its associated optical switch runs an IP routing protocol allowing it to exchange full reachability information across the OUNI.

Many routing protocols already in existence may be adapted to provide adaptive routing in WDM. In the next section we evaluate some of those protocols.

3.2 Candidate Adaptive Routing Protocols for WDM Networks

Routing protocols can be divided into two broad classes, *Interior Gateway Protocols (IGPs)*, and *Exterior Gateway Protocols (EGPs)*. IGPs are intended for use within an *Autonomous System*, i.e., a set of networks that are under a single or closely coordinated management organizations. Routing between different autonomous systems is maintained by the EGPs, of which the *Border Gateway Protocols (BGPs)* are one example. While IGPs are designed to maintain dynamically information about network topology, produce optimal routes, and respond to changes quickly, EGPs emphasize stability and administrative control above route optimality. As the networking community moves towards further decentralization of control, EGPs are viewed as serving to protect one system of networks against errors or intentional misrepresentation by other systems.

Along a different axis, routing protocols, can also be classified into *distance vector routing* protocols, which implement distributed algorithms to compute all pairs of shortest paths, and *link state routing* protocols, which propagate information about network topology to all routers.

In distance vector routing protocols, such as the *Routing Information Protocol (RIP)*, each router maintains the distance from itself to each possible destination, and the distance vector routing protocols relax these values down to their shortest value. In doing so, RIP is essentially a distributed implementation of the Bellman–Ford all-pairs shortest path algorithm.

In link state routing protocols, such as the *Open Shortest Path First (OSPF)* protocol, every router maintains a representation of the network's topology, and this representation is updated regularly on the basis of the information being exchanged between the network routers. In particular, each router is responsible for discovering its neighboring routers and then advertising the identities of its adjacent neighbors and the "cost" to reach each. This information is advertised between routers by periodic exchange of link state packets. A link state protocol arms each router with a dynamic map of the network's topology; using this, the router can compute paths to any destination.

In the subsections that follow, we evaluate several candidate routing protocols and assess their suitability for operation in the WDM environment.

3.2.1 Routing Information Protocol (RIP)

The Routing Information Protocol (see [16,17,22]) is a distance vector routing protocol for small networks. Each RIP-enabled router is classified as either active or passive: active routers advertise their routes to other routers; passive routers listen and update their routes based on the advertisements they receive but do not advertise any information themselves. Typically, routers run RIP in active mode, while hosts use passive mode. RIP is attractive because of its simplicity and elegance, but it suffers from several undesirable features:

Network size limitations: RIP was designed as a routing protocol for small networks of diameter ≤ 15. The convergence time of RIP limits its applicability for larger networks.

Slow convergence: The essential reasons behind the slow convergence are (1) the information that RIP needs to propagate is itself routed using the same routing tables that RIP strives to build, and (2) the accuracy of these routing tables increases only gradually because of the nature of distributed Bellman–Ford.

Count to infinity: The failure of a link or node failure may result in two routers continually exchanging update vectors because each is claiming to have the right information. Many solutions (e.g., "Split Horizon," "Flush Updates," "Poison Reverse Update") have been proposed to address this issue. But count to infinity may still occur even when using these techniques.

Looping: Incorrect routing information available in some nodes may result in forwarding loops within the network.

Restricted view of the network: A RIP router does not have a view of the topology of the entire network and so cannot detect problems such as looping or count to infinity.

Limitations on the metrics that can be used for shortest path computation: RIP attempts to determine the shortest path by relaxing the count of intermediate hops between each source–destination pair. It is unclear how to extend RIP to support computation of the "best path" using multiple interdependent (particularly non additive) link metrics such as those required by TE.

Inability to support load balancing: Occasionally the need arises to split traffic load between two parallel paths between two nodes, or to cache alternate paths for use in case of failures. RIP presently supports only the computation of the unique shortest path between nodes.

The undesirable characteristics described above make RIP a poor candidate to serve as the foundation for an optical routing protocol.

The next three protocols we consider are all link state routing protocols. By sequencing successive updates of link state information, these protocols sidestep the count-to-infinity problem exhibited by RIP. In addition, by enabling each switch to acquire a complete view of the network topology, they implicitly avoid RIP's routing loop problem.

3.2.2   Intermediate System to Intermediate System (IS-IS)

In ISO terminology, a router is referred to as an *Intermediate System* (IS). The *IS-IS* protocol (see [10,18,22]) is a link state IGP that was originally developed for routing ISO/CLNP* packets. IS-IS has many desirable features, including

Scalability to large networks: IS-IS intradomain routing is organized into two levels of hierarchy, allowing large domains to be administratively subdivided.

Multipath forwarding: IS-IS uses a "shortest path first" algorithm to determine routes, and the new IS-IS protocol supports multipath forwarding.

Genericity: IS-IS can support multiple communication protocols and is therefore generic.

IS-IS does not flood any routing information across level 1 areas. Thus internal IS-IS routers always send traffic destined to other level 1 areas to the nearest level 2 router Area Border Router (ABR). If there is more than one ABR, IS-IS may not pick the optimal route. In this manner, IS-IS trades off route optimality in exchange for lower control traffic; this allows IS-IS to scale to larger networks. Unfortunately, IS-IS is not an IETF standard and has only one RFC, which is not up to date with commercial implementations. As a consequence, IS-IS is no longer an open IGP routing protocol and so is a risky choice for extension to the optical domain.

3.2.3   Private Network to Network Interface (PNNI)

PNNI is a link state routing protocol standardized by the ATM Forum [7]. Intended for use in ATM networks, PNNI offers many features that would make it a good starting point for developing a WDM routing protocol. These include

---

* International Organization for Standardization/Connectionless Network Protocol.

Scalability to large networks: PNNI implements hierarchic partitioning of
networks into Peer Groups (up to 104 levels) and supports aggregation
of information from within each Peer Group. Together, hierarchy and
aggregation provide a flexible trade off curve between route optimality
and protocol traffic. The PNNI hierarchy is configured automatically
based on switch addresses using dynamic election protocols.

Support for traffic engineering and QoS: PNNI distributes link state information in PNNI Topology State Elements (PTSEs); these come in many
flavors, including Horizontal Link InfoGroup (HLinkIG) packets for advertising link characteristics and Nodal InfoGroup (NodalIG) packets for
advertising switch state information. HLinkIG packets include resource
availabilities for a link on a per service-class basis; it would be quite easy
to extend this description to include wavelength information. Similarly, it
would be fairly straightforward to extend the NodalIG to include information about the wavelength conversion capabilities of a switch.

Standardization: PNNI was approved by the ATM Forum and has since
been adopted by most ATM vendors and shown to work in the field.

In terms of the technical advantages outlined above, PNNI is a strong candidate
to serve as the foundation for an optical routing protocol. PNNI supports policy-
based QoS aware routing by permitting the source switch (and each border
switch) to compute a partial route through its peer group towards the destination.
This makes it difficult for PNNI to support multipath forwarding. PNNI is a
relatively complex protocol, and perhaps as a consequence the optical switch
industry has turned away from considering PNNI-based WDM routing solutions.
Interoperability with existing optical switch technology makes PNNI a problem-
atic choice.

3.2.4   Open Shortest Path First (OSPF)

Open Shortest Path First Routing Protocol [17,18,21] is a link state IGP that
distributes routing information between routers within a single autonomous system. OSPF has many desirable features, including

Easily extended for TE and wavelength information. OSPF distributes link
state information in Link State Advertisement (LSA) packets. In the RFC
[12], Coltun proposes Opaque LSAs to extend the OSPF protocol.
Opaque LSAs have a standard LSA header followed by an application
specific information, which may be used directly by OSPF or by other
applications. Interoperability was achieved by introducing three new
Link State (LS) types, each with fixed flooding regimes: (1) LS type 9
packets are not flooded beyond the local subnetwork, (2) LS type 10

packets are not flooded beyond the borders of their associated area, and (3) LS type 11 packets are flooded throughout the Autonomous System. Routers that are not opaque capable may receive Opaque LSAs because they cohabit broadcast networks with opaque: capable routers; in this case the LSAs are discarded, because they have unknown LS types. Opaque LSAs make it possible to extend OSPF to incorporate new domain-specific TE information, as needed; a lot of effort has been expended to specify the exact format of these packets for TE in the IP domain.

Scalability: OSPF supports two levels of hierarchy and groups routers into areas. Because OSPF advertises a summary of the information describing the areas inside the Autonomous System, the protocol is more likely (compared to IS-IS) to compute an optimal path when routing between different areas. OSPF converges reasonably quickly even for large networks.

Simplicity and Acceptance: OSPF is a relatively simple protocol (compared to PNNI) and enjoys widespread commercial adoption.

Standardization: OSPF is a full IETF Standard (unlike IS-IS).

In the next section, we present some of the OSPF-based initiatives for WDM routing.

3.3 OSPF-Based Initiatives for WDM Routing

Currently, most of the routing protocols being proposed for WDM are based on OSPF. We consider two of the most promising proposed extensions now.

3.3.1 The QoS Extension to OSPF

In the RFC [6], Apostolopoulos et al. propose an extension to OSPF to support QoS. In the same reference, the authors propose that the Type of Service (TOS) field within the protocol packet options field should be used to specify whether the originating router is QoS capable. The bandwidth and delay of each link will be encoded into the metric field of the LSA packet by QoS capable routers. By using three bits for the exponent part and the remaining thirteen bits for the mantissa, the scheme can be used to represent bandwidths ranging into Gbit/s.

The main advantage of this proposal is that it does not require the addition of any additional LSAs. This makes it simple to integrate QoS-capable routers running the extended protocol with routers running older versions of OSPF (i.e., non-QoS-capable routers). However, the proposed scheme suffers from the drawback of not being able to advertise additional information such as a complete list of available wavelengths for routing inside WDM networks.

3.3.2   OSPF Extension Using Opaque LSAs

Here we give a brief outline of ongoing efforts of proposals using Opaque LSAs to extend OSPF for routing in the optical domain.

Recent Internet drafts by Chaudhuri et al. [11] and Basak et al. [9] present requirements for optical bandwidth management and restoration in a dynamically reconfigurable network. Both papers expand on the list of optical network characteristics that must be maintained in the OXC routing database. The information is classified into two categories. First there is the information that must be advertised using OSPF, e.g., the total number of active channels, the number of allocated channels, the number of preemptable channels, the risk groups throughout the network. The second category consists of information that is kept locally and is not advertised for protocol scalability reasons. Examples of such information are the available capacity, the preemptable capacity, the association between fibers and wavelengths. The proposals choose to put information about the wavelength availabilities in the second category, arguing that available wavelength changes occur frequently, so advertising the changes in the available wavelength does not yield a performance increase proportionate to the costs in terms of control traffic.

In the internet draft [19], Kompella et al. present an extension to OSPF for optical routing. In particular, they specify some of the parameters needed to be advertised by the Opaque OSPF LSAs. Like the author of [9,11], Kompella et al. propose that OSPF should not advertise any information pertaining to wavelength availability, and postpone wavelength assignment to the time when lightpaths are signaled.

In the internet draft [24], Wang et al. propose to use the OSPF protocols to advertise both the available wavelength per fiber and the available bandwidth. To address the objection that the advertisement of the available bandwidth is impractical because of rapid changes in network link usage, the authors also introduce the notion of ''significant change.'' In their extension, OSPF readvertises only if the available bandwidth changes by a factor higher than a certain tunable threshold.

Some of the TE and QoS parameters change very frequently, raising the issue of when to advertise the changes of the network characteristics throughout the whole network. The original OSPF standard mandates a variety of tunable parameters controlling the flooding of LSAs, including the ''MinLSInterval,'' which specifies the time between any two consecutive LSA originations, and the ''MinLSArrival,'' which limits the frequency of accepting newer instances of LSAs.

Apostolopoulos et al. [5] present other policies dealing with the issue of when a router should flood a new LSA to advertise changes in its link metric. Some of the proposed policies include

> *Threshold-based policies* that trigger updates when the difference between the previously flooded and the current value of available link bandwidth is larger than a configurable threshold.
>
> Class-based policies, which partition the capacity of a link into a number of classes and readvertise when a class boundary is crossed.
>
> *Timer-based policies*, which generate updates at fixed time intervals or enforce a minimum spacing between two consecutive updates.

In the next section, we present a design of a routing protocol in an optical network: Toolkit for Routing in Optical Networks (TRON). TRON combines both models for routing information exchange: the peer model in the sense that TRON makes the router aware of the optical network details, and the overlay model in the sense that TRON uses the collected information to compute source routes.

3.4 Toolkit for Routing in Optical Networks (TRON)

The Toolkit for Routing in Optical Networks (TRON) is a simulation/emulation library developed to facilitate experiments on OSPF-based routing protocols for optical networks. Currently, TRON implements the *LightWave-OSPF* routing protocol, which is an adaptation of the optical extensions to OSPF proposed in [19] and [24]. TRON may be used both to simulate *LightWave-OSPF* routing in large optical networks and to emulate routing on a live optical switch.

In the next section, we present the LightWave-OSPF.

3.4.1   LightWave-OSPF

LightWave-OSPF follows the internet draft [19] of Kompella et al., which presents an extension to OSPF for routing in optical networks. The proposal makes use of the Opaque OSPF LSAs, which were first introduced by Coltun [12] as a way to extend the OSPF protocol. While *LightWave-OSPF* largely adopts the proposal of [19], there is one crucial exception, which we now describe.

Like the earlier drafts of Chaudhuri et al. [11] and Basak et al. [9], Kompella et al. say that an optical adaptation of the OSPF protocol should not advertise any information pertaining to wavelength availability in link state advertisements. They argue that the set of available wavelengths changes frequently on each link, so advertising these changes would not yield a performance increase proportionate to the communication cost of increased control traffic. Instead, they postpone the problem of wavelength assignment to a later stage when the lightpaths are being signaled/provisioned. Here, the authors break from the proposal of Kompella et al.

Instead, *LightWave-OSPF* takes the approach of Wang et al. as presented in Internet draft [24]. There the authors propose that an optical adaptation of the OSPF protocol should advertise both the number of available wavelengths per fiber and the total available bandwidth. To address the objection that advertisement of the available bandwidth is impractical because of rapid changes in network link usage, the authors of the draft incorporate a tunable threshold to determine when a change is "significant" enough to mandate readvertisement.

The rationale for selecting the approach of Wang et al. is the following: In an heterogeneous optical network where a significant fraction of the switches are not wavelength-conversion capable, the absence of wavelength information in the description of a network link will cause the route computation algorithm to be operating in the dark. In particular, as wavelength utilization of network links increases, the probability of selecting an effective source route that can actually be provisioned decreases dramatically, because while many routes exist from source to destination, wavelength continuity constraints at the transit non-wavelength-conversion-capable switches render most of these routes infeasible. In the proposals [9,11,19], the infeasibility of the computed routes would not be detected until signaling was actually attempted and failed. Each lightpath setup experiences a large number of "crankbacks" or reattempts, with the cumulative effect being that as the load on the network increases, the lightpath setup latency would increase prohibitively (even in the absence of contention between concurrent setup requests!). We remark that the above problem gets worse as the fraction of nonwavelength-conversion-capable switches in the network increases.

LightWave-OSPF also incorporates the recommendations presented by Apostolopoulos et al. [5] by implementing a range of policies to determine when a router should flood a new LSA advertising a change in its link metric. The policies presently implemented include (1) *threshold-based policies* that trigger updates when the difference between the previously flooded and the current value of available link bandwidth is larger than a configurable threshold, and (2) *timer-based policies* that generate updates at fixed time intervals or enforce a minimum spacing between two consecutive updates.

In the next section, we present the TRON architecture.

3.4.2 TRON Architecture

The TRON architecture can be decomposed functionally into a set of cooperating modules, as shown in Fig. 13. The modular design facilitates future modifications and extensions to the TRON software. Because each OXC switch has several interfaces, certain modules such as the "Adjacency Protocol," the "Hello Protocol," the "Network Interface," and the "Hardware Abstraction" modules have to be reproduced for each interface. In contrast, the "Routing," "Flooding," and "Switch Database" modules are instantiated once for each OXC switch. We briefly describe each of these components.

1. **Network Interface** and **Hardware Abstraction** modules. The TRON stack is built over these two modules. Together, they are responsible for reporting

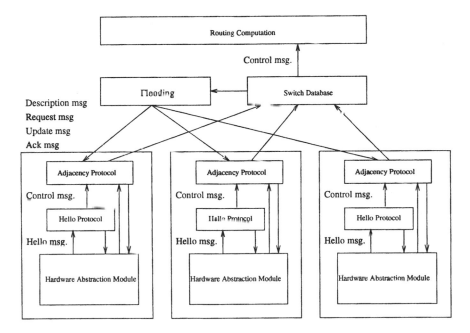

Figure 13   TRON architecture.

any failure or change in local switch resources, by generating events and forwarding them upwards for the rest of the TRON stack to respond to.

The Network Interface module represents the connection between the OXC router and the network. This module informs the router whether the interface to the physical network is up or down and maintains information about the type of network to which the interface attaches, the interface's IP address, the area ID, a list of neighboring routers, the router priority, and the IP address of the Designated Router (DR) and Backup Designated Router (BDR).

The Hardware Abstraction module monitors the OXC for alarms and alerts triggered by the failures or changes to the OXC's internal components. The Hardware Abstraction module maps these physical hardware dependent signals to a set of generic messages that are processed by the Hello and Flooding module. Examples of generic messages are the OXCPortUp and the OXCPortDown messages, which indicate a port's operation and failure, respectively. Extending TRON to work with other optical switches then requires only modification to the Hardware Abstraction module.

2. **Hello** module. The Hello module implements the Hello protocol, which is responsible for maintaining neighbor relationships between OXCs. Specifi-

cally, the Hello protocol serves to certify that the links connecting neighboring routers correctly support bidirectional communication. The protocol specifies that every HelloInterval seconds, each router sends a Hello packet containing the identities of the neighbors that it knows about. TwoWay connectivity is attained when the local router's ID is listed in the Hello packet received from a remote router; otherwise, the connectivity is declared to be only OneWay. The Hello protocol is also responsible for electing a Designated and Backup Designated Routers, by carrying election priority information in the Hello packets.

The Hello module is located directly above the Network Interface and Hardware Abstraction modules, and the Hello protocol is sensitive to messages generated by these modules.

3. **Adjacency** module. The Adjacency module lies above the Hello module within the TRON stack and implements the Adjacency/Flooding (A/F) protocol. The A/F FSM (see Fig. 3) is quite similar to the adjacency FSM of the original OSPF. The A/F FSM is triggered by the Hello protocol. Upon reaching the TwoWay state the Hello protocol sends a Begin control message to the A/FFSM triggering it to become operational by entering the Extstart state.

The Adjacency protocol is responsible for the synchronization of Switch Databases of neighboring routers. The exchange protocol is asymmetric: the first step is to determine a ''master'' and a ''slave'' relationship. After the master–slave relationship negotiation, the new state is Exchanging. The two routers then exchange Description packets containing a summary of their respective databases that lists the advertisements within it. Based on this, the routers will request and exchange the information as necessary to synchronize their databases. This is achieved using Request packets, Update packets, and Acknowledgment packets. The routers use checksums to ensure that all protocol packets have valid contents. Once all necessary exchanges have been completed, the protocol enters the Full state, and the Flooding module is notified. Subsequently, the Adjacency module is subservient to the Flooding module, which may periodically give it LSAs to flood. The state transition diagram of the A/F FSM is in Fig. 14.

The Adjacency protocol is terminated if the Hello module sends a message up withdrawing certification of TwoWay connectivity with the peer. Such failures are detected by the Hardware Abstraction and Network Interface modules, which inform the Hello protocol, which in turn notifies the A/F FSM by sending a Halt control message. The latter message causes the interruption of the A/F FSM and its transition to the initial state WaitMessageFromHello.

In *LightWave-OSPF*, the Update packets carry a list of Link State Advertisements (LSAs). These LSAs are encapsulated into Opaque LSAs. The Opaque LSAs carry different types of Type/Length/Value (TLVs) and SubTLVs packets. The TLVs and SubTLVs include the following: Traffic Engineering Metric, Maximum Bandwidth, Unreserved Bandwidth, Maximum Reservable Bandwidth, Re-

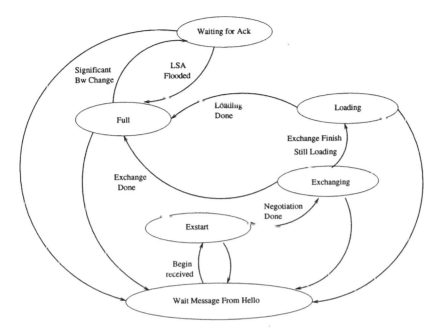

Figure 14   A/F FSM for LightWave-OSPF.

source Class/Color, Link Media Type, Shared Risk Link Group, Available Wavelength, and Number of Active as well as Preemptable Channels.

4.   **Flooding** module. The Flooding module may interact with all instances of the Adjacency modules. This interaction begins only after the Adjacency module has fully synchronized with the remote peer. The Flooding module is invoked whenever one of the following three events occurs:

1.   When there is a failure in one of the incident links as detected by the Hello module. In this scenario, the Flooding module must flush the LSA for the failed link from the network. It does this by inserting a new LSA for the link into its Switch Database (with a higher sequence number and age set to MaxAge) and forwarding it out on all interfaces, via their respective Adjacency modules.

2.   In case of a significant change in the available bandwidth of an incident link. In this scenario, the Flooding module inserts a new, more accurate LSA describing the link into its Switch Database—with a higher sequence number—and forwarding the new LSA out on all interfaces, via their respective Adjacency modules.

3. When a new LSA arrives with more up-to-date information about the
    network's topology/attributes. In this scenario, the Flooding module
    inserts the newer LSA into the Switch Database, acknowledges the
    receipt of the update, and then forwards the new LSA on all interfaces
    via their Adjacency modules *with the exclusion of the Adjacency mod-
    ule that reported the update to the Flooding module.*

5. **Switch Database** module. This module maintains a collection of LSAs
that have been collected by the Adjacency protocol and subsequent operation of
the Flooding module. Upon the receipt of a new LSA, the SwitchDatabase determines whether the LSA is a newer version of an LSA that it already owns. If
so, the newer LSA replaces the older one; otherwise the LSA is discarded. The
SwitchDatabase is also responsible for continuously aging the LSAs and removing those whose age reaches MaxAge. The Switch Database also periodically
reoriginates LSAs (with higher sequence number and age set to zero) that have
been previously generated by the router, thereby ensuring that they do not disap-
pear from the Databases of other routers in the network.

6. **Routing** module. The Routing module has the main task of building a
graph representation of the network based on the contents of the Switch Database
module, and it responds to requests for the computation of routes to destination
routers. This route computation may require consideration of Traffic Engineering
and QoS requirements for the connection and reconciling them with the corresponding information within the LSAs. Traditionally, the next-hop along these
routes is cached in a routing table for hop-by-hop routing. In the case of *Light
Wave-OSPE*, however, the entire route is used as a source route; the lightpath
provisioning protocol attempts to provision the path along the specified route.

3.4.3   Integration of Provisioning/Signaling with TRON

Presently, when running in emulation mode, TRON sends all its control messages
out of band, on an auxiliary control network operating IP over ATM. Each optical
switch running a TRON stack must therefore be configured with information
about the identities of its neighboring optical switches. Specifically, it must be
told the location and listening port of the TRON stacks of each of its neighbors,
so that it can establish out-of-band connectivity to these peers prior to booting.
In future releases of TRON, we hope to consider in-band transport of routing
control messages.

In simulation mode, TRON operates multiple stacks within the same process, so all control messages are exchanged within the simulation process and
no out-of-band (or in-band) connectivity is required.

The TRON software can be considered as the routing computation component of the provisioning protocols, which are in turn viewed as clients of TRON.
These clients use the information available computed by TRON to set up

lightpaths physically. In our present scheme, the lightpath is determined by source routing establishment, unlike hop-by-hop schemes such LSP establishment in MPLS.

Once the provisioning protocol gets the necessary information from the Routing module, it starts physically setting up the path by establishing bindings at each of the transit switches. If it succeeds, the provisioning protocol informs the Hardware Abstraction modules of every switch along the route about the change in bandwidth availability on affected links.

The Hardware Abstraction modules report this change, and if the change is determined to be significant, a new link state advertisement of the affected link is reoriginated and flooded to all neighbor routers. If the change is not significant, then the notification is ignored. Figure 15 represents the interaction between the TRON and the provisioning protocol

3.4.4 The CASiNO Framework

The TRON software was implemented in C++ using the Component Architecture for Simulating Network Objects (CASiNO) [20]. In this section, we present a brief overview of the CASiNO framework.

The Component Architecture for Simulating Network Objects (CASiNO)

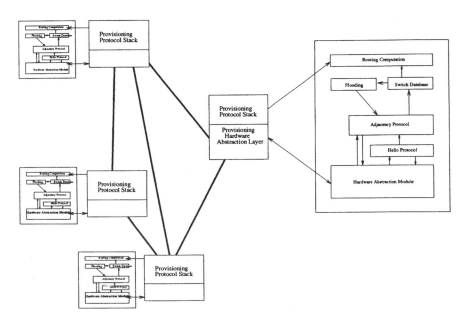

Figure 15 Interaction of optical OSPF and provisioning protocols.

is useful for the implementation of communication protocol stacks and network simulators. CASiNO framework implements a rich, modular coarse-grained dataflow architecture, with an interface to a reactor kernel that manages the application's handlers for asynchronous I/O, real timers, and custom interrupts. In the subsequent sections, we introduce the principal components of CASiNO.

The **Reactor Kernel** is the name given to the collection of classes that provide applications with access to asynchronous events, such as the setting and expiration of timers, notification of data arrival, and delivery and receipt of intraapplication interrupts. CASiNO can be used to implement both live network protocols and their simulations. Both cases require precise coordination of program execution relative to some notion of time. In the case of network simulation, however, the notion of time is quite different, because the time within the simulation need not remain in lockstep with the real wall clock time. The Reactor Kernel can switch between these paradigms transparently, and user programs are not affected in any way (in fact are oblivious to) whether they are operating in simulated or "live" mode.

The **Data-Flow Architecture** is the name given to the collection of classes that define the behavior of the modules within a protocol stack. CASiNO has three tiers in its design: (1) The Conduit level of abstraction represents building blocks of the protocol stack; the function of the Conduit class is to provide a uniform interface for joining these blocks together and injecting Visitors into the resulting networks. (2) The Behavior level of abstraction represents broad classes, each capturing different semantics regarding the Conduit's purpose and activity. Finally, (3) at the Actor level of abstraction, the user defines concrete implementations. There are five subclasses of the Actor class: Terminal, Accessor, State, Creator, and Expander. These subclasses of the Actor correspond respectively to the Behavior's subclasses, Adapter, Mux, Protocol, Factory, and Cluster. We now describe each of these Behaviors and their associated Actor.

1. **Protocol Behavior—State Actor**. A communication protocol is often specified by a finite-state machine. A Protocol behavior implements a finite-state machine using a specialized State Actor. When a Visitor arrives at a Protocol, it is handed to the State actor. The State then queries the Visitor to determine dynamically what type of Visitor it is. Once the State determines the type of the Visitor, the State transforms the Visitor into the appropriate type casting and operates on the Visitor accordingly. The State Actor is implemented using the state pattern defined in [15].

2. **Adapter Behavior—Terminal Actor**. The Adapter Behavior, as its name indicates, converts or adapts the outside world to the Conduits and Visitors composing the CASiNO application. It is used typically for testing, where a Protocol stack is capped at both ends with two Adapters, one to inject Visitors and the other to sink them. The *Terminal* actor is used to configure the Adapter with a simple interface to allow Visitors to be injected and absorbed.

3. **Mux Behavior—Accessor Actor**. The Conduit with a Mux Behavior allows developers to implement multiplexing and demultiplexing. The demultiplexing occurs when a Visitor arrives and needs to be routed to one of the several other Conduits on the basis of some aspect of the Visitor's contents, which we refer to as the Visitor's key or address. The Accessor Actor is responsible for examining the incoming Visitor's contents and deciding which Conduit the Visitor should be sent to.

4. **Factory Behavior—Creator Actor**. The Factory Conduit allows dynamic instantiation of new Conduits at Muxes. Thus a Factory Behavior is always found with Muxes; see Fig. 16. A Conduit configured with Factory Behavior is connected to the B side of a Conduit configured with Mux Behavior. Upon receipt of a Visitor, the Factory may make a new Conduit. If it decides to do this, the Factory makes the newly instantiated Conduits known to the Accessor of the Mux. In this manner, Visitors can be sent to a Mux to install new Conduits dynamically, provided that a Factory is attached to the Mux. The Factory behavior is specified with a Creator Actor. After the installation of the newly created Conduit object, the incoming Visitor, which prompted the installation, is handed to the freshly instantiated Conduit by the Factory.

5. **Cluster Behavior—Expander Actor**. In addition to the four primitive Behaviors of Protocol, Factory, Mux, and Adapter, CASiNO has generic extensibility via the Cluster Behavior. Just as the primitive behaviors mentioned above are configured with a particular Actor (i.e., a concrete State, Creator, Accessor, or Terminal object), the programmer can fully specify a Cluster's behavior by configuring it with a concrete Expander Actor. An Expander may be thought of as a black-box abstraction of a network of Conduits.

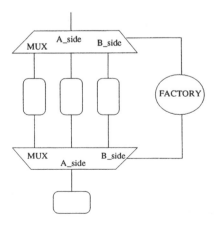

Figure 16 Common usage of two muxes and a factory.

3.4.5   The TRON Architecture in Terms of the
CASiNO Framework

1.   **Router Expander**. The Router Expander is a collection of different types of Conduits: the SwitchDB Conduit, the SwitchProtocol Expander, and the HardwareAccessor Conduit. The SwitchProtocol Expander is attached to the B-side of the SwitchDB Conduit via its A-side, and to the A-side of the HardwareAccessor Conduit via its B-side. Figure 17 shows two Router Expanders attached to each other through a LinkSim Conduit, and two Hardware Abstraction Module (HAM). We now describe each of these Conduits in further detail.

The **SwitchDB** Conduit maintains the switch network topology database. The operations it supports include searching for LSAs, inserting LSAs, flooding LSAs, aging, and flushing expired LSAs. For example, when the Adjacency protocol running over the link reaches the Full state, the LSA corresponding to that link is instantiated and inserted into the SwitchDB. Once the LSA has been inserted, the SwitchDB calls the flooding protocol, which ages the LSA by adding the hop transmission delay time to the LSA age's field, then appends the LSA to an OSPF update Visitor. A copy of this Visitor is given to all Adjacency FSM running in the SwitchProtocol Expander.

The **SwitchProtocol** Expander represents the set of protocols running on the ports of the switch. More details about the internal architecture of the SwitchPort Expander will be provided in the next section.

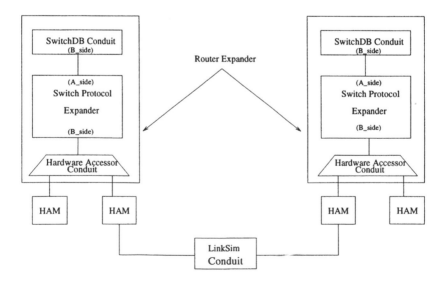

Figure 17   Router expander architecture.

The **HardwareAccessor** Conduit is responsible for routing the outgoing Visitors to their respective interfaces and HAMs.

The **HAM** is located between the Router Expander and the LinkSim Conduit. The main task of the HAM is to map the hardware signals and alarms returned by the OXC switch into a set of generic events that will be handled by the routing protocol. This module insures that the routing protocol is hardware independent, because the protocol relies only on the events returned by the **HAM**, and the HAM implements the adapter pattern [15]. In case of different types of hardware, only the HAM has to be updated so that the signals and alarms returned by the switch will be mapped to the same set of protocol events.

The **LinkSim** Conduit represents an optical link. It is a means to simulate link failure, since it can be tuned so that it passes through a first set of Visitors, then suicides the next coming visitors (showing that the link is down), and finally passes through the next incoming Visitors (indicating that the link has come up.)

We now examine the internal structure of the SwitchProtocol Expander.

2. **SwitchProtocol Expander.** The SwitchProtocol Expander (Fig. 18) is a collection of several types of Conduits: a Factory Conduit, TopMux and BottomMux Conduits, and as many PortStack Conduits as are needed.

The TopMux and the BottomMux each maintains a dictionary that allows

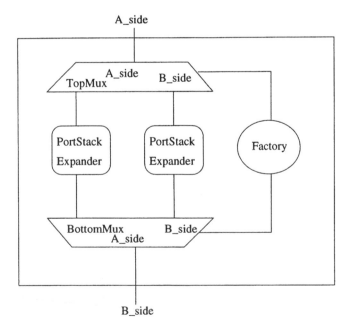

Figure 18 SwitchProtocol expander.

the incoming Visitors to be forwarded to their appropriate PortStack Expander. Upon the reception of a Visitor by either of these muxes, the Accessor searches in its dictionary for the Conduit PortStack that should receive the Visitor. In case this search fails, the Visitor is forwarded to the factory Conduit, causing the instantiation of a new PortStack Expander between the TopMux and the BottomMux. Hence the newly created PortStack Conduit becomes known to both muxes, and the Visitor is forwarded through the newly made Conduit PortStack.

When this occurs, the above mechanism is used dynamically to create Port-Stack Expanders. As soon as two routers are joined by LinkSim Conduit, two instances of InstallerVisitors are made inside the LinkSim Conduit and one is sent to each of the attached routers. This special Visitor is always sent by the BottomMux to the factory, thus causing a new PortStack Conduit to be made.

We now describe the internal structure of the PortStack Expander.

3.   **PortStack Expander**. The PortStack architecture (Fig. 19) is a collection of two Conduits build on top of each other. The top Conduit contains a State Actor implementing the Adjacency (**FSM**.) The bottom Conduit contains a State Actor implementing the Hello FSM. The top Conduit is attached to the A-side of the bottom one through its B-side. Both Conduits are grouped together into a single Expander. The A-side of this Expander as a whole is the A-side of the Adjacency FSM Conduit; the B-side of the Expander is the B-side of the Hello FSM Conduit.

All incoming Visitors entering the PortStack from the network are first seen by the Hello protocol Conduit. If the Visitor is not a HelloVisitor, then it is just passed upwards to be handled by the Adjacency Protocol Conduit. The Hello FSM and the Adjacency FSM operate with the help of various types of timers.

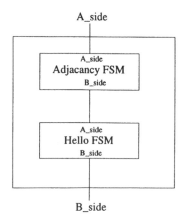

Figure 19   PortStack expander.

For example, in the Hello FSM, whenever a router sends a hello Packet to its peer, the FSM calls Register ( ) to start an InactivityTimer. This timer is aborted by calling Cancel ( ) only if the Hello FSM receives a hello packet in response to the sent message. On the other hand, if the InactivityTimer expires, the hello FSM concludes that the connection over this link is broken and therefore it has to notify the upper running protocol about this change. Likewise, during the database exchange, the Adjacency FSM uses timers to cope with packet loss. For example, a DB_InactivityTimer is always registered whenever a database description packet is sent. Expiring of the timer indicates a failure event in the Adjacency FSM, which triggers the necessary corrective actions.

REFERENCES

1. *Sun Microsystems Inc., Java Management Extensions; see http://java.sun.com/ products/JavaManagement/wp/.*
2. *Sun Microsystems Inc., Jiro Technology Release 1.0.1A; see http://www.sun.com/ jiro/.*
3. M. Ali, B. Ramamurthy, and J. Deogun. Routing and wavelength assignment (RWA) with power considerations in optical networks. IEEE Globecom '99 Symposium on High-Speed Networks, December, 1999.
4. J. Anderson, J. Manchester, A. Rordiguez-Moral, and M. Veeraraghvan. Protocols and architectures for IP optical networking. Bell Labs Technical Journal, January, 1999.
5. G. Apostolopoulos, R. Guerin, S. Kamat, and S. Tripathi. Quality of service based routing: a performance perspective. Proceedings of SIGCOMM, September, 1998.
6. G. Apostolopoulos, D. Williams, S. Kamat, A. O. R. Guerin, and T. Przygienda. QoS routing mechanisms and OSPF extensions. Request for Comments No. 2676, August, 1999.
7. ATM-Forum. *Private Network–Network Interface Specification, Version 1.0.* ATM-Forum, 1996.
8. D. Awduche, J. Malcolm, J. Agogbua, and J. M. M. O'Dell. Requirements for traffic engineering over MPLS. Request for Comments No. 2702, September, 1999.
9. D. Basak, D. Awduche, J. Drake, and Y. Rekhter. Multi-protocol lambda switching: issues in combining MPLS traffic engineering control with optical crossconnects. Internet draft, January, 2000.
10. R. Callon. Use of OSI IS-IS for routing in TCP/IP and dual environments. Request For Comments No. 1195, December, 1990.
11. S. Chaudhuri, G. Hjalmtysson, and J. Yates. Control of lightpaths in an optical network. Internet draft, February, 2000.
12. R. Coltun. The OSPF opaque LSA option. Request for Comments No. 2370, July, 1998.
13. D. Fedyk and A. Ghanwani. Metrics and resource classes for traffic engineering. Internet draft, October, 1999.

14. T. O. Foundation. *OpenLDAP; see http://www.openLDAP.org/*.
15. Gamma, Helm, Johnson, and Vlissides. *Design Patterns: Elements of Reusable Object-Oriented Software*. Addison Wesley/Professional Computing Series, April, 1997.
16. C. Hedrick. Routing information protocol. Request for Comments No. 1058, June, 1988.
17. C. Huitema. *Routing in the Internet*. Prentice Hall.
18. IETF. The MPLS operations mailing list. IETF's MPLS Working Group.
19. K. Kompella, Y. Rekhter, D. Awduche, J. L. G. Hjalmtysson, S. Okamoto, D. Basak, G. Bernstein, J. Drake, N. Margalit, and E. Stern. Extensions to IS-IS/OSPF and RSVP in support of MPL(ambda)S. Internet draft, October, 1999.
20. S. Mouncastle, D. Talmage, S. Marsh, B. Khan, A. Battou, and D. C. Lee. CASiNO: a component architecture for simulating network objects. Proceedings of SIG-COMM, September, 1999.
21. J. Moy. OSPF Version 2. Request for Comments No. 2178, July, 1997.
22. R. Perlman. *Interconnections Bridges and Routers*. Addison Wesley, 1993.
23. R. Ramaswami and K. Sivarajan. *Optical Networks: A Practical Perspective*. Morgan Kaufmann, 1998.
24. G. Wang, D. Fdyk, V. Sharma, K. Owens, G. Ash, M. Krishnaswamy, Y. Cao, M. Girish, H. Ruck, S. Bernstein, P. Nquyen, S. Ahluwalia, L. Wang, A. Doria, and H. Hummel. Extensions to OSPF/IS-IS for optical routing. Internet draft, March, 2000.
25. H. Zang, J. Jue, and B. Mukherjee. A review of Routing and wavelength assignment approaches for wavelength-routed optical WDM network. August, 1999.

6

Routing and Path Establishment for Point-to-Point and Point-to-Multipoint Communications over WDM Networks

Taieb F. Znati and Rami Melhem
University of Pittsburgh, Pittsburgh, Pennsylvania

1   INTRODUCTION

The next generation internet (NGI) technology will, for the most part, be driven by the increasing need for high throughput and low latency. Optical networks have the potential for achieving these two goals by offering unprecedented bandwidth in a medium that is free from inductive and capacitive loadings, thus relaxing the limitations imposed on the bandwidth-distance product [37,42]. Optical fibers may be used for transmitting terahertz signals in low attenuation passbands while maintaining low error rates and low sensitivity to noise [48,38].

Although different types of photonic switching networks have been reported and demonstrated, wavelength division multiplexing (WDM) has emerged as one of the most attractive approaches for data transfer in interconnection networks [32,38,34,46]. WDM divides the optical fiber link spectrum into several channels, each corresponding to a different wavelength.[1] An incoming wavelength in one input port can be routed to one or more output ports. Due to electromagnetic interference, however, the same wavelength coming from two separate input ports cannot be routed to the same output port.

---

[1] Current commercial WDM systems support up to 16 channels, at 2.5 Gbps per channel, for an aggregate capacity of 40 Gbps per fiber and are expected to scale to 32 channels at 10 Gbps per channel within the next few years.

In *single-hop* WDM networks, the optical layer provides a *lightpath* to the higher layers. A lightpath is an all-optical transmission path between a source and a destination that uses the same wavelength between the intermediate nodes along the path. Each lightpath provides a full wavelength's worth of bandwidth to the higher layer. Lightpaths enable an efficient utilization of the optical bandwidth since data transmission through a lightpath does not require wavelength conversion or electronic processing at intermediate nodes. Therefore, an all-optical path between two access nodes can exploit the large bandwidth of optics without the overhead of buffering and processing at intermediate nodes.

In optical networks, however, the number of wavelengths available at the intermediate nodes and the tunability of the optical transceivers at these nodes are both limited. These constraints, combined with the fact that optical networks are usually large in size, make all-optical networks practically infeasible [31]. To increase the probability of successfully establishing a connection between a source and a destination, *multihop* WDM networks allow conversion between different wavelengths to take place at the intermediate nodes.[2] The resulting path is referred to as a *semilightpath*.

Despite the renewed popularity of WDM, it has not been easy to bridge the gap between the Internet and optical networks. Much of the focus of optical communications research has been on transmission, a link property, rather than on networking. The feasibility of economical all-optical wideband amplifiers and multichannel links has been demonstrated, thus proving WDM to be an effective transmission solution. However, optical networking, a property of the overall system and its dynamics, still remains a challenge to the research community. Addressing this challenge requires the identification of the synergies between the two technologies that will enable the wide deployment of Internet protocols and services on advanced optical networks. One very important aspect of the development of efficient routing and path establishment protocols for point-to-point and point-to-multipoint communications.

In order to provide dynamic connection establishment in WDM networks, where a high rate of connection setups and tear-downs is expected, connection protocols should combine path and wavelength selection with resource reservation. Consequently, adaptive and global algorithms are necessary to augment connection establishment protocols in all-optical WDM networks with the capabilities of adapting to changing wavelength availability on the network's links. This adaptivity is accomplished by exploring alternate paths as well as alternate wavelengths dynamically during connection setup.

Similarly, multicasting, which is currently supported in most common commercial operating systems and routers, is expected to play a major role in the Internet. It provides the means to deliver messages to multiple destinations as a network service without requiring message replication by the end systems. As

---

[2] The physical realization of wavelength conversion can be either opto-electronic or fully optical.

such, support for *native* WDM multicast is essential if WDM networks are to play an effective role in the next generation Internet.

Multicasting requires building a *point-to-multipoint tree* of paths between the source node and a *group* of destination nodes. The focus of the early research work on multicasting was aimed at extending the LAN-style multicasting across a wide area network by developing heuristics for approximate solutions to the Steiner problem. In this context, the objective was to find a minimum cost tree that connects all the multicast nodes. The heuristics developed were aimed at finding a low-cost multicast tree, given a connectivity graph and a single cost metric; the goal was to minimize the sum of link metrics in the tree.

The primary limitation of using the previous approaches to support high data rate applications rests with their sole consideration of low cost. In addition to low cost, multimedia applications have different demands in terms of bandwidth, reliability, delay, and jitter. A key property of multimedia data is its time dependency. The support of sustained streams of multimedia objects, over a period of time, requires the establishment of reliable, low delay, and low cost source-to-destination routes. Nevertheless, the objective is not to develop a strategy that produces the lowest possible end-to-end delay but a strategy to ensure that the data traffic arrives within its delay bound, thereby allowing a tradeoff between delay and cost. Thus the objective is to produce a minimal cost tree that guarantees a bounded number of wavelength conversions along the path from the source to all destinations.

Finding the minimum delay paths from a source to a set of destination nodes can be achieved in polynomial time using one of the well-known shortest path algorithms. However, finding a delay-bound, minimum-cost multicast tree is at least as complicated as the classical Steiner minimum tree (SMT), which is known to be NP-hard. Finding an approximation solution for the SMT problem has been the focus of research in multicasting, and several algorithms for constructing multicast trees have been developed for traditional networks.

It must be mentioned, however, that there is a fundamental difference between the traditional networks and the WDM networks regarding the multicasting issue. This difference reflects in their optimization objectives. Multicasting in traditional networks has focused on bounding the end-to-end delay while keeping the cost minimum. In WDM networks, wavelengths constitute critical resources and the cost of wavelength conversion plays an important role in the optimization objectives. Furthermore, the optimization objectives in WDM networks involve more than one cost metric. The tradeoff between the desire to reduce wavelength convergence to a minimum and the need to increase the likelihood of successfully establishing a multicast tree brings about challenges that are unique to WDM networks.

The focus of this chapter is twofold: (1) We develop adaptive and global algorithms to augment connection establishment protocols in all-optical WDM networks with the capability of adapting to changing wavelength availability on the network's links. This adaptivity is accomplished by exploring alternate paths as well as alternate wavelengths dynamically during connection setup. (2) We discuss algorithmic issues related to establishing trees of semilightpaths between a source node and a group of destination nodes and develop adequate heuristics for point-to-multipoint communications, more specifically, a heuristic that can be used to build minimum-cost, conversion-bounded multicast trees in WDM networks. Contrary to many solutions proposed in the literature, our approach decouples the cost of building the multicast tree from the data transmission delay due to wavelength conversion and processing at intermediate nodes along the semilightpath.

The rest of this chapter is organized as follows: In the first part, we first discuss a connection establishment protocol that combines adaptive wavelength selection with resource reservation along a given selected path. We show how to enhance the connection establishment protocol with the capability of dynamically changing the route according to the availability of wavelengths. We also explore a variation of the protocol that attempts to balance the wavelength utilization on the links of the network. A simulation study is then used to compare the performance of the adaptive route selection techniques that rely on local knowledge with route selection techniques that rely on global knowledge of wavelength utilization. In the second part, we formalize the multicast problem in WDM, in terms of minimizing the cost and minimizing the delay by bounding the number of wavelength conversions along a path from the sender to a receiver, and we describe a graph model to characterize this problem. We then present a heuristic to build such a multicast tree in WDM networks. In the last section, we summarize the contributions of this chapter and discuss future work.

2 ROUTING AND PATH ESTABLISHMENT FOR POINT-TO-POINT COMMUNICATION

In WDM networks, it is desirable to avoid wavelength conversion along a connection path in order to improve the quality of service, in terms of bandwidth and end-to-end delay; an all-optical path between two access nodes can exploit the large bandwidth of optics, without the overhead of buffering and processing at intermediate nodes. However, the ability of converting between wavelengths along the same path increases the probability of successfully establishing a connection [1,11,12,16,20,23,51], even when such ability is limited [22,24]. The tradeoffs that result from using wavelength converters have been extensively studied in the literature and will not be considered in this chapter. Rather, we will study the effect of adaptive routing and wavelength selection on path establishment in the absence of wavelength conversion capabilities. The absence of such capabilities requires that the same channel be used on all the links of a given path and thus distinguishes the problem at hand from that of routing in electronic networks, where different channels can be used on individual links of a path.

Connection establishment in high-speed wide-area networks is accomplished by first selecting a path in the network and then reserving the resources along this path [9]. However, when the rate at which connections are established and released is high, the resource availability may change rapidly, and the likelihood of establishing a connection along the selected path decreases. This difficulty can be avoided by combining route selection and resource reservation [8]. A few techniques have been proposed to reserve wavelengths dynamically in WDM networks along a selected path [17,19,26,25]. These techniques ensure that the same wavelength is used along all the links of a path. Adding the capability of adaptively selecting the route to these techniques [2,10,13] should reduce the blocking probability, thereby reducing the the connection establishment delay.

In addition to adaptively selecting a route, dynamic reservation protocols can take advantage of another degree of adaptivity that is possible in WDM networks, namely, the selection of a wavelength. Specifically, in order to establish a connection, a path in the network has to be selected, a wavelength on this path has to be chosen, and a connection using this path and wavelength has to be established. A closer look at this process, however, reveals that combining the above three steps is likely to lead to increased effective network bandwidth, especially in environments where resource availability changes frequently and the information about this availability prior to path establishment cannot be accurately obtained.

In this discussion, a network is represented by a graph, $G = (V, E)$, where V is a set of nodes, corresponding to the network nodes, and E is a set of edges, corresponding to the network links. An edge from a node $u \in V$ to a node $v \in V$ is denoted by $\langle u, v \rangle$. A path, $P(s, d)$, from a source node, $s \in V$, to a destination node, $d \in V$, is an ordered set of nodes $\{v_0, v_1, \ldots, v_k\}$, such that $v_0 = s$, $v_k = d$; and for $i = 0, \ldots k - 1$, $\langle v_i, v_{i+1} \rangle \in E$. The function $\text{next}_{P(v_0, v_k)}(v_i) = v_{i+1}$, $0 \leq i \leq k - 1$, will be used to indicate the node following v_i on $P(v_0, v_k)$. Only shortest-path routing will be used, in the sense that, connections will be established on paths with the least number of intermediate nodes between a source and a destination.

It is assumed that W wavelengths, $\lambda_1, \ldots, \lambda_W$, are available on each link and that wavelength-sensitive switches are available at each node to route signals either to the next link toward the destination or to the local processor. The switching is performed in the optical domain, but the switches are controlled by an electronic controller, as depicted in Fig. 1. A control network, which is separate from the optical data networks but shadows its topology, is used for exchanging control messages. The traffic on the control network consists of small packets and thus is lighter than the traffic on the data network. Given that, switch controllers at intermediate nodes will need to examine the control packets and update local

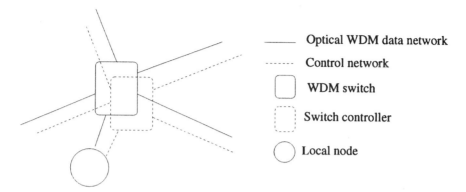

Figure 1   The data and control networks.

bookkeeping information accordingly; the control network can be implemented as an electronic network. It is possible, however, to reserve a wavelength on the data network to exchange control messages exclusively as proposed in [19].

In the remainder of this section, we describe a connection establishment protocol that combines adaptive wavelength selection with resource reservation along a given selected path. We then show how to enhance the connection establishment protocol with the capability of dynamically changing the route according to the availability of wavelengths. We also explore a variation of the protocol that attempts to balance the wavelength utilization on the links of the network. A simulation study is then used to compare the performance of the adaptive route selection techniques that rely on local knowledge with route selection techniques that rely on global knowledge of wavelength utilization.

2.1   Dynamic Wavelength Selection and Path Establishment

A distributed reservation protocol is needed in WDM networks with rapidly changing resource availability. Such a protocol should efficiently handle the situations that arise when resources that were available at the time of path selection become unavailable during connection setup [4,7,9,15]. In WDM networks, wavelengths are the main resources to be reserved along a path.

In a number of dynamic path establishment protocols for time-division multiplexed networks were studied [18,27]. These protocols, which were applied to WDM networks in [17,26], distributively reserve the resources along a given path and dynamically select the time slots for the connection. Two classes of protocols are presented in [27]: a forward reservation protocol and a backward reservation protocol. In this section, we will consider only forward reservation protocols, which adaptively select a wavelength while dynamically establishing a connection along a given path, and show that they can lead to a higher rate of connection establishment in WDM networks.

Given a *selected route*, the process of distributively establishing a connection along this path requires the exchange of control messages. The switch controller at each node in the network maintains a state for each wavelength on each link emerging from that node. For a wavelength λ on link $\mathscr{L}$ the state can be one of the following:

> *AVAIL* indicates that λ is available and can be used to establish a new connection,
> *LOCK* indicates that λ is locked by some request in the process of establishing a connection.
> *BUSY* indicates that λ is being used in some connection to transmit data.

For a link, $\mathscr{L}$, the set of wavelengths that are in the *AVAIL* state is denoted by *Avail* ($\mathscr{L}$). When a wavelength, λ, is not in *Avail* ($\mathscr{L}$), an additional field is maintained to identify the connection request locking λ, if λ is in the *LOCK* state, or the connection using λ, if λ is in the *BUSY* state.

Each connection request is assigned a unique identifier, *id*, which consists of the identifier of the source node and a serial number issued by that node. Each control message related to the establishment of a connection carries its *id*, which becomes the identifier of the connection, when successfully established. Four types of packets are used to establish a connection.

Reservation packets (*RES*), each carrying a set, *cset*, of wavelengths. This set, which can be represented as a *W*-bit vector, is used to keep track of the set of wavelengths that can be used to establish a connection. These wavelengths are locked at intermediate nodes while the *RES* message progresses toward the destination node.

Acknowledgment packets (*ACK*), used to inform source nodes of the success of connection requests. An *ACK* packet contains a *channel* field, which indicates the wavelength selected for the connection. As an *ACK* message travels from the destination to the source, it changes the state of the wavelength selected for the connection to *BUSY* and unlocks (change from *LOCK* to *AVAIL*) all other wavelengths that were locked by the corresponding *RES* message. Also, when a switch controller receives an *ACK*, it reconfigures the WDM switch to accommodate the corresponding connection.

Fail or Negative ack packets (*FAIL/NACK*), used to inform source nodes of the failure of connection requests. While traveling to the source node, a *FAIL/NACK* message unlocks all wavelengths that were locked by the corresponding *RES* message.

Release packets (*REL*), used to release connections. A *REL* message traveling from a source to a destination changes the state of the wavelength reserved for that connection from *BUSY* to *AVAIL*.

The protocol requires that control messages from a destination, *d*, to a source, *s*, follow the reverse path of their corresponding messages from *s* to *d*. We will denote the fields of a packet by *packet. field*. For example, *RES.id* denotes the *id* field of the *RES* packet.

The reservation proceeds as follows: when the source node, v_0, wishes to establish a connection to node v_k on a given path $P(v_0, v_k)$, it composes a *RES* message with *RES.cset* set to the wavelengths that the node may use. This message is then routed to the destination. When an intermediate node, v_i, receives the *RES* packet, it determines the next outgoing link, $\mathcal{L} = next_{P(v_0, v_k)} (v_i)$, for the *RES* packet, and updates *RES.cset* to *RES.cset* $\cap$ *Avail* $(\mathcal{L})$. If the resulting *RES.cset* is empty, the connection cannot be established, and a *FAIL/NACK* message is sent back to the source node. This process of failed reservation is shown in Fig. 2(a). Note that if *Avail* $(\mathcal{L})$ is represented by a bit-vector, then *RES.cset* $\cap$ *Avail* $(\mathcal{L})$ is a bit-wise *and* operation.

If the resulting *RES.cset* is not empty, the node reserves all the wavelengths in *RES.cset* by changing their state to *LOCK*. It then forwards *RES* to the next node. This way, as *RES* approaches the destination, the path is reserved incrementally. Once *RES* reaches the destination with a nonempty *RES.cset*, the destination selects from *RES.cset* a wavelength to be used for the connection and informs the source node of the wavelength to be used by sending an *ACK* message with

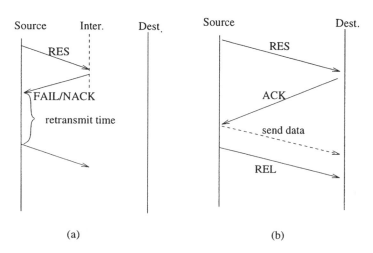

Figure 2　Control messages in forward reservation.

ACK.channel set to the selected wavelength. The source can start sending data on the all-optical path once it receives the *ACK* packet. After all data is sent, the source node sends a *REL* packet to tear down the connection. This successful reservation process is shown in Fig. 2(b).

The protocol described above does not specify the initial set of wavelengths *RES.cset* chosen by the source node. The size of that set specifies the aggressiveness of the reservation process. An aggressive scheme choses *RES.cset* to include all the wavelengths. This can be indicated by setting $RES.cset = \{1, \ldots, W\}$. It allows the maximum flexibility for adaptively selecting the wavelength but has the disadvantage of unnecessarily locking resources during the reservation process. A very conservative scheme choses *RES.cset* to include only one wavelength, say λ_j, $1 \le j \le W$. In this case, no dynamic adaptation is possible in selecting the wavelength, and the reservation is successful only if λ_j is available on all the links along the path.

Finally, it should be mentioned that, instead of aborting the reservation process as soon as *RES.cset* becomes empty at an intermediate node (*RES* can no longer move forward on the next link, $\mathcal{L}$), a holding mechanism can be used to keep the wavelengths on the partial path locked for some "time-out" period, hoping that during this period some wavelengths on $\mathcal{L}$ will be released, thus allowing *RES* to progress forward and the reservation to continue. If after the time-out period *RES* still cannot move forward, the reservation fails, and a *FAIL/NACK* message is sent to the source node, to release the reserved partial connection.

2.2 Combining Adaptive Routing, Wavelength Selection, and Path Establishment

The protocols discussed in the previous section aimed at reserving wavelengths for establishing a connection along a given path $P(v_0, v_k)$. The protocol dynamically selects the wavelength for the connection but assumes a fixed path from the source v_0 to the destination v_k. The connection establishment fails at an intermediate node, v_i, when no wavelength is available for the connection on the link v_i, $\text{next}_{P(v_0, v_k)}(v_i)$).

Given that $P(v_0, v_k)$ is a shortest path between v_0 and v_k, the path $\{v_i, v_{i+1}, \ldots, v_k\}$, denoted by $P(v_i, v_k)$, is necessarily a shortest path between v_i and v_k. However, $P(v_i, v_k)$ may not be the only shortest path from v_i to v_k, and there may be another path, $\overline{P}(v_i, v_k) = \{v_i, \bar{v}_i, \ldots, \bar{v}_{d-1}, v_k\}$, with the same number of intermediate hops as $P(v_i, v_k)$.

Definition: At any node v_i, one of the shortest paths to a destination, d, will be called the *principal shortest path* to d, and the other shortest paths will be called the *alternate shortest paths* to d. The next node on the principal shortest path to d will be denoted by $\text{next}_P(v_i, d)$, and the next node on the jth alternate shortest path to d will be denoted by $\text{next}_{AP_j}(v_i, d)$, where $j = 1, \ldots, q$, and q is the number of alternate shortest paths.

The fixed route path establishment algorithm described in the previous section will be denoted by "**Fixed**." It may be viewed as one in which, en route to the destination d, each intermediate node v_i sends the *RES* packet forward to $\text{next}_P(v_i, d)$. However, the connection establishment does not progress if $RES.cset \cap Avail(\text{next}_P(v_i, d)) = \phi$. Instead of stopping the progress of the path establishment, an adaptive routing scheme would dynamically change the principal route and forward *RES* on an alternate route if possible. That is, instead of aborting the connection establishment process, the *RES* packet can be forwarded to node $\text{next}_{AP_j}(v_i, d)$ for some j, such that $RES.cset \cap Avail(\text{next}_{AP_j}(v_i, d)) \neq \phi$. In other words, the path to the destination is selected adaptively (at each node) among alternate shortest paths, with priority given to the principal shortest path. Because the path reservation protocol combined with the route selection mechanism described here takes an alternate route when the principal route is blocked, we call the resulting protocol the "**Detour**" adaptive path establishment protocol.

The Detour adaptive protocol deviates from a given route only when wavelength reservation on that route is blocked. Given that the route adaptation decision is based only on local information, the chances of completing the reservation successfully may be improved if v_i forwards *RES* to $\bar{v}_{i+1}$ rather than v_{i+1} when $RES.cset \cap Avail(\bar{v}_{i+1})$ is larger than $RES.cset \cap Avail(v_{i+1})$, even when the route through v_{i+1} is not blocked. This choice follows a greedy strategy to maximize the size of $RES.cset$ at each intermediate node (within shortest paths), and thus is called "**Greedy**."

A different strategy may be to try to increase the chance of successfully completing future reservations rather than the chance of successfully completing the current reservation. In order to accomplish this goal, it may be appropriate to try to balance the wavelength utilization on the links of the network. For this, an intermediate node v_i may choose to forward *RES* to $\bar{v}_{i+1}$ rather than v_{i+1} if $Avail(\bar{v}_{i+1})$ is larger than $Avail(v_{i+1})$. We call this strategy "**Balanced**."

In order to illustrate the difference between the four strategies, consider the network shown in Fig. 3(a). Assume that the path $\{s, t, u, v, z, d\}$ shown in Fig. 3(b), which is a shortest path from s to d, is the principal path from s to d. Clearly, at each of the nodes s, t, u, and v, an alternate shortest route can be taken. For example, assume that the *RES* packet reaches node u with $RES.cset = \{1,3,6,7\}$ and assume that $Avail(<u, v>) = \{2,5\}$ and $Avail(<u, w>) = \{1,2,3,6\}$. At node u, the "Fixed" protocol cannot forward *RES* to node v along the principal path since $RES.cset \cap Avail(<u, v>) = \phi$. Hence "Fixed" will fail and will send back a FAIL/NACK message to s. The "Detour" adaptive policy, however, will send *RES* to node w, which is on an alternate shortest path to d as long as it can do so. In fact, all of "Detour," "Greedy," and "Balanced" will send *RES* to w with $RES.cset = \{1,3,6\}$.

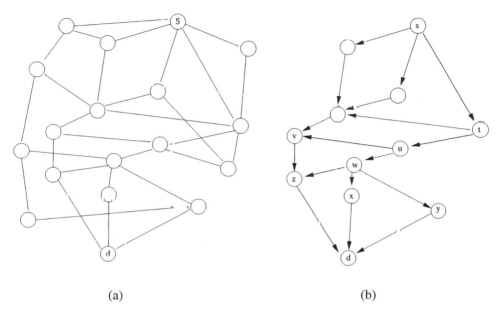

(a)   (b)

Figure 3   (a) A graph and (b) the shortest paths from s to d.

Continuing with the above example, assume that the principal shortest path from w to d is through z, and that $Avail(\langle w, z \rangle) = \{1,7\}$, $Avail(\langle w, x \rangle) = \{1,3,7\}$, and $Avail(\langle w, y \rangle) = \{2,4,6,7\}$. The "Detour" policy will send RES to node z since it follows the principal path whenever possible. However, the "Greedy" policy will send RES to node x, since $RES.cset \cap Avail(\langle w, x \rangle) = \{1,3\}$ is larger than $RES.cset \cap Avail(\langle w, y \rangle) = \{6\}$ and larger than $RES.cset \cap Avail(\langle w, z \rangle) = \{1\}$. The rationale is that with two wavelengths in $RES.cset$ rather than one, RES has a better chance of reaching d. Finally, the "Balanced" policy will send RES to node y since $\langle w, y \rangle$ has more available wavelengths than the other links and thus is least loaded.

A simulation program to compare the "Fixed" route strategy with the "Detour," the "Greedy," and the "Balanced" strategies was developed. The simulated network assumes that the control network is separate from the optical data network and that data is packetized such that the time to route a data packet on the all-optical path from a source to a destination is equal to the time to buffer, process, and route a control message between two neighboring nodes. This time is defined as "a time unit." Messages are generated at individual nodes according to uniform distributions for both the message interarrival rate and the message size (in packets). The load that each node exerts on the network is $\gamma = L/\mu$, where μ is the average interarrival rate and L is the average message size. That is, γ is the average demand, in terms of number of data packets, that each node exerts on the network every time unit. The destination of each message generated at a node is chosen randomly among the other nodes in the network.

We are primarily interested in comparing the routing policies in terms of the network throughput, defined as the average number of data packets delivered to their destinations in a time unit. Given that we fix the average generation rate in each experiment, the throughput inversely correlates with the number of blocked connections and thus with the probability of call blocking. We will report on the throughput for two network topologies, namely an irregular network with 80 nodes and a 10×10 two-dimensional mesh-connected network. The irregular network was generated using a random graph generator. It has a diameter of 5, a maximum node degree of 15, and an average node degree of 6.7. The principal route for the mesh network is the X–Y route while one of the shortest paths is chosen as the principal route for the irregular network. We will assume that 10 wavelengths are available for data transmission on each link. We have experimented with different networks, regular and irregular, and different number of wavelengths. The results that we observed are similar to the results that we will report in this section for the two representative networks.

Figure 4 compares the four routing policies and shows the effect of the protocol aggressiveness (initial size of *RES.cset*, denoted by *agg*) on the throughput for the 10×10 mesh network. The load γ is taken to be 2.5 in this experiment. Two different message sizes are chosen for this experiment, namely 32 and 128 (Figs. 4(a) and 4(b), respectively). These two values are chosen such that the former represents a case in which the control network is saturated, while the latter represents a case in which the control network is not saturated as indicated by the following approximate calculation. Given that each node generates, on average, $2.5/L$ messages per time unit, the average number of control packets generated every time unit by all 100 nodes in the network is $3(250/L)$ (three control packets per message). Assuming that, on average, each control packet travels about 10 links on the mesh, the average number of control packets in the network in each time unit is $7500/L$. Hence the 400 links of the control network will saturate when $7500/L > 400$, that is, when $L < 19$. Due to the nonuniformity of the traffic, it is reasonable to assume that $L = 32$ will still saturate the control network while $L = 128$ will not.

From Fig. 4, it is clear that choosing the right size of *RES.cset* has a large effect on the performance of the protocols. Although neither $agg = 1$ nor $agg = 10$ gives the maximum throughput, it is not trivial to find the optimal value for *agg*. When the control network is saturated, $agg = 1$ outperforms $agg = 10$ with the optimal performance at $agg = 3$, while when the control network is not saturated, $agg = 10$ outperforms $agg = 1$ with the optimal performance at $agg = 5$. The figure also shows that, in general, Greedy and Balanced actually

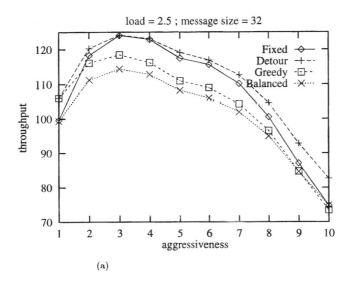

(a)

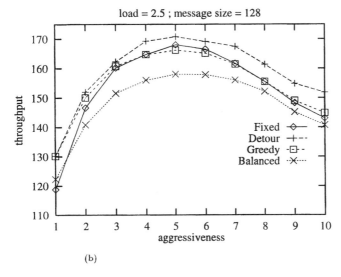

(b)

Figure 4   Effect of aggressiveness on throughput for a 10 × 10 mesh. (a) Control network is saturated and (b) is not saturated.

perform worse than Fixed, while Detour outperforms Fixed, especially when $agg = 1$ and $agg = 10$.

Figure 5 studies the effect of the average message size and the load on the throughput of Fixed and Detour when $agg = 1$, 4, and 10. The figure confirms that Detour, in general, outperforms Fixed for different loads and message sizes, and that the advantage of Detour increases when agg is not optimal. Noting that choosing the optimal agg maximizes the benefit of adaptively selecting the wavelength and that Detour adaptively selects the route, one interpretation of the above results is that route adaptively is not very beneficial when wavelength adaptivity is maximized.

Figure 6 presents results for the 80-node irregular network described earlier. Note that the network is more dense than a mesh (average node degree = 6.7 as opposed to 4), and thus the control traffic will saturate at message sizes smaller than the ones that saturate the mesh. An interesting observation is that, although Detour still outperforms Fixed in the irregular network, Greedy and Balanced are the most efficient policies, especially when the control network is saturated (small message sizes).

The inferiority of Greedy and Balanced in the mesh case and their superiority in the irregular network case may be attributed to the observation that, for the mesh case, Greedy and Balanced disturb the uniformity inherent in the Fixed X–Y routing, while for the irregular case they add uniformity and balance to an otherwise irregularly distributed load.

2.3   Adaptive Routing Based on Global Information

The adaptive route selection algorithm described in the previous section assumes that nodes have only local knowledge about the wavelength availability. That is, each node knows only about the wavelengths available on its incident links. If each node has global knowledge about wavelength availability on all the links of the network, then it is possible for the source node to select a route to the destination such that at least one wavelength is available on this route. Once the route and a wavelength are selected by the source node, the path reservation algorithm of Sec. 2 can be used to establish the connection, with *RES.cset* set to contain the selected wavelength. That is, conservative resource reservation can be used.

In order to confine the selection of routes to shortest paths, we define for each source, s, and destination, d, a subgraph, $G_{s,d} = \{V_{s,d}, E_{s,d}\}$, of G whose nodes and edges lie on some shortest path from s to d. Moreover, if the shortest distance between s and d is k, then we divide the nodes in subgraph $G_{s,d}$ into $k + 1$ disjoint layers as follows:

$$V_0 = \{s\}$$

$$V_{i+1} = \{v; \langle u, v \rangle \in E_{s,d}, u \in V_i, v \notin \cup_{j=0}^{j=i} V_j\} \quad \text{for} \quad i = 0, 1, \ldots, k - 1$$

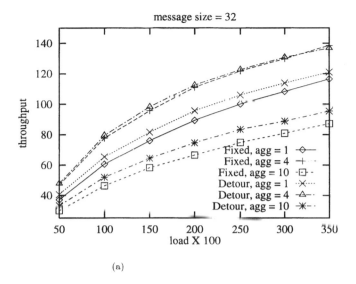

(a)

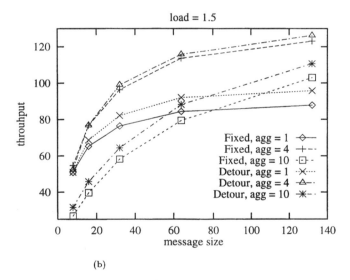

(b)

Figure 5 Effect of load and message size on throughput for a 10 × 10 mesh. (a) Throughput versus load and (b) throughput versus message size.

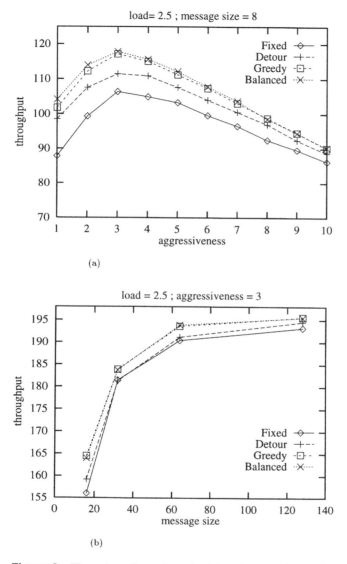

Figure 6 Throughput for an irregular 80-node network. (a) Throughput versus aggressiveness and (b) throughput versus message size.

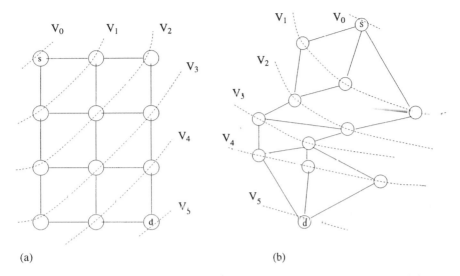

Figure 7 Layering the graph $G_{s,d}$ hr: (a) a two-dimensional mesh network and (b) an irregular network.

Figure 7 shows $V_0, \ldots, V_5$ for two examples where the distance between s and d is $k = 5$. The example in Fig. 7(a) is for a regular two-dimensional graph, and the example of Fig. 7(b) is for the irregular subgraph shown in Fig. 3(b).

An efficient algorithm is proposed to find a shortest path from s to d such that at least one wavelength is available on all the links of that path. For this, let $CAN\_USE(w)$ be the set of wavelengths that can be used to reach any node w in $G_{s,d}$ from s. Given the sets $Avail(\ )$ for the links of $G_{s,d}$, a dynamic programming approach can be used to compute $CAN\_USE(w)$ for each node in the graph. The set $CAN\_USE(d)$ will then contain the wavelengths that can be used to reach d from s. Specifically, if w is in V_{i+1} and can be reached from two nodes u and v in V_i, then $CAN\_USE(w)$ can be computed from $CAN\_USE(u)$, $CAN\_USE(v)$ as follows:

$$CAN\_USE(w) =$$
$$CAN\_USE(u) \cap Avail \langle u, w \rangle) \cup (CAN\_USE(v) \cap Avail \langle v, w \rangle)$$

Algorithm Global
    $CAN\_USE(s) = \{1, \ldots, W\}$
    For each level V_i, $i = 1, \ldots, k$ Do
        For each node $w \in V_i$ Do
            Let H_w be the set of nodes in V_{i-1} that are neighbors to w,
            $CAN\_USE(w) = \bigcup_{u \in H_w} \{CAN\_USE(u) \cap Avail(\langle u, w \rangle)\}$
        End For

End For
If $CAN\_USE(d) = \phi$, then no path is available–exit
let λ be any wavelength in $CAN\_USE(d)$
$w = d$
Repeat
    let u be a node in H_w such that λ is in $CAN\_USE(u) \cap Avail(\langle u, w \rangle)$
    $next_{P_g}(u) = w$
    $w = u$
Until $w == s$

    We illustrate the algorithm in Fig. 8. $CAN\_USE(u) = \{1,4,6\}$ means that there is a shortest path from s to u that can use wavelengths λ_1, λ_4, or λ_6. Given that $Avail \langle u, w \rangle = \{1,2,4\}$, then any path from s to w that goes through u can use any of the wavelengths in $CAN\_USE(u) \cap Avail(u, w) = \{1,4\}$. Similarly, any path from s to w that goes through v can use any of the wavelengths in $CAN\_USE(v) \cap Avail(v, w) = \{1,5\}$. Hence wavelengths λ_1, λ_4, or λ_5 can be used to connect s to w. That is, $CAN\_USE(w) = \{1,4,5\}$.

    After finding $CAN\_USE(d)$ and selecting a wavelength that can be used for the connection, the path from s to d that uses the selected wavelength can be determined by backtracking through $G_{s,d}$. The following algorithm finds a wavelength λ that can be used to connect s and d, if any is available, and the path, P_g, that uses this wavelength. The complexity of the algorithm is equal to the number of edges in $G_{s,d}$.

    After selecting the wavelength, λ, and the path, P_g, the protocol of Sec. 2

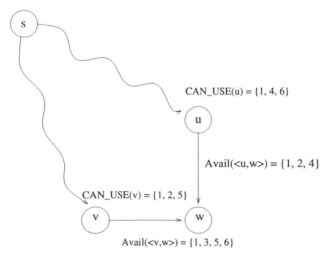

Figure 8 Computation of CAN_USE().

can be used to establish a connection. However, due to the dynamic changes in the wavelength availability, the protocol may still fail if λ becomes unavailable on some link on P_g when the *RES* packet reaches that link. This may be due to one of two factors:

    1. The information in the sets *Avail*( ) used by Global may not be up to date. Typically this is because gathering information about the state of the network cannot be done instantaneously, and thus the information about the availability of wavelengths that Global uses may not reflect the current state of the network.

    2. Even if Global uses the most current information about wavelength availability, this availability at a given link may change by the time the *RES* packet reaches that link.

    Using Global to select a path from s to d has thus a possible drawback, namely, it deprives the path establishment protocol from the dynamic adaptivity for selecting the route and the wavelength by fixing the route during path establishment and forcing *RES.cset* to contain only one wavelength. This may be rather harmful if the wavelength availability changes frequently. Note that although *CAN _USE(d)* found by Global may contain more than one wavelength, it is possible that these wavelengths are available on different paths and thus cannot all be used on the same path from s to d. If only one *RES* packet is used for path establishment (rather than parallel *RES* packets sent on parallel paths), then this *RES* is associated with a selected path and thus a selected wavelength.

    In order to compare the effectiveness of the adaptive routings described in Sec. 3 with the adaptive global path selection described in this section, we make the reasonable assumption that each node in the network has a copy of $Avail(\mathfrak{L})$ for every link $\mathfrak{L}$ in the network, but that this copy is updated using the actual state of the network every Δ time units. We simulated the path establishment with global route selection for different values of Δ and show in Fig. 9 the results of comparing Global and Detour on the 10×10 mesh network. For Detour, the value of *agg* is taken to be 4; for Global, Δ is taken to be 1, 5, and 10. As expected, the performance of Global degrades with larger values of Δ. However, even with perfect global knowledge of the network state ($\Delta = 1$), Detour outperforms Global, except when the load is very high on either the data network (high γ) or the control network (small message sizes). In such cases, global management of resources becomes crucial.

3 POINT-TO-MULTIPOINT COMMUNICATIONS OVER WDM NETWORKS

The minimum-cost, conversion-bounded multicast tree problem, hereafter referred to as *conversion-bound multicast tree* (CBMT), can be viewed as a derivative of the Steiner minimal tree (SMT) problem in which the delay from the

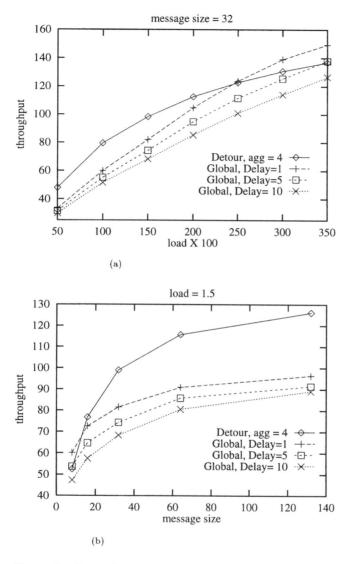

Figure 9 Comparing the throughputs of Global with Detour for the 10×10 mesh. (a) Throughput versus load and (b) throughput versus message size.

source to the destinations is bounded [39]. Both problems are NP-hard, and several heuristics have been proposed to address these problems [49,45,36, 28,40,56,47,44,43]. The metric used in building point-to-multipoint paths must take into consideration the cost of using a specific wavelength on a given link, the cost of converting an incoming wavelength into one or more outgoing wavelengths, and the time required to convert an incoming wavelength into one or more outgoing wavelengths. Most of the proposed solutions that address WDM multicast communication combine these costs to produce a weighted cost structure, which is then used to produce a multicast tree. The CBMT problem, however, consists of two subgoals, namely minimizing the multicast tree cost and bounding the delay due to switching and wavelength conversion. Our approach decouples the cost optimization from bounding the delay. Furthermore, given that in WDM networks, the need for wavelength conversion is expected to be low and must be avoided as much as possible, our approach is first to build a close-to-optimal minimal cost tree and deal with delay violations whenever they occur.

3.1 Network Model

The network model considered in this paper consists of routers attached to an optical core network and connected to their peers over dynamically established switched lightpaths. The optical network is assumed to consist of multiple optical subnetworks interconnected by optical links in a general topology.

In such a network, an all-optical path between two network nodes can exploit the availability of a large bandwidth, without the overhead of buffering and processing at intermediate nodes. In general, however, it is worth noting that WDM links are affected by several factors that can introduce impairments into the optical signal path. These factors include the serial bit rate per wavelength, the amplification mechanism, and the type of fiber being used. Furthermore, the number of wavelengths on a single fiber are often limited, which in turn limits the number of all-optical paths connecting a source to a destination. However, it has been shown that the ability of converting between wavelengths along the same path increases the probability of successfully establishing a connection [51].

Unfortunately, current technology does not allow wavelength conversions to be performed efficiently and cost effectively in the optical domain. Hence it is often the case that the switching fibers of many optical networks are not capable of performing wavelength conversions. In such networks, semioptical paths may still be established if intermediate nodes are used as relays to receive a message on one wavelength and retransmit it on another. A typical intermediate node in a network with two wavelengths is shown in Fig. 10. In this simple example, the node is connected to its neighbors by two input links and two output links. The optical input and output signals to the local node are either multiplexed (Fig. 10a) or demultiplexed (Fig. 10b).

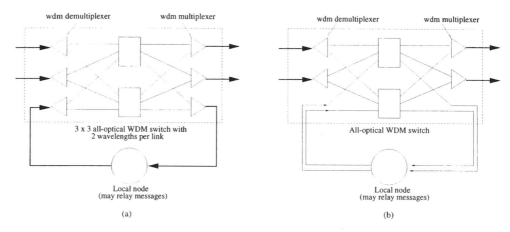

Figure 10   Node model in a WDM network with all-optical switches.

When the node receives several signals, coming from different links at the same wavelength, which need to be transmitted over their respective output ports, the incoming wavelength must be converted to a number of desirable outgoing wavelengths. Due to speed and other physical limitation of electronic wavelength converters, the time required by the last conversion to wait for its turn may be significant. Consequently, a message that is relayed at such a node may suffer a relatively large delay because of buffering, processing, and optics/electronic conversion.

Thus wavelength conversion in networks with all-optical switches introduces delays at intermediate nodes which affect the quality of service on the connections. For this reason, it is desired to minimize the number of wavelength conversions along a given connection. Moreover, if a certain quality of service guarantees is required in terms of delay and jitter, then a limit on the number of wavelength conversions needs to be imposed on a connection in order to meet these requirements. The multicast heuristics proposed in this chapter aim to achieve this goal.

3.2   Problem Formulation

What makes the problem of building WDM multicast trees fundamentally different from the multicasting problem in traditional networks is the cost structure associated with the underlying network. The metrics used in building point-to-multipoint paths in WDM networks must take into consideration the cost of using a specific wavelength on a given link, the cost of converting an incoming wavelength into one or more outgoing wavelengths, and the delay required for wavelength conversions.

Consider a WDM optical network with N nodes, M optical links, and K wavelengths. This network can be modeled by a graph $G = (V, E, \Omega)$, where

V denotes a set of vertices, corresponding to the network nodes.

E denotes a set of edges, corresponding to the network optical links. A link incident from node $u \in V$ onto node $v \in V$ is denoted as (u, v).

$\Omega = \{v_1, v_2, \ldots, v_K\}$ represents the set of all available wavelengths in the network.

Each link $l = (u, v) \in E$, connecting node $u \in V$ to $v \in V$, is associated with a set $\Omega_l \subseteq \Omega$ of available wavelengths. Furthermore, each link $l = (u, v) \in E$, is associated with the following costs and delays:

A *wavelength usage cost*, $\omega_l(v)$, for each $v \in \Omega_l$, which represents the cost of using v on link l.

A *wavelength conversion cost*, $c_l(v_i, v_j)$, representing the cost of converting a wavelength, v_i, on some incoming link $e = (x, u)$ into a wavelength v_j on the outgoing link $l = (u, v)$, where $v_i \in \Omega_e$ and $v_j \in \Omega_l$. This cost is infinity if such a conversion is not possible at u and is reduced to zero if $v_j = v_i$. Furthermore, we assume that $c_l(v_i, v_j) = c_{l'}(v_i, v_j)$ for all l and l'.

A *wavelength conversion indicator*, $\delta_l(v_i, v_j)$, representing the need to convert a wavelength, v_i, on incoming link $e = (x, u)$ into a wavelength, v_j, on outgoing link $l = (u, v)$, where $v_i \in \Omega_e$ and $v_j \in \Omega_l$. This indicator is zero if $v_j = v_i$.

The wavelength usage cost can be used to enforce a specific wavelength assignment strategy. The First-Fit policy, for example, selects the wavelength with the lowest index to assign to a new connection request [41]. The MaxSum policy, on the other hand, assigns the wavelength that maximizes the wavelength path capacity (WPC), defined as the capacity of the most congested link along the path, over all potential paths and all wavelengths [29]. The wavelength usage cost can be used to enforce either of these two policies. In the case of the First-Fit policy, the cost of available wavelengths on a given link can be assigned in such a way that the wavelength with the lowest index bears the lowest cost. Similarly, setting the cost of the wavelength that maximizes the WPC to be the lowest enforces the MaxSum wavelength assignment policy.

The wavelength conversion cost can be used to minimize the number of wavelength conversions along a path. In some cases, it can also be used to discourage wavelength conversion at nodes that perform critical functionalities within the network, such as nodes interconnecting different networks, or at nodes that are heavily loaded. The wavelength conversion cost can be also used to adapt to changing network environments and to reflect the availability of wavelengths at a given node. Links incident from interconnection nodes, heavily loaded nodes, or nodes with a reduced number of available wavelengths incur higher wavelength conversion costs.

The wavelength conversion indicator reflects the significance of the time required to convert an incoming wavelength into an outgoing wavelength in comparison to the time required to transmit the signal over the optical link. Furthermore, in multicast configuration, wavelength conversion can take place between an incoming wavelength on a given input port and a number of outgoing wavelengths on several output ports, as required by the underlying multicast tree. Unless the wavelength conversions proceed in parallel, the required wavelength conversion delay can become even more significant in comparison with the transmission delay.

The cost structure proposed in this chapter decouples the conversion cost from the conversion delay. This is necessary to account for the fact that in a multicast tree two connections from the source to two different destinations that share some nodes along the path face conversion jointly at the shared node but suffer the conversion delays separately. Therefore the conversion cost must be accounted for only once, while the end-to-end delay requirement must be verified for each connection separately. Consequently, the CBMT problem can be viewed as consisting of two subgoals, namely minimizing the multicast tree cost and bounding the delay. Given a source and a set of predefined multicast nodes, the CBMT problem consists of finding the minimal cost tree that contains all selected multicast nodes. In addition to finding the optimal multicast tree that satisfies the delay requirements, a procedure to assign a specific wavelength $v_i \in \Omega_l$ on each link $l = (u, v)$ along the path is required. More specifically, the CBMT problem can be stated as follows:

Let $\mathcal{G}$ be a multicast group composed of a source $s \in V$ and a set of multicast nodes $D \subseteq V$. Every destination node $d \in D$ has a delay bound Δ_d. The objective is to find

1. A multicast tree $T = (V', E')$, $V' \subseteq V$ and $E' \subseteq E$, connecting s to all $d \in D$
2. A wavelength assignment function, $\nabla_P \colon E' \to \Omega$, which assigns a wavelength, $\nabla_P(l) = v \in \Omega_l$, to link l for each path, $P(s, d)$, connecting source s to a given destination $d \in D$, such that the total cost of the tree is minimum and the delay from the source s to any multicast node $d \in D$ does not exceed Δ_d.

In order to formalize the cost of the multicast tree and the delay on a given path, define $P(s, d)$ as the sequence of links $(s, v_1), \ldots, (v_i, v_{i+1}), \ldots, (v_k, d)$ forming the multicast tree path from source s to node d. Furthermore, let $Pred(l) = (v_{i-1}, v_i)$, be the predecessor of link $l = (v_i, v_{i+1})$ on path $P(s, d)$.

Note that if two paths, $P(s, d_1)$ and $P(s, d_2)$, share the same link, l, the $Pred(l)$ on $P(s, d_1)$ is the same as $Pred(l)$ on $P(s, d_2)$. This follows directly from the tree property of the multicast graph.

Further, define $\nabla(l) = \{\nabla_P(l) | l \in P(s, d) \ \forall d \in D\}$ to be the set of all wavelengths currently used on link l by all the source-to-destination paths that include l. Finally, let $\mathcal{W}_l = \{(\mu, \nu) \mid \exists P \supseteq l: \nabla_P(Pred(l)) = \mu$ and $\nabla_P(l) = \nu\}$ be the set wavelength pairs used on link l and its predecessor, respectively, along the same path. The following observations can be made regarding paths of a multicast tree in WDM networks:

Two paths, $P(s, d_1)$ and $P(s, d_2)$, can share some link, l, without necessarily sharing the same wavelengths on l. This may be required if at one node along either one of the two paths, further conversions are no longer possible without violating the delay requirements.

If two paths share the same wavelength on a given link, l, then they share the same wavelength on $Pred(l)$. This is required to eliminate unnecessary wavelength conversions, which can only increase the cost and delays of the final multicast tree.

Based on the above, the CBMT problem can be formalized as follows:

Find

A multicast tree $T(V', E')$ and a wavelength assignment function, ∇_P for each each path P, such that

$$Cost(T) = \sum_{l \in E'} \sum_{\nu \in E\nabla(l)} \omega(l, \nu) + \sum_{l \in E'} \sum_{(\mu, \nu) \in \mathcal{W}_l} c_l(\mu, \nu) \qquad \text{is minimum} \qquad (1)$$

Subject to

$$Delay(P(s, d)) = \sum_{l \in P(s,d)} \delta_l(\nabla_P(Pred(l)), \nabla_P(l)) < \Delta_d \qquad \forall d \in D \qquad (2)$$

The CBMT problem can be viewed as a derivative of the Steiner minimal tree (SMT) problem in which the delay from the source to the destinations is bounded. Minimizing the tree cost is inherently NP-hard, while bounding the delay can be achieved in polynomial time. When the two parameters are combined, the problem is NP-hard and therefore has to be approached by a heuristic method. What makes the CBMT problem in WDM networks inherently more complex than in traditional networks is that path selection in WDM networks is compound with wavelength assignment along the path. A heuristic for WDM networks, which takes delay, cost, and wavelength assignments into consideration, is discussed next.

3.3 Multicast Heuristics for WDM Networks

Given that in a WDM network, the number of delay violations is expected to be low, an efficient approximation approach to the CBMT problem is first to build a close-to-optimal least-cost tree and deal with delay violations as a special case. In the following, we show how we transform the original graph representing the network into an expanded graph to facilitate route computation and wavelength assignment. We then describe three new heuristics that can be used to build a low-cost, bounded-delay multicast tree for WDM networks.

3.4 Graph Transformation

Route selection and wavelength assignment are difficult to achieve based on the original graph. This is mostly due to the fact that the cost metrics and the delay computation must be considered dynamically during the path selection process. To make path selection and delay computation more tractable, an expanded directed and weighted graph, $\mathcal{A}(\mathcal{M}, \mathcal{L}, \Omega)$, derived from the original graph, is used [31]. The expanded graph, $\mathcal{A}(\mathcal{M}, \mathcal{L}, \Omega)$, is defined as follows:

> $\mathcal{M}$ represents a set of *nodes*, (m, ν), where $m \in V$ is a node in the original graph, and $\nu \in \Omega$ is a wavelength in the network.
> $\mathcal{L}$ represents a set of links connecting pairs of nodes in M.
> Ω represents a set of all available wavelengths in the network.
> An edge between two nodes (m_1, ν) and (m_2, ν) exists in the expanded graph, $\mathcal{A}(\mathcal{M}, \mathcal{L}, \Omega)$, only if there exists a link l from node m_1 to node m_2 in the original graph, $G(V, E, \Omega)$, and the wavelength ν is available on link l.
> An edge between (m, ν_1) and (m, ν_2) exists only if wavelength conversion from ν_1 to ν_2 is available at node m.

The costs and delay associated with the links of the original graph, namely wavelength usage cost, wavelength conversion cost, and wavelength conversion delay, are directly mapped into the corresponding costs and delay of the links of the expanded graph. Figure 11a shows the expanded graph resulting from the original graph in Fig. 11b.

Using the expanded graph, we describe a new heuristic for multicast communication in WDM. The heuristic decouples the cost optimization from bounding the delay by first building a low-cost tree and then handling any delay violations that may occur in the tree.

3.5 Conversion Bound Multicast (CBM) Heuristic

The basic idea of the *CBM* is first to build a least-cost tree using a shortest path heuristic (*SPH*) approach [57]. Using the resulting least-cost tree, conversion bounds are checked for every multicast node in the tree, and violations are removed accordingly.

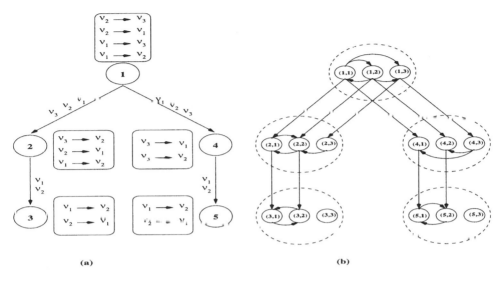

Figure 11 Graph transformation from (a) original to (b) auxiliary graph.

In the *SPH* heuristic, the link cost takes into consideration both the cost of wavelength conversion and the cost of using a specific wavelength on the link.

The *SPH* heuristic is similar to Prim's minimum spanning tree algorithm. The time complexity of *SPH* is $O(n^2 \log(n))$, where n is the number of nodes in the network. However, the need to compute the least-cost path at each stage raises

---

Algorithm 1 PROCEDURE Shortest Path Heuristic

INPUT:
 G: Network graph.
OUTPUT:
 T: Least-cost multicast tree.
Shortest Path Heuristic (G, T)
 Let $PATH(d, T)$ represent the least-cost path from a multicast node $d \in D$ to a
  tree T.
Step 1:
 Set $i = \text{i}$ and $M_1 = \{\emptyset\}$.
 Construct a subtree, $T_1 = \{s\}$, consisting of the source node s.
Step 2:
 Set $i = i + 1$.
 Find the least-cost node, d_i, to T_{i-1} such that $d_i \in (D - M_{i-1})$ (ties are broken
  arbitrarily).

Step 3:

Construct a new subtree, T_i, by adding all edges and nodes in $PATH(d_i, T_{i-1})$ to T_{i-1}.

Set $M_i = M_{i-1} \cup \{d_i\}$.

Step 4:

    if $|M_i| < |D|$

       Go to **Step 2**

    else

       DONE: $T = T_i$ is the final least-cost multicast tree.

    end if

---

the total complexity of the algorithm to $O(|Z| \times n^2)$, where $Z = D \cup \{s\}$ s is the source node and D is the set of multicast nodes.

The *CBM* heuristic uses the expanded graph, $\mathcal{A}(\mathcal{M},\mathcal{L},\Omega,)$ and invokes *SPH* to produce a least-cost multicast tree. The *SPH*, however, does not guarantee that the end-to-end delays on all paths of the least-cost tree meet the delay bounds required by the multicast nodes. To address potential delay bound violations, *CBM* identifies each node, v, whose delay exceeds its end-to-end delay bound, and replaces its current path with a new least-cost path that meets the end-to-end delay requirement of node v, if such a path exists. Notice, however, that adding the new path to the tree may cause nodes to have two incoming paths. This is undesirable, as it causes the same information to be transmitted twice to the same node over different paths.[3] To illustrate this case, consider Fig. 12, where S is a source node and M, R, and L are multicast nodes, which depicts the least-cost tree produced by SPH. Assume, however, that in this tree the path from S to R, namely $((S, E) (E, H), (H, R))$, does not satisfy the end-to-end delay requirement of R. *CBM* finds a new least-cost path, from S to R, which meets the delay-bound requirement of node R. The addition of this path to the tree causes R to have two incoming paths. *CBM* removes the path with the higher delay up to the closest intermediate node of degree three or greater, as depicted in Fig. 13 (link (H, R) is removed in this case). The resulting tree satisfies the delay requirements of all multicast nodes. The addition of this new path to the tree, however, causes an intermediate node, N in this case, to have two incoming paths. This is again undesirable, as node N receives the same information over two different paths. Furthermore, the presence of two incoming paths may unnecessarily increase the overall cost of the tree.

In general, the problem of intermediate or multicast nodes with two incoming paths can be resolved by removing the *relay-path* on the larger-delay path. A relay-path is characterized by the following two properties: (1) all internal

---

[3] It may be acceptable to have a node with two incoming paths if the main objective is to find a path from the source to all multicast nodes. The final multicast graph, however, will not be a tree.

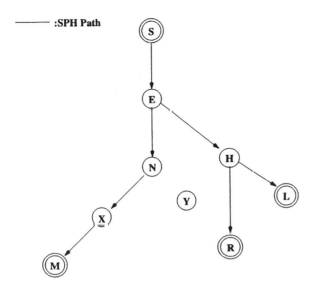

Figure 12   SPH low-cost multicast tree.

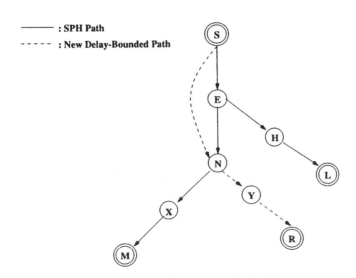

Figure 13   Resulting tree after removal of the path with larger delay.

nodes on the path are neither multicast nor source nodes and are exactly of degree two; (2) the end nodes of the relay-path are either a source node, a multicast node or a node of degree three or higher.[4] In Fig. 13, the relay-path, namely link (E, N), is redundant and may be removed. Notice, however, that because of potential wavelength conversions that can take place between the incoming link to node N on the new path and the outgoing links from node N along the paths to the multicast nodes attached to N, the removal of link (E, N) may not be feasible. For example, the removal of relay-path, (E, N), is only feasible if the end-to-end delay on the new path to multicast node M does not exceed M's delay bound.

Based on the above, one possible way to verify the feasibility of removing a *redundant relay-path* at a given node, N, is by checking that its removal does not cause the end-to-end delay requirement of any multicast node attached to N to be violated. This, however, may increase the time complexity of the algorithm, since it requires revisiting all nodes in the subtree attached to N. In the following, we derive a *sufficient* condition which can be used to verify the feasibility of removing a *redundant relay-path* without causing any multicast node to violate its end-to-end delay requirements. The condition involves quantities that can be obtained locally at the node and its verification can be achieved in a constant amount of time.

Consider an intermediate node, N, and let $D_N = \{d \in D \mid \exists P(N, d)\}$ represent the set of multicast nodes of the subtree rooted at node N, as depicted in Fig. 14. Let l_N, and l'_N, represent the incoming link to node N on $P(S, N)$ and $P'(S, N)$, respectively. We use v_N and v'_N to denote the respective wavelengths associated with l_N and l'_N. Further, denote the wavelength associated with the outgoing link, l_R, at node N along $P(S, R)$ through node X as v_R. Finally, let v_d, $\forall\, d \in D$, denote the wavelength associated with the outgoing link, l_d, at node N along $P(N, d)$, to multicast node d.

Assume, without loss of generality, that $P(S, R)$ meets the delay requirements of node R, while the delay on $P'(S, R)$, denoted as $Delay(P'(S, R))$, exceeds the delay bound of node R. Then the following expression holds:

$$Delay(P(S, N)) + c_{l_R}(v_N, v_R) < Delay(P'(S, N)) + c_{l_R}(v'_N, v_R) \qquad (3)$$

which can be rewritten as

$$Delay(P(S, N)) - Delay(P'(S, N)) < c_{l_R}(v'_N, v_R) - c_{l_R}(v_N, v_R) \qquad (4)$$

where $c_{l_R}(v_N, v_R)$ represents the conversion cost on link l_R along $P(S, R)$, and $c_{l_R}(v'_N, v_R)$ represent the conversion cost on link l_R along the concatenated path $P'(S, N \cup P(N, R)$. Notice, however, that the relay-path on $P'(S, N)$ can be

---

[4] Notice that according to the above definition, a relay-path can be composed only of one edge.

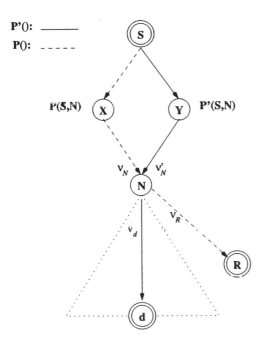

P'(): ———
P(): - - - -

P(S,N)

P'(S,N)

Figure 14 Redundant relay paths.

removed if

$$Delay(P(S, N)) + c_{l_d}(v_N, v_d) < Delay(P'(S, N)) + c_{l_d}(v'_N, v_d) \qquad (5)$$

which can be rewritten as

$$Delay(P(S, N)) - Delay(P'(S, N)) < c_{l_d}(v'_N, v_d) - c_{l_d}(v_N, v_d) \qquad (6)$$

where $c_{l_d}(v_N, v_d)$ and $c_{l_d}(v'_N, v_d)$ represent the conversion cost on link l_d along $P(S, d)$, and $P'(S, d)$, respectively. Combining Eq. (4) and Eq. (6), a sufficient condition for a relay-path removal can be expressed as

Relay-path $P'(S, N)$ can be removed if

$$c_{l_R}(v'_N, v_R) - c_{l_R}(v_N, v_R) \leq c_{l_d}(v'_N, v_d) - c_{l_d}(v_N, v_d) \qquad (7)$$

3.5.1 CBM Basic Steps

Using the *SPH* heuristic, *CBM* first builds a low-cost tree. It then handles any delay violations that may have occurred in the process of building the tree by finding a new delay-bounded path to the delay-bound violating node, if such a path exists. Finally, the *CBM* heuristic removes redundant relay-paths that may cause a node to have more than one incoming edge. The basic steps of the *CBM* heuristic are described in Algorithm 2.

Algorithm 2 PROCEDURE Conversion Bound Multicast

INPUT:
   $\mathcal{A}(\mathcal{M}, \mathcal{L}, \Omega)$: Extended graph.
OUTPUT:
   T: Minimal-cost, delay-bounded multicast tree.
Least Delay Replacement $(\mathcal{A}(\mathcal{M}, \mathcal{L}, \Omega), T)$
   Step 1:
      Find the minimum cost multicast tree, *T*, using *SPH*.
   Step 2:
      Let $P_\delta(s, d)$ be the end-to-end delay on *Path*(s, d) connecting node *s* to *d* on the tree.
      Find $W = \{ d \in D, | P_\delta (s, d) > \Delta_d \}$.
   Step 3:
      for each node $v \in W$
         Invoke the *Delay Constrained Path* procedure, *DCP*$(\mathcal{A}, T, v)$, described in Algorithm 3, to find the minimal-cost, delay-bounded path from *s* to *v*, whose addition to the tree and the removal of the resulting redundant relay-path do not cause any tree node to violate its delay bound.
         if such a path exists
            Attach node *v* to the tree using this new path and remove the redundant relay-path to node *v*;
         else
            Node *v* cannot be attached to tree. Exit.
         end if
      end for
   Step 4:
      Output $T = T_i$ as the final minimal-cost, delay-bound multicast tree.

The *DCP* procedure searches for the least-cost, delay-bounded path that connects the delay violating node, $v \in D$, to the current tree, *T*. This is achieved by building a new graph G', obtained by reversing the links of *G* while keeping all its nodes. *DCP* then builds a set of paths composed of the shortest delay paths from *v* to the tree nodes and the shortest cost paths from *v* to nodes in T', in that order. Prior to computing the shortest cost paths, *T*'s links are removed from G' to create different independent least-cost paths. This is necessary to avoid creating paths that are mutually derived from each other. Using this set of paths, *DCP* selects the path that meets the following three criteria: (1) the end-to-end delay on the path meets the delay bound of the multicast node, (2) the addition of this path to the tree and the removal of the redundant relay-path associated with *v* do not cause any other multicast node currently attached to the tree to violate its delay bounds, and (3) the path is the least-cost path that meets criteria (1) and (2). The basic steps of the *DCP* procedure are described in Algorithm 3.

The *CBM* heuristic uses the shortest delay paths, so it finds a solution if one exists. Furthermore, if there is no violation, or the number of violations is low, then the cost of the resulting tree is close to the cost of the *SPH* tree. The complexity of the algorithm is bounded by the complexity of *SPH*, which is $O(|Z| * n^2)$, where *n* is the total number of nodes in the graph. When there is a delay violation, then for each violating node the shortest delay path can be found in $O(n^2)$. In the worst case, we have $|D|$ violating nodes. Hence, the total time complexity of *CBM* is $O(|Z| * n^2)$.

Algorithm 3 PROCEDURE Delay Constrained Path

INPUT:
   G: Network graph.
   T: Current multicast tree.
   v: Delay-violating node.
OUTPUT:
   $p(u, v)$: Bounded-delay path, if one exists.
Delay Constrained Path (G, T, v)
   Create a new graph G' consisting of G's nodes and G's edges reversed.
   Using G', create P, the set of the least-cost delay paths, $p(u, v)$, from v to each
      node u in $T \in G'$.
   Remove the edges of T from G'.
   Using G', find the set of the least-cost paths from v to each node in T and add them
      to P.
   Out of all paths in P select path, $p(u, v)$, such that
      $Delay(p(u, v)) < \Delta_v$.
      $Cost(p(u, v))$ is minimum.
      Addition of $(p(u, v))$ to the tree only creates a redundant relay-path (i.e., the
         removal of the redundant relay-path does not cause any other node to violate its
         delay bound).
   if no such path exists **then**
      return ''no solution''
   else
      return $p(u, v)$.
   end if

---

4   CONCLUSION

In recent years, WDM networks have been the subject of research and development. Dense WDM technology, which uses minimum spacing between channels, can accommodate up to one hundred optical channels per fiber without coherent detection techniques. Furthermore, the ability to use tunable transmitters and receivers in WDM networks allows greater network flexibility, as channels can be allocated dynamically according to traffic requirements. Both of these advantages make WDM an attractive technology to support the anticipated traffic of the next generation Internet. One specific aspect of future Internet services is multicasting.

High-speed networking research efforts have concentrated on developing frameworks to support QoS guarantees for a wide range of real-time applications. A critical aspect of these frameworks is their ability to support routing and path establishment both for unicast and multicast communication. As the networks grow in size and user requirements increase in complexity the determination of the ''best'' route or multicast tree for the required level of QoS becomes a difficult task. Further, it is no longer sufficient to select routes with minimum cost; what is required is efficient mechanisms for building multicast trees that are not only of minimal cost but also support the application required level of QoS.

Given that WDM networks are fundamentally different from traditional networks when it comes to optimization objectives, it is imperative to explore the impact of the WDM networks' characteristics, in terms of wavelength availability, light-wave conversion cost, and link delays, on the path selection process, both for unicast and for multicast communication, and to understand the impact of the inaccuracy in the available network state information on the efficiency of the route establishment schemes.

This chapter addresses the above issues. First, the focus was on developing adaptive and global algorithms to augment connection establishment protocols in all-optical WDM networks with the capabilities of adapting to changing wavelength availability on the network's links. This adaptivity is accomplished by exploring alternate paths as well as alternate wavelengths dynamically during connection setup. The chapter then discussed algorithmic issues related to establishing trees of semilightpaths between a source node and a group of destination nodes and developed adequate heuristics for point-to-multipoint communications. A cost structure that takes into consideration both link cost and conversion delays was proposed. Based on this cost structure, an optimization problem was formulated and a heuristic to this problem was proposed.

Further research investigating multicast tree establishment protocols that can adapt to resource availability is required, and different options for dynamically adapting to resource availability need to be developed. This includes allowing wavelength conversion when permitted by the QoS requirement, switching to overflow routes, and selecting alternate routes dynamically. These options must be carefully addressed in order to produce efficient schemes that will not only produce nearly optimum paths but most importantly can be implementable in large WDM based optical networks.

REFERENCES

1. R. Barry and P. Humblet. Models of blocking probability in all-optical networks with and without wavelength changers. In *Proceedings of the IEEE InfoCom*, 402–412, 1995.

2. A. Birman. Computing approximate blocking probabilities for a class of all-optical networks. *IEEE J. on Selected Areas in Communication*, 13:852–857, June 1996.

3. I. Chlamtac, A. Ganz, and G. Karmi. Lightpath communications: an approach to high-bandwidth optical wans. *IEEE/ACM Trans. on Communications*, 40:1171–1182, July 1992.

4. I. Cidon, I. S. Gopal, and A. Segall. Connection establishment in high-speed networks. *IEEE/ACM Trans. on Networking*, 1(4):469–482, 1993.

5. P. Dowd and K. Sivalingam. A multi-level wdm access protocol for an optically interconnectred parallel computer. In *Proceedings of the IEEE InfoCom*, 400–408, 1994.

6. A. Aggarwal et al. Efficient routing and scheduling in optical networks. In *Proc. of the ACM-SIAM Symp. on Discrete Algorithms*, 412–423, 1993.

7. Awerbuch et al. Distributed control for paris. In *Proceedings of 9th Annual ACM Symp. on Principles of Distributed Computing*, 145–160, 1990.

8. I. Gidon and R. Rom. Multi-path routing combined with resource reservation. In *Proceedings of the IEEE InfoCom*, 1997.

9. A. Girard. *Routing and Dimensioning in Circuit-Switched Networks*. Addison-Wesley, 1990.

10. H. Harai, M. Murata and H. Miyahara. Performance analysis of wavelength assignment policies in all-optical networks with limited range wavelength conversions. In *IEEE JSAC*, 16(7):1051–1060, 1998.

11. E. Karasan and E. Ayanoglu. Effect of wavelength routing and selection algorithms on wavelength conversion gain in WDM optical networks. In *IEE/ACM Trans. on Networking*, 6(2):186–196, 1998.

12. M. Kovacevic and A. Acampora. On wavelength translation in all-optical networks. In *Proceedings of the IEEE InfoCom*, 413–422, 1995.

13. A. Mokhtar and M. Azizoglu. Adaptive wavelength routing in all-optical networks. *IEEE/ACM Trans. on Networking*, 6(2):197–206, 1998.

14. B. Mukherjee, D. Banerjee, S. Ramamurthy, and A. Mukherjee. Some principles for designing wide area optical networks. *IEEE/ACM Trans. on Networking*, 4(10):684–696, 1996.

15. S. Nugent. The iPSC/2 direct-connect communications technology. In *Proceedings of the 3rd Conference on Hypercube Concurrent Computers and Applications*, 1988.

16. R. Pankaj and R. Gallager. Wavelength requirements of all-optical networks. *IEEE/ACM Trans. on Networking*, 3(3):269–280, 1995.

17. C. Qiao and Y. Mei. Wavelength reservation under distributed control. In *IEEE/ LEOS Summer Topical Meeting: Broadband Optical Networks*, August 1996.

18. C. Qiao and R. Melhem. Reconfiguration with time division multiplexing MINs for multiprocessor communications. *IEEE Transactions on Parallel and Distributed Systems*, 5(4):337–352, 1994.

19. R. Ramaswami and A. Segall. Distributed network control for wavelength routed optical networks. In *Proceedings of the IEEE InfoCom*, 138–147, 1996.

20. R. Ramaswami and K. Sivarajan. Optimal routing and wavelength assignment in all-optical networks. In *Proceedings of the IEEE InfoCom*, 970–979, 1994.

21. B. Ramamurthy and B. Mukherjee. Wavelength conversion in WDM networking. In *IEEE JSAC*, 16(7):1061–1073, 1998.

22. V. Sharma and E. Varvarigos. Limited wavelength translation in all-optical WDM networks. *Proc. of Infocom 98*, 1998.

23. S. Subramanian, A. Somani, M. Azizoglu, and R. Barry. A performance model for wavelength conversion with non-Poisson traffic. In *Proceedings of the IEEE Info-Com*, 1997.

24. S. Subramaniam, M. Azizoglu, and A. K. Somani. Connectivity and sparse wavelength conversion in wavelength-routing networks. In *Proceedings of the IEEE Info-Com*, 148–155, 1996.

25. E. Varvarigos and V. Sharma. The ready-to-go virtual circuit protocol: a loss-free protocol for multigigabit networks using fifo buffers. *IEEE/ACM Trans. on Networking*, 5(5), 1997.

26. X. Yuan, R. Gupta, and R. Melhem. Distributed control in optical wdm networks. In *Proceedings of the IEEE MILCOM*, 100–104, 1996.

27. X. Yuan, R. Melhem, and R. Gupta. Distributed path reservation algorithms for multiplexed all-optical interconnection networks. In *Proceedings of the IEEE Symp. High Performance Computer Architecture*, 1997.

28. T. Ballardie, P. Francis, and J. Crowcroft. Core-based trees (cgt) an architecture for scalable inter-domain multicast routing. *Computer Communication Review*, 23(4): 85–95, 1993.

29. A. Barry and S. Subramanian. The max_sum wavelength assignment algorithm for wdm ring networks. In *Proceedings of OFC'97*, Feb. 1997.

30. B. Beauquier, J. C. Gargano, S. Pereness, P. Hall, and U. Vaccaro. Graph problems arising from wavelength routing in all. *Proceedings of the 2nd Workshop on Optics and Computer Science*, 1997.

31. I. Chlamtac, A. Faragó, and T. Zhang. Lightpath (wavelength) routing in large wdm networks. *IEEE/ACM Journal on Selected Areas in Communications*, 14:909–913, 1996.

32. H. Choi, H. A. Choi, and M. Azizoglu. Efficient scheduling transmissions in optical broadcast networks. *IEEE/ACM Transactions on Networking*, Dec. 1996.

33. E. W. Dijkstra. A note on two problems in connection with graphs. *Numeriskche Mathematik*, 1:269–271, 1959.

34. P. Dowd and K. Sivalingam. A multi-level wdm access protocol for an optically interconnected parallel computer. *Proceedings of IEEE INFOCOM'94*, 1, June 1994.

35. T. Erlebach and K. Janson. Scheduling of virtual connections in fast networks. *Proc. of the 4th Workshop on Parallel Systems and Algorithms*, 13–32, 1996.

36. D. Estrin, D. Farinacci, A. Helmy, D. Thaler, S. Deering, M. Handley, V. Jacobson, C. Liu, P. Sharma, and L. Wei. Protocol independent multicast-sparse mode (PIM-SM): protocol specification. *Internet RFC 2117*, http://ds.internic.net/rfc/rfc2117.txt, June 1997.

37. M. Feldman, S. Esener, C. Guest, and S. Lee. Comparison between optical and electrical interconnects based on power and speed considerations. *Applied Optics*, 29: 1742–1751, 1990.

38. G. Gravenstreter, R. G. Melhem, D. M. Chiarulli, S. P. Levitan, and J. P. Teza. The partitioned optical passive stars (POPS) topology. *IEEE Proceedings of the Ninth International Parallel Processing Symposium*, April 1995.

39. V. Kompella, J. Pasquale, and G. Polyzos. Multicast routing for multimedia communication. *IEEE/ACM Transactions on Networking*, 1(3):286–292, June 1993.

40. L. Kou, G. Markowsky, and L. Berman. A fast algorithm for Steiner trees. *Acta Informatica*, 15:141–145, 1981.

41. M. Kovaceivc and A. S. Acampora. Benefits of wavelength translation in all-optical clear channel networks. *IEEE Journal on Selected Areas in Communication*, 14(5): 868–880, June 1996.

42. P. Lalwaney, L. Zenou, A. Ganz, and I. Koren. Optical interconnects for multiprocessors: cost performance analysis. *Proceedings of the Symposium on Frontiers of Massively Parallel Computation*, 278–285, 1992.

43. W. Liang and H. Shen. Efficient multiple multicast in wdm networks. *Proceedings of the 1998 Intern. Conf. on Parallel and Distributed Processing Techniques and Applications*, 1998.

44. W. Liang and H. Shen. Multicast broadcasting in large wdm networks. *Proceedings of the 12th Intern. Conf. on Parallel Processing Symposium*, 1998.

45. J. Moy. Multicast routing extensions for OSPF. *Communications of the ACM*, 37(8): 61–66, August 1994.

46. B. Mukherjee. WDM-based local lightwave networks, Part I: single-hop systems. *IEEE Networks*, 12–27, May 1992.

47. Y. Ofek and B. Yener. Reliable concurrent multicast from bursty sources. *Proceedings of IEEE Infocom'96*, 1433–1441, 1996.

48. P. Prucnal, I. Glesk, and J. Sokoloff. Demonstration of all-optical self-clocked demultiplexing of tdm data at 250 gb/s. *Proceedings of the Int. Workshop on Massively Parallel Processing Using Optical Interconnections*, 106–117, 1994.

49. T. Pusateri. Distance vector multicast routing protocol. *Internet Draft*, Feb. 1997.

50. B. Rajagopalan, D. Pendarakis, D. Saha, R. S. Ramamoorthy, and K. Bala. Ip over optical networks: architectural aspects. *IEEE Communication Magazine*, 38(9):94–103, Sept. 2000.

51. B. Ramamurthy and B. Mukherjee. Wavelength conversion in wdm networking. *IEEE J. on Selected Areas in Communications*, 16(7), July 1998.

52. S. Ramanathan. Multicast tree generation in networks with asymmetric links. *IEEE/ACM Transactions on Networking*, 4(4):558–568, Nov. 1996.

53. V. J. Rayward-Smith and A. Clare. On finding Steiner vertices. *Networks*, 16:283–294, 1986.

54. A. Somani and B. Ramamurthy. Issue on optical communication networks for the next-generation internet. *IEEE Network*, 14(6), Nov./Dec. 2000.

55. J. Strand, A. L. Chui, and R. Tkach. Issues for routing in the optical layer. *IEEE Communication Magazine*, 39(2):81–88, Feb. 2001.

56. Q. Sun and H. Langendoerfer. Efficient multicast routing for delay sensitive applications. *Proceedings of Second Workshop Protocols Multimedia Systems*, 452–458, April 1995.

57. H. Takahashi and A. Matsuyama. An approximate solution for the Steiner problem in graphs. *Mathematica Japonica*, 24:573–577, 1980.

58. B. M. Waxman. Routing of multipoint connections. *IEEE Journal on Selected Areas in Communications*, 6(9):1611–1622, Dec. 1988.

7
Routing in Multihop Optical WDM Networks with Limited Wavelength Conversion

Hong Shen and Susumu Horiguchi
Japan Advanced Institute of Science and Technology, Tatsunokuchi, Japan

Yi Pan
Georgia State University, Atlanta, Georgia

1 INTRODUCTION

An emerging key technology in communications networks is optical networks that deliver promises for various applications that require data transmission rates much higher than traditional electronic networks can provide [1,29,39,32]. Optical networks are often implemented by *Wavelength-Division Multiplexing* (WDM) [40], which divides the optical spectrum fiber-optically into many channels, each corresponding to a different optical wavelength, and thus allows multiple laser beams carrying different data streams to be transferred concurrently along a single fiber-optic provided that each beam uses a distinct wavelength. All nodes (stations) in a WDM network are interconnected by point-to-point fiber-optic links, where each link can support a certain number of wavelengths. At each node a set of input ports receiving incoming data and output ports delivering outgoing data are attached. A WDM network allows an incoming signal to be routed to one or more output ports, but not multiple signals simultaneously to the same output port on the same wavelength. Multiple incoming signals are allowed to use the same output wavelength in the same output port at different times through a queue (buffer) [34]. In *switched* (also known as *reconfigurable*) *multihop* WDM networks, signals on input ports at each node are routed to appropriate output ports directing to their destinations via a set of switches, and wavelength conversion during the course of transmission may happen at some intermediate nodes on the communication path. At each of these intermediate nodes, the signal is converted from optic form to electronic form and then retransmitted on another wavelength (converted from electronic to optic). In multihop networks, a communication path between a source–destination pair is called a *semilightpath* [7], which is obtained by establishing and chaining several lightpaths together.

Routing in a WDM network requires one to set up a communication path between each pair of given nodes by chaining a set of optical channels together, with all channels on each path being assigned a number of wavelengths, and channels of different paths sharing the same optical link having different wavelengths. An important topic of practical significance is to set up minimum cost paths for fastest routing of a set of requests when the available wavelengths in a network are fixed and given by the network. In order to solve various application problems on a WDM network, mechanisms must be developed to handle not only point-to-point routing but also group (collective) communication involving transporting information from a group of nodes to another group of nodes in the network. A typical group communication is *multicast*, which transports information from one source node to a set of destination nodes. A more general version of group communication is *multiple multicast*, which contains more than one multicast group, each having its own source node and destination set [33]. *Off-line* routing constructs all paths before routing actually takes place, whereas *on-line* routing is carried out simultaneously while the underlying path is being constructed. Routing can be carried out on a WDM network that is either *reliable*, where no faults can occur, or *unreliable*, if it allows the existence of hardware faults including optical channel and wavelength conversion faults. There is a considerable body of work for routing in both single-hop (all-optical) and multihop optical networks [1,2,3,6,10,28,30,31,11,14,17,25,7,20,41]. While most work is for off-line point-to-point routing, research on group communication, particularly multicast, and on-line routing has become active recently [21, 22,26,27,32,35,34,36,37].

This chapter gives an overview on some new results on efficient routing in multihop optical WDM networks with limited wavelength conversion. Section 2 is an introduction to the new routing cost model proposed recently, which is more general than other existing models. This model takes into consideration not only the cost of wavelength access and conversion but also the delay for queuing signals arriving at different input channels that share the same output channel at the same node. Section 3 presents a set of algorithms for efficient off-line communication in reliable WDM networks. Section 4 presents algorithms for efficient off-line communication in unreliable WDM networks. Sections 5 and 6 present algorithms for on-line communication in reliable and unreliable networks, respectively. Section 7 addresses dynamic group membership maintenance for on-line communication. The contents of this chapter are extracted from the results of [34] for Secs. 2–4, of [36] for Secs. 5 and 6, and of [37] for Sec. 7.

2 A GENERAL ROUTING COST MODEL

We first describe the model of multihop WDM network with limited wavelength conversion and output signal queuing proposed in [34]. For a network of n nodes using k wavelengths, a node v contains a set of n' input ports and n'' output ports. Each input port contains a dedicated electronic *receiver* with buffering capability for each incoming (signal) wavelength, which converts the signal from optical to electronic, and each output port contains a dedicated laser *transmitter* for each outgoing wavelength that converts the signal from electronic to optical on that wavelength and transmits the optical signal. To handle the situation of the simultaneous arrival of multiple incoming signals routed to the same optical channel (wavelength) in the same output port, we further assume that these signals are queued in a (electronic) buffer before they are sent to the corresponding transmitter at the output port one by one. There are n'' sets of buffers, where set i is dedicated to output port i and contains $o_i(v)$ buffers if there are $o_i(v)$ wavelengths available on output port i. Incoming signals are switched to the buffers through a crossbar-like switch after the receivers. The switch is dynamically set according to the wavelength conversion cost from incoming wavelength to outgoing wavelength and the length of the queue on the outgoing wavelength for the purpose of minimizing the total waiting time of the signal at the output port. That is, a signal with incoming wavelength λ_i chooses outgoing wavelength λ_j so that $\min_j \{c_v(\lambda_i, \lambda_j) + q_j(e)\}$, where $q_j(e)$ is the length of queue j on channel (edge) e in the output port. Figure 1 shows the procedure of wavelength conversion at a node.

The above model uses a combination of *circuit switching*, to reserve each communication path for each entire session when it is established, and *packet switching*, to transmit physically the messages in each buffer along these paths.

Let $\Gamma = \{\lambda_1, \lambda_2, \ldots, \lambda_k\}$ be the set of available wavelengths in a WDM network. A WDM network can be represented by a directed graph $G = (V, E, \Gamma)$ with $|V| = n$ and $|E| = m$, where $\Gamma = \{\Gamma_{e0}, \Gamma_{e_1}, \ldots, \Gamma_{e_{m-1}}\}$, $\Gamma_e \subseteq \Gamma$, is the set of wavelengths available at edge $e \in E$ with $w(e, \lambda)$ associated with wavelength λ as the cost required to access λ. Converting a particular (incoming) wavelength (λ_i) to another (outgoing) wavelength (λ_j) at node v causes a fixed cost $c_v(\lambda_i, \lambda_j)$ for all available λ_j on all outgoing edges, where $c_v(\lambda_i, \lambda_i) = 0$ indicates no wavelength conversion is incurred.

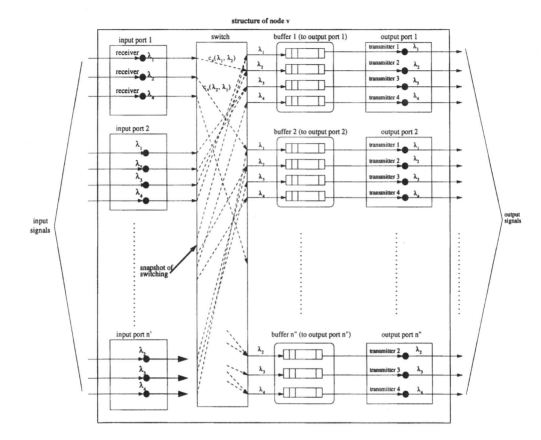

Figure 1 Wavelength conversion at a node.

Let P be a semilightpath connecting a pair of nodes in the network to fulfill a point-to-point routing request. Clearly P consists of a sequence of optical channels $e_1, e_2, \ldots, e_l$, where e_i carries wavelength λ_{pi}, $1 \leq p_i \leq k$. All channels e_1, $e_2, \ldots, e_l$ are chained together so that the tail of e_{i+1}, $t(e_{i+1})$, coincides with the head of e_i, $h(e_i)$, for all $1 \leq i < l$.

For point-to-point routing, the cost structure for constructing semilightpath P is defined as [34]

$$C(\mathcal{P}) = \sum_{i=1}^{l} w(e_i, \lambda_{p_i}) + \sum_{i=1}^{l-1} (c_{h(e_i)}(\lambda_{p_i}, \lambda_{p_{i+1}}) + d_{\lambda_{p_{i+1}}}(e_i)) \tag{1}$$

For multicast, we are required to construct a *multicast tree MT* rooted at *s* that connects all destination nodes within the multicast group. Assume that $\{e_1, e_2, \ldots, e_{|MT|}\}$ is the sequence of edges obtained by left-first traversal (left-visit-right) on *MT* that enumerates semilightpaths in *MT*, and *L* is the set of leaf nodes in *MT*. We denote the queuing delay for transmitting any incoming signal using wavelength λ on edge *e* by $d_\lambda(e)$, which is proportional to the number of signals in the queue that buffers this signal, that is, the queue length. All signals in the same queue will follow the same queuing delay because signals are transmitted in packet switching along the optical channel of wavelength λ on edge *e*. The following cost model $C(MT)$ for traversing *MT* was defined in [34]:

$$C(MT) = \sum_{i=1}^{|MT|} w(e_i, \lambda_{p_i}) + \sum_{1 \le i \le |MT|, h(e_i) \notin L} (c_{h(e_i)}(\lambda_{p_i}, \lambda_{p_{i+1}}) + d_{\lambda_{p_{i+1}}}(e_i)) \qquad (2)$$

To support routing, the following method was proposed in [34] that transforms $G = (V, E, \Gamma)$ into another auxiliary $G_M = (V_M, E_M)$.

Let $\delta[1..k](e)$ represent the queue I/O delays (time) required for queuing all incoming signals on link *e*, where $\delta[i](e)$ is the queue I/O delay for incoming signals using output wavelength λ_i on *e*, which is mainly defined by the speed of the underlying buffer of the queue. Here we consider the general case that different incoming wavelengths may have different queuing delays subject to the length of the queue and speed of the buffer for each wavelength.

Call all original nodes in *V node*, all auxiliary nodes in G_M *vertex*, optical channels on all links in *E* and auxiliary edges in G_M *edge*. G_M is a directed and weighted graph with both fixed edge weights and dynamically changing edge weights with initial value zero.

1.   For each $v \in V$, construct a bipartite graph $G_v = (A_v \cup B_v, E_v)$, where vertex sets A_v and B_v represent the input wavelengths and output wavelengths at $v \in V$, and E_v represents all possible wavelength conversions at v—$(a \in A_v, b \in B_v) \in E_v$ iff wavelength *a* can be converted to wavelength *b* at *v* (i.e., $c_v(a, b)$ exists). Set $c_v(a, b) = 0$ if $a = b$. Assign weight $c_v(a, b)$ to edge (a, b). Connect all vertices in A_v to *v* through introducing *k* new edges E'_v. Assign zero weight to each of these new edges. Vertices in B_v are connected to the appropriate nodes in *V* by edges transformed from links in *E* described in the following step.

2.   Replace each $e \in E$ with $|\Gamma_e| \le k$ parallel edges (channels), E_e. For each $e' \in E_e$ carrying wavelength λ_i, assign edge weight $w(e, \lambda_i)$ to it. These edges connect vertices in B_v to the corresponding vertices in $A_{v'}$ of *v*'s neighbor(s) v'.

3.   Assign an edge weight $d_{\lambda i}(e')$, initially zero, to edge e' (representing outgoing wavelength λ_i) in E_e, indicating the queuing delay for sending message from $h(e') \in B_v$ to $t(e') \in A'_v$ using wavelength λ_i. This edge weight is dynamically changing—it is increased by a queuing delay $\delta[i](e)$ when an incoming signal arrives at the queue for this wavelength.

4. Let $V_M = \bigcup_{v \in V} A_v \cup B_v$, and $E_M = (\bigcup_{v \in V}(E_v \cup E'_v)) \cup (\bigcup_{e \in E} E_e)$.

Since $|A_v| = |B_v| = k$, $|E_v| \leq k^2$ and $|E_e| \leq k$, we can easily obtain the equations

$$|V_M| \leq 2kn \tag{3}$$

$$|E_M| \leq k^2 n + km \tag{4}$$

An example of G_M is given in Fig. 2.

For general G_M, which is a directed graph, we define the (edge) *connectivity* of G_M to be the minimal number of edge-disjoint directed paths from any node to any other node in G_M.

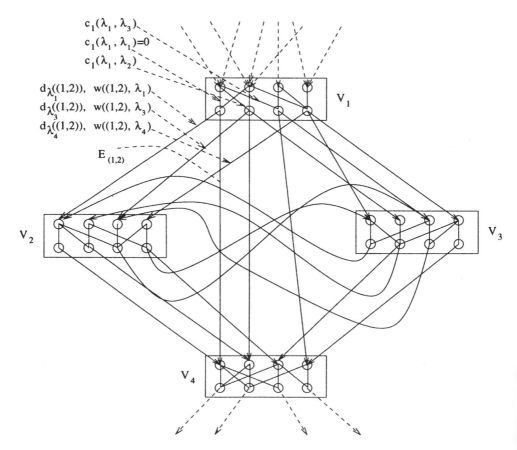

Figure 2 An example of a component of $n = 4$, $m = 7$, and $k = 4$ in G_M.

We equivalently say that G_M is t-edge connected if G_M has a connectivity of t. Whenever appropriate, we use *weight* and *cost* interchangeably.

3 OFF-LINE ROUTING IN RELIABLE NETWORKS

In this section we study routing in a reliable WDM network. We present a set of efficient algorithms for point-to-point requests, multicast, and multiple multicast, respectively, on the proposed cost models of Eqs. (1) and (2).

3.1 Multiple Point-to-Point Routing

Multiple point-to-point routing is defined thus: given r requests $\mathcal{R} = \{(s_1, t_1), (s_2, t_2), \ldots, (s_r, t_r)\}$, we are required to establish a communication path from s_i to t_i for each request (s_i, t_i), where $1 \le i \le r$ and r is smaller than the connectivity of G_M. Let SP_i be the shortest path for request (s_i, t_i) in G_M without considering the queuing delay caused by the existence of SP_j, $1 \le i \ne j \le r$. When queuing delay is taken into account, an optimal solution producing minimum total cost of all the paths needs to compute the alternative shortest paths of SP_1, $SP_2, \ldots, SP_r$ in G_M without passing through any edge in $E(SP_{i_1}) \cap E(SP_{i_2}) \cap \ldots \cap E(SP_{i_q})$ for all different combinations of $1 \le i_1 \ne i_2 \ne \ldots \ne i_q \le r$ and $2 \le q \le r$, and then take those to replace their corresponding SP's if such replacement results in the minimum total cost for all the r paths. Clearly the total number of these alternative paths is prohibitively large for large r, so it is unrealistic to expect an optimal solution. We use a greedy approach to design our algorithm that employs a modified shortest-path algorithm to construct paths one by one in their length-increasing order, where the length of a path is the number of edges on the path. During each step of including an edge to a path, the weight of the edge is increased by an appropriate additive factor of δ to reflect the increment of queuing delay for all signals in the same queue when being transmitted.

Let the index of the ith path in length-increasing order be π_i. For any edge e on path P under construction in G_M, we call the probability of e falling on an edge occupied by other paths the *edge overlapping probability* of e. We define the *path overlapping probability* of P to be the average edge overlapping probability that equals the aggregated edge overlapping probabilities of all edges on P divided by the length of P. Path overlapping probability shows how likely edges on the path are to fall on occupied edges in G_M and consequently gives an indication on the quality of the path, that is, the ratio of its total cost versus its optimal cost. It was shown [34] that the path overlapping probability of SP_i, $SP_2, \ldots, SP_r$ is minimized if we establish paths in the order of $SP_{\pi_1}, SP_{\pi_2}, \ldots, SP_{\pi_r}$, and hence the proposed heuristic is optimal in this regard in the expected case when every edge in G_M has an equal probability to be used by all shortest paths.

Once all paths are established, signals are sent along the paths using packet switching, observing the delays calculated in path establishment. Our modified shortest path algorithm for path establishment takes into consideration the dynamic edge weights for edges on the path under construction and uses *source routing* in which path establishment is initiated by the source node and the path is extended step by step from one node to the next toward the destination. Source routing can be viewed as a limited version of centralized routing without global central control. Our multiple point-to-point routing algorithm has the following structure.

Algorithm **A.1**
{*Construct a shortest path for each request (s_i, t_i), $1 \leq i \leq r$.*}
1. **for** i: $= 1$ **to** r **do**
   Use Dijkstra's algorithm to compute the length L_i of s_i-t_i shortest path;
   Insert L_i into the sorted list of increasing order $\{L_{\pi_1}, L_{\pi_2}, \ldots, L_{\pi_i}\}$;
2. **for** i: $= 1$ **to** r **do**
   Find the shortest path from $s_{\pi i}$ to $t_{\pi i}$ using source routing,
   where for any edge (channel) e using wavelength λ_j the distance from $h(e)$
   to $t(e)$ is $w(e, \lambda_j) + c_{h(e)} (\lambda'_j, \lambda_j) + d_{\lambda_j}(e)$;
   {*λ'_j is the wavelength on the preceding edge of the path from which λ_j is converted.*}
   Add $\delta[j](e)$ to $d_{\lambda_j}(e)$ for each e of wavelength λ_j on the path constructed;
   {*Increase the queuing delay of all signals in the same queue by a prespecified I/O cost, where all queuing delays are initialized to 0.*}

In Algorithm **A.1**, the shortest path from s_{π_i} to t_{π_i}, when all paths from s_{π_j} to t_{π_j}, have been established for all $j < i$, is constructed using a modified Dijkstra's algorithm [9] with the change in distance formula to include dynamic edge weight as stated above. At the first step of routing, source (s_{π_i}) will pick up a channel (wavelength) to send the signal, so there is only channel cost at the source node. Wavelength conversion may occur in all subsequent steps. Using adjacent lists to represent G_M, Dijkstra's algorithms can be implemented in $O(|E_M|+|V_M|\log|V_M|)$ time. By Eqs. (3) and (4) we have the following theorem:

Theorem 1 *The problem of multiple point-to-point routing for r requests in a WDM network of n nodes and m links with k available wavelengths can be solved in $O(rk(kn + m + n \log(kn)))$ time.*

3.2 Multicast

Multicast requires to transport information from source s to a set of destinations $D = \{t_1, t_2, \ldots, t_g\}$. Multicast can be realized by first constructing a multicast tree MT rooted at s including all nodes $\{t_1, t_2, \ldots, t_g\}$ in G, and then transmitting information from the root to all destinations along the tree edges using appropriate wavelengths. We are interested in finding an optimal MT in which the total cost for multicast is minimum. Putting the dynamic edge weight aside, when G is transformed into G_M, it is clear that finding an optimal MT is equivalent to finding a minimum directed Steiner tree in G_M, which is unfortunately NP-complete. Our generalized cost in Eq. (2) requires us to consider both static and dynamic edge weights in calculating the cost, which is even harder than the minimum directed Steiner tree problem on G_M. We therefore have to turn to approximate solutions that produce an MT whose cost is close to that of an optimal MT. We use an approach based on that of [16] to construct a *minimum spanning tree* (MST) rooted at s on an induced graph $I(\{s\} \cup D)$ of G_M to approximate the Steiner tree in G_M, where $I(\{s\} \cup D)$ is a complete graph on vertex set $\{s\} \cup D$ with the distance (length of shortest path) being the edge weight between any pair of vertices. Note that the algorithm of [16] works only for undirected graphs and may not work for general directed graphs, as $I(\{s\} \cup D)$ may not exist for general directed graphs. However, due to the special properties of G_M, we can make this approach work also for our case. Since our graph G_M is a special directed graph in which there is a path between any pair of nodes v_i, $v_j \in V$ (not those in A_v and B_v), the induced graph $I(\{s\} \cup D)$ exists and is a completely directed graph. We can therefore apply the approach of [16] for approximate multicast in our case as follows. We construct $I(\{s\} \cup D)$, which is a completely directed graph with vertex set $\{s\} \cup D$ and edge weight dist (u, v) in G_M for all u, $v \in \{s\} \cup D$. After we have constructed $I(\{s\} \cup D)$, we find the directed MST rooted at s instead of the undirected MST in the undirected case [16]. Because we are seeking for a Steiner tree in G_M rooted at s covering D, we can use the approximation ratio of the undirected MST on $I(\{s\} \cup D)$ to the undirected Steiner tree on $\{s\} \cup D$ to estimate that of the directed MST rooted at s to the directed Steiner tree on $\{s\} \cup D$. The directed MST rooted at s in $I(\{s\} \cup D)$ can be constructed as follows. Extend a most economic path from s to every node in D one by one, where the most economic path adds a least weight edge to the MST under construction. It expands the MST from originally only containing s to finally covering all nodes in D by repeatedly adding an edge of direction outwards from the MST with the least weight in the neighborhood of the MST. This construction can be completed in $O(|I(\{s\} \cup D)|^2) = O(g^2)$ time, because each step needs to consider at most g neighbors. It was shown [34] that the MST constructed by the above method is the minimum cost-directed spanning tree that connects s to every other node in $I(\{s\} \cup D)$.

With the help of the above algorithm, we can now present our algorithm for multicast in the WDM network. Let dist(u, v) be the shortest distance from u to v in G_M that is the summed edge weight on the shortest path from u to v. We also keep the shortest path corresponding to dist(u, v) in $P[u, v]$ accordingly.

The induced graph $I(\{s\} \cup D)$ is a complete graph on $g + 1$ nodes ($\{s\} \cup D$) with cost dist(u, v) associated with edge (u, v). The algorithm is:

Algorithm **B.1**
{*Multicast routing for $\mathcal{M} = (s, D)$, where $D = \{t_1, t_2, \ldots, t_g\}$.*}
1. **for** each ordered pair of $u, v \in \{s, t_1, t_2, \ldots, t_g\}$ **do**
   Compute the shortest path from u to v, $P[u \to v]$, and dist(u, v) in G_M using a modified Dijkstra algorithm to include dynamic edge weight updating as used for point-to-point source routing;
   For each $e \in P[u \to v]$ add $\delta[j](t(e))$ to $d_{\lambda_j}(e)$ if e uses wavelength λ_j;
2. Construct $I(\{s\} \cup D)$;
3. Compute the MST_I rooted at s in $I(\{s\} \cup D)$ using the algorithm described before;
4. Replace each edge in MST_I with the corresponding path in G_M, that is dist (u, v) with $P[u \to v]$, and break all cycles at their maximum weighted edges (removal) so that the resulting subgraph is a Steiner tree ST;
5. For each edge e of wavelength λ_j in ST, add $\delta[j](e)$ to $d_{\lambda_j}(e)$.
{*Increase the queuing delay of all signals in the same queue by a prespecified I/O cost.*}

For a completely directed graph $G_I = I(\{s\} \cup D)$, let ST be the Steiner tree constructed in G_M and MST be the minimum spanning trees constructed by Algorithm B.1, both rooted at s. We have the following lemma showing their approximation ratio.

Lemma 1 [34] *Approximating ST on g destination nodes by MST using Algorithm* **B.1** *has the following approximation ratio in the expected case when all edge weights in G_M are randomly and independently chosen from a fixed range. That is,*

$$\frac{\text{Cost}(MST)}{\text{Cost}(ST)} \leq 2 - \frac{2}{g + 1} \tag{5}$$

From Lemma 1, we know that the MST_I obtained by Algorithm **B.1** is $(2 - 2/g)$-OPT. Since $|V_M| = 2kn$ and $|E_M| = k^2n + km$ by Eqs. (3) and (4), Step 1 of the algorithm requires $O(g^2k^2n + g^2 km + g^2kn \log (kn))$ time. Steps 2 and 3 can be done in $O(g^2)$ time. Step 4 requires $O(gkn)$ time. Therefore we have the following theorem:

Theorem 2 *A $(2 - 2/g)$-OPT approximate multicast tree for multicast of a group of size g in a WDM network of n nodes and m links can be computed in $O(g^2k(kn + m + n \log(kn)))$ time in the expected case, where k is the number of available wavelengths in the network.*

Note that after Step 4 replacement of each dist (u, v) with its corresponding path $P[u \rightarrow v]$, MST_I may contain $|V_M|$ nodes because all these shortest paths may span over the entire G_M.

3.3 Multiple Multicast

When several groups of multicast wish to take place concurrently, a more general communication pattern, namely *multiple multicast*, is formed. Given r groups of multicast $\mathcal{M}_i = (s_i, D_i)$, where s_i is a source and $D_i = \{t_i^1, \ldots, t_i^{g_i}\}$ are the destinations, $1 \leq i \leq r$, r smaller than the connectivity of G_M, assume that $\mathcal{M}_i$ alone (without considering the existence of other groups) can be realized by a multicast tree MT_i. Let multicast forest $MF = \cup MT_i$. It is clear that several edges of different MT_i in MF may fall onto the same edge of G_M and hence attempt to use the same wavelength at the same node in the network. This will possibly cause *contention* on a particular wavelength when these requests arrive simultaneously at a node. Figure 3 shows an example of wavelength contention caused by three multicast trees.

While wavelength contention is forbidden in the most conventional optical models, our optical model in Fig. 1 does allow it to happen by buffering all signals using the same wavelength on the same physical link in a queue and then transmitting them out in packet switching in different time slots. However, as with multiple point-to-point routing, in order to result in a minimal cost MF we need to minimize the aggregated wavelength contention probability on all optical channels. Clearly, wavelength contention probability on an optical channel in G is the edge overlapping probability on that channel's corresponding edge in G_M. Therefore we take the same greedy approach as for multiple point-to-point routing, that is, find an approximate optimal multicast tree for each multicast MT_i one by one employing Algorithm B.1 in their size-increasing order. From Sec.

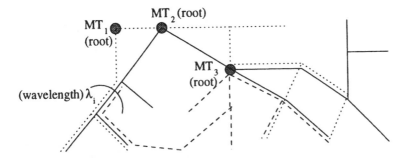

Figure 3 Wavelength contention caused by three multicast trees.

3.1 we know that the above approach will minimize the tree overlapping probability, which is the average edge overlapping probability over all edges in the tree, for all trees in MF in the expected case when every edge in G_M has an equal probability to be used by all the trees. Our algorithm for multiple multicast is described as follows:

Algorithm **C.1**
{*Multiple multicast for $\mathcal{M}_1, \mathcal{M}_2, \ldots, \mathcal{M}_r$, where $\mathcal{M}_i = (s_i, D_i)$.*}
1. Sort $\{\mathcal{M}_1, \mathcal{M}_2, \ldots, \mathcal{M}_r\}$ into increasing size order $\{\mathcal{M}_{\pi_1}, \mathcal{M}_{\pi_2}, \ldots, \mathcal{M}_{\pi_r}\}$.
2. **for** $i = 1$ **to** r **do**
   Construct multicast tree MT_{π_i} for $\mathcal{M}_i$ using Algorithm **B.1**.

The correctness of the algorithm is seen clearly from the greedy approach. The time complexity of the algorithm is $O\,(r \log r + \sum_{i=1}^{r} t_{MTi})$, where t_{MT_i} is the time complexity required for constructing the multicast tree for $\mathcal{M}_i$. With our result for multicast in the previous section we have the following theorem.

Theorem 3 *The problem of multiple multicast for r groups of sizes $g_1, g_2, \ldots, g_r$, respectively, in a WDM network can be solved in $O((\sum_{i=1}^{r} g_i^2)k(kn + m + n \log(kn)))$ time, where n, m, and k are the number of nodes, links, and available wavelengths in the network, respectively.*

The probability of edges of MT_{π_j} falling to those of MT_{π_i} is the probability of wavelength contention and hence queuing delay increase caused by MT_{π_i} and MT_{π_j} both wanting to access the wavelength represented by this edge (channel). As shown in Sec. 3.1, in the expected case when all edges in G_M have an equal probability to be used by all multicast trees, our heuristic is optimal in the sense that it minimizes the average probability of edges of MT_{π_j} falling to those of MT_{π_i} for $j > i$. Therefore Algorithm C.1 has the same approximation ratio as Algorithm B.1 in the above expected case.

4 OFF-LINE ROUTING IN UNRELIABLE NETWORKS

In this section we consider the routing problem in an unreliable WDM network in which both optical channel (wavelength) and wavelength conversion faults may occur. The optical channel fault occurs in cases such as when the designated wavelength on the channel is accidentally lost, distorted, or insufficiently amplified. The wavelength conversion fault occurs when the corresponding wavelength conversion within a node cannot be completed correctly due to a hardware fault in the receiver or switch. We call the faulty hardware blocking of a single wavelength conversion a *faulty gate*. From our optical model described in Sec. 2 we know that a channel of any link and a wavelength conversion in the WDM network G corresponds to an edge in G_M, respectively. By transforming G into G_M, we have effectively converted the channel faults and wavelength conversion faults in the WDM network into only edge faults in G_M. We give in this section a set of efficient algorithms for point-to-point requests, multicast, and multiple multicast on our cost models of Eqs. (1) and (2) in an unreliable WDM network. We assume that G_M in this section is $(f + 1)$-edge connected so that any f faulty edges of the same direction at one node will not disconnect G_M.

4.1 Multiple Point-to-Point Routing

We consider routing in an unreliable WDM network with f edge faults in G_M. Let $F = \{e_1^*, e_2^*, \ldots, e_f^*\}$ be the set of edges that are faulty. The whole process of routing consists of the following consecutive three stages: (1) finding the path, (2) establishing the found path, and (3) transmitting a message along the established path. F can be known locally at each associated node in G_M at different stages of routing, requiring different strategies for fault tolerance. Note we do not require global state consensus. We consider three cases:

Case 1: F is known before routing stage (1).
Case 2: F is known after (1) but before (2).
Case 3: F is known after (2) but before (3).

For Case 1, since F is known before path finding, simply assigning infinitely large weight to each faulty channel will convert the unreliable network to a reliable network, and hence algorithms described in the previous section will apply.

For Case 2, which is more realistic and general and hence of greater interest, we establish multiple paths for each edge in G_M such that for any portion of F falling to a path we are able to choose an available alternative path from them to skip the faulty edges. This approach is better and more practical than the two straightforward solutions. The first finds f edge-disjoint shortest paths from s_i to t_i for each i. This however can generate paths of unpredictable large weight. The second computes and stores the shortest path from s_i to t_i for each of the cases when taking f edges out from the s_i–t_i shortest path computed by Algorithm **A.1**, in order to achieve a minimum cost. This will however require clearly exponential time, since the number of such cases is exponential to f.

For Case 3, different strategies can be applied to obtain a solution. One method is that routing for each request is realized by implementing a reliable communication protocol that requires the receiver to acknowledge receipt of the message to the sender for each step of routing. This requires considerable overhead but guarantees reliability. Another method requiring less overhead is to construct two or more edge-disjoint paths for each request and send the message along these paths independently. This method is more efficient but does not guarantee reliability. The message is sent along the shortest path established by Algorithm **A.1**. At any step if a sender u doesn't receive an acknowledgment from a receiver v, it should assume that there is an edge fault on the path from u to v, and as a result an alternative path from u to v is sought and the message is sent along that path.

Our purpose here is to present a simple and efficient solution with the above approach for fault-tolerant routing in Case 2. Given r requests $\mathcal{R} = \{(s_1, t_1), (s_2, t_2),\ldots,(s_r, t_r)\}$, for each request (s_i, t_i) we are required to establish a robust communication path from s_i to t_i that can tolerate any (up to) f edge faults in each of the above two cases. Note since our routing involves circuit switching, a physical path from source to destination must be established before the message transmission takes place. We propose the following approach. Let $P_i = \{e_i^1, e_i^2, \ldots, e_i^{p_i}\}$ be the shortest path from s_i to t_i, where $e_i^j = (h(e_i^j), t(e_i^j))$. Assume that the size of F, f, is known before routing stage (1) (but not the content of F). We compute f alternative paths across two endpoints after removal of each $e \in E_M$, that are edge-disjoint shortest paths in length increasing (non decreasing) order and have the same orientation as edge e. This is done as preprocessing only once without knowing F. During the course of establishment of paths for the routing requests when F is known before routing stage (2), we chain several such paths together to skip all faulty edges in F if they are encountered. The edge-disjointness of the f redundant paths for each edge on P_i guarantees that an alternative path can always be found in routing stage (2) in the presence of any f faulty edges.

We assume that both preprocessing (Algorithm **A.2**) and path establishment (Algorithm **A.3**) are executed in the way of source routing, where **A.2** finds alternative routes from $h(e)$ to $t(e)$ when e is removed and **A.3** establishes physical paths to skip all faulty edges on each path found by Algorithm **A.1** in the previous section.

Algorithm **A.2**
{*Construct alternative paths for every edge $e \in E_M$.*}
for every edge $e \in E_M$ **do**
    Find f shortest paths connecting $h(e)$ to $t(e)$ in $E_M - \{e\}$
    that are edge-disjoint with each other;
    Store these paths in $\mathcal{P}(e)$ according to length increasing order in $\mathcal{P}(e)[1,f]$.

When all redundant paths are computed by Algorithm **A.2**, path establishment from s_i to t_i along each path found by **B.1** in the presence of $F = \{e_1^*, e_2^*, \ldots, e_f^*\}$ is carried out as follows. For any path $P = \{e_1, e_2, \ldots, e_p\}$, let $t(P) = t(e_1)$ and $h(P) = h(e_p)$. We say that node u precedes node v in P, denoted by $\acute{u}\ v$, if there exist e_i and e_j in P, $i \leq j$, such that $t(e_i) = u$ and $h(e_j) = v$. That is, node u precedes node v in P if u appears before v on path P. We use $P[t(e_i), h(e_j)]$ to denote the segment path $\{e_i, e_{i+1}, \ldots, e_j\}$ in P, and $R \parallel P$ to denote the update to path R with P such that $R \parallel P$ contains the shortest path from $t(R)$ to $h(P)$. Here is the algorithm:

Algorithm **A.3**
{*Establish a physical fault-free path for every request (s_i, t_i) along the already found path $P_l = \{e_i^1, e_i^2, \ldots, e_i^{p_i}\}$ by **A.1**, $1 \le i \le r$.*}
for $i = 1$ **to** r **do**
(a) $R_i := e_i^1$;
(b) **for** $j = 1$ **to** p_i **do**
    if $(e_i^j \in F) \wedge (h(R_i) < h(e_i^j))$ **then**
{*If $h(e_i^j) < h(R_i)$ removal of e_i^j will have no effect on R_i.*}
    $u := 1$;
    while $\mathcal{P}(e_i^j)[u] \cap F \ne \phi$ **do**
    $u := u + 1$;
{*Choose a shortest path across $V^1[P_i - \{e_i^j\}]$ and $V^2[P_i \quad \{e_i^j\}]$ that contains no faulty edges.*}
    $R_i := R_i \parallel \mathcal{P}(e_i^j)[u]$;
{*Include the chosen path into the route.*}
(c) $R_i := R_i \parallel P_i[h(R_i), t_i]$;
    {*Include the last segment of P_i that connects $h(R_i)$ to $h(e_i^{p_i}) = t_i$.*}
    Add $\delta[k](t(e))$ to $d_{\lambda_k}(e)$ for each $e \notin P_i$ using wavelength λ_k.
    {*Increase the queuing delay for each replacement

Note that since we do not know which replacement path will be used for each edge in P_i, we do not assign any queuing delay for edges on replacement paths when we construct the redundant paths in Algorithm **A.2**. Instead, we assign each such edge a queuing delay when the replacement path is physically included in the path for routing message in Algorithm **A.3**. This is different from the queuing delay calculation for the edges in P_i in **A.1** for which we know beforehand that all edges on P_i will be used for the ith request routing. It was shown [34] that Algorithm **A.3** finds a path correctly for each request (s_i, t_i) when there are f faulty edges in G_M. The quality of the path found by Algorithm **A.3** is given by the following lemma.

Lemma 2 [34] *If $F \subset P$, Algorithm **A.3** finds a path connecting the first node to the last node of path P whose length is no more than f times the optimal (shortest) one.*

In the general case $F \not\subset P$, the quality of the paths found by Algorithm **A.3** does not follow the above f-factor approximation. In fact, in order to achieve f-approximation, an algorithm needs to compute alternative paths that skip every edge on the previously computed alternative path recursively. This will clearly require exponential time.

Using Dijkstra's algorithm to find a shortest path in G_M in $t_{SP} = O(|E_M| + |V_M|\log|V_M|)$ time, we can complete Algorithm **A.2** in time

$$f|E_M|t_{SP} = O(f|E_M|(|(E_M| + |V_M|\log|V_M|)).$$

For Algorithm **A.3**, we use a bit-vector B to represent all edges in F: $B[i] = 1$ if $e_i \in F$ and $B[i] = 0$ otherwise; we record at each node its rank in path P_i at the same time when P_i is constructed in **A.2**. With this we can check $(e_i^j \in F) \wedge (h(R_i) < h(e_i^j))$ in $O(1)$ time and $\mathcal{P}(e_i^j)[r] \cap F \neq \phi$ in $O(|E_M|/\log|E_M|)$ time. Consequently we can complete Algorithm **A.3** in $O(\sum_{i=1}^{r} p_i f|E_M|/\log|E_M|) = O(fr|V_M||E_M|/\log|E_M|)$ time. By Eqs. (3) and (4)) we have the following theorem:

Theorem 4 *When f redundant paths are precomputed for each edge in G_M, the problem of multiple point-to-point routing for r requests in an unreliable WDM network with up to f faulty optical channels and wavelength conversion gates can be solved in $O(frk^2 n(kn + m)/\log(kn))$ time, where n, m, and k are the number of nodes, links, and available wavelengths in the network, respectively.*

Clearly, if multiple point-to-point routing in a reliable WDM network requires time t_{MPP} (Algorithm **A.1**), the same problem in an unreliable WDM network with f faulty edges can be solved in $O(fkn/\log(kn)t_{MPP})$ time.

Precomputing all redundant paths by Algorithm **A.2**, executed only once, requires $O(k^2 f(kn + m)(kn + m + n \log(kn)))$ time. All redundant paths are stored permanently for use by different routing requests.

4.2 Multicast

Now we consider the problem of multicast in an unreliable WDM network modeled by G_M. For multicast $\mathcal{M} = (s, D)$ that transports information from source node s to all destination nodes in set $D = \{t_1, t_2, \ldots, t_g\}$, we know that in a robust G_M $\mathcal{M}$ can be accomplished through sending a message along the tree edges of a multicast tree MT rooted at s. From Sec. 3.2 we also know that there is an MT that can be constructed efficiently whose cost is within an approximation ratio of less than 2 to that of the optimal Steiner tree in the expected case. Our task in this section is to construct a communication structure for $\mathcal{M}$ that can tolerate any f edge faults in an unreliable G_M. Let $F = \{e_1^*, e_2^*, \ldots, e_f^*\}$ be the set of faulty edges in G_M which are known after routing stage (1) and before routing stage (2). For other cases solutions are the same as given in the previous section. The basic idea to achieve fault-tolerant multicast is to enhance every edge in multicast tree MT with multiple alternative paths so that MT is always connected via at least one of these paths in the case that all edges in F are broken for any F. To achieve this, a trivial solution is to compute $(f + 1)$ edge-disjoint minimum spanning trees of G_M. Another straightforward approach is to establish k edge-disjoint alternative paths for each edge in MT that connects the two endpoints of the edge so that the two endpoints are always connected via one of these paths in case of k faulty edges. These two approaches, although both feasible, do not provide a low cost to the modified MT. In order to maintain the cost of MT as low as possible, a better approach is to reconnect the two connected components, not necessarily the two endpoints of the faulty edge, when an edge in MT is faulty. For a faulty edge $e = (u, v)$, let $MT^{(u)}$ and $MT^{(v)}$ be two connected components (trees) after the removal of e, where $MT^{(u)}$ and $MT^{(v)}$ contain endpoints u and v, respectively. Our approach first calls Algorithm **A.2** to enhance each edge in G_M with f replacement paths (redundant edges) so that an MT constructed in G_M can tolerate any F edge faults. As with **A.2**, this is done only once as a preprocessing procedure. We then find a shortest replacement path connecting node u and any node in $MT^{(v)}$ for any faulty $e = (u, v) \in E(MT)$ after MT has been found by **B.1**.

When each edge in G_M has been enhanced with f alternative paths by Algorithm **A.2**, after the multicast tree MT has been found by Algorithm **B.1** for multicast request $\mathcal{M}$, path establishment on MT in the presence of up to f faulty edges $F = \{e_1^*, e_2^*, \ldots, e_f^*\}$ is carried out as follows. Let $\{e_1, e_2, \ldots, e_{|E(MT)|}\}$ be level-by-level ordered edges of MT. The multicast proceeds by sending a message originated at root s along edges e_i for $i = 1$ to $|E(MT)|$ in MT. If edge e is faulty, then an alternative path of the shortest length that does not contain any faulty edge is chosen from $\mathcal{P}(e)$ to deliver the message. To support faulty edge detection, each path in $\mathcal{P}(e)$ uses a bit vector of $|E_M|$ bits to store the presence of each edge of G_M in the path: ''0'' for nonpresence and ''1'' for presence. To facilitate alternative path selection, all paths in $\mathcal{P}(e)$ are stored in the order of increasing length. We use an array of $f \times |E_M|$ bits for $\mathcal{P}(e)$ and let F store the global indices of all faulty edges, that is, faulty edge $e^*_i = e_{F[i]}$ for $1 \leq i \leq f$. Thus we have immediately the following multicast path establishment algorithm, which is called for each multicast request after the multicast tree MT has been found by **B.1** and executed in the way of source routing.

Algorithm **B.2**
{*Establish physical paths for messages multicast from the root in MT found by **B.1**.*}
for $i = 1$ **to** $|E(MT)|$ **do**
  if $e_i \in F$ **then**
    Deduct $\delta[k](e_i)$ from $d_{\lambda_k}(e_i)$ if e_i uses wavelength λ_k;
  {*Reduce its queue length by 1 to reflect release of channel e_i.*}
    $j: = 1$; $alt: = FALSE$;
    while $(j \leq f) \wedge (alt = TRUE)$ **do**
      $q: = 1$; $alt: = TRUE$;
      while $(q \leq f) \wedge (alt = TRUE)$ **do**
        if $\mathcal{P}(e_i)[j][F[q]] = 1$ **then**

        alt: = *FALSE*;
          q: = *q* + 1;
       j: = *j* + 1;
{*Choose a shortest path in $\mathcal{P}(e)$ that contains no faulty edges.*}
    if the above replacement path contains a node $u' \in MT^{(u)}$ **then**
       Delete the edge pointing to u' in $MT^{(u)}$;
{*Eliminate 'loop' while maintaining the path connecting from u to $MT^{(v).*}$}
   $MT = MT \parallel \mathcal{P}(e)[j]$;
   Add $\delta[k](t(e'))$ to $d_{\lambda_k}(e')$ for each $e' \in \mathcal{P}(e)[j]$ using wavelength λ_k.
{*Update *MT* and the edge weight for each edge on the new path.*}

Note that 'loop' in the above means more than one incoming edge to a tree node. It is not a loop in the directed sense.

Lemma 3   [34] *Algorithm* **B.2** *correctly implements multicast* $\mathcal{M}$ *in an unreliable WDM network with up to f faulty edges in* G_M.

We know that Algorithm **A.2** can be completed in time $O(f|E_M|(|E_M| + |V_M|\log|V_M|))$ which is the preprocessing time. Using Algorithm **B.1** to construct *MT* in time t_{MT} and Dijkstra's algorithm to find a shortest path, Algorithm **B.2** requires $O(|E_{MT}|f^2 + |V_M|\log|V_M|)$ time. With $|V_M| = 2\,kn$ and $|E_M| = k^2n + km$, we have the following theorem:

Theorem 5   *The problem of multicast of group size g in an unreliable WDM network with up to f faulty optical channels and wavelength conversion gates can be solved in* $O(kf^2(kn + m) + kn\log(kn))$ *time, with preprocessing support of* $O(k^2f(kn + m)(kn + m + n\log(kn)))$ *time, where n, m, and k are the number of nodes, links, and available wavelengths in the network, respectively.*

The preprocessing here is the same as for multiple point-to-point routing implemented by Algorithm **A.2** and runs only once for all routing requests.

Let multicast in a reliable WDM network require time t_{MC} (Algorithm **B.1**). From the above discussion it is clear that multicast in an unreliable WDM network with f faulty edges requires $O\,(f^2/g^2t_{MC})$ time.

4.3 Multiple Multicast

We now consider the general group communication pattern of multiple multicast. Let $\mathcal{M}_1, \mathcal{M}_2, \ldots, \mathcal{M}_r$ be r groups of multicast, and let $\mathcal{M}_i = (s_i, D_i)$, $1 \leq i \leq r$, where s_i is a source and $D_i = \{t_i^1, \ldots, t_i^{g_i}\}$ are the destinations; G_M must be at least $(f + r + 1)$-edge connected. In an unreliable WDM network with up to f faulty edges in G_M that are known after routing stage (1) and before routing stage (2), $\mathcal{M}_i$ alone can be realized by a multicast tree MT_i constructed by Algorithm

B.2. As we stated in Sec. 3.3, since all MT_i's are constructed concurrently and independently, edges of different MT_i in $MF = \cup MT_i$ may fall onto the same edge of G_M and hence possibly cause wavelength contention on the same optical link of the network. So our task here is to construct all MT_i's in such a way that results in a minimal wavelength contention for all the trees in MF. We use the same greedy approach as in Sec. 3.3 to achieve the above: construct an edge-enhanced G_M for fault tolerance by Algorithm **A.2** as preprocessing; then, after approximate multicast tree MT_i for each multicast $\mathcal{M}_i$ has been found by Algorithm **C.1**, establish physical paths in each MT_i one by one in size-increasing order in the presence of any F applying Algorithm **B.2**. This will ensure that the tree overlapping probability is minimum for all trees in MF, and hence the probability of wavelength contention is minimal. Our algorithm for multiple multicast in an unreliable WDM network is

Algorithm **C.2**
{*Establish physical paths for multicast trees MT_{π_1}, MT_{π_2}, . . ., MT_{π_r} sorted in size-increasing order found by **C.1** in an unreliable WDM network with up to f faulty edges.*}
for $i = 1$ **to** r **do**
    Call Algorithm **B.2** to establish a set of physical routes $\mathcal{R}(MT_{\pi_i})$ for MT_{π_i} that skips all faulty edges in MT_{π_i}.

The correctness of the above algorithm is obvious. The time complexity of the algorithm can be directly obtained from that of algorithms **A.2** and **B.2**. This results in the following theorem:

Theorem 6 *The problem of multiple multicast of r groups of maximum size g in an unreliable WDM network with up to f faculty optical channels and wavelength conversion gates can be solved in $O(rk(f^2(kn + m) + n \log(kn)))$ time under the same preprocessing as for multiple point-to-point and multicast routing, where n, m, and k are the number of nodes, links, and available wavelengths in the network, respectively.*

Let the time for multiple multicast routing in a reliable WDM network be t_{MMC} (Algorithm **C.1**). It is clear that multiple multicast in an unreliable WDM network with f faulty edges would require $O(f^2/\sum_{i=1}^r g_i^2 t_{MMC})$ time.

5 ON-LINE ROUTING IN RELIABLE NETWORKS

In this section we study on-line routing in a reliable WDM network and give a set of efficient algorithms for on-line multiple point-to-point routing and multiple multicast on the general cost models of Eqs. (1) and (2). We say that a routing problem is solved if physical path(s) to realize the routing are found and established.

5.1 Multiple Point-to-Point Routing

In general, routing for multiple point-to-point requests is not always synchronized for the reason that some may start earlier than others, and each path, once established, should be able to be used repeatedly for the same request throughout a session. For simplicity and without loss of generality, we ignore the timing differences among different paths arriving at the same queue. Though these differences may result in some differences in their queuing delay, this is insignificant in comparison with the total routing cost in Eq. (1) and yet difficult to incorporate in the measurement of queuing delay. Our queuing delay therefore reflects only the number of signals arriving at the same node within a small time interval, but not their actual timing differences.

For on-line multiple point-to-point routing, requests are generated and terminated dynamically, and request-activated paths are constructed concurrently, each ignoring the existence of the others. A physical route for each request is constructed by shortest path routing, where updating the edge weight for each edge on the path is coordinated through a mutual exclusion mechanism.

For request (s_i, t_i), our algorithm [36] uses a greedy approach to construct the shortest path that in addition to adding the edge weight to the path distance (cost) it also adds an appropriate additive factor δ to the edge weight of each edge on the path when it is reached. The algorithm has the following structure. Concurrent adding by multiple paths is prohibited by the mutual exclusion mechanism *semaphores* [38], which we use here.

Algorithm **OMP**
{*Construct a shortest path for request (s_i, t_i), $1 \leq i \leq r$.*}
1. Find the shortest path from s_i to t_i, where for any edge (channel) e using wavelength λ_j the distance from $h(e)$ to $t(e)$ is $w(e, \lambda_j) + d_{\lambda_j}(t(e))$;
2. For each new edge added to the path do
   wait(*mutex*);
{*Mutual-exclusion, *mutex* = 1 initially.*}
   Add $\delta[j](t(e))$ to $d_{\lambda_j}(t(e))$;
   signal(*mutex*).

The shortest path from s_i to t_i is established using a modified Dijkstra algorithm [9] with the change in distance calculation to include edge weight as stated above.

The time complexity of the algorithm is the maximum of that required for routing for a single request assuming that all requests arrive simultaneously because routes are constructed concurrently.

Using an adjacent list to represent G_M, Dijkstra's algorithm can be implemented in $O(|E_M| + |V_M|\log|V_M|)$ time. Updating node weight at each node requires $O(r)$ time in the worst case, yielding $O(r|V_M|)$ time in total for the whole path. By Eqs. (3) and (4) we have the following theorem:

Theorem 7 *The problem of on-line multiple point-to-point routing for r requests in a WDM network of n nodes and m links with k available wavelengths can be solved in O(kn(k + r + log(kn)) + km) time.*

5.2 Multiple Multicast

In off-line multiple multicast, all MT_i's are constructed one by one as described in Sec. 3.3.

We now consider the case of on-line multiple multicast in which all groups of multicasts are carried out concurrently, that is, all MT_i for $1 \leq i \leq r$ are constructed concurrently. In this case, as with on-line multiple point-to-point routing, multiple MT_i's may want to update the edge weight of each common edge they are using simultaneously. This can be resolved by imposing mutual exclusion to edge weight updating. However, this will result in the above method for the off-line case not directly usable. The reason is that each step of extending each MT_i will update the edge weight at common edges and hence change the distances of many pairs of nodes in V_M, resulting in different edge weights of many edges in $I(\{s\} \cup D)$. Our main task here is to find an effective way to update the edge weights in $I(\{s\} \cup D)$ in correspondence to each edge weight update in G_M.

We observe that it is difficult to accomplish the above task if we use the same data structure as used in the off-line case because each edge weight in $I(\{s\} \cup D)$ corresponds to the accumulated path weight in G_M, which is difficult to update with respect to an edge weight change. We therefore use an auxiliary graph G_I to represent $I(\{s\} \cup D)$. G_I is obtained by replacing each edge in $I(\{s\} \cup D)$ with its corresponding path in G_M, and its edge weight with the edge weight on the path in G_M. Let G_I^i be the G_I used for constructing MT_i, $1 \leq i \leq r$. With the above replacement, any edge weight update obtained by path extension of MT_i in G_I^i will be immediately reflected in G_I^j for all $j \neq i$ and thus affect other path extensions of MT_j.

Our algorithm for constructing MT_i is [36]

Algorithm **OMM**
{*Multicast tree construction for $\mathcal{M}_i = (s_i, D_i)$, $D_i = \{t_i^1, \ldots, t_i^{g_i}\}$ in multiple multicast of r groups.*}
1. **for** each pair of $u, v \in \{s_i, t_i^1, t_i^2, \ldots, t_i^{g_i}\}$ **do**
   Compute the shortest path from u to v, $P[u \rightarrow v]$, in G_M;
   {*This shortest path will be used to connect u to v in G_I^i.*}

2. Construct a "complete" graph G_I^i, where the edge (u, v) is the path $P[u \rightarrow v]$ (with all edge weights preserved);
3. Compute a shortest path tree MT_i rooted at s_i reaching all destinations (D_i) in G_I^i, where for each new edge e added to ST_i do the following:
{*Update the corresponding edge weight in all $G_I^{i'}$ mutual-exclusively.*}
$wait(mutex)$;   {*Mutual-exclusion, where $mutex = 1$ initially.*}
For each $1 \leq i' \leq r$ if $t(e) \in V(G_I^{i'})$ then
    Add $\delta[j](t(e))$ to $d_{\lambda_j}((t(e))$ in $G_I^{i'}$ if e uses wavelength λ_j;
$signal(mutex)$.

The correctness of the algorithm can be seen clearly from the greedy approach. We use an ordered data structure in node indices to store the nodes used in G_I^i, that is, use an array B_i of size $|V_M|$ initialized to 0 for G_I^i and add "1" to $B_i[j]$ "1" if an edge pointing to node j occurs in G_I^i. The time complexity of constructing G_I^i and updating edge weights in Steps 2 and 3 is $O(r^2 g_i^2)$, because adding each edge to MT_i requires us to examine all r B_i's and update the weight of the edges pointing to nodes in B_i used in other groups when necessary, and each of these may need to wait for other group updates in case of concurrent updating, which brings in another r factor. Step 1 constructing shortest paths can be done in $O(g_i^2 k(kn + m + n \log (kn))) = O(g_i^2 k(kn + m + n \log(kn)))$ using Dijkstra's algorithm [9] (single source all destinations). Since $|MT_i| \leq |V_M|$, we have the following theorem.

Theorem 8   *On-line multiple multicast for r groups in a WDM network of n nodes and m links with k available wavelengths can be completed in $O(g^2 k(r^2/ k + kn + m + n \log (kn)))$ time, where g is the maximal group size.*

6   ON-LINE ROUTING IN UNRELIABLE NETWORKS

In this section we consider the on-line communication problem in an unreliable WDM network in which both optical channel (wavelength) and switch gate (conversion) faults may occur. We give a set of efficient algorithms for point-to-point requests and multiple multicast on our cost models of Eqs. (1) and (2) in an unreliable WDM network.

6.1   Multiple Point-to-Point Routing

We consider an unreliable WDM network with f edge faults in G_M. Let $F = \{e_1^*, e_2^*, \ldots, e_f^*\}$ be the set of edges that are faulty. We consider the case that F is known before establishment of the physical path but after the path has been found. Other cases are less interesting than this one, because faults occurred before path finding can be handled by simply getting around them during path finding, and faults occurred after path establishment can be handled by applying a reliable communication protocol. For on-line multiple point-to-point routing, as in the reliable case, since all requests arrive on line, they are generated and terminated dynamically, and therefore paths are constructed concurrently, each ignoring the existence of others.

Assume that path finding finds path $P_i = \{e_i^1, e_i^2, \ldots, e_i^{p_i}\}$ for request (s_i, t_i) before the occurrence of F, where $e_i^j = (h(e_i^j), t(e_i^j))$, and that we know the size of F, f, beforehand at the beginning. Our basic idea for handing F when it occurs before path establishment is to augment edges on P_i so that any edge when it becomes faulty can be replaced quickly by an alternative path connecting its head to its tail. Since there can be f faulty edges, we need to compute f edge-disjoint alternative paths to augment each edge on the path to guarantee that an alternative path can always be found. In order to save the cost of computing edge augmentation for the path, we use the same approach as for off-line routing [34] and compute the edge augmentation, f edge-disjoint alternative paths across the two parts, for each $e \in E_M$ beforehand; and store them in length-increasing (non-decreasing) order. This is done as preprocessing only once so that it can handle any P_i containing faulty edges. After that, during the course of establishment of any path for the routing when F is known, we then chain several such paths together to skip all faulty edges in F if they are encountered.

Our algorithm combines the Algorithm **OMP** and the fault-tolerant off-line routing algorithm in [34]. As OMP, it adds an appropriate additive factor δ to the edge weight for each edge on the shortest path during construction of the shortest path. Like fault-tolerant off-line routing, for each edge in E_M it precomputes f alternative paths across its two endpoints prior to the start of routing, and then chains several such paths together to skip all faulty edges in F when they occur during the establishment of physical paths. Our algorithms for preprocessing of edge augmentation and physical path establishment are presented as follows [36].

Algorithm **EdgeAugment**
{*Construct alternative paths for every edge $e \in E_M$.*}
for every edge $e \in E_M$ **do**
    Find f shortest paths connecting $h(e)$ to $t(e)$ in $E_M - \{e\}$
    that are edge-disjoint with each other;
    Store these paths in $\mathcal{P}(e)$ according to length-increasing order in $\mathcal{P}(e)$
    $1, f]$.

Algorithm **FOMP**
{*Establish a physical fault-free path for on-line request (s_i, t_i) along the already found path $P_i = \{e_i^1, e_i^2, \ldots, e_i^{p_i}\}$ by OMP, $1 \leq i \leq r$.*}
1. $R_i := e_i^1$;
2. **for** $j = 1$ to p_i **do**

if $(e_i^j \in F) \wedge (h(R_i) < h(e_i^j))$ **then**

{*If $h(e_i^j) < h(R_i)$ removal of e_i^j will have no effect on R_i.*}

   $u: = 1;$

   while $\mathcal{P}(e_i^j)[u] \cap F \neq \phi$ **do**

   $u: = u + 1;$

{*Choose a shortest path across $V^1[P_i - \{e_i^j\}]$ and $V^2[P_i - \{e_i^j\}]$ that contains no faulty edges.*}

   $R_i: = R_i \parallel \mathcal{P}(e_i^j)[u];$

{*Include the chosen path into the route.*}

3. $R_i: = R_i \parallel P_i[h(R_i), t_i];$

{*Include the last segment of P_i that connects $h(R_i)$ to $h(e_i^{p_i}) = t_i$.*}

wait(mutex);   {*Mutual-exclusion, where *mutex* $= 1$ initially.*}

Add $\delta[k](t(e))$ to $d_{\lambda_k}(e)$ for each $e \notin P_i$ using wavelength λ_k;

{*Increase the queuing delay for each replacement edge.*}

signal(mutex).

Algorithm **EdgeAugment** requires time $f|E_M|t_{SP} = O(f\,|E_M|(|E_M| + |V_M|$ $\log|V_M|))$, using Dijkstra's algorithm to find a shortest path in G_M. For algorithm FOMP, let faulty edges be represented by their head endpoints, that is, F contains only head endpoints of faulty edges. We use a bit-vector B_i of $|V_M|$ bits to represent the heads of all edges in P_i and mark $B_i[j]$ "1" if node j occurs on path P_i. Thus we can complete the algorithm in $O(p_i f^2|V_i|)$ time. Since updating $d_{\lambda_k}(e)$ by multiple paths must be mutually exclusive, an additional $O(r^2 kn)$ cost is needed for the last step. Since **EdgeAugment** is executed only once as preprocessing, it is not a part of the routing cost. We therefore have the following theorem with Eqs. (3) and (4).

Theorem 9   *When f redundant paths are precomputed for each edge in G_M, the problem of on-line multiple point-to-point routing for r requests in an unreliable WDM network with up to f faulty optical channels and wavelength conversion gates can be solved in $O(kn(f^2 kn + r^2))$ time, where n, m, and k are the number of nodes, links, and available wavelengths in the network, respectively.*

If on-line multiple point-to-point routing in a reliable WDM network requires time t_{MPP} (Algorithm **OMP**), the same problem in an unreliable WDM network with f faulty edges can be solved in $O((f^2 n + r)t_{MPP})$ time.

6.2   Multiple Multicast

We now consider on-line multiple multicast in unreliable WDM networks. As with point-to-point routing, the basic idea to achieve fault tolerance in multicast is to augment every edge in E_M with multiple alternative paths such that any MT constructed in G_M is always connected via at least one of these paths for any possible F. For multiple multicast, as for the off-line case [34], since edge weight is shared, the above updating must also be mutually exclusive. Our algorithm is given as [36]

Algorithm **FOMM**
{ *Establish physical paths for multicast tree MT_i for on-line multicast $\mathcal{M}_i$ found by OMM in an unreliable WDM network with up to f faulty edges.*}
for $i = 1$ **to** $|E(MT)|$ **do**
    if $e_i \in F$ then
        $wait(mutex1)$; {*Mutual-exclusion, where $mutex = 1$ initially.*}
        Deduct $\delta[k](e_i)$ from $d_{\lambda_k}(e_i)$ if e_i uses wavelength λ_k;
    {*Reduce its queue length by 1 to reflect release of channel e_i for its future use.*}
    $signal(mutex1)$;
    $j: = 1$; $alt: = FALSE$;
    while $(j \leq f) \wedge (alt = TRUE)$ **do**
        $q: = 1$; $alt: = TRUE$;
        while $(q \leq f) \wedge (alt = TRUE)$ **do**
            if $\mathcal{P}(e_i)[j][F[q]] = 1$ **then**
                $alt: = FALSE$; $q: = q + 1$;
        $j: = j + 1$;
    {*Choose a shortest path in $\mathcal{P}(e)$ that contains no faulty edges.*}
    if the above replacement path contains a node $u' \in MT^{(u)}$ **then**
        Delete the edge pointing to u' in $MT^{(u)}$;
    {*Eliminate 'loop' while maintaining the path connecting from u to $MT^{(u)}$.*}
    $MT = MT \| \mathcal{P}(e)[j]$;
    $wait(mutex2)$;
    {*$mutex2 = 1$ initially.*}
    Add $\delta[k](t(e'))$ to $d_{\lambda_k}(e')$ for each
    $e' \in \mathcal{P}(e)[j]$ using wavelength λ_k;
    $signal(mutex2)$;
    {*Update MT and the edge weight for each edge on the new path.*}
    $wait(mutex3)$;   {*$mutex3 = 1$ initially.*}
    For each $e \in \mathcal{R}(MT_i) - MT_i$ mark $w(e)$ with weight ∞;
    $signal(mutex3)$.

MT_i can be constructed by Algorithm **OMM** in time t_{MT}. Because each mutual exclusion for updating causes r^2 factor delay for the reasons explained in Algorithm **OMM**, and inside the **for** loop the computation takes $O(f^2 + |MT_i|)$ $= O(f^2 + |V_M|)$ time, Algorithm **FOMM** requires $O(|E_{MT}|r^2(f^2 + |V_M|))$ time. With $|V_M| = 2kn$ and $|E_M| = k^2n + km$, we have the following theorem:

Theorem 10 *On-line multiple multicast or r in an unreliable WDM network with f faculty channels can be completed in $O(kr^2(f^2 + kn)(kn + m)$ time, with* *preprocessing support of $O(k^2 f(kn + m)(kn + m + n \log(kn)))$ time, where n, m, and k are the number of nodes, links, and available wavelengths in the network, respectively.*

From the above discussion it is clear that multicast in an unreliable WDM network with f faulty edges requires $O(r^2(f^2 + kn)/g^2 t_{MC})$ time.

7   GROUP MEMBERSHIP UPDATING

On-line communication allows dynamic membership changes in the designated communication groups during the course of communication. In this section we present efficient algorithms for updating communication groups to accommodate dynamic group membership changes such as insertion and deletion of requests and destinations, group splitting and merging during the course of on-line communication of multiple point-to-point routing, multicast and multiple multicast. Our algorithms work for both reliable and unreliable WDM networks on the general cost model.

7.1   Group Membership Updating for On-Line Multiple Point-to-Point Routing

Our first task is to present efficient algorithms for updating group membership for dynamic insertion and deletion of point-to-point routing requests.

As discussed in Sec. 5.1, requests in on-line multiple point-to-point routing are generated and terminated dynamically, and request-activated paths are constructed concurrently, each ignoring the existence of others. We consider the cases of inserting a new request and deleting an existing request.

For inserting a new request (s, t), the algorithm employs a shortest path construction algorithm from s to t, and each step (edge) expanding of the shortest path to include a new edge increases the queuing delay of all edges sharing the same queue by a predefined factor. Since routing for all requests is carried out concurrently, several factors may be added to the queuing delay of the same queue simultaneously when multiple paths arrive at the same edge at the same time during their expansion. In this case, in order to ensure that concurrent value additions to the same queuing delay are carried out correctly, a suitable synchronization mechanism available in the system needs to be employed.

Using an adjacent list to represent G_M, Dijkstra's algorithm [9] for shortest path construction can be implemented in $O(|E_M| + |V_M| \log |V_M|)$ time. Taking consideration of the waiting time at each edge on the path due to mutual exclusion on edge weight updating, the total waiting time on the path in the worst case is $O(r|V_M|)$, since in total r paths may be concurrently constructed as the new path.

Clearly the total time required for updating group membership when new request (s, t) is inserted into the group is the same as that for on-line routing for all the requests after (s, t) is inserted, in the worst case, because we assumed that all paths are constructed concurrently.

For deleting an existing request, we need to delete the shortest path constructed for that request. More specifically, we need to release the edges (optical channels) used by the path and decrease the edge weight (queuing delay) for each edge on the path by the prespecified factor. This requires $|V_M|$ (path length) updates to edge weights on the path. Since each update may take $O(r)$ time when all r paths are concurrently updating each note weight, the algorithm requires in total $O(r|V_M|)$ in the worst case.

By Eqs. (3) and (4) we have the theorem [37]

Theorem 11 *For on-line multiple point-to-point routing for r requests in a WDM network of n nodes and m links with k available wavelengths, insertion and deletion of a request require $O(kn(k + r + \log(kn)) + km)$ and $O(rkn)$ time, respectively.*

7.2 Group Membership Updating for On-Line Multicast

We now consider the problem of updating the group membership—multicast tree—for on-line multicast where destination nodes can be dynamically inserted to or deleted from the multicast tree.

Assume that MT is the current multicast tree rooted at s and spans to all nodes in D. We use $p(v)$ to denote the precedent (parent) node of v in MT. When a node d is inserted to MT, we first compute dist(u, d) and dist(d, u) for every node $u \in MT$ in G_M; then we update MST with the path that has the minimal dist(u, d) + dist(d, v) − dist$(p(v),v)$. If dist$(d, v) \geq$ dist$(p(v),v)$, we include path $u \to d$ into MT. Otherwise we include path $u \to d \to v$ and delete path $p(v) \to v$. Our algorithm is

Algorithm **B⁺**
{*Insert a new destination d to the multicast group.*}
1. For every node $u \in MT$ compute dist(u, d) of path $P[u \to d]$ and dist(d, u) of path $P[d \to u]$;
2. Compute $\min_{u \neq v \in MT}$ {dist(u, d) + dist(d, v) − dist$(p(v),v)$} and let the found nodes be u^* and v^*;
3. If dist$(d, v^*) \geq$ dist$(p(v^*),v^*)$ then $MT = MT \cup P[u^* \to d]$
   else $MT = (MT − P[p(v^*) \to v^*]) \cup P[u^* \to d] \cup P[d \to v^*]$.

In the case of deleting a destination from the multicast group, we compute the shortest cycle connecting the two parts of MT, $MT'(s)$, and MT'' that are disconnected due to removal of node d. That is, for all $u, u' \in MT'(s)$ and $v,v' \in$

MT'', we compute dist(u, v) and dist(v',u') and take the minimal dist(u, v) + dist(v',u') −dist$(p(u'),u')$ to update MT. Here is our algorithm:

Algorithm **B⁻**
    {*Delete a destination node d from the multicast group.*}
    1. For every node $u \in MT'(s)$ and $v \in MT''$ compute dist(u, v) and dist(v, u);
    2. Compute $\min_{(u,u' \in MT'(s)) \wedge (v,v' \in MT'')}$ {dist(u, v) + dist(v',u') − dist$(p(u'),u')$ and let these nodes be $u^* = u$, $u'^* = u'$, $v^* = v$, $v'^* = v'$;
    3. If dist$(v'^*,u'^*) \geq$ dist$(p(u'^*),u'^*)$ then $MT = MT's \cup P[u^* \rightarrow v^*] \cup MT''$ else $MT = (MT'(s) - P[p(u'^*) \rightarrow u'^*]) \cup P[u^* \rightarrow v^*] \cup P[v'^* \rightarrow u'^*]$ $\cup MT''$.

A direct implementation of the above algorithms requires clearly $O(|V_M||MT|^2)$ time, because there are $\Theta(|MT|^2)$ pairs of nodes and for each we need to compute two or three distances of shortest paths of length at most $|V_M|$. A more careful implementation suggests to precompute all pairs' shortest paths in G_M, which takes $O(|V_M|^3)$ time, and store them in a table for later retrieval. With this scheme, a distance can be obtained in $O(1)$ time by a table lookup, and therefore the total time for the above algorithms becomes $O(|MT|^2)$. Since $|MT| \leq |V_M| = (2k + 1)n$, we have the following theorem.

We can also see that the resulting MT after updating has the same approximation ratio as the original MT. This is because the updating in both cases of insertion and deletion adds a minimum possible weight to incorporate the changes.

Theorem 12 [37] *For on-line multicast in a WDM network of n nodes and k available wavelengths, dynamically inserting and deleting a destination requires $O(k^2n^2)$ time, preserving the same approximation ratio as the multicast tree before updating, provided that a precomputation of all pairs' shortest paths in G_M is given.*

7.3 Group Membership Updating for On-Line Multiple Multicast

Our final task is to consider the group membership maintenance problem for on-line multiple multicast in WDM networks.

When multiple multicast is carried out in on-line fashion, we are concerned with how to maintain MF with respect to the following dynamical changes:

1. Destinations may dynamically join and leave multicast groups.
2. A group may be split into two (or more), with one (or more) of its destinations being a new source(s).
3. Two (or more) groups may be merged together, where the source of

a designated group becomes the common source of all the groups, whereas sources of other groups become destinations.

For (1), we employ algorithms $\mathbf{B}^+$ and $\mathbf{B}^-$ of the previous section to dynamically maintain the multicast tree MT_i for each multicast group $\mathcal{M}_i$, where adding an edge to MT_i also updates the queuing delay at the corresponding edge accordingly for all i. Concurrent updates on queuing delay to the same edge are coordinated with a suitable synchronization mechanism. Time complexity for this case is at most r times of that required for on-line maintaining a single multicast tree due to the waiting time for edge weight updating, that is, $O(rk^2n^2)$ if a precomputation is done.

For (2), we reconstruct a multicast tree for each new group after splitting. Construction of different multicast trees is carried out concurrently without their knowing each other using an on-line multicast tree construction algorithm described in Sec. 5.2. Concurrent updates to queuing delay are handled in the same way as in (1). The time complexity for this case is thus $O(g^2k(r^2/k + kn + m + n\log(kn))$ by the on-line multiple multicast time complexity [36], as each step of updating edge weight requires $O(r)$ time waiting for total r groups of multicast.

For (3), we need to merge two (or more) multicast trees MT_i and MT_j. This can be done by finding out the shortest path joining them into a single multicast tree rooted at the root of MT_i by the following algorithm.

Let $MT_i = \{s_i\} \cup D_i$ and $MT_j = \{s_j\} \cup D_j$, and s_i the designated root for $MT_i \cup MT_j$.

Algorithm $\mathbf{C}^{\cup}$
{*Merge multicast groups MT_i and MT_j into one group, with root s_i of MT_i being the common root.*}
1. For every node $u \in D_i$ and every node $v \in MT_j$ compute u^* and v^* such that dist(u^*, v^*) + dist(v^*, s_j) −dist(s_j, v^*) = $\min_{u \in D_i, v \in MT_j}\{$dist$(u, v)$ + dist(v, s_j) −dist$(s_j, v)\}$; Keep the corresponding path of dist(x, y) in $P[x \rightarrow y]$;
2. If $v^* \neq s_j$ then merge $MT_i \cup MT_j \cup P[u^* \rightarrow v^*] \cup P[v^* \rightarrow s_j]$ − $P[s_j \rightarrow v^*]$, {*dist$(v^*, s_j) <$ dist(s_j, v^*).*}
else merge $MT_i \cup MT_j \cup P[u^* \rightarrow v^*]$.

end.

The time complexity for $C^{\cup}$ is $O(r |MT_i\|MT_j|)) = O(rk^2n^2)$, where r is the factor for waiting time for each node updating.

Summarizing the above cases, we have the following theorem:

Theorem 13 [37] *In on-line multiple multicast of r groups in a WDM network of n nodes and k available wavelengths with maximal group size g, a single group membership change and merging require $O(rk^2n^2)$ time, and a group splitting requires $O(g^2k(r^2/k + kn + m + n\log(kn))$ time.*

With the time complexity of maintaining a single multicast tree $t_{MC} = O(k^2 n^2)$ by Theorem 2, we know that a single group membership change and merging in multiple multicast of r groups require $O(r t_{MC})$ time, and a group splitting requires $O(g^2/k\ t_{MC})$ time.

8 CONCLUDING REMARKS

We have presented some recent developments on routing in multihop WDM networks with limited wavelength conversion [34,36,37]. The contents we have covered in this chapter include off-line and on-line routing in both reliable and unreliable WDM networks for point-to-point, multicast, and multiple multicast communication patterns on general and realistic cost models. For on-line routing, efficient algorithms for updating group membership to accommodate dynamic membership changes arisen during the course of routing have also been presented. All the algorithms run efficiently in time polynomial to the network size and the number of wavelengths.

We hope that the materials presented in this chapter will provide useful information for future research on routing in optical WDM networks.

REFERENCES

1. A. Aggarwal, A. Bar-Noy, D. Coppersmith, R. Ramaswami, B. Schieber, and M. Sudan. Efficient routing in optical networks. *J. of the ACM* 46:973–1001, 1996.
2. Y. Aumann and Y. Rabani. Improved bounds for all optical routing. In *Proceedings of the 6th Annual ACM-SIAM Symp. on Discrete Algorithms (SODA'95)*, 567–576, 1995.
3. R. A. Barry and P. A. Humblet. On the number of wavelengths and switches in all-optical networks. *IEEE Trans. on Communications (Part I)*, 583–591, 1994.
4. B. Beauqier, J. C. Gargano, S. Perenees, P. Hell, and U. Vaccaro. Graph problems arising from wavelength-routing in all-optical networks. In *Proc. of 2nd Workshop on Optics and Computer Science (WOCS)*, on CD-Rom, 1997.
5. K.-M. Chan and T.-S. Yum. Analysis of least congested path routing in WDM lightwave networks. In *Globecomm*, 962–969, 1994.
6. K. W. Cheung. Scalable, fault tolerant 1-hop wavelength routing. In *Globecom'91* 1240–1244, 1991.
7. I. Chlamtac, A. Faragó, and T. Zhang. Lightpath (wavelength) routing in large WDM networks. *IEEE Journal on Selected Areas in Communications* 14:909–913, 1996.
8. I. Chlamtac, A. Ganz, and G. Karmi. Lightpath communications: a novel approach to high bandwidth optical WAN's. *IEEE Trans. on Communications*. 40:1171–1182, 1992.
9. E. W. Dijkstra. A note on two problems in connexion with graphs. *Numerische Mathematik*, 1:269–271, 1959.

10. T. Erlebach and K. Jansen. Scheduling of virtual connections in fast networks. In *Proc. of 4th Workshop on Parallel Systems and Algorithms (PASA '96)*, 13–32, 1996.

11. J. C. Gargano, P. Hell, and S. Perenees. Colouring all directed paths in a symmetric tree with applications to WDM routing. In *Lecture Notes on Computer Science 1099 (Proceedings of ICALP'97)*, 505–515, 1997.

12. L. Gargano. Limited wavelength conversion in all-optical networks. *Proc. of 25th International Colloquium on Automata, Languages and Programming*, 544–555, 1998.

13. P. E. Green. *Fiber-Optic Communication Networks*. Prentice Hall, 1992.

14. K. Kaklamanis, G. Persiano, T. Erlebach, and K. Jansen. Constrained bipartite edge coloring with applications to wavelength routing. In *Lecture Notes on Computer Science 1099 (Proc. of ICALP'97)*, 460–470, 1997.

15. L. Kou, G. Markowsky, and L. Berman. A fast algorithm for Steiner trees. *Acta Informatica*, 15:141–145, 1981.

16. K. Bharath-Kumar and J. M. Jaffe. Routing to multiple destinations in computer networks. *IEEE Trans. on Communications*, COM-31:343–351, 1983.

17. E. Kumar and E. Schwabe. Improved access to optical bandwidth in trees. In *Proc. of the 8th Annual ACM-SIAM Symp. on Discrete Algorithms (SODA '97)*, 437–444, 1997.

18. H.-M. Lee and G. J. Chang. Set-to-set broadcasting in communication networks. *Discrete Applied Mathematics*, 40:411–421, 1992.

19. K. Li, Y. Pan, and S. Q. Zheng, eds. *Parallel Computing Using Optical Interconnections*. Kluwer Academic, 1998.

20. W. Liang, G. Havas, and X. Shen. Improved lightpath routing in large WDM networks. To appear in *Proc. of 18th Intern. Conf. on Distributed Computing Systems*, Amsterdam, The Netherlands, IEEE Computer Society Press, May 26–29, 1998.

21. W. Liang and H. Shen. Multicast broadcasting in large WDM networks. *Proc. of 12th Intern. Conf. on Parallel Processing Symp.* (IPPS/SPDP), Orlando, Florida, USA. IEEE Computer Society Press, 1998.

22. R. Malli, X. Zhang, and C. Qiao. Benefit of multicasting in all-optical WDM networks. *Conf. on All-Optical Networks*, SPIE Vol. 3531, 209–220, 1998.

23. G. De Marco, L. Gargano, and U. Vaccaro. Concurrent multicast in weighted networks. Manuscript.

24. A. D. McAulay. *Optical Computer Architectures*. John Wiley, 1991.

25. M. Mihail, K. Kaklamanis, and S. Rao. Efficient access to optical bandwidth. In *Proc. of FOCS'95*, 548–557, 1995.

26. B. Mukherjee. *IEEE Comm Magazine* (special issue), Jan/Feb 1999.

27. Y. Ofek and B. Yener. Reliable concurrent multicast from bursty sources. *Proc. of IEEE INFOCOM'96*, 1996, 1433–1441.

28. P. Raghavan and E. Upfal. Efficient routing in all-optical networks. In *Proc. of STOC'94*, 133–143, 1994.

29. R. Ramaswami. Multi-wavelength lightwave networks for computer communication. *IEEE Communications Magazine* 31:78–88, 1993.

30. G. N. Rouskas and M. H. Ammar. Analysis and optimization of transmission schedules for single-hop WDM networks. In *Infocom'93*, 1342–149, 1993.

31. G. N. Rouskas and M. H. Ammar, Multi-destination communication over tunable-receiver single-hop WDM networks. TR-96-12, Department of Computer Science, North Carolina State University.

32. L. H. Sahasrabuddhe and B. Mukherjee. Light-trees: optical multicasting for improved performance in wavelength-routed networks. *IEEE Communications*, 37(2): 67–73. 1999.

33. H. Shen. Efficient multiple multicasting in hypercubes. *J. System Architectures* 43(9):655–662, 1997.

34. H. Shen, F. Chin, and Y. Pan. Efficient fault-tolerant routing in multihop optical networks. *IEEE Trans. on Parallel and Distributed Systems*, 10(10), 1999.

35. H. Shen and W. Liang. Efficient multiple multicast in WDM networks. *Proc. of 1998 Intern. Conf. on Parallel and Distributed Processing Techniques and Applications*. Las Vegas, USA, 1998.

36. H. Shen, J. Sum, G. H. Young, and S. Horiguchi. Efficient algorithms for on-line communication in WDM networks. *Proc. of 2000 Intern. Conf. on Parallel and Distributed Processing Techniques and Applications*. Las Vegas, USA, 2000.

37. H. Shen, J. Sum, G. H. Young, and S. Horiguchi. Efficient dynamic group membership updating for on-line communication in optical WDM networks. *Proc. of 2000 Intern. Conf. on Parallel and Distributed Processing Techniques and Applications*. Las Vegas, USA, 2000.

38. Silbertcharze. *Operating Systems Concepts*. Addison-Wesley, 1998.

39. R. J. Vitter and D. H. C. Du. Distributed computing with high-speed optical networks. *IEEE Computer* 26:8–18, 1993.

40. S. S. Wagner and H. Kobrinski. WDM applications in broadband telecommunication networks. *IEEE Communications*, 27, 3:22–30, 1989.

41. Z. Zhang and A. S. Acampora. A heuristic wavelength assignment algorithm for multihop WDM networks with wavelength routing and wavelength re-use. *IEEE Journal on Networking* 3:281–288, 1995.

8
Optical Switching Architectures

Mohsen Guizani
University of West Florida, Pensacola, Florida

1 MOTIVATION

Telecommunications and computer networks are evolving rapidly from the narrow-band towards the broadband integrated services digital networks and the Internet. The directions taken by the Internet are influenced by the emergence of large number of service applications with different requirements. These applications include video on demand, video conferencing, videophony, and high-speed data transfer. Each of these services generates a large number of requirements. These introduce the need of flexible and universal networks. Both the need for flexible networks and progress in hardware technology led to the development of ATM.

Two other parameters are influencing the directions taken by the Internet: fast evolution of semiconductor and optical technologies. The combination of optical and electronic components appears capable of dramatically enhancing and extending the capabilities of current electronic systems.

One of the most decisive factors for making ATM switching a potential solution is that it has been defined totally independently of an underlying transmission technology. Since these emerging applications require increased bandwidth capacity, the vision of using photonic technology in communication channels, signal processing, and switching fabric is very promising. A communication network can be seen as switching elements combined with communication links. Optics present several advantages in all parts of the networks. Fiber optics has matured enough and is now a proven technology. Optical communications introduces a new era of interconnections. Furthermore, optical networks arise from their potential to offer unprecedented low-cost communications capacity.

Currently, fiber optics is used as the physical medium for different networks such as Synchronous Optical Network (SONET)/Synchronous Digital Hierarchy (SDH), Fiber Distributed Data Interface (FDDI), Dual Queue Dual Bus (DQDB), to name few. SONET is a transmission protocol over fiber and the Internet backbone that is widely used in all point-to-point links. One of the major factors of developing optical ATM switching is the standardization of the SONET transmission system to carry ATM cells. The development of optical switching is improving as the optical technology advances. This is expected to ease the interfacing between ATM cells and SONET frames.

2   ADVANTAGES OF OPTICAL SWITCHING

There are many advantages for designing switching elements using optical components. These advantages include decreased switching time (less than 1/10 of a picosecond (10^{-12})), less cross talk and interference, increased reliability, increased fault tolerance, enhanced transmission capacity, economical broadband transport network construction, enhanced cross-connect node throughput, and flexible service provisioning.

Most of these advantages stem from the fundamental properties of light. Light is composed of photons, which are neutral bosons unaffected by mutual interactions and coulomb forces. Thus multiple beams of photons (with different frequencies) can cross paths without significant interference. The speed of optical links is limited only by the switching speed and capacitance of transmitters and receivers. This property allows holographic interconnects to achieve a three-connection density with only two-dimensional optical elements [24]. These characteristics are highly desirable in developing a very high capacity and flexible communication systems. However, using optical frequencies still has some difficulties, especially with optical component design, which requires its own technology. This is because the extremely small wavelength requires micrometric manipulations.

3   OPTICAL TECHNOLOGY

During the last few years, a new type of optical wave has been discovered that is called a soliton wave. The technology uses a nonlinear optical effect to create waves that have special characteristics. A soliton pulse can propagate for a long distance without distortion. The word soliton was first coined by Zabusky and Kruskal in dealing with the wave motion of a one-dimensional nonlinear lattice vibration. It is found that the shape of a nonlinear wave remains unchanged even

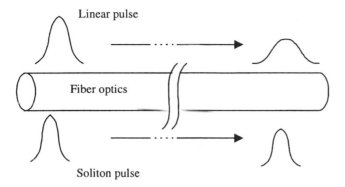

Figure 1 Comparison between an optical soliton pulse and a linear optical pulse.

when two such waves collide. Solitons were demonstrated experimentally in a single mode fiber by Mollenaucr et al. in 1980 [27].

Progress in optical soliton communication has been very rapid, since the invention of the 1.48 μm laser diode and erbium-doped optical fiber amplifiers (EDFAs). The main difference between a soliton pulse and a linear pulse is shown in Fig. 1.

This figure shows a comparison of a soliton pulse in an optical fiber to a linear optical pulse. An optical soliton does not change its waveform as it propagates, but the pulse in the linear transmission system broadens due to the group velocity dispersion. Thus solitons offer the potential of sending extremely clear pulses of light through very long runs of fiber. This means that solitons can be used to utilize the full bandwidth of a fiber over almost arbitrary distances.

Remarkable progress has been accomplished in the development of information transport networks based on optical technologies. Transmission systems with data rates exceeding 2.5 Gbps that use optical fiber have been introduced. SDH based on digital cross-connect systems and add-drop multiplexers (ADMs) are expected to be widely deployed, resulting in reduced transport network cost and enhanced reliability [24]. Subscriber networks have been gradually changing toward the introduction of optical techniques into subscriber loops, fiber to the home (FTTH), and fiber to the curb (FTTC). These advances are expected to strengthen in the near future. Once optical fibers are introduced into subscriber networks, the bit rate restriction of local access networks will be practically independent of the transmitted bit rate. In this context, trunk transport network throughput enhancement and cost reduction will be the keys to providing end-to-end broadband capacity at low cost.

So far, optical technologies have been introduced only at the physical layer to enhance transmission capability. To increase cross-connect throughput, photonic technology should be introduced into higher levels of the network. Recent technical advances in wavelength division multiplexing (WDM)/frequency division multiplexing (FDM) techniques have reached a level at which their practical application seems feasible. When the available number of wavelengths (frequencies) on a single fiber is an order of 10, the WDM technique is applicable to the path layer. But when it exceeds 10^2, the technique could be applicable to the circuit layer using wavelength routing [22].

Progress toward realization of optical networks manifests a need to develop the system, network, and device concepts required to expand the utilization of optical technology in a manner that will provide networks that support a variety of advanced applications.

4   OPTICAL NONLINEAR DEVICES

In this section, we briefly discuss nonlinear optical devices. The classical devices are discussed thoroughly in the literature, so they are not described in this chapter. Because of the advantages described above, semiconductor optical technology is emerging as a leading technology for building high-speed systems. Based on this technology, a number of high-speed optical devices such as optical bidirectional couplers, self-electrooptic effect devices (SEEDs), optoelectronic integrated circuits (OEICs) and interference filters using logic etalons (OLEs) have been experimentally demonstrated. These devices can provide extremely high data rates and a very large number of parallel channels. The projected data rate of optical etalon systems is well into 100–500 Gbps with the number of interconnections exceeding one million.

Interference filters (IFs) are used as optical bistable devices which are nonlinear devices. The technology of these optical filters is based on the Fabry–Perot resonator.

A Fabry–Perot resonator is a device made by sandwiching a nonlinear electrooptic material such as lithium niobite (LiNbO$_3$), between two partially reflective surfaces (mirrors). Light entering the cavity bounces back and forth between the mirrors. If the spacing between the mirrors is related to the wavelength of the optical signal in precisely an integer number ratio, then the combination of mirrors will act as a transparent window. Changing the spacing results in switching the device from reflective mode to transparent mode. The length of the gap is dependent upon the index of refraction of the nonlinear material and can be changed by applying a control optical beam that causes the device to switch. With low-intensity input light, the cavity will not resonate, but when the input intensity exceeds a certain threshold, the device switches and most of the light is transmitted instead of reflected. It has been experimentally demonstrated that these interference filters can be used for storage elements since they have the capability of transmission, reflection, and absorption based on the intensity of the incident light. The switching speed of these optical devices is of the order of picoseconds.

A few examples of using optical devices to design switching elements are discussed in this chapter.

5 CLASSIFICATION OF OPTICAL SWITCHES

Optical switches are mainly classified into three types:

Space-division switches, which includes also free-space switches
Time-division switches
Wavelength (frequency)-division switches

5.1 Space-Division Switches

In space-division switching networks, inputs and outputs are connected via physical paths. Different connections use different paths, and each new connection requires additional physical space in the switching fabric. Photonic space-division switching is an attractive way for constructing flexible broadband networks. The Polarization independent LiNbO$_3$ switch is a good example of this type, so it is discussed next.

Polarization Independent LinbO$_3$ Switch. Lithium niobate (LiNbO$_3$) switch matrices show the best characteristics with regard to matrix size and insertion loss. Several kinds of LiNbO$_3$ photonic switch matrices have been reported using different architectures. The number of lines needed in small switches in practical applications for space division switches should be more than 100. A 128-line prototype photonic switch/network was proposed in [19,20]. The photonic network uses different switch sizes as shown in Fig. 2.

This figure shows the structure of the 8 × 8 switch matrix used in the network with 64 polarization-independent directional coupler switch elements on one LiNbO$_3$ chip using a simplified tree structure (*STS*) architecture. The advantage of the *STS* architecture is its reduced switching voltage, insertion loss, and cross talk. The switching voltage of the 8 × 8 switch is approximately 70 volts. This is acceptable for a circuit switching application. For an 8 × 8 switch, the insertion loss and the cross talk value are about 12 dB and 18 dB, respectively.

Free-space Switches. Recently, free-space photonic switching has attracted a large number of researchers. In free-space photonic switches, light travels through free space or some medium (glass, vacuum, or air). The signals are typically directed (through free space) to their destinations using bulk optical components, such as lenses, mirrors, beam splitters, and retardation plates.

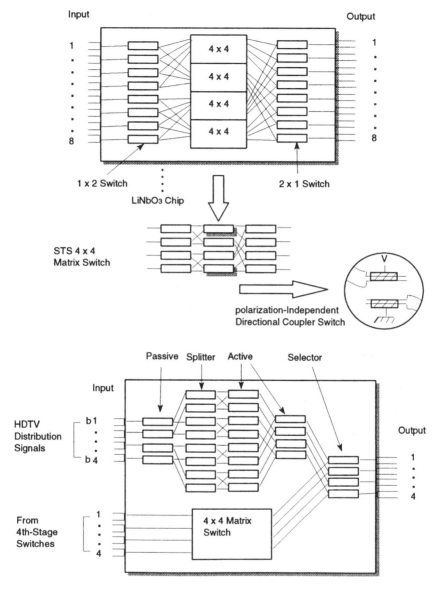

Figure 2 Polarization independent LiNbO$_3$ switch matrix. (a) 8 × 8; (b) 8 × 4 with distribution function.

A salient feature of the free-space switch is its high-density three-dimensional structure that enables compact large-scale switches to be created. The free-space switches are classified according to the following criteria:

Logical network configuration. Logical network configuration is implemented using small-scale guided switches (e.g., 8-input, 8-output switch) and fiber interconnections. These kinds of switches are not compact due to the large number of interconnections.

Path-establishment method. Free-space switches can be categorized into path-shifting or branching-and-gating switches similar to guided-wave switches.

Number of physical stages. Three-dimensional free-space switches can be classified as either single-stage or multistage physical structures as shown in Fig. 3. Single-stage switches are better than multistage switches because there is no accumulation of loss and cross talk due to cascaded connections.

Signal waveform transmission. Free-space switches can be classified by the form of signal waveform transmission into analog and digital switches.

Interconnection optics. Usually, there are two types of interconnection systems: imaging interconnection systems and microbeam interconnection systems. In the first type, a large lens transmits light signals from the source plane to the destination plane. In the second type, the beams on the destination plane are focused by a microlens array. This is suitable for relatively low-density interconnections because the minimum beam diameter is limited by diffraction. Figure 4 shows an example of these two types of interconnections.

5.2 Time Division Switches

Photonic time division (TD) switching systems are based on the same principles as conventional digital TD switching systems. The main features are expandability and small physical size. TD multiplexed (TDM) signals in the optical fiber transmission line are switched by interchanging the time slot in the switching network. However, the line's required operating speed increases significantly with TDM multiplexing. A photonic TD switching system, using 150 Mbps full-motion video terminals instead of telephones requires a switching speed of about 32 Gbps. This is 2,000 times higher than that of the conventional digital switching system. Recently, a 10 Gbps optical transmission with intensity-modulation direct detection has been demonstrated [20].

Figure 5 shows a switch architecture that has N input and N output channels carrying electrical signals (or optical signals). It consists of input modules, TDM modules, an optical star broadcast network, output grouping modules, optical splitters, output buffering modules, and cell expanding modules. The electrical signals entering or leaving the switch have bit rates of V Gbps and a cell of a duration of T seconds. The N input / N output channels are divided into M channel groups. Each group has the same number of K channels time-multiplexed over

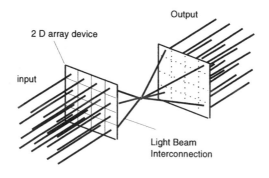

(a)

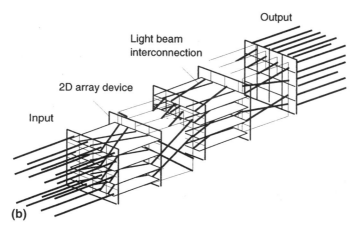

(b)

Figure 3 Classification based on the number of physical stages. (a) Single-stage switch. (b) Multistage switch.

an optical highway by a TDM module where cells with a duration of T seconds are optically time-compressed to cells with a duration of T/K seconds. The output time-multiplexed optical signals from individual TDM modules have different wavelengths. These outputs are combined by the star broadcast network and then evenly broadcast to M individual output grouping modules. Each of the output grouping modules wavelength-demultiplexes the combined signals and selects the cells destined for it. The selected cells within a group are buffered and broadcast via an optical splitter to the output buffering modules of that group where the address matched cells are finally identified. The cells emerging out of an

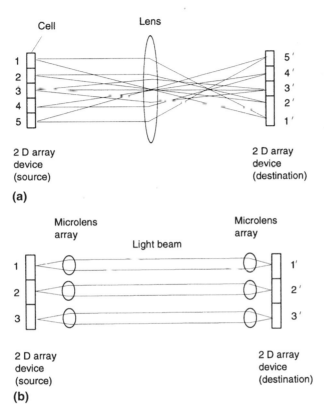

Figure 4 Interconnection optics for free-space switches. (a) Imaging interconnection system. (b) Microbeam interconnection system.

output buffer which are time-compressed are converted back into a cell expanding module.

5.3 Wavelength Division Switches

The wavelength (or frequency) switch in the wavelength division (WD) network corresponds to the time switch in the TD network. The wavelength switch combines wavelength filters and wavelength converters. It also interchanges the wavelength of signals multiplexed on the input highway. One prototype of WD switches is shown in Fig. 6.

In this WD switch, the optical signal selected by the collinear acousto-optic switch is led to the photodiodes (PD); then the electrical signals from the PD drive the laser diode (LD) [6].

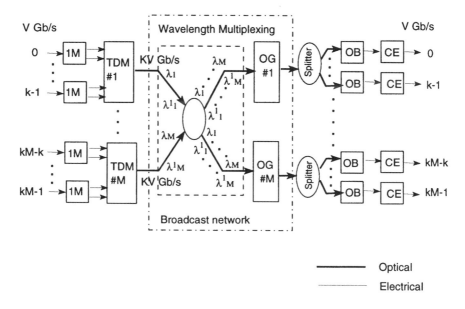

| Optical |
| Electrical |

IM: Input Module
TDM: Time-division multiplexing module
OG: Output Grouping module
OB: Output Buffering module
CE: Cell expanding module

Figure 5   Block diagram of the proposed switch architecture using hierarchical photonic TDM and WDM multiplexing.

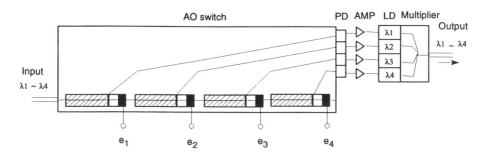

Figure 6   Wavelength conversion switch.

6 EXAMPLES OF OPTICAL SWITCHES AND NETWORKS

Many different photonic switching systems have been proposed in the literature, and few experimental prototypes have appeared in laboratories as testbeds [1]. Only few of these systems will be cited in this text. The list presented here is by no means exhaustive. For more information, the reader is referred to [22,23].

6.1 Recent Proposals

Zhong et al. [9] proposed a new photonic ATM switch structure that uses ultra-short optical pulses and a TDM scheme. In their structure, the bit pattern of the ATM cell is used to modulate a sequence of ultrashort optical pulses generated by a gain-switched distributed-feedback laser diode (DFB-LD). Shrinking the interval between two consecutive pulses allows the data cells from the individual input channels to be time-multiplexed and broadcast to all the outputs, each of which selects the address-matched cells. This optical TDM switch has a simple structure, but its capacity is less than 100 Gbps. This is because the number of channels that can be time-multiplexed on an optical highway is restricted by the technologies for generating and detecting ultrashort optical pulses. Using optical TDM and WDM technologies, the architecture discussed in [11] achieves a switch capacity of more than 1 Tbps (tera = 10^{12}), but it requires a large amount of hardware because it does not make use of the advantage of hierarchical multiplexing.

In [15], a photonic ATM switch architecture is described that has a new cell buffering scheme that combines the advantages of a large bandwidth of photonic circuits and advanced logical functions of electronic circuits. The switch buffers ultrafast optical cells by using optical delay lines that eliminate the need for very high-speed electronic first-in first-out (FIFO) memory. The new structure can thus serve as the basis of a modular, expandable, high-performance ATM switching system that requires a very large capacity of the order of terabits per second. This ATM switch architecture uses hierarchical TDM and WDM (or SDM) multiplexing schemes (see Fig. 7).

The idea behind this hierarchical multiplexing in the switch design is to provide a switching capacity of the order of terabits per second with an acceptable complexity of the switch fabric. The optical switch technology based on the ultra-short optical pulses reduces the required number of wavelengths by a factor of K, where K is the number of channels time-multiplexed on a wavelength. As a result, the required number of wavelengths and complexity of the switch fabric are reduced.

A novel photonic frequency division multiplexing switching named frequency assign photonic switching system (FAPS) is discussed in [11]. The basic concept of FAPS is to assign a single frequency for each call. Network configura-

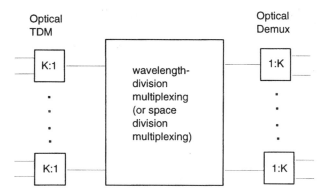

Figure 7   A generic photonic ATM switch structure using hierarchical TDM and WDM (or SDM) multiplexing.

tion, subscriber network, and multiplicity of frequencies are considered. The idea is that the system consists of n independent subnetworks, with n being a multiplicity of the frequencies. Each network topology is identical. When a connection is required by a terminal, the network selects a subnetwork that can afford it and assigns a frequency representing the selected subnetwork to the terminal. This system eliminates frequency converting devices and traffic concentration equipment, which reduces the size and cost of the system.

Multiplexing in ultrafast photonic ATM (ULPHA) switch. An ultrafast photonic ATM switch based on time-division multiplexing and on a "broadcast-and-select" network with optical output buffers was proposed. The ULPHA switch has an ultrahigh throughput and excellent traffic characteristics since it utilizes ultrashort optical pulses for cell signals and avoids cell contentions by novel optical output buffers. Feasibility studies show that an 80×80 ULPHA switch with 1-Gbps input/output is possible by applying the present technology. In addition, more than 1 Tbps is possible by making a three-stage network using such switches. As an experimental demonstration, 4 bit, 40 Gbps optical cells are generated, and certain cells are selected at an output on a self-routing basis. With its high throughput and excellent traffic characteristics, the ULPHA switch is a strong candidate for the future large-capacity optical switching node [21]. All the cells are time-multiplexed on the optical *TD* highways as ultrafast optical cells. Certain cells addressed to each output are selected from the highways with their throughput controlled by the optical output buffers. All these operations are processed on a cell basis. Cell multiplexing is performed by the cell coder and state coupler. First of all, it is assumed that electrical cells with a duration T arrive at the inputs from various transmission links at a data rate V. Then, in each of the input interface modules, the cells are regenerated to suppress frequency to phase jitter and reclocked by the system clock at frequency slots as shown in Fig. 8.

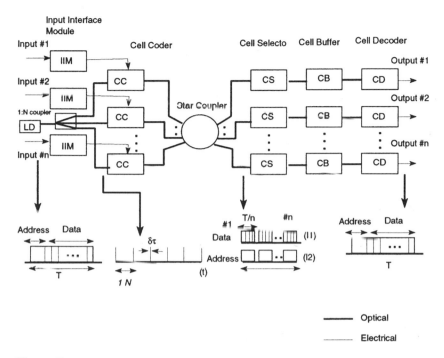

Figure 8 Block diagram of ULPHA switch.

Simultaneously, ultrashort optical pulses having a wavelength x_1 and a width β (where β is extremely small) are generated by a gain-switched DFB-LD at a repetition rate of V synchronously with the system clock and distributed to n cell coders. This single DFB-LD may be replaced by n synchronized DFB-LDs to reduce optical power distribution losses. Next, using the optical pulses, the cell coder converts the data fields of the electrical cells into ultrafast optical data cells with a duration of around T/n and a wavelength x_2. It also wavelength-multiplexes the ultrafast data and slow address cells at the cell coder output. Then the star coupler time-multiplexes all the cells from all the cell coders and distributes them on the optical TD highways, where the multiplexed data cells are separated by a certain margin time, enough for the subsequent processing.

6.2 Terahipas Switch

The switch architecture using TDM and WDM schemes is shown in Fig. 5. It has N input/output channels carrying electrical signals (or optical signals) [15]. The functional units shown in this figure are

Input Modules: this has four functions, which are

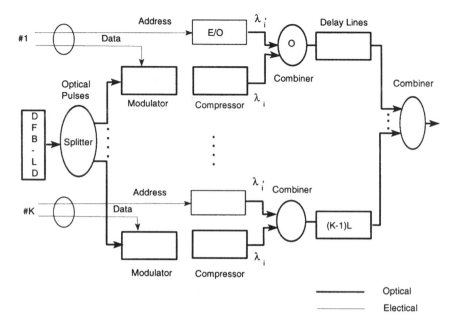

Figure 9   Configuration of the TDM module.

1. To convert the incoming optical cells into electrical cells
2. To obtain the destination address of a cell by electrically analyzing its header and attaching a new header to the cell
3. To synchronize the cells and address data over individual input channels
4. To send cells and address data simultaneously to the TDM module

TDM module: The block diagram of the TDM module is given in Fig. 9. Its function is to multiplex optically a group of K channels.

6.3   Shared-Medium Switches

In this type of switching network, the k-fold multiplexed signals on each of the L input lines are switched onto L output lines with respect to frequency space position. Figure 10 shows two types of shared-medium switches. Any permutation of frequencies at the entrance of the star coupler could be chosen to implement a crossbar switch using the shared-medium principle shown in Fig. 10 (top).

If the same permutation is chosen, for example, $f_1, f_2, \ldots, f_L$, at all shared-medium switches, we obtain the network shown in Fig. 11 with k' identical switches.

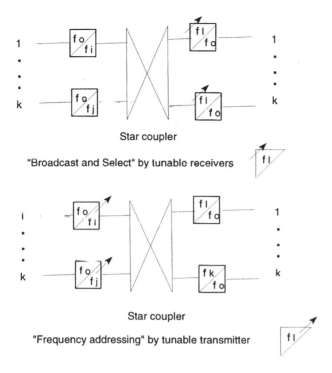

Figure 10 Basic types of shared-medium switches. The arrow denotes frequency tunability.

Since this architecture is derived from Clos configuration, the condition for absolute freedom from internal blocking can be given as $k' \geq 2k - 1$. To ensure that the channel at the input of the star coupler can still be distinguished, the number of input lines, L, must not exceed the number of carrier frequencies. The number of channels is determined by the product $L \times k$. The number of channels is around 1000 if the values of L are 10 or 25, and k consequently is 100 or 40, respectively.

6.4 Parallel λ Switch

A parallel λ switch is used to accomplish wavelength interchange. Figure 12 shows a three-stage photonic wavelength division switching network (λ^3 switching network), using wavelength multiplexers and demultiplexers (MUXs, DMUXs) in interstage networks.

The interstage switch can provide each λ switch with potential connectivity

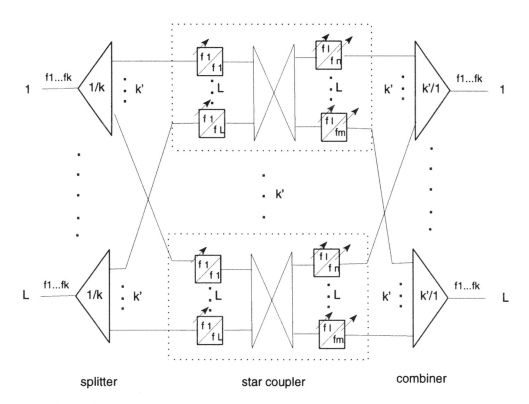

Figure 11 Switching network, realized exclusively with tunable transmitters and receivers. Inputs of the outlined switching modules are composed of tunable wavelength filters and fixed wavelength transmitters. Outputs comprise tunable-wavelength filters and converters.

to every next-stage λ switch. Since the switching network shown in Fig. 12 is equivalent to a multistage Clos network using $n \times n$ switch matrices, a nonblocking switching network can be constructed using this network. Utilizing wavelength interchange function of λ switches, this kind of network allows the reuse of wavelength channels in different switching stages. This leads to a large-capacity switching system with fewer wavelength channels. Consequently, all WDM input lines to the switching network can use the same wavelength channels $\lambda_1, \lambda_2, \ldots \lambda_n$. The line capacity for a single λ switch is expressed as n, which is the same for the WDM passive-star network. Line capacities for λ^3 and λ^5 switching networks are expressed as n^2 and n^3, respectively. Extremely large line capacity can be achieved using multistage connection with λ switches even with a limited number of WD channels.

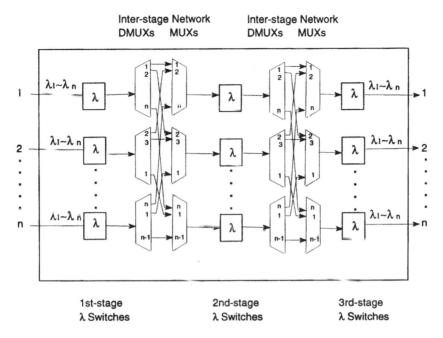

Figure 12 Multistage WD switching network.

6.5 Electrocapillary Optical Switch

The *electrocapillary optical* (ECO) switch is based on an electrochemical effect whereby a mercury droplet is driven by an electric current in an electrolyte. The electrically produced force is a shear force tangential to the interface. Therefore it is called the *continuous electrowetting* (CEW) effect.

This kind of switch is successful in switching optical paths between multimode fibers (acting as a 1×2 switch) with an attractive level of performance, as it has

> Fast response time (≈ 20 ms)
> Low driving power (≈ 25 μW)
> Low driving voltage (0.5 to 1.0 V)

Since the surface of the mercury is used as a mirror, this switch is independent of polarization and wavelength. Apart from these properties, it is bistable (i.e., no power is required to maintain the switching state). The bistability is required for route reconnection switches, which are installed in the subscriber network as the switching matrices between optical cables. Figure 13 shows the concept of an ECO switch on a planar lightwave circuit (PLC).

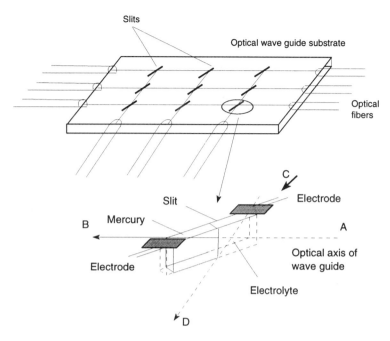

Figure 13 Concept of electrocapillary optical switch.

Slits are etched into a glass substrate at the crossing points of buried-type optical waveguides. There is an electrolyte in the slit that acts as an index matching liquid and also a memory droplet that functions as a micromirror. A voltage is applied between the electrodes at each end of the slit to drive the mirror. The guided optical powers are transmitted straight (from **A** to **B** and from **C** to **D**) when the mirror is facing away from the crossing point of the waveguides (OFF state). If the mirror is at the cross point (ON state), the optical power from **A** is reflected into the cross guide **D**. In the latter state, waveguides **C** and **B** are not used, but this causes no problem because the switch elements in the ON state do not share the same column numbers or the same row numbers in the matrix switches.

Figure 14 shows the structure of an ECO switch for an optical susbscriber network. The switch system is used to connect together the fibers in the three cables.

The two cables containing *N* fibers are upper and lower main route cables, and the cable with 2*N* fibers is a branch cable. The switch consists of two matrices:

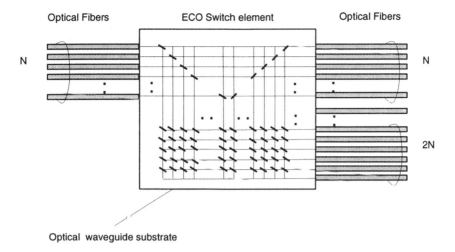

Optical Fibers   ECO Switch element   Optical Fibers

N

N

2N

Optical waveguide substrate

Figure 14   Circuit structure of an ECO switch for use in optical subscriber network.

A matrix with switch elements in diagonal positions
A full $2N \times 2N$ cross-connect matrix

The first part enables the main route transmission to avoid the switching power loss caused by reflection. It also keeps transmission loss at a minimum through the slits. Branching signals are reflected twice and transmitted through at most $2(N-1)$ OFF state slits. Therefore the total transmission loss in the network is kept at a low level. This switch has the potential for use in bistable waveguide matrix switches, which are suitable for route reconnection in the optical subscriber network.

6.6   Recent Projects

A self routing fault-tolerant optical ATM switch (SEROS) has been proposed by Guizani and Memon [29].

The contention resolution by delay-lines (CORD) project is an ATM optical packet-switched networking consortium of GTE Laboratories, Stanford University, and the University of Massachusetts. CORD focuses on the problems of resource contention, signaling, and local and global synchronization. With CORD, an optical solution to resource contention, based on the use of optical switches and delay lines, is utilized. Signaling is achieved with subcarrier multiplexing of packet headers. Synchronization issues are resolved by means of clock tone multiplexing, new ultrafast digital bit clock recovery techniques, and new distributed techniques for global packet-slot alignment. The principles and technologies developed by the CORD consortium are flexible and are expected to find numerous future applications, together or separately, in a variety of topologies and network architectures. A passive star-topology network with two nodes was built. Each node has one fixed wavelength transmitter, operating at $\lambda_1 = 1300$ and $\lambda_2 = 1320$ nm. Each node transmits ATM-sized packets at 2.5 Gbps, addressed both to the other node and to itself. This gives rise to receiver contentions. At each receiver, a WDM demultiplexer separates the two wavelengths so that either one can be received.

LAMBDANET was proposed by Bellcore in the mid-1980s [26]. The network interconnects n nodes using a fixed transmitter(s) and fixed receivers(s) approach. Each node on the network had a single fixed transmitter and n receivers, one for each sending channel. The receiver listens to all the channels all the time and pulls out the transmission it wishes. The prototype network connects sixteen hosts, each channel having a bandwidth of 2 Gbps.

The obvious drawback of the LAMBDANET design is the requirement for n receivers at each node. This approach is clearly expensive. However, the approach has some very important virtues such as easy multicasting without the need of transmission scheduling. When LAMBDANET was built, tunable receivers were still very primitive. By using multiple receivers, LAMBDANET was able to test some of the ideas that would later be used in fixed transmitter(s) and tunable receivers(s) networks.

One way to avoid the need for fast tunable transmitters or receivers is to build a multihop WDM network. In multihop WDM systems, transmitters and receivers typically are tuned to a fixed frequency. They are capable of being tuned slowly and returned in case of a failure in the network.

Multihop networks can be designed in a variety of ways. The essential idea is to build a connectivity graph among the nodes, so that the number required to get between two nodes is kept small. Few multihop networks have been built. TERANET is a multihop WDM network being built as part of the Advanced Communications Organizations for Research Networks (ACORN) project at Columbia University in New York.

7 SUMMARY

Optical networks can be viewed as potential candidates for a national telecommunications infrastructure supporting a low-cost universal network. This network will allow access to supercomputers and vast multimedia databases and more sophisticated point-to-point connections. Finally, advanced optical network technologies will also accelerate the research process in all disciplines and enable education to integrate new knowledge and methodologies into course curricula.

REFERENCES

1. M. Guizani. ATM architectures using optical technology. an overview on switching, buffering and multiplexing. *International Journal of Network Management*. Vol. 7, No. 4. July–August 1997.
2. M. Yamaguchi and K. Yukimatsu. Recent free-space photonic switches. *IEICE Transactions on Communication*, Vol. E77-B, No. 2, pp. 128–138, February 1994.
3. M. Yamaguchi and T. Matsunaga. Comparison of a novel photonic frequency-based switching network with similar architectures. *IEICE Transactions on Communication*, Vol. E77-B, No. 2, pp 147–154, February 1994.
4. M. Fujiwara and T. Sawano. Photonic space-division switching technologies for broadband networks. *IEICE Transactions on Communication*, Vol. E77-B, No. 2, pp. 110–118, February 1994.
5. K. Murakami and S. Kuroyanagi. Overview of photonic switching systems using time-division and wavelength-division multiplexing. *IEICE Transactions on Communication*, Vol. E77-B, No. 2. pp. 119–126, February 1994.
6. S. Suzuki, M. Nishio. T. Numai, M. Fujiwara, M. Itoh, S. Murata, and N. Shimosaka. A photonic wavelength-division switching system using tunable laser diode filters. *Journal of Lightwave Technology*, Vol. 8, No. 5, pp. 660–666, May 1990.
7. M. Sato. Electrocapillary optical switch. *IEICE Transactions on Communication*, Vol. E77-B, No. 2, pp. 197–203, February 1994.
8. M. Nishio, S. Suzuki, K. Takagi, I. Ogura. T. Numai, K. Kasahara, and K. Kaede. A new architecture of photonic ATM switches. *IEEE Communication*, pp. 62–68, April 1993.
9. W. D. Zhong, M. Tsukada, K. Yukimatsu, and Y. Shimazu. Terahipas: a modular and expandable terabit/second hierarchically multiplexing photonic ATM switch architecture. *Journal of Lightwave Technology*. Vol. 12, No. 7, pp 1307–1315, July 1992.
10. A. Himeno, R. Nagase, T. Ito, K. Kato, and M. Okuno. Photonic inter-module connector using 8 × 8 optical switches for near-future electronic switching systems. *IEICE Transactions on Communications*, Vol. E77, No. 2, pp. 155–162, Feb. 1994.
11. T. Yasui and A. Uemurai. A proposal of a new photonic FDM switching system FAPS (frequency assign photonic switching). *IEICE Transactions on Communications*, Vol. E77, No. 2, pp. 174–183, Feb. 1994.
12. W. D. Zhong, Y. Shimazu, M. Tsukuda, and K. Yukimatsu. A modular tbit/s TDM/WDM photonic ATM switch using optical output buffers. *IEICE Transactions on Communications*, Vol. E77, No. 2. pp. 190–196, Feb. 1994.
13. H. Haga, M. Izutsu, and T. Sueta. An integrated 1 × 4 high-speed optical switch and its applications to time demultiplexers. *Journal of Lightwave Technology*, Vol. LT-3, No. 1, pp. 116–120, 1992.

14. N. Yokoto, Y. Hakamata, and Takemoto. Time division 8:1 multiplexing and demultiplexing system adapting 7BIC code operating at 6 Gbps. *Electronics Letters*, Vol. 26, No. 13, pp. 923–925, June 1990.
15. T. J. Cloonan and F. B. McCormick. Photonic switching applications of 2-D and 3-D crossover networks based on 2-input, 2-output switching nodes. *Applied Optics*, Vol. 30, pp. 2309–2323, 1991.
16. E. Kerbis, T. J. Cloonan, and F. B. McCormick. An all-optical realization of a 2 × 1 free-space switching node. *IEEE Phot. Tech. Letters*, Vol. 2, pp. 600–602, 1990.
17. T. J. Cloonan and A. L. Lentine. Self-routing crossbar packet switch employing free-space optics for chip-to-chip interconnections. *Applied Optics*. Vol. 30. pp. 3721–3733, 1991.
18. T. J. Cloonan, M. J. Herron, F. A. Tooley, G. W. Richards, F. B. McCormick, E. Kerbis, J. L. Brubaker, and A. L. Lentine. An all-optical implementation of a 3-D crossover switching network. *IEEE Photo. Tech. Letters*, Vol. 2, pp. 438–440, 1990.
19. T. Matsunaga, M. Okuno, and K. Yukimatsu. Large-scale space division switching system using silica-based 8 × 8 matrix switches. *Proceedings OEC'92*, (Makuhari, Chiba), pp. 256–257, July 1992.
20. M. Goto, K Hironishi, A. Sugata, K. Mori, T. Horimatsu, and M. Sasaki. A 10 Gbps optical transmitter module with a monolithically integrated electro-absorption modulator with a DFB laser. *IEEE Photo. Techn. Lett.*, Vol. 2, p. 896, 1990.
21. Y. Shimazu and M. Tsukada. Ultrafast photonic ATM switch with optical output buffers. *Journal of Lightwave Technology*, Vol. LT-10, No. 2, pp. 265–272, Feb. 1992.
22. Ken-Ichi Sato. *Advances in Transport Network Technologies*. Artech House, 1996.
23. IEEE Journal on Selected Areas in Communications, Vol. 14, No. 5. June 96.
24. C. Tocci and H. J. Caulfield. *Optical Interconnection, Foundations and Applications*. Artech House, 1994.
25. M. Sexton and A. Reid. *Broadband Networking: ATM, SDH, and SONET*. Artech House, 1997.
26. M. S. Goodman, H. Kobrinski, M. P. Vecchi, R. M. Bulley, and J. L. Gimlet. The LAMBDANET multiwavelength network: architecture, applications, and demonstrations. *IEEE Journal of Selected Areas in Communications*. Vol. 8, No. 6, pp. 995–1004, August 1990.
27. L. F. Mollenaur, R. H. Stolen, and J. P. Gorden. Experimental observation of picosecond pulse narrowing and solitons in optical fibers. *Phys. Rev. Lett.*, Vol. 45, No. 13, pp. 1095–1098, 1980.
28. Masataka Nakasawa. Soliton transmission in telecommunication network. *IEEE Communications*, pp. 34–41, March 1994.
29. M. Guizani and A. Memon. *SEROS*: a self-routing optical ATM switch. *International Journal of Communications Systems*, Vol. 9, No. 2, pp. 115–125, March–April 1996.

9

Scalable Electro-Optical Clos Switch Architectures

Fotios K. Liotopoulos
Computer Technology Institute, Athens, Qrooce

1 INTRODUCTION

Progress in all forms of today's information society is fundamentally based on the information exchange between producers and consumers. With the amounts and rates of exchanged information traffic exploding exponentially in the last few years, the problem of switching a multitude of diverse data streams between several sources and destinations, at very high rates, has become a demanding challenge [8,24,42,53,55,59]. Typical areas where this challenge is apparent are

Internet core switching at large internet service providers and internet exchange points

Multimedia service distribution networks (such as video-on-demand distribution networks and archive distribution and storage area networks (SAN))

Telecommunication provider switching networks (fixed or mobile telephony core networks)

Supercomputing interconnection networks [for multiprocessors and cluster of workstations (COW, NOW)]

The explosion in the use of information services and the Internet during the 1990s is expected to continue in the first decade of the new millennium. To accommodate this explosion, the underlying communications and internetworking infrastructure must be able to provide the extra capacity and resources required. This new reality thus poses several challenging questions to communications engineers: What are the characteristics of this explosion that need to be accounted for? How much scalability do we need to design for? What is the asymptotic behavior of this information service proliferation?

Statistical study of information service trends indicates that the expected increase will mostly be attributed to increases in the data rates of the various information services and less to increases in the number of users and services themselves.

The profile of the upcoming explosion can be characterized by the rates of increase in *users*, *services*, and *data*. Information users are expected to increase by no more that one order of magnitude, as current penetration in the global population is in the order of 10%. As information services grow in diversity, content, and usefulness and users are becoming more and more exposed to them, the daily frequency and duration of their use will also increase. This will effectively lead to an overall increase in the average service rates by one to two orders of magnitude. On the other hand, due to the improving quality and content of multimedia services and the transition from sound-based to video-based services, the rate of data transmitted over communications networks is expected to increase by three to four orders of magnitude.

These growing demands for internetworking and switching capacity have stimulated the development of high-speed optical networks, [9–11,15–21,32,33,37,57]. Currently, wavelength division multiplexed (WDM) networks with WDM channel rates of up to approximately 10 Gbps per channel are being aggressively developed and deployed [2,10,36,62]. Ultra-fast optical time-division multiplexed (OTDM) networks with single-channel rates in excess of 100 Gbps may offer advantages over WDM networks, depending on the profile of the users and the network traffic. Such advantages include flexible bandwidth-on-demand (BoD) service, packet/cell switching and self-routing, statistical multiplexing, variable quality of service (QoS) support, digital signal regeneration, and buffering.

Based on combinations of WDM and OTDM optical technologies as well as electronic packet or cell switching technologies (e.g., ATM), switch engineers can utilize single multigigabit channels to transfer data and design very large capacity core switches [37,45,46].

Optical switches are critical elements in all-optical networks [2]. They make up the fabric of the network by joining different links with each other, controlled by either a centralized network routing controller or the routing information carried by individual packets in their headers. There is a distinction between switches used in passive and active optical networks. Passive optical networks employ passive switching elements. For example, the star couplers and wavelength division multiplexers/demultiplexers used in broadcast and select networks and wavelength routed networks, respectively, are fixed and nonconfigurable switches [34,35]. The optical switches employed in active optical networks are configurable switches whose state is set by the routing information carried by the packets traveling through the switch.

However, the technological achievements in single-channel optical transport are not the most important factors in developing scalable switches, which are essential to support multiterabit (Tbps) switching capacities. Scalable high-performance switch architectures are also essential to be studied and evaluated that will effectively utilize optical channels and components to produce switching fabrics with low call/cell blocking, minimum resource conflicts, and high reliability. Furthermore, certain switch functions, such as intelligent resource management and control, quality of service, call admission, service charging, and policing need to be implemented using electronics [40,41,47,48]. Therefore both high-performance electronic and high-performance optical technologies need to be applied in optimum balance and with similar scalability properties in order to avoid localized bottlenecks that will hinder the overall switch performance and weaken the advantages of some technologies because of the deficiencies of others.

In these directions, we advocate the use of three-stage Clos network architectures for the development of scalable high-performance, high-capacity, fault-tolerant, active core switches. During the late 1990s, several photonic ATM switch architectures have been proposed and prototypes have been developed [33,37,46]. In particular, a significant portion of them involved the implementation of such switch fabrics and interconnection schemes based on three-stage Clos networks [5–7,11,26,30,61]. A comparative evaluation of photonic ATM switch architectures, with respect to their performance and the number of required optical switching devices to be implemented, can be found in [37].

In this chapter, we advocate the combination of optical and electrical technologies and components in the development of switching devices. In particular, the current trends seem to favor the combination of fast optical cell transmission and switching functions (e.g., at rates of 10 Gbps) and intelligent electrical control with relatively low speed (e.g., 42.4 nsec per cell) parallel processing [37].

2 THREE-STAGE CLOS NETWORKS

Three-stage Clos networks [1,13,14] have been used for multiprocessor interconnection networks as well as for data-switching fabrics. Examples include the Memphis switch of the IBM GF-11 parallel system [3], Fujitsu's FETEX-150, NEC's ATOM, Hitachi's hyperdistributed switching system, the OPTIMA [61, 62] and TORUS electro-optical Tbps switch, the ATMOS optical switch, [5–07,30], and Lucent's Atlanta™ chipset [53–56].

Clos networks can be used to implement low latency high-bandwidth connection-oriented ATM switching. Clos-type switching fabrics have also been found to be economically and technically attractive. Combined with their inherent fault tolerant and multipath routing properties, they pose as a very appealing choice for reliable broadband switching [1,29,43,44,53].

A three-stage Clos network (see Fig. 1) consists of three successive stages of switching elements, which are interconnected by point-to-point links. In a symmetric three-stage network, all switching elements in a stage are uniform. In a symmetric Clos network such as this, there are r switches of size $n \times m$ in the first stage, m switches of size $r \times r$ in the second, and r switches of size $m \times n$ in the third stage. This network thus interconnects $n*r$ input ports with $n*r$ output ports in one direction.

In a three-stage Clos network architecture, we can apply a variety of blocking or nonblocking call admission control (CAC) algorithms, proposed by numerous researchers (e.g., [25,40,41]) for generalized three-stage Clos switching networks in multirate environments. Note at this point that the term "nonblocking" is used in this context to refer to the nonblocking property of the switching network *at call setup time*.

We distinguish the following four nonblocking CAC modes characterizing the switching operation and CAC algorithm of a three-stage Clos network (see also Fig. 2):

> *Strictly Nonblocking (SNB)*, where no particular CAC algorithm is applied, and link capacity allocation is performed ad hoc. As a result, it is very

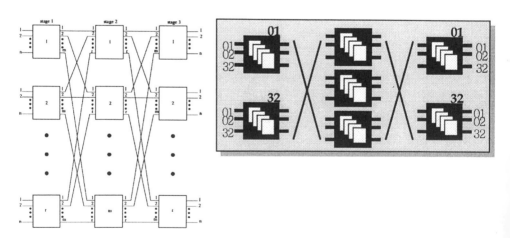

Figure 1 (Left) A symmetric [r, m, r] three-stage Clos switch architecture. (Right) (up) A 1024 × 1024 i/o symmetric configuration.

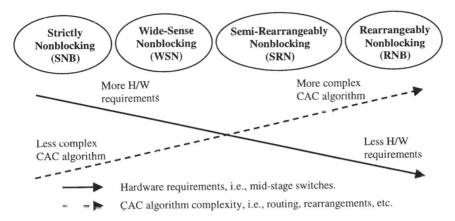

Figure 2 Trade-offs between the four main nonblocking CAC modes.

demanding in terms of middle-stage switches, in order to guarantee non-blocking call setup.

Wide-Sense Nonblocking (WSN), which uses a special connection placement algorithm, in order to reduce the hardware requirements, compared to the SNB mode.

Semirearrangeably Nonblocking (SRN), where connections are established based on a placement algorithm (usually aiming at call balancing or call stacking). A limited number of connection rearrangements are performed only after call disconnection.

Rearrangeably Nonblocking (RNB), which is similar to SRN but allows an unlimited number of rearrangements to be performed. This mode thus requires the least hardware resources but may be prohibitive in terms of its practical implementation.

The implementation of most intelligent and efficient CAC algorithms requires global knowledge of the status of the switch resources, namely the capacity utilization of all i/o and interstage links. If we assume a centralized call admission controller, which maintains all global information and applies the appropriate CAC algorithm, then this controller becomes a severe bottleneck at moderate to high traffic loads, depending on the complexity of the CAC algorithm.

Since call requests arrive randomly at the input ports of the switch, the CAC decisions regarding call placement can be partially parallelized to alleviate the aforementioned centralized bottleneck. In order to exploit this parallelism, one needs to explore distributed architectures and implementations of the call admission controller.

3 OPTICAL SWITCH COMPONENTS AND TECHNOLOGIES

Interconnection and *packaging* are the two main technological aspects that are crucial for the effective scaling of the switching capacities into the multiterabit (Tbps) range. As the physical size of the switching fabric increases in order to achieve higher capacities, the connectivity requirements, in terms of the number of links and the speed between stages, grow higher.

Higher node capacities require higher levels of packaging and eventually impose the requirement for a large number of i/o ports at both the IC and the board level. In addition, high-speed multiplexed data streams should be transmitted to reduce the number of interconnection links. In this direction, the introduction and development of *multichip module* (MCM) and *optical interconnection* (OIC) technologies allow for the design and feasible implementation of high-capacity, high-performance switching devices [4,39,60].

3.1 Multichip Modules

Multichip modules are a promising technique that allow the reduction of device "real estate" area, improve integration density, and increase the overall system performance. MCMs involve the mounting and small-scale interconnection of bare dice on a single substrate. As an example, we report a 16 × 16 switching element, designed by CSELT [4] and realized as an 11-layer MCM-C (cofired ceramic) by IBM. This MCM uses four 8 × 8 switching modules properly interconnected to avoid blocking and takes up one-sixth of the space it would normally take if packaged with standard components. It has a complexity of about 2.8 million transistors, operates at 78 MHz, and dissipates about 10 watts. The output drivers can drive 48 balanced lines (ECL or PECL) at up to 320 MHz, with an area of about 800 mm^2 and a power dissipation of 2.4 watts. The input receiver adapts 48 balanced PECL inputs to 48 single TTL outputs at up to 100 MHz, with an area of about 800 mm^2 and a power dissipation of 2 watts.

3.2 Optical Interconnections

Optical interconnection technologies are becoming more and more available due to the progress in low-power, small-dimension, highly reliable electro-optical devices, such as very low threshold (sub-mA) vertical cavity surface emitting lasers (VCSEL), which have made practical and relatively cheap the development of parallel optical interconnections [4]. The combination of optic fibers and microelectronics plays an important role in this direction [12,22,23,39,49,50, 60,63].

Point-to-point optical interconnection focuses on the development of two basic components:

1. A *parallel-to-serial* and a *serial-to-parallel* converter. As an example, we mention the ATM serial transmitter/receiver integrated circuit for data links (ASTRID) [ref.Merging]. ASTRID has a parallel interface (4/8 data, data clock, cell clock) at 77.8/38.9 MHz and a serial interface at 311 Mbps (PECL levels), and the conversion includes cell alignment and cell start detection. Implemented using 0.7 μm CMOS technology, it dissipates about 1.1 watts.

2. A *laser driver* and a *diode receiver* circuit. As an example, we mention the multipurpose interconnection unit for electro-optic transceivers (MINUET) [04]. An implementation of MINUET by CSELT (Italy) has been reported operating at 311 Mbps, using PECL electrical interfaces. The receiver sensitivity was −20 dBm with a dynamic range of 12 dB and the power dissipation was 100 mW for the transmitter and 150 mW for the receiver.

In the following paragraphs we examine various optical interconnection components, which can be applied in the development of a high-capacity, high-performance switching fabric, based on three-stage Clos networks. The way these components are proposed to be used is in the form of interconnecting input ports and output ports within a switch module, in a similar way as remote stations are optically interconnected in a LAN or WAN.

3.3 Broadcast-and-Select (Star and Ring Couplers)

Broadcast-and-select (B&S) WDM is widely used in optical networking. It consists of a *star coupler*, to which a number of remote stations are interconnected [34,35,59]. Each station transmits information using a multiplexer and a laser diode and receives information via an optical amplifier, an optical splitter, and multiple selectors. A typical station multiplexes its local channels and transmits the multiplexed optical signal using a preassigned wavelength. The star coupler combines all optical signals and redistributes the combined WDM signal to all stations. Each station receives the signal and splits it accordingly, based on its wavelength selectors. Subsequently, further electrical demultiplexing of the specific wavelength can be used to split the single optical signal into smaller information channels.

The star-coupler-based WDM B&S network can be replaced by a *ring*-type WDM B&S network, where all stations are placed in a ring configuration (like add/drop multiplexers) [18]. Each station adds its own wavelength and it splits and receives all of the wavelengths multiplexed on the ring fiber. Thus the ring-type WDM B&S network is logically equivalent to the star-type network.

3.4 Arrayed Waveguide Grating Filters

Arrayed waveguide grating filters (AWGF) are optical modules consisting of N inputs and N outputs [51,52,60,63]. In order to perform $N \times N$ switching, the AWGF requires only N wavelengths, $\lambda_1, \ldots, \lambda_N$, as shown in Table 1.

According to Table 1, this AWGF is capable of simultaneously switching up to N cells of different wavelengths to a given output.

Yamanaka et al. [60,61] have developed an efficient AWGF component, which was used as a modular optical interconnection device in the OPTIMA project. The proposed AWGF had an 80 Gbps switching capacity, wide channel spacing (525 GHz, centered at 1555.2 nm), and a wider passband (160 GHz) than conventional AWGFs (30 GHz). This means that the system allows a temperature range five times wider than that of a conventional device. As a result, the total system size is reduced. The transmit/receive module is only $125 \times 85 \times 23$ mm. The key technologies applied are MCMs (multichip modules) and EA lasers (electronic absorption lasers).

3.5 Photonic Cell Buffering

Earlier proposed photonic switches have used high-performance *electronic RAMs* for packet buffering. However, electronic RAMs have a limited access speed, which poses speed and capacity constraints to photonic packet switching systems. The electronic approach has the disadvantage of optical-to-electrical (O/E) and electrical-to-optical (E/O) conversions, which increases its complexity and access delay. *All-optical RAMs* avoid the speed bottleneck and O/E and E/O conversions. There have been several efforts to study and develop all-optical RAMs, but without practical usefulness.

An alternative to all-optical RAMs for packet buffering is to use *optical fiber delay (OFD)* lines [31,58]. OFDs incorporate various optical components, including optical gate switches, optical couplers, optical amplifiers, and wave-

Table 1 Wavelength Assignment for an 8×8 AWGF Switching Module

WL	ICh1	ICh2	ICh3	ICh4	ICh5	ICh6	ICh7	ICh8
OCh1	λ_1	λ_2	λ_3	λ_4	λ_5	λ_6	λ_7	λ_8
OCh2	λ_2	λ_3	λ_4	λ_5	λ_6	λ_7	λ_8	λ_1
OCh3	λ_3	λ_4	λ_5	λ_6	λ_7	λ_8	λ_1	λ_2
OCh4	λ_4	λ_5	λ_6	λ_7	λ_8	λ_1	λ_2	λ_3
OCh5	λ_5	λ_6	λ_7	λ_8	λ_1	λ_2	λ_3	λ_4
OCh6	λ_6	λ_7	λ_8	λ_1	λ_2	λ_3	λ_4	λ_5
OCh7	λ_7	λ_8	λ_1	λ_2	λ_3	λ_4	λ_5	λ_6
OCh8	λ_8	λ_1	λ_2	λ_3	λ_4	λ_5	λ_6	λ_7

length converters to realize photonic packet buffering. OFD line-based buffers can be classified into two categories: (1) the *travelling type* and (2) the *recirculating type*. The former usually consists of multiple fiber delay lines with lengths equivalent to multiples of a packet transfer time and optical space switches to select the delay lines. The latter is more flexible, in the sense that the storage time is adjustable, based on the circulation number [52].

In the context of photonic packet switching, WDM has been considered to increase the buffering capacity, resulting in several proposals for traveling type or recirculating type photonic buffers. These include random wavelength accessible recirculating loop buffers, cascaded recirculating loop buffers, and wavelength-routed buffers.

3.6 Limitations

Currently, the feasibility of implementing all-optical packet switching within a high-capacity switch is still limited by the lack of adequate optical switching and buffering elements. In electronically switched networks, static memory is cheap and readily available. Electronic data can be stored statically and accessed at later times. This buffering, or store and forward function, is key to packet switching. Protocols for switching, flow control, and contention resolution are already in place. However, since optical data cannot be stored statically, any optical data through the switches must be processed and switched on the fly. For this to be feasible, tuneable sources and filters in the switches need response times that are fast enough to access and process packet headers. Response times for tuning the sources are currently in the order of milliseconds, which are inadequate for packet processing and switching. The limited amount of buffering which an all-optical switch can provide is not capable of handling the typical packet-traffic of data networks, given the need for flow control and contention resolution. The optical buffers are either passive fiber delay lines or recirculatory buffers, in which the packet loops around until needed. Optical (''RAM'') memories are still in the experimental stages and are currently expensive and impractical. These component limitations make all-optical switches difficult to implement at the present time, but as more progress is made in the development of optical component devices, such switches may be common ground in the near future.

4 A SURVEY OF PROPOSED OPTICAL CLOS SWITCH ARCHITECTURES

4.1 WDM Burst Switching Project

WDM burst switching [54–56] is an approach to building very high capacity switches, based on optical channels and electronic control. In a burst switched network, each link comprises a number of WDM data channels and one (or more) control channels, carrying burst header cells (BHC). This separation of data and burst-level control information allows the switches to propagate data without ever converting it to electronic form, while still allowing control information to be processed electronically.

The scalable burst switch architecture proposed by Turner [55] consists of a set of I/O modules (IOM), which interface external links and a three-stage Clos (or multistage Beneš) interconnection network of burst switch elements (BSE). A three-stage configuration, consisting of d port switch elements, can support up to d^2 external links (each carrying h WDM channels). Each IOM contains a control section to process BHCs received on the control channel. The address information in the BHC is looked up in a 4096-entry VPI/VCI lookup table, and an output port, where the associated burst will be forwarded, is selected. If the required output link has an idle channel then the burst data is switched through this link. If no channel is available, the burst is stored in a burst storage unit (BSU) within the BSE.

Regarding the applied Call Admission Control, the selection of the middle-stage switch to route the burst through is done dynamically on a burst-by-burst basis to balance the traffic load throughout the interconnection network.

In general, a burst switching system such as the one proposed, supporting $n = d^k$ external links (each with h channels), requires $2k - 1$ stages and each stage has n/d BSEs. Such a system requires $n(2k - 1)$ optical space switches, $nh(2k - 1)$ wavelength selectors (WS), and $nh(2k - 1)$ wavelength converters (WC). For each optical output channel, the system requires $2k - 1$ of these cost-dominant components. As an example of the scalability of this switch it is calculated that a system with $n = 4096$, $h = 128$, $d = 16$, will have $k = 3$ and will require 5 wavelength selectors and converters per output channel. With each channel operating at 10 Gbps, the total switching capacity reaches approximately 5 Pbps.

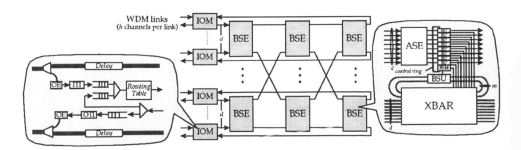

Figure 3 WDM burst switching.

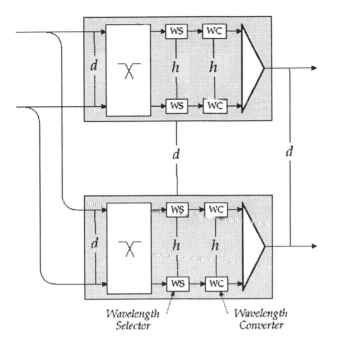

Figure 4 Burst switching system.

Assuming that the combined cost of a WS and a WC is C, then the total (parts) cost per Gbps of total system bandwidth is around $C/2$. For an optical implementation of such a switch to be cost-competitive with an electronic implementation, it is estimated that we need $C < \$1000$.

A small experimental prototype of the burst switch is under development at Washington University, St. Louis, MO, USA by J. S. Turner and his group. The system will support seven external links, with 32 channels of 1 Gbps each, totaling a raw switching capacity of 224 Gbps and scalable to 10 Tbps.

4.2 The OPTIMA System

In 1997, N. Yamanaka et al. proposed optically interconnected multistage ATM switch (OPTIMA), a 640 Gbps, quasi-nonblocking ATM switch based on the three-stage Clos architecture [60]. The switch is using advanced MCM-C (ceramic multichip modules), 0.25 μm CMOS, and optical WDM interconnection technologies.

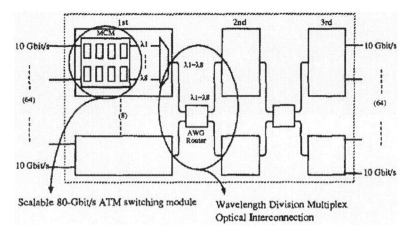

Scalable 80-Gbit/s ATM switching module   Wavelength Division Multiplex
                                           Optical Interconnection

Figure 5   A quasi-nonblocking ATM switch.

A 40-layer, 160 by 114 mm, ceramic multichip module comprises the basic ATM switching element of 80 Gbps switching capacity and 230 watt power dissipation. These switching elements are interconnected in a three-stage Clos network using optical WDM. The WDM interconnection is implemented by means of compact 10 Gbps 8 wavelength optical transmitters and receivers as well as optical WDM routers based on AWGF optical routing elements [62].

In terms of call admission control, OPTIMA has two interesting features [40,41]:

1. The selection of the middle stage switch to route a VC connection through is performed randomly for more even distribution of traffic.
2. Self-rearrangement of existing connections can be triggered by excessive second-stage traffic load and it is realized in hardware.

The scalability and cost-effectiveness of the OPTIMA architecture has been further improved by the introduction of the concept of optical WDM grouped links and dynamic bandwidth sharing [38].

4.3   The ATMOS Photonic Switch

In the framework of the European RACE-II program, called ATMOS, Callegati et al. [5–7,30–32] presented the design of a 256 × 256 ATM switch operating at 2.5 Gbps with electronic control and an all-optical cell path. The switch has been shown by simulation to have cell loss probability (CLP) less than 10^{-9} with load up to 0.8.

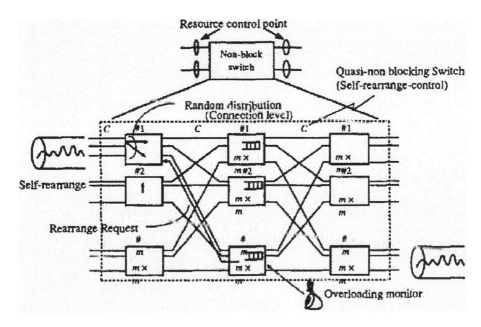

Figure 6 A 40-layer ceramic multichip module.

The architecture of the switch is organized in parallel planes, followed by a two-stage output concentration stage, which collects cells addressed to the same output in the same time slot. Switching elements on each stage are interconnected in a three-stage Clos configuration. Interstage flow control to prevent cell loss was considered but was not implemented.

The feasibility of the switching elements has been proven by the development of a 16 input, 16 output optical switching element with a buffer capacity of up to 16 cells (per output), scalable to 32 cells in the near future.

Besides proving the feasibility of all-optical ATM switches, one of the major conclusions of the ATMOS project was that a new cell transfer format, specifically designed for the requirements and limitations of today's optical packet switching networks, is necessary.

Starting in 1995 and in continuation to the ATMOS project, a follow-up project in the framework of the ACTS program, called keys to optical packet switching (KEOPS), was initiated. The project aimed at identifying a general structure for an optical packet switched network, called an optical transparent packet network (OTP-N) [6].

Key optical components for the development of OTP-Ns are all-optical wavelength converters (AOWCs), multiwavelength optical gates, semiconductor optical amplifier (SOA) gates, and wavelength selectors (WSs). The switching architecture relies on the broadcast-and-select (BS) scheme. Output queuing is achieved through suitable management of a set of WDM fiber delay lines (FDLs).

Although 16×16 switching elements were demonstrated at 2.5 Gbps and at 10 Gbps, for arbitrary $N \times N$ configurations of larger capacity switches, three-stage Clos configurations were proposed in a way similar to the ATMOS project. This time, back-pressure signaling and several flow control policies were suggested and evaluated, including random choice (RC), delay based choice (DBC), and delay based choice with skip (DBCS).

5   A CASE STUDY

In this section we present a case study of an electro-optical ATM switch architecture, proposed as a three-stage Clos network of switching elements. The so-called all-optical switches are all optical in the sense that no data regeneration occurs in the switches. Thus the switch is transparent to payload format and data rates. The electronics cannot be completely eliminated in the all-optical switch, since at least the routing control functions, such as the header processing and resource control, still rely on electronics.

The electrical part of the proposed architecture mainly comprises the ''intelligent'' part of the switch and includes resource management functions, cell header processing, routing, protocol layer implementation, etc. Optical components and technologies are applied along the data transport paths of the physical layer, including input/output port interfaces and the core switching fabric.

5.1   Switch Functionality

Typical functionality of the input and output modules interfacing with the basic switching elements (SE) of the Clos switching network is depicted in Fig. 7.

Assuming that ATM cells enter the switch in optical form (either bit-serial or bit-parallel), the first action to be performed is the separation between the cell header and the payload. For this purpose, a front unit of the SE's input port first splits off a fraction of the incoming optical signal to determine the routing wavelength. While the header wavelength analysis is being performed, the main signal is temporarily buffered using optical delay fiber lines. Once the header wavelength is determined, the main (data) signal is released. For bit-parallel packet switching, the header and payload are already on separate wavelengths, so header stripping is as simple as filtering out the header wavelengths.

Other functions (mostly electrical) that are performed at the input and output ports are briefly outlined here:

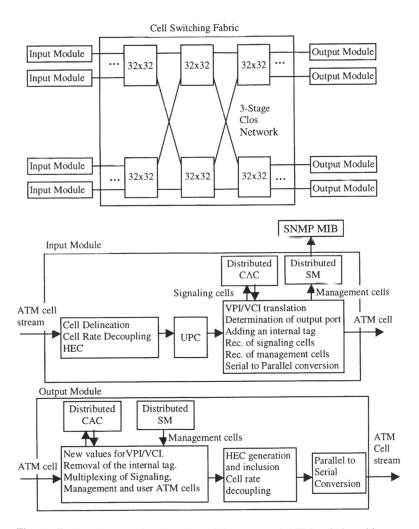

Figure 7 Input/output functionality of the proposed ATM switch architecture.

5.1.1 Input Modules

The block structure of the input module consists of the following elements or block functions:

1. Cell delineation
2. Cell rate decoupling (discarding empty cells)
3. Checking the HEC field

4. Usage parameter control (UPC)
5. Translation of VPI/VCI values
6. Determination of the target output port
7. Addition of a tag in the cell header for internal use
8. Recognition of signaling cells and routing to CAC
9. Recognition of management cells and routing to SM
10. Serial to parallel conversion
11. Distributed CAC
12. Distributed SM functions.

The degree of distribution of CAC and SM functions is a design open issue. It should be noted that all the above make the input modules rather complex components and a very costly part of the overall switching system. Moreover, it is obvious that a fully optical implementation of these functions is not foreseen in the next few years.

5.1.2   Output Modules

The output modules are responsible for preparing ATM cells for physical transmission on the output links. Many of the physical- and ATM-layer functions in the input modules must be performed in the output modules in reverse order.

As with the input modules, the block structure of the output module consists of the following elements or block functions (assuming that pure ATM cell streams are sent to the output ports):

1. Removal of the internal tag/wrapper from each ATM cell.
2. New values for VPI/VCI.
3. HEC generation and inclusion in the ATM cell headers.
4. Multiplexing of CAC, SM, and user data ATM cells.
5. Cell rate decoupling (adding empty cells as frequently as 26 user data cells).
6. Parallel to serial conversion.
7. Distributed call admission control (CAC) will be investigated.
8. Distributed system management (SM) functions will be investigated.

5.1.3   Cell Switching Fabric

The architectural requirements and design issues regarding the core switching fabric mainly concentrate on the definition of the following:

1. A symmetric, scalable, electro-optical Clos interconnection fabric
2. Self-routing (to internally route user data cells, only, from some input to some other output port)
3. Internal cell buffering (both input and output)

4. Buffer management (FIFO with cell discarding only based on CLP bit of the header)
5. Redundancy (middle-stage SEs) for fault tolerance.

5.1.4 Call Admission Control

One of the most critical issues (mainly in terms of scalability) in the implementation of the switch is the definition of the call admission control (CAC) algorithm. This important task involves the following issues:

1. Distributed CAC implementation, suitable for a large-scale switch (design open-issue.) A semidistributed version will be implemented.
2. Allocation of the switch resources for VPCs/VCCs.
3. Admission/rejection of requested VPCs/VCCs.
4. Generation of UPC parameters.

Centralized CAC becomes a call-processing bottleneck for large switches. To overcome this bottleneck we have to distribute the functions of CAC to the input and eventually to the output modules, depending on the actual implementation. However, a trade-off is involved between performance and consistency. As the amount of distributed CAC functions increases, the switch is more scalable to larger sizes because no processing bottleneck appears. On the other hand, in order to ensure that the distributed CAC functions behave consistently (as they affect common resources of the switch), we have to install additional functions for coordination.

In general, a call may consist of multiple connections, and any connection can be established, modified, or terminated during a call. However, for the implementation we assume single-connection calls (with Q2931 signaling protocol at the UNI only), whereas CAC supporting multiple-connection calls will be investigated during the second project year.

5.1.5 System Management

Central system (switch) management (SM) is another very critical part of the overall design. A central control unit will be providing this functionality. Each switching element (or group of switching elements) will also be performing distributed system management control functions, and it will be communicating with the central control unit, thus implementing a two-level hierarchical and semidistributed scheme for effective switch management functionality. The following issues will be considered in both levels of the control hierarchy:

1. Distributed management functions (design open-issue). A semidistributed version will be implemented.

2. Fault management (through the loss of cell delineation and F4-F5 OAM cells).
3. Performance management (through OAM performance management cells and HEC).
4. Accounting management (by counting user data cells at the output ports, called- and calling-party numbers, date, time, and elapse time).
5. Congestion management (through measures of queue lengths, buffer overflows).
6. Management information: SNMP MIB.

As with centralized CAC, centralized SM becomes a processing bottleneck for large switches, because a great amount of data collection and processing is involved. Distributed SM alleviates this bottleneck. Besides, some SM functions (e.g., performance management) must be distributed among the input and output modules, because of the nature of these functions. The degree of distribution depends on the trade-offs between performance and cost of duplication (the same SM functions will appear in both input and output modules).

5.2 The Switching Fabric Architecture

The proposed switching fabric architecture is based on a novel feature, proposed by Liotopoulos [26]. This architectural feature introduced a two-level hierarchy of busses, within each switching element of the Clos switching network, aiming to increase the scalability of the overall architecture and to achieve higher switching capacities. Considering that simple optical switching modules, such as AWGF routers, cannot scale to more than 8 or 16 inputs very easily, due to the limited number of simultaneously usable wavelengths, as well as other cost and complexity issues, we propose a similar architecture, featuring a two-level hierarchy of optical cross-connects (OXC) (e.g., AWGF optical routers).

5.2.1 The Core Switch Module (CSM)

Figure 8 shows the modular design of a typical switching element (SE) (i.e., a stage-switch) of the Clos switching fabric. It consists of K core switch modules (CSMs) in parallel, interconnected via a (global) optical cross-connect (GOXC), such as $K \times K$ AWGF.

As already mentioned, in this design, we break down the single cross-connect "pool" of the SE into a more scalable, two-level hierarchy of simpler cross-connects, one local to the module (LOXC) and one global for the entire switching element (GOXC). Modules of the same switch can communicate each other, via the GOXC, by means of a pair of (global) input and output FIFOs (GiQ, GoQ).

Each CSM has $K - 1$ serial optical input ports and $K - 1$ serial optical output ports (e.g., ATM ports operating at 155 Mbps (OC-3), also referred to as

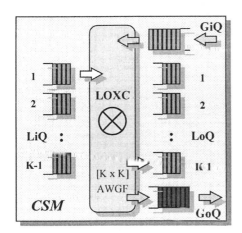

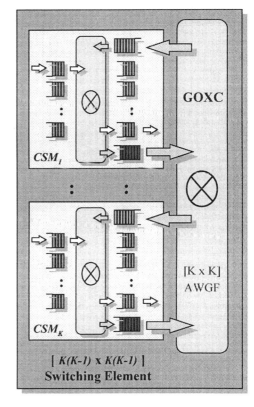

Figure 8   The architectural description of a typical switching element (SE), consisting of a number of core switch modules (CSM).

synchronous transport signal level 3 (STS-3c), or even OC-192 (~10 Gbps)). The serial data stream at each input port, after being separated from the header, may optionally be transformed, stored, and transferred in parallel (e.g., 32-bit electrical words), by means of a serial-in-parallel-out (SIPO) shift register. Subsequently, the cell data is stored in a (local) input FIFO queue (LiQ). Cells can be switched from any input FIFO to any output FIFO within the same module or across different modules of the same switch. All system FIFOs can be implemented using optical delay fiber lines.

A cell destined to an output port of the same module is transferred directly to the corresponding local FIFO LoQ, via the local bus, LOXC. If the destination of the cell is an output port of a different module, within the same switch, then the cell is first transferred to the GoQ FIFO, and through the GOXC to the GiQ of the target module. At the remote module, the cell is transferred from the GiQ

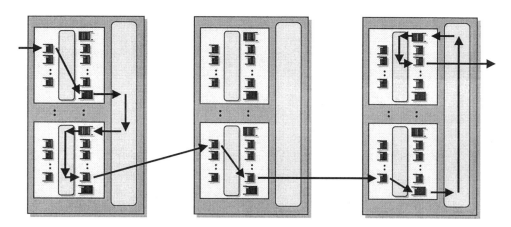

Figure 9   An example of cell routing through three switching elements, one per stage.

to the appropriate LoQ before it exits the current switching element and moves on to the next stage (see also the example depicted in Fig. 9).

If they are processed in parallel form, cells from the output FIFOs are then transformed into a serial stream by a parallel-in-serial-out (PISO) shift register and subsequently to a compatible optical signal (with single or multiple wavelengths) in order to be switched to the next stage via an internal link or to an output port. It is often desirable for internal links to have higher capacity than the input or output ports. This is usually implemented with either wider data paths or higher transfer rates.

5.3   Sizing Considerations

Assuming optical 8 × 8 AWGFs as both LOXC and GOXC (i.e., $K = 8$), each SE can then scale up to 56 inputs and 56 outputs. Given that each optical input can carry traffic at up to 10 Gbps, the total switching capacity of the switch is 560 Gbps. A 3-stage Clos network consisting of such switching elements can therefore achieve:

Up to 28*560 = 15.68 Tbps (28 SEs, 1568 input ports) (in general: fK^2 $(K - 1)^2/2$ for $f = 10$ Gbps, $K = 8$) of strictly nonblocking (SNB) switching capacity (in a [28 × 56 × 28] Clos configuration). This configuration requires 112 switches of 9 OXCs each, or a total of 1008 OXCs. Assuming that an OXC dissipates about 10 W and will eventually cost about $1000, the power dissipation and cost of this switch is estimated in the order of 10 kW and $1 M, respectively.

Up to 56*560 = 31.36 Tbps (56 SEs, 3136 input ports), (in general: fK^2 $(K - 1)^2$ for $f = 10$ Gbps, $K = 8$) of rearrangeably nonblocking (RNB) switching capacity (in a [56 × 56 × 56] Clos configuration). This configuration requires 168 switches of 9 OXCs each, or a total of 1512 OXCs. The cost and power dissipation of this switch is roughly 50% higher than the previous one, while it can achieve very low call blocking at up to 90% of its capacity (=28.22 Tbps).

It can be easily shown that, if the number of inputs/outputs of the OXC is changed by a factor a, the total switch capacity is changed by a factor of approximately a^4. Therefore, for $f = 10$ Gbps, $K = 16$, the resulting switching capacity is 576 Tbps, or half a petabit per second. Such a switch would have 57,600 input ports and would require three stages of 240 SEs each, and a total of 4080 OXCs. Assuming that the new OXC ($K = 16$) costs about 2.5 times more than the OXC with $K = 8$, the corresponding cost and power dissipation of the larger switch would then be about an order of magnitude higher, or roughly 150 kW and $15 M.

5.4 Simulation Study

For the purposes of evaluating the performance of the proposed switch architecture we have modified a functional simulator, first introduced in [27] for the architecture in [28], to include also the functionality of OXCs. The methodology, characteristics, and operation of the simulator is described next.

5.4.1 Simulator Features

Due to memory and programming limitations, the simulator can simulate three-stage Clos networks of up to 14 switching elements per stage, four modules per switching element and seven i/o ports per module (i.e., 392 input ports, or roughly 4 Tbps switching capacity) and 25–50 VCs per i/o port. Each FIFO can buffer four to eight ATM cells. For our case study, the message unit was chosen to be one ATM cell and the base time unit to be the transfer time of one ATM cell, over an input port, e.g., 42.4 ns for 10 Gbps input ports. Cell destinations (i.e., output ports) are chosen randomly over the entire switching fabric, and resource allocation is performed in such a way as to balance the available capacity over the internal switch links.

5.4.2 Input Parameters

The user can assign values to the following input parameters (interactively or in batch form):

Simulation duration
Number of switches per stage
Number of modules per switch
Number of input ports per 1st-stage switch
Number of output ports per 3rd-stage switch

Size of local and global FIFOs
Relative speedup of the LOXC and GOXC
Choice of single or multiple wavelength buffering within the OXCs
Relative speedup of internal, interstage links
Number of virtual channels per input/output port
Number of virtual channels per internal link
Choice of CAC algorithm
Switch capacity utilization limit (for performance reasons)
Input traffic parameters (call inter-arrival time, average call duration).

5.4.3 Output Results

The simulator produces both aggregate and averaged performance results, as well as snapshots of the network's status at any point of time at the user's request.

Aggregate results: number of blocked cycles 1 for all i/o ports and for all local and global busses.

Averaged results: steady-state averages for the absolute and relative two throughput, cell latency, switch population, LiQ, LoQ, GiQ, and GoQ lengths.

Snapshots: all the above (current) aggregates and averages, plus a spatial FIFO-length dump of the entire switch. This spatial dump can directly produce (e.g., via a Matlab macro) insightful "temperature surface" graphs, which can be useful in determining switching bottlenecks. Errors and exceptions are recorded in a log file.

5.4.4 Simulation Methodology

The simulator implements a reverse parse algorithm, scanning and servicing the network's resources from the output ports to the input ports. This direction is the reverse of the regular cell flow, and it is chosen so that there is no cell propagation through more than one resource (OXCs) during the same network cycle. Local and global OXCs as well as internal links are serviced at rates that are multiples of the input (first stage) cell service rate. OXCs are serviced in a circular manner, starting from a random input. The service of an OXC or internal link consists of a single cell transfer from one FIFO to another. An OXC can be serviced more than one time during the same network cycle, depending on its bus-speedup factor.

Initially, random calls, consisting of CBR traffic, are generated and each one is assigned a pair of input and output ports (i.e., source and destination). Subsequently, these calls are routed by the CAC algorithm of choice, and the routing information is recorded in a routing table and in the header of each cell.

```
for stg:=1 to 3 do begin
  forall switches in stage(stg) do
    parbegin
      forall out_ports in switch do
        Service (out_port) ;
      Service (GOXC) ;
      forall modules in Switch do
        Service(LOXC) ;
      forall inp_ports in switch do
        Service(inp_port) ;
    parend ;
    Barrier_Synchronization ;
end ;
```

Figure 10 Parallel pseudo code

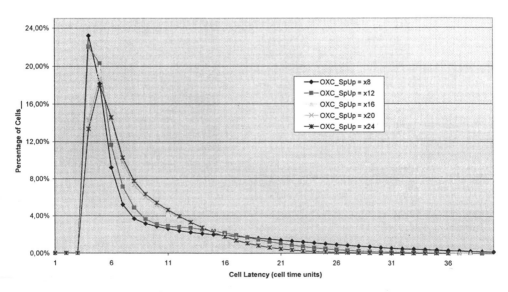

Figure 11 Relative throughput (or goodput) achieved for various internal speedup factors.

From a first-stage input port to a third-stage output port, cells are self-routed within the switching fabric, using the information loaded in their VPI field.

According to the reverse parse algorithm, resources are serviced on a per-stage basis (from third to first) in the following order: first, the internal links (output ports) are serviced, then the GOXCs, and finally the LOXCs. GiQ FIFOs are interlocked, so that a cell cannot propagate from a GOXC through a LOXC to an output port within the same cycle.

5.4.5 Simulator Parallelization

The simulator has been written in such a way that its parallelization, in order to achieve faster results, can be straightforward. Since the code section simulating a single switch does not have any interswitch dependencies, within the same stage, it can be executed in parallel, as in the self-explanatory parallel pseudo-code shown in Fig. 10.

5.4.6 Simulation Results

Figures 11–13 show sample results from simulation experiments run for 500,000 network cycles. (Note at this point that, within the given experimental parameters,

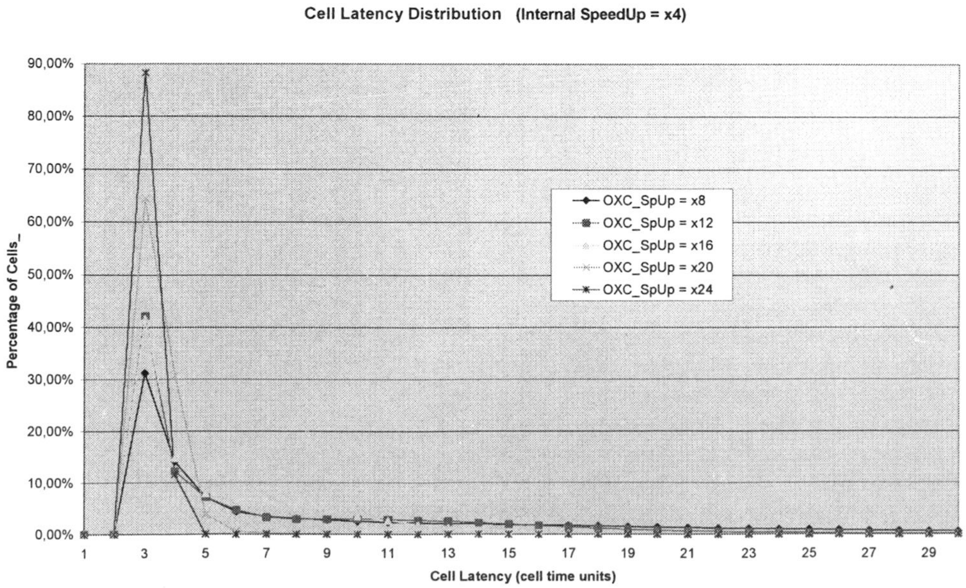

Figure 12 Relative throughput (or goodput) achieved for various internal speedup factors.

Relative Throughput vs. OXC Speedup

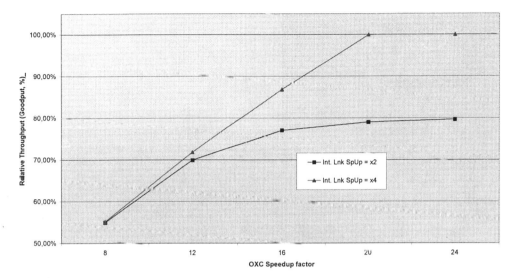

Figure 13   Relative throughput (or goodput) achieved for various internal speedup factors.

a steady state is reached at about 300,000 to 400,000 cycles.) All figures refer to the same [8 × 14 × 8] Clos configuration, with each SE consisting of four [7 × 7] CSMs. Therefore the total simulated switching capacity was: 8*4*7*(10 Gbps) =2.24 Tbps. The traffic was assumed to be continuous, CBR, with 25 to 50 VCs per link and only one-time call arrivals.

Figure 11 and 12 show the cell latency distribution for two values (×2 and ×4, respectively) of the internal (inter-stage) link speed-up factor. In each graph we can see how the cell latency distribution varies as the OXC speed-up factor varies from ×8 to ×24. We observe that, as the internal photonic circuits become faster than the i/o links, either by applying bit-parallel operations or by using ultra high speed bit-single links, the cell latency and cell latency variance becomes smaller. As expected, the higher the speedup factor is, the more "steep" the distribution curves become. Steeper latency distribution curves indicate smaller latency variance.

Figure 13 shows how the switch throughput is affected as the internal link and OXC speedup factors are varied. The graph shows the relative throughput (or goodput), as percent of the offered load. The offered load during these experiments was 1 cell per port per time unit, or a total of 112 million cells during the duration of each run. We observe that for a link speed-up value of ×2 a goodput of 55% up to 80% is achieved, while for a link speed-up value of ×4 the goodput lies between 55% and 100%. In the last case, (i.e., goodput = 100%) the internal cell blocking probability is near zero.

CONCLUSIONS

Our review of the state-of-the-art research in the area of high-speed photonic switching indicates that there is a continuously growing interest in: a) improving the currently applicable optical technologies and components, used to develop all-optical networks and b) studying the performance and feasibility of scalable network and switch architectures that can accommodate the ever growing user traffic.

An ATM switch architecture, based on a three-stage Clos network of simpler switching elements, in turn consisting of optical WDM routers (such as AWGF routers) structured in a two-level hierarchy of optical cross-connect modules is proposed and evaluated. The use of fixed length ATM cells can significantly simplify the implementation of cell contention resolution and buffering, cell routing, as well as cell synchronization.

We claim that such active core switches must be *hybrid* (i.e., electro-optical and not purely optical). The core switching fabric can be mostly optical and use combined multi-WDM and OTDM technologies and optical components, such as AWGFs, optical delay fibers for buffering, optical amplifiers, splitters, selectors, laser transmitters, tunable receivers, etc. The "intelligent" part of the switch, concerning resource management functions, call-admission control, congestion and traffic control, routing, cell-header processing, etc. cannot but be electrical.

Simulation results indicate that the proposed architecture can achieve relative throughput (or goodput) from 50% to 100%, depending on the internal (link and OXC) speedup. The cell latencies vary from 2 to 30 cell cycles for 90% of the transferred cells. The feasibility of functions, such as call admission control and resource management is not considered in this work and it is currently under investigation using parallel and distributed approaches. With today's standards, electro-optical switches of this scale and magnitude may dissipate as much as 10 to 100 KW and cost from $1M to $15M. These figures, however, are expected to drop as opto-electronic technologies become more mature.

REFERENCES

1. H. Ahmadi and W. E. Denzel, "A Survey of Modern High-Performance Switching Techniques", *IEEE J. Sel. Areas Comm.*, 7, no. 7, pp. 1091–1103, Sep. 1989.

2. R. Barry, "The AT&T/DEC/MIT All-Optical Network Architecture" in *Photonic Networks*, Edited by G. Prati, Springer Verlag, 1997.

3. J. Beetem, M. Denneau and D. Weingarten, "The GF11 supercomputer," *Proc. of the 12th Ann. Intl. Symp. on Computer Arch.*, pp. 108–115, June 1985.

4. B. Bostica and L. Licciardi, "Merging Electronics and Photonics towards the Terabit/s ATM Switching," *IEICE Trans. Comm.*, E81-B, no. 2, pp. 459–465, Feb. 1998.

5. M. Casoni, D. Chiaroni, G. Corazza, J. B. Jacob, F. Masetti, P. Parmentier and C. Raffaelli, "Performance analysis of a 256 × 256 TM photonic modular switching fabric", *Proc. 1st IEEE BSS Workshop*, pp. 101–107, Poznan, Poland, April 1995.

6. M. Casoni, et al., "System design and evaluation of a large modular photonic ATM switch", *European Trans. Telecomm.*, Nov. 1996.

7. M. Casoni, G. Corazza, J. B. Jacob, F. Masetti and C. Raffaelli, "Clos Architecture for the design of large photonic ATM switches", *Proc. EFOC&N'94*, Heidelberg, June 1994.

8. T. Channey, J. A. Fingerhut, M. Flucke and J. S. Turner, "Design of a Gigabit ATM Switch", *Proc. IEEE INFOCOM'97*, pp. 2–11, 1997.

9. D. Chiaroni, et al., "A novel photonic architecture for high-capacity ATM switching applications", *Proc. Photonics in Switching*, P.Th.C3, Salt Lake City, UT, March 1995.

10. D. Chiaroni, et al., "Feasibility issues of a high-speed photonic packet switching fabric based on WDM sub-nanosecond optical gates", *Proc. ECOC'96*, Oslo, Sept. 1996.

11. D. Chiaroni, et al., "Theoretical feasibility analysis of a 256 × 256 ATM optical switch for broadband applications", *Proc. ECOC'93*, pp. 485–488, Montreux, Sept. 1993.

12. D. Chiaroni, et al., "Optical packet switching systems based on optical amplifier gates", *Proc. Int'l W. Photonic Net. & Tech*, Lerici, Italy, Sept. 1996.

13. C. Clos, "A study of non-blocking switching networks", *Bell Syst. Tech. J.*, pp. 406–424, March 1953.

14. P. Coppo, M. D'Ambrosio and R. Melen, "Optimal Cost/Performance Design of ATM Switches", *IEEE Transactions on Networking*, 1, no. 5, pp. 566–575, Oct. 1993.

15. P. Gambini, "State of the Art of Photonic Packet Switched Networks", in *Photonic Networks*, Edited by G. Prati, Springer Verlag, 1997.

16. K. Genda and N. Yamanaka, "TORUS: Terabit-per-second ATM switching system architecture based on distributed internal speed-up ATM switch", *IEEE J. Sel. Areas Comm.*, 15, no. 5, 1997.

17. M. Gustavsson, "Technologies and Application for Space Switching in Multi-Wavelength Networks", in *Photonic Networks*, Edited by G. Prati, Springer Verlag, 1997.

18. K. Habara, T. Matsunaga and K. Yukimatsu, "Large scale WDM star-based photonic ATM switches", *IEEE J. Lightwave Tech.*, 16, no. 12, pp. 2191–2201, Dec. 1998.

19. K. Habara, Y. Yamada, A. Misawa, K. Sasayama, M. Tsukada, T. Matsunaga and K. Yukimatsu, "Demonstration of frequency-routing type photonic ATM switch (FRONTIERNET) prototype", *Proc. 22nd ECOC*, 4, no. ThC.3.4, pp. 41–44, Oslo, Sept. 1996.

20. S. Hino, S. Urushidani, J. Nishikido and K. Yamasaki, "Implementation issues for very large capacity ATM switching fabrics optically intraconnected", *Proc. GLOBECOM'92*, vol. 1, pp. 208–212, 1992.

21. S. Hino, S. Yamaguchi, Y. Ohtomo, S. Yasuda, S. Urushidani and K. Yamasaki, "An implementation of a high-performance ATM switching fabric", *Proc. ISS'95*, vol. 1, pp. 399–403, 1995.

22. K. Hirabayashi, T. Yamamoto, S. Hino, Y. Kohama and K. Tateno, "Optical beam direction compensating system for board-to-board, free-space optical interconnection in high capacity ATM switch", *IEEE J. Lightwave Tech.*, *15*, no. 5, pp. 874–882, May 1997.

23. T. Ikegami, "WDM Devices, State of the Art", in *Photonic Networks*, Edited by G. Prati, Springer Verlag, 1997.

24. Y. Kamigaki, et al., "160 Gbit/s ATM Switching System for Public Network", *Proc. IEEE GLOBECOM'96*, pp. 1380–1387, 1996.

25. F. Liotopoulos and S. Chalasani, "Semi-Rearrangeably Nonblocking Operation of Clos Networks in the Multirate Environment", *IEEE Trans. on Networking*, *4*, no. 2, pp. 281–291, April 1996.

26. F. Liotopoulos, "A Modular, 160 Gbps ATM Switch Architecture for Multimedia Networking Support, based on a 3-Stage Clos Network", *Proc. 16th Int'l. Teletraffic Cong. (ITC-16)*, Edinburgh, UK, June 7–11, 1999.

27. F. Liotopoulos, "Issues on Gigabit Switching, using 3-Stage Clos Networks", *J. Informatica*, special issue on "Gigabit Networking", *23*, no. 3, pp. 335–346, Sept. 1999.

28. F. Liotopoulos, "Simulator-Assisted Performance Evaluation of a High-Capacity 3-Stage Clos Network, used for Broadband ATM Switching", *3rd IEEE ISPDBN'98*, Quebec, Canada, Oct. 1998.

29. F. Masetti, "System Functionalities and Architectures in Photonic Packet Switching", in *Photonic Networks*, Edited by G. Prati, Springer Verlag, 1997.

30. F. Masetti, et al., "ATMOS (ATM Optical Switching): Results and Conclusions of the RACE 2039 Project", *ECOC'95*, Bruxelles, pp. 645–652, Sept. 1995.

31. F. Masetti, et al., "Fiber delay lines optical buffer for ATM photonic switching applications", *Proc. INFOCOM'93*, pp. 935–942, S. Francisco, March 1993.

32. F. Masetti, et al., "High-speed, high-capacity ATM optical switches for future telecommunication transport networks", *IEEE J. Sel. Areas. Comm.*, *14*, no. 5, June 1996.

33. T. Matsunaga, K. Habara, A. Misawa, Y. Yamada, T. Ogawa, M. Tsukada, S. Hino, K. Sasayama and K. Yukimatsu, "Design of a photonic ATM switch and rack mounted prototype", *Proc. ICE'97* pp. 1292–1297, Montreal, June 1997.

34. A. Misawa, M. Tsukada, Y. Yamada, K. Sasayama, K. Habara, T. Matsunaga and K. Yukimatsu, "40 Gbit/s broadcast-and-select photonic ATM switch prototype with FDM output buffers", *Proc. 22nd ECOC*, *4*, no.THD.1.2, pp. 107–110, Oslo, Sept. 1996.

35. A. Misawa, M. Tsukada, Y. Yamada, K. Sasayama, K. Habara, T. Matsunaga and K. Yukimatsu, "A prototype broadcast-and-select photonic ATM switches with a WDM output buffer", *IEEE J. Lightwave Tech.*, *16*, no. 12, pp. 2202–2211, Dec. 1998.

36. T. Miyazaki, T. Kato and S. Yamamoto, "A demonstration of an optical switch circuit with bridge-and-switch function in WDM four-fiber ring networks", *IEICE Trans. Comm.*, E82-B, no. 2, pp. 326–334, Feb. 1999.

37. Y. Nakahira, H. Sunahara and Y. Oie, "Comparative Evaluation of Photonic ATM Switch Architectures", *IEICE Trans. Comm.*, E81-B, no. 2, pp. 473–481, Feb. 1998.

38. K. Nakai, E. Oki and N. Yamanaka, *Optical WDM Grouped Links and Dynamic Bandwidth Sharing for Scalable 3-Stage ATM Switching Systems*.

39. Y. Ohtomo, S. Yasuda, M. Nogawa, J. Inoue, K. Ymakoshi, H. Sawada, M. Ino, S. Hino, Y. Sato, Y. Takei, T. Watanabe and K. Takeya, "A 40 Gb/s 8×8 ATM switch LSI using 0.25mm CMOS/SIMOX", *Proc. ISSCC'97*, pp. 154–155, 1997.

40. E. Oki and N. Yamanaka, "A high-speed ATM switch based on scalable distributed arbitration", *IEICE Trans. Comm.*, E80-B, no. 9, pp. 1372–1376, 1997.

41. E. Oki and N. Yamanaka, "High-speed connection admission control in ATM networks by generating virtual requests for connection", *Proc. IEEE ATM'98*, pp. 295–299, May 1998.

42. T. Ozeki, "Ultra opto-electronic devices for photonic ATM switching systems with Terabits/sec throughput", *IEICE Trans. Comm.*, E77-B, no. 2, pp. 100–109, Feb. 1994.

43. S. Sabesan, W. A. Crossland and R. W. A. Scarr, "Nonblocking ATM switching networks composed of ATM switching modules", *Proc. GLOBECOM'97*, pp. 232–236, Phoenix, 1997.

44. S. Sabesan, W. A. Crossland and R. W. A. Scarr, "Reliable opto-electronic ATM switching networks with X out of Y sparing modules at the centre stage", *Proc. MICC'97*, Kuala Lumpur, Malaysia, Nov. 1977.

45. K. Sasayama, et al., "FRONTIERNET: Frequency routing-type time-division Interconnection Network", *IEEE, J. Lightwave Tech.*, 15, no. 3, 1997.

46. K. Sasayama, Y. Yamada, F. Fruh, K. Habara, K. Yukimatsu, S. Suzuki and H. Ishii, "Demonstration of a photonic frequency routing type, time division interconnection network FRONTIERNET and performance analysis of FDM output buffers", *Proc. ISS'95, C8.3*, pp. 452–456, Berlin, June 1995.

47. Special issue on "Bandwidth allocation in ATM networks", *IEEE Comm. Mag.*, 35, no. 5, May 1997.

48. Special issue on "Flow and congestion control", *IEEE Comm. Mag.*, 34, no. 11, Nov. 1997.

49. K. Stubkjaer, et al., "Semiconductor Optical Amplifiers as Linear Amplifiers, Gates and Wavelength Converters", *Proc. ECOC'93*, TuC5.1, Montreux, Switzerland, Sept. 1993.

50. K. Stubkjaer, et al., "Wavelength conversion devices and techniques", *Proc. ECOC'96*, ThB2.1, Oslo, Sept. 1996.

51. H. Takahashi, et al., "Transmission characteristics of Arrayed Waveguide NxN Wavelength Multiplexer", *IEEE J. Lightwave Tech.*, 13, no. 2, 1995.

52. R. S. Tucker and W. D. Zhong, "Photonic Packet Switching: An Overview", *IEICE Trans. Comm.*, E82-B, no. 2, pp. 254–264, Feb. 1999.

53. J. S. Turner and N. Yamanaka, "Architectural Choices in Large Scale ATM Switches", *IEICE Trans. Comm.*, E81-B, no. 2, pp. 120–137, Feb. 1998.

54. J. S. Turner, "Terabit Burst Switching", *Washington Univ. Tech. Report, WUCS-98-17*, 1998.
55. J. S. Turner, "WDM Burst Switching for Petabit Capacity Routers", *Proc. MIL-COM*, 1999.
56. J. S. Turner, "WDM Burst Switching", *Proc. INET*, June 1999.
57. S. Urushidani, M. Yamaguchi and T. Yamamoto, "A high-performance switch architecture for free-space photonic switching", *IEICE Trans. Comm., E82-B*, no. 2, pp. 298–305, Feb. 1999.
58. Y. Yamada, K. Sasayama, K. Habara, A. Misawa, M. Tsukada, T. Matsunaga and K. Yukimatsu, "Optical output buffered ATM switch prototype based on FRONTIERNET architecture", *IEEE J. Sel. Areas Comm., 16*, no. 7, pp. 1298–1308, Sept. 1998.
59. M. Yamaguchi, K. Yukimatsu, A. Hiramatsu and T. Matsunaga, "Hyper-Media Photonic Information Networks as Future Network Service Platforms", *IEICE Trans. Comm., E82-B*, no. 2, pp. 222–230, Feb. 1999.
60. N. Yamanaka, "Interconnection and packaging technologies in ultra high speed ATM switching system", *Proc. PS'96*, no. PTuA2, pp. 44–45, Sendai, Japan, April 1996.
61. N. Yamanaka, S. Yasukawa, E. Oki, T. Kawamura, T. Kurimoto and T. Matsumura, "OPTIMA: Tb/s ATM switching system architecture based on highly statistical optical WDM interconnection", *Proc. IEEE ISS'97*, Systems Architecture 1997.
62. S. Yasukawa, N. Yamanaka, E. Oki and R. Kawano, "High-Speed Multi-Stage ATM Switch based on Hierarchical Cell Resequencing Architecture and WDM Interconnection", *IEICE Trans. Comm., E82-B*, no. 2, pp. 271–280, Feb. 1999.
63. M. Zirngibl, et al., "Digitally tuneable channel dropping filter/equalizer based on Waveguide Grating Router and Optical Amplifier Integration", *IEEE Photonics Tech. Lett., 6*, no. 4, pp. 513–515, April 1994.

10

Parallel Free-Space Optical Interconnection

Guoqiang Li

University of California, San Diego, La Jolla, California, and Shanghai Institute of Optics and Fine Mechanics, Shanghai, People's Republic of China

Sadik Esener

University of California, San Diego, La Jolla, California

1 INTRODUCTION

The rapid progress in high-performance very-large-scale-integrated (VLSI) technology has made possible the integration of more than 10 million transistors onto a single silicon chip with on-chip clock rates at more than 1 GHz. For complementary metal-oxide-semiconductor (CMOS) technology, as the Semiconductor Industry Association predicted [1], the number of transistors available for logic circuits and memory chips will continue to double and quadruple every three years, respectively. By the year 2006, transistors will be made with 0.1 micron line width technology, the on-chip clock speed will be approximately 3.5 GHz, and each chip may have up to 3000 I/O pins. By the year 2009, more than 5000 chip I/O pins will be required, and the off-chip clock speed will reach 2–3 GHz. The shrinking feature size used in VLSI chips results in a higher density of gates per unit area. As the integration density and clock rate improve, the communication bandwidth at different levels (chip level, multiple chip module level, and board level) needs to keep pace. In high-performance computer and communication systems only 30% of the total dissipated power is consumed by processing, whereas 70% is used to support the interconnections. Furthermore, present interconnection and packaging technologies are expected to approach their performance limits at 2000 I/O connections. For higher clock rates, power dissipation concerns and cross talk will limit the system performance. To overcome this bandwidth bottleneck, advanced interconnection technologies must be developed.

Optical interconnection [2–8], it is currently believed, may provide a solution to this bandwidth problem by offering advantages over all-electronic interconnection, such as high bandwidth, high density, low cross talk, low power consumption, and global communication. To perform the optical interconnection, an additional photonics layer is inserted into the conventional electronic system to provide electro-optic and optoelectronic signal conversion and proper optical signal routing. The transmission medium can be either free space or guided-wave optics such as waveguides or optical fibers. Here we have focused on free-space optical interconnection (FSOI). FSOI allows communication between two-dimensional (2D) arrays to be realized in the third spatial dimension. A generalized view of 3D free-space optical interconnection is shown in Fig. 1. A photonics layer consists of an array of optical transmitters and their corresponding drivers, the optical routing elements, and an array of optical detectors and their corresponding amplifiers. The performance and packaging of these components ultimately determines whether optical interconnection can outperform electronic interconnection. Recently many advances have been made in surface-normal transmitters and hybrid integration of the transmitters to CMOS circuits. These transmitters include lead-lanthanum-zirconium-titanate (PLZT) modulators [9], gallium-arsenide (GaAs) multiple-quantum-well (MQW) electroabsorption modulators [10], and vertical-cavity surface-emitting lasers (VCSELs) [11]. Integration techniques such as solder-bump flip-chip bonding has been used to fabricate high-density optically interconnected submicrometer CMOS integrated circuits by bonding directly above active silicon gates. The feasibility of integrating sev-

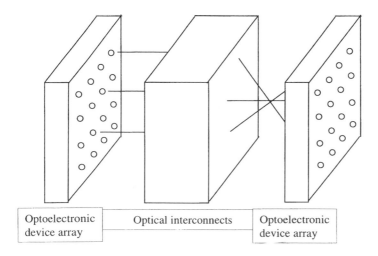

| Optoelectronic device array | Optical interconnects | Optoelectronic device array |

Figure 1 Conceptual view of three-dimensional optical interconnects.

eral thousand optoelectronic devices on a single chip has been demonstrated [12], and the data rate per device can be as high as 1 Gb/s [10,11]. These results indicate that an aggregate bandwidth in excess of 1 Tb/s is possible.

The characteristics of surface-normal operation of the optical devices has two interesting consequences in building parallel optical system architectures [13,14]. First, the surface-normal optical interconnection permits the construction of a pipelined architecture by cascading arrays of devices or processing elements where the output from an array is directed straightforwardly to the next array. Second, a fixed interconnection pattern can be simply realized by imaging one array to another in parallel.

The optical routing elements can include traditional macrolenses, prisms, mirrors, polarization components, holographic elements, or photorefractive crystals. With the development of micro-optics technology, microlenses, diffractive optical elements with small feature size, and microelectromechanical (MEM) mirrors are now available. A hybrid optical interconnection system can be constructed using both macro- and micro-optical elements. The optoelectronic system can be packaged through optomechanic approaches, stack approaches [15–17], or reflective planar approaches [18].

For parallel multiprocessor computer systems and telecommunication systems, high-speed dynamically reconfigurable communication linking processors, memories, and input/output ports are required [19–22]. Again, although continuous improvements have been made in processors and memory, the performance of such systems is increasingly limited by the communication problem. When the number of processors is small, the simplest solution is to use a high bandwidth bus. For large systems with hundreds of processors or more, a complex interconnection network, defined as a system of switches and links that accepts an array of inputs and produces an array of outputs, must be used. Such an interconnection network could be a crossbar with minimum latency, or a multistage interconnection network (MIN) with a reasonable number of switches but with larger latencies.

One novel way of packing multiple silicon chips into a smaller volume is to use 3D packaging technology by stacking electronic chips together. Each chip in the stack communicates with its neighbors using either electrical or optical interconnections, while the global interconnections between the stacks can be implemented using free-space optical technology. This packaging approach has the potential to significantly reduce the interconnection length between chips and chip modules, therefore leading to a new design methodology for high-performance 3D parallel optoelectronic computing and communication systems allowing improvements in terms of bandwidth, density, power consumption, and volume [23,24].

In this chapter, some of the most recent advances are reviewed. Section 2 gives a description of the advantages and limitations of FSOI compared to electronic interconnections. Options for the transmitter and receiver are presented in Sec. 3. Imaging systems with macro-, mini-, and microlenses and their trade-offs are discussed in Sec. 4, which is followed by a review of the spot array generator. Section 6 presents different optoelectronic switches and the implementation of optical interconnection networks. A hybrid system composed of optically interconnected 3D VLSI stacked processors is introduced in Sec. 7. Alignment and packaging technologies regarding optomechanics, planar optics, and stacked optics are briefly described in Sec. 8. A summary is given in Sec. 9.

2   ADVANTAGES AND LIMITS OF OPTICAL INTERCONNECTION

Electrical interconnection [3–7] is physically limited in its time-bandwidth product due to capacitance and resistance associated with the wires. Both capacitance and resistance increase proportionally to the wire length L; the required interconnect energy also increases with L; and the interconnect time (RC constant) increases with L^2. The dependence of bit-rate capacity of electrical interconnects on the aspect ratio, i.e., the ratio of the interconnection length to the total cross-sectional dimension $\sqrt{A}$, has been analyzed [7]. For RC lines, the bit-rate capacity is approximately $\sim 10^{16} A/L^2$ (bits/s). As the feature size of the device scales down, the interconnect performance becomes worse. When the dimensions are reduced by a factor F, the RC delay of a fixed length interconnection increases by a factor of F^2. Moreover, with increase of the chip size, longer connection wires are required and the interconnection bottleneck becomes even more severe. Similarly, the bandwidth limitations are also apparent for off-chip communications. For example, the speed of the transistor is on the order 1 ps, while the delay at the chip level is about 10 ns and that of the bus is about 100 ns. The delay variations in separate clock and signal paths result in clock and signal skew. Moreover, electrical interconnection suffers from signal cross talk and electromigration. Cross talk is caused by electromagnetic interference when lines are put closer. Scaling of the feature size will increase both cross talk and electromigration, which has an additional influence on interconnection.

Electrical interconnection is topologically limited by the 2D planar nature of VLSI circuits. With constant improvement of VLSI technology, more and more devices are integrated onto a single chip as a result of the scaling, and correspondingly more interconnections are required to exploit the performance of the processors. However, the number of devices and interconnections scale differently with chip area. The number of devices grows in proportion to the area, whereas the number of interconnections available grows with the dimension of the chip. The number of input/output pins is also limited.

In contrast, FSOI can offer a large number of parallel interconnections with high bandwidth and can be used to alleviate many of the problems encountered in electronic systems. The advantages of FSOI include

High interconnection density. Parallel interconnections between the 2D electronic circuits interfaced with transmitters and receivers can be achieved by inserting an optical imaging system between them. With high resolution in the optical imaging system, it is possible to connect several thousand outputs on one plane to several thousand of inputs on another, with spot size on the order of a few wavelengths. Interconnection of more than 8000 optical beams into a chip with an area of about 1.7 mm^2 has been demonstrated [25].

High utility of the chip area. Because optical interconnection is implemented on top of the electronic circuits, 3D FSOI has a high utility of the chip area. The number of interconnections can scale with the chip area A.

Less cross talk and global interconnections. These advantages come from the fact that multiple beams can propagate separately without interference. Both regular and irregular interconnections can be achieved by various routing elements.

Efficient power utilization. Photons can travel through air or transparent mediums without appreciable absorption or power dissipation. The power consumption is basically connection-length independent.

Less signal skew. For two fixed optoelectronic arrays, different links have basically equivalent path length. Neglecting the time delay of the beam propagation, the speed of the optical link is determined by the speed of the switches and the response time of the transmitters and detectors. Therefore the speed of FSOI is also length independent. In other words, the bit-rate capacity is not dependent on the aspect ratio. Furthermore, mode-locked lasers can generate picosecond or femtosecond ultrashort pulses, and we can propagate and deliver such pulses with a high degree of synchronization over a quite large system [5]. We can also use short pulses to remove the signal skew from different signals, e.g., read out all modulators with synchronized short optical pulses during the time when all the modulator signals are valid, generating perfectly synchronous optical outputs.

More degrees of freedom for signal multiplexing. The nature of optics offers more degrees of freedom such as wavelength, spatial position, angle, and polarization. All of these can be used for multiplexing to enhance the throughput of a communication channel.

A comparison has been made of 3D integration of conventional electronic circuits interconnected by optical, normally conducting, and other approaches by analyzing the signal delay and bandwidth [26], and it is shown that optical interconnection is superior to normally conducting interconnection in such a system.

We should point out that since most electrical interconnection problems worsen when the connection length increases and optical interconnection is basically length independent, in the near future a compromise could be made to perform local connections by electronics and global interconnections by optics. However, in order to exploit the potential of FSOI in a practical system, several issues need to be studied:

Thermal dissipation. Since 2D optoelectronic input/output device arrays with high speed and high density are desirable, a large amount of functionality will be concentrated in a small area, and hence an effective solution must be developed for thermal management.

Energy utilization. The efficiency of diffractive and holographic optical elements must be improved, and the energy losses in beam combination and surface reflection from sequences of optical components must be minimized.

Packaging and volume. The alignment of the FSOI system should have micrometer or even submicrometer scale positioning accuracy. Mechanical stability and temperature tolerance have to be guaranteed. Conventional optomechanics have difficulty in satisfying these requirements simultaneously. Prototype research demonstration systems are still bulky and expensive. For FSOI to play any role in computing applications, it is crucial to explore alternative, more cost effective, and lighter weight packaging techniques. An alternative to optomechanics technology is to use micro-optics integration technology. Two approaches are currently being studied: the planar approach [18] and the stacked approach [16]. The idea of the planar approach is to fold 3D optical systems into a 2D layout and the micro-optical elements are integrated on one or two sides of an optical flat. This approach is compatible with the standard VLSI fabrication technology. The stacked approach is to put the micro-optical elements behind each other in a 3D fashion.

3 OPTOELECTRONIC DEVICE ARRAY

As described above, an optoelectronic interconnection link is defined as the routing of an optical signal from the transmitter through the optical system to the destination detector. Both the transmitter and the detector are integrated with the corresponding electronic circuits. To show the advantages of an FSOI link over its electronic counterpart, diverse requirements on the optoelectronic device have to be addressed. These requirements include efficient high-speed, low-power, high-density, low-cost packaged transmitters and receivers and high-complexity optoelectronic integration technology that combines the functional devices for light detection, logic, and light emission (modulation). Two important parameters used to evaluate the performance of an optical link are the areal data throughput and the energy required for a transmitted data bit. The two competing technologies, MQW modulators and VCSEL sources, are considered as the potential candidates for the transmitters. Hybrid integration technology such as flip-chip bonding has been well developed to combine these transmitters and silicon CMOS circuits. Smart pixel arrays have been successfully fabricated by integrating the transmitter, the receiver, and the associated electronic circuits into a single device. In this section, we will review the advances of the optoelectronic device arrays based on the two transmitter technologies and compare the performance between the optical interconnection with these technologies and the electronics.

3.1 Smart Pixel Array

A smart pixel [27] is an optoelectronic cell that consists of a photodetector, an electronic processing circuit (CMOS, bipolar, etc.), and an optical transmitter. The optical signals input to the smart pixel are detected by p-i-n or metal-semi-conductor-metal (MSM) photodetectors. The detected signals are then amplified to the required digital electrical levels, which will be further processed by the following electrical circuits. Finally, the processed signals are used to drive the transmitter as the outputs, either a source such as a VCSEL or a modulator such as an MQW modulator. As shown in Fig. 2, SPAs can be classified into three categories in terms of the partitioning of the optical I/Os: uniformly distributed, centralized, and clustered modes [28]. The uniformly distributed SPA (Fig. 2a)

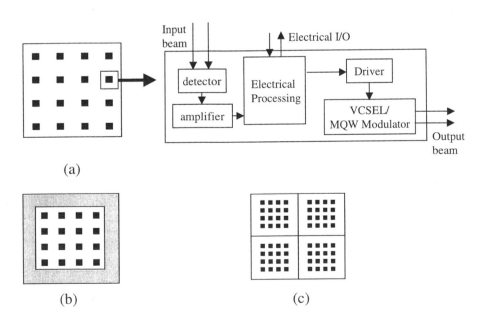

(a)

(b)   (c)

Figure 2 Partitioning of a smart pixel array: (a) uniformly distributed smart pixels, (b) centralized I/Os, and (c) clustered smart pixels.

is formed by creating a 2D array of similar smart pixels. The electronics in each pixel is localized to a small area, avoiding long wire lines that may cause reduced electrical performance due to skew and cross talk as described above. This architecture requires a relatively large field of view for the routing optics. The sizes of the transmitter and detector are usually much smaller than the pitch, and this kind of distribution is called a dilute array. In the centralized mode, the SPA are concentrated in the central area (Fig. 2b). The field of view is reduced and the optical I/Os are more compact. Because the size of the device is comparable with the pitch, it is referred to as a densely packed array. However, the electrical interconnections inside the circuits may be limited to the number of paths that one can get into the lasers/modulators and detector arrays from the surrounding SPA electronics, much like today's electronic chips. Clustered SPA (Fig. 2c) is a combination of both of the previous architectures. In this approach, an SPA is separated into several small arrays, and each small array is a cluster. The smart pixels in the same cluster can share common interconnection optics.

According to the device technology used, SPAs can be categorized as modulator-based or source-based smart pixels. These categories have different integration technologies, which will be reviewed in the following subsections. These SPAs have been used in various demonstration systems.

3.2 Multiple-Quantum-Well (MQW) Modulator

A quantum well [29] is formed by a very thin ($\sim$100 Å) layer of a semiconductor material with a low bandgap such as GaAs sandwiched between a higher bandgap material such as Al_xGa_{1-x} As, as shown in Fig. 3a. The lower bandgap material is called the well, and the higher bandgap material is called the barrier. Multiple quantum wells are repeating layers of the two materials. The structure can be

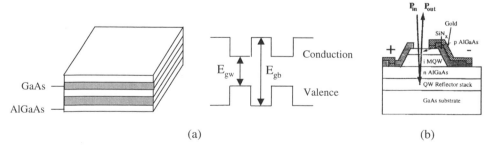

(a)   (b)

Figure 3 (a) Structure of multiple quantum wells. E_{gb}, energy gap of the barrier material (AlGaAs); E_{gw}, energy gap of the well material (GaAs). (b) Reflective MQW modulator. (From Ref. 33, © 1990 IEEE.)

grown with molecular beam epitaxy (MBE) or metal organic chemical vapor deposition (MOCVD), which can precisely control the layer thickness to one atomic layer. At room temperature, the absorption spectrum of the quantum well exhibits peaks at certain wavelengths. When a photon is absorbed at these wavelengths, a bound electron–hole pair (exciton) is created, confined by the wells, and it remains intact until the carriers escape from the wells. When an electric field is applied perpendicular to the plane of the quantum wells, the electron and hole tend to move toward the opposite side of the well. The field reduces the photon energy required to generate the electron and hole, and there is a red shift in the absorption spectrum. This effect is called the quantum confined Stark effect. If a quantum well is put in the intrinsic region of a reverse-biased p-i-n diode, the optical transmission will change with the externally applied voltage, and hence an input optical signal of a wavelength near the excitonic peak can be modulated. Among various materials, GaAs/AlGaAs MQW-pin structures are suitable for surface-normal optical modulators because of their large electroabsorptions, and they give the highest extinction ratios with short light-absorbing lengths. Originally, applications of this type of modulator were limited due to the relatively low saturation intensity. Concepts such as resonant tunning [30] and high intensity phase modulation [31] have been proposed to increase the saturation intensity.

There are two types of MQW modulators, transmissive and reflective [32]. For the transmissive modulator, light illuminates from one side, transmits through the device, and comes out from the other side. The device is placed on top of a transparent material. If the device is fabricated on a GaAs substrate operating near 850 nm, there should be an open window because GaAs is not transparent for this wavelength. A reflective modulator is shown in Fig. 3(b) [33], in which a reflection mirror is placed on top of the substrate. Light enters and exits from the same side. Polarization optical elements can be used to separate the input and output beams. The advantages of the reflective modulator lie in that the substrate need not be removed to open a window and the modulator has a higher extinction ratio for a lower driving voltage because the light propagates through the device twice. Furthermore, this structure has better heat conduction because the device can be mounted directly to a heat sink, and it also makes easier the integration of the modulators with CMOS circuits. For these reasons, the reflective modulator is chosen in most applications.

This device can also be used as a detector, generating a photocurrent in response to the incident light. If the current generated from a photodetector is used to change the electrical field across the quantum well region of the modulator, this kind of device can perform some Boolean logic functions and is called a self-electro-optic effect device (SEED). This device has optical input and optical output. There are different kinds of SEED structures, for example, resistor-biased SEEDs [34], diode-based SEEDs [35], symmetric SEEDs [36], and transistor-biased SEEDs [37]. Based on MQW, an exciton absorption reflection switch (EARS) has been developed [38].

MQW modulators are of particular interest due to their high speed and low electrical drive power requirements. The disadvantage of the modulator is that it requires an external optical source and a corresponding optical system, which may add system complexity, cause energy loss, and complicate alignment. However, this disadvantage can be turned to an advantage because the laser source can be used together with an optical array illuminator to generate a large beam array. This enables the optimization of the laser power, wavelength, and modulation properties. In addition, simple and skewless clock distribution can be achieved by modulating the laser at the system clock rate. In this case, a short-pulse laser source can be used for the signal retiming and also for delivering a sharp signal to the receiver, thus the performance of the receiver can be improved.

To take advantage of the spatial bandwidth available in optics, optical modulators and detectors must be integrated with VLSI circuits. This mixture of the processing capabilities of electronics and communication capabilities of optics will allow the implementation of connection-intensive architectures with more complex nodes than simple switches. Monolithic integration of MQW modulators with GaAs field-effect transistors (FETs) has been demonstrated [39,40]. Integration enables optical and electrical devices with higher reliability, lower cost integration, and potentially high performance. For example, a design for the integration of InAlGaAs/InGaAs MQW modulators and InAlAs/InGaAs modulation-doped FETs (MODFETs) has been reported [39]. The FET-SEED SPA [40] was fabricated by monolithically integrating MQW reflection modulators, p-i-n photodetectors using the same MQW stack as the modulators, doped-channel metal-insulator-semiconductor (MIS)-like FET, and optional integrated resistors. A 4 × 4 FET-SEED SPA has been used in the AT&T System5 photonic switching system demonstrator [41] and other demonstration systems [42,43]. However, integration of MQW modulators with silicon VLSI has an advantage over integration with GaAs electronic circuits because silicon circuits can be fabricated with higher density. The ideal goal is to integrate the modulators and detectors monolithically with silicon VLSI circuits. GaAs MQW modulators have been grown on silicon subtrates with long lifetimes ($>$10,000 h) [44] but not yet integrated with silicon circuits. This is due to the fact that monolithic growth of GaAs devices on silicon is not very compatible with existing silicon processes, e.g., Ga makes SiO_2 conducting, and it is a highly undesired material in any silicon process. Such technical problems have to be solved to make monolithic integration practical. A more mature scheme for dense integration of optical modulators with silicon VLSI is to use a hybrid technique [45] which allows minimum changes of CMOS processing and independent optimization of the optical and electrical device. The most flexible approach is solder-bump flip-chip bonding. In this technique, solder is deposited on one or both of the chips to be bonded,

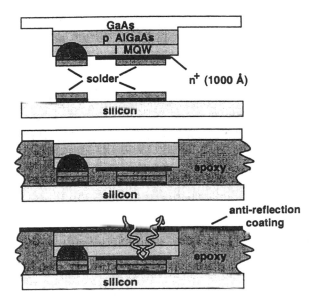

Figure 4 Illustration of the three-step solder bonding process used for MQW modulators and silicon VLSI circuits. (From Ref. 45, © 1995 IEEE.)

and the two chips are brought together with controlled temperature. Assume an array of modulators is fabricated on a GaAs substrate. The flip-chip bonding can be accomplished in three steps as shown in Fig. 4 [45]: (1) Modulators are fabricated in the GaAs chip with coplanar p and n pads, and matching contact pads are designed on a silicon VLSI circuit. The two chips are then bonded together. (2) Etch-protectant is flowed into the remaining space between the two chips. (3) The GaAs substrate is removed by a selective chemical etch, and an antireflection coating is deposited onto the top of the modulators. The electrical connections between the optical and electrical devices are realized with the two kinds of devices isolated. Alignment to submicron tolerances is possible. A high-speed optoelectronic VLSI switching chip with more than four thousand devices has been made using this technique [12], and 1-Gb/s two-beam transimpedance smart-pixel optical receivers made from hybrid GaAs MQW modulators bonded to 0.8 μm silicon CMOS have been demonstrated [42].

3.3 Vertical-Cavity Surface-Emitting Laser (VCSEL)

The first VCSEL was invented in Japan in 1978 [46], and it has aroused much interest throughout the world. After more than a decade of research, VCSELs are moving from research into the manufacturing arena [47]. VCSEL radiation, as reflected in its name, travels in the direction perpendicular, rather than parallel, to the wafer surface, as shown in Fig. 5 [48]. A high-finesse Fabry–Perot cavity is formed by two high-reflectivity distributed Bragg reflection (DBR) mirrors with an active region in between. The active medium contains one or more quantum wells as described in the previous subsection. Different working wavelengths require different quantum well materials, e.g., InGaP, AlGaAs, GaAs, InGaAs, and InGaAsP are appropriate for lasers at 650, 780, 850, 980, and 1300 nm, respectively. Typically the length of the cavity is one wavelength, resulting in a single longitudinal mode. The thickness of the optical cavity is critical, because the threshold current is at a minimum when the optical resonance matches the peak of the laser gain. Variation of the thickness will cause variation of the resonance wavelength, and spectral misalignment of the optical resonance with the laser gain leads to an increase in the threshold current. The DBR mirrors consist of alternating layers of $\lambda/4$-thick high and low refractive index materials, comprising either monolithically grown semiconductors or other dielectric materials. Since multiple high-to-low index interfaces are separated by a distance $\lambda/2$, the reflected beams at each interface interfere constructively to produce mirrors with maximum reflectance of greater than 99%. Dielectric DBR mirrors have a large refractive index difference between the high and low index layers and can generate high reflectivity with only a few pairs, whereas monolithic semiconductor DBR mirrors have a lower index difference and require more pairs for high reflectivity. The semiconductor materials are selected to allow maximum index difference, and they are doped so that the active medium is located at the p-n junction of the laser. A VCSEL has two possible emission forms: top-emitting and bottom-emitting.

The monolithic VCSEL was first grown by MBE [49], but recently, metalorganic vapor phase epitaxy (MOVPE) or MOCVD [50] is preferred for VCSEL growth due to its advantages: rapid growth rate, high wafer throughput, high

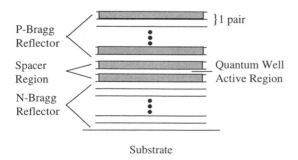

Figure 5 Structure of a VCSEL.

wafer uniformity, and broad flexibility of materials and dopants. For transverse optical and electrical confinement, three approaches have been used, namely, air-post etching, ion implantation, and selective oxidization [51]. In comparison, oxide confined VCSELs are superior, and they have been demonstrated for a wide range of wavelengths from 640 nm to 1550 nm.

In recent years, a lot of effort has been made to improve the performance of VCSELs toward lower threshold current (voltage), lower power dissipation, higher power conversion efficiency, larger modulation bandwidth, higher density, and better uniformity. Ultralow threshold current ($<$ 10 μA) [52] and low threshold voltage (1.33 V) [51] have been reported for oxide confined VCSELs. Power conversion efficiency greater than 50% has also been shown [53]. VCSELs with a 3 dB bandwidth of 15.3 GHz at a low bias current of only 2.1 mA have been achieved [54]. A wavelength uniformity of $\pm0.05\%$ for a two-inch diameter has been realized [55]. Device pitch as low as 3 μm has been demonstrated [56]. However, optical, electrical, and thermal cross talk, physical dimensions of electrical contacts, and optical alignment tolerances limit the pitch in practical designs. A 2D VCSEL array can now be easily fabricated.

The dependence of performance on temperature is one of the important parameters of VCSELs. During operation, part of the electrical power is dissipated as heat, which will change the performance of the laser and the surrounding electronics. With the increase of temperature, both the optical cavity resonance and the laser gain shift to longer wavelengths because of the dependence of the refractive index and bandgap on temperature. However, they do not shift at the same speed. Usually the laser gain shifts faster than the cavity resonance. This mismatch will cause spectral misalignment and thus deteriorate the VCSEL performance. The laser gain profile and amplitude also change with temperature. To compensate for the mismatch, the cavity resonance is usually designed for a slightly longer wavelength relative to the peak gain at room temperature so that the shifted laser gain overlaps the shifted cavity resonance [47]. Figure 6 shows the temperature dependence of the L-I and I-V curves of a flip-chip bonded CMOS/VCSEL [11]. At room temperature, the threshold current and voltage of this particular VCSEL was approximately 1 mA and 1.5 V, with peak power in excess of 4 mW. As the temperature of the chip is increased, the threshold current of the VCSEL increases, and its output power decreases. I-V curves show no dependence on temperature. Oxide confined VCSELs with low threshold currents of less than 200 μA, operating voltage of less than 2 V for 1 mW of optical power, and operation over a wide temperature range, up to 190°C, have been shown [57].

There has been considerable progress in integration of VCSELs to various substrates [11] such as silicon, AlGaAs, and fused silica glass for signal delivery and testing purposes. Monolithic integration of VCSELs with GaAs MESFETs has been realized. A monolithic integration of VCSELs and photodetectors has

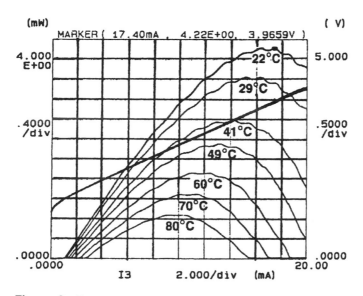

Figure 6 Temperature dependence of the I-V and L-I curves of a flip-chip bonded CMOS/VCSEL. (From Ref. 11, © 1999 IEEE.)

been presented. Substrate removal for both top-emitting and bottom-emitting VCSELs have been accomplished for hybrid integration. Hybrid integration of VCSELs with GaAs-AlGaAs heterojunction bipolar transistors, GaAs MESFETs, and an NMOS circuit have been demonstrated. A novel hybrid integration technology that uses polyimide bonding [58] was proposed recently to bond VCSEL arrays to silicon substrates, where the epitaxial layer for the VCSELs and the photodetectors on a GaAs substrate is bonded to the Si substrate with a polyimide. Another integration technology is based on epitaxial liftoff (ELO) [58]. This technique fully processes the wafer into VCSELs, temporarily attaches the VCSELs to another substrate, removes the substrate from the VCSEL wafer, and then transfers individual VCSELs to the electronic chip. Flip-chip bonding is a low-cost choice for packaging 2D optoelectronics with electronics. For bottom-emitting and detecting devices, the VCSEL emits through the substrate and the detector is illuminated through the substrate. This geometry is compatible with flip-chip bonding without the need for substrate removal [59]. This also allows monolithic integration of the microlenses on the backside of the substrate. However, the substrate must be transparent to the laser wavelength. For GaAs material, wavelengths greater than 900 nm must be employed. For high-speed optical links, the final goal is to integrate dense VCSEL arrays with CMOS circuits. To this end, three hybrid integration techniques [60], namely, coplanar flip-chip bonding, top-bottom contact bonding, and top contact bonding, have been used to integrate 4×4 VCSEL arrays to CMOS chips. The rise and fall times of the optical output are <1 ns, indicating that the VCSEL/CMOS combination is capable of operating at >500 Mb/s per channel. The first demonstration of a flip-chip bonded CMOS/VCSEL array technology capable of gigabit-per-second performance per laser has been shown [11]. VCSELs are bonded directly to the top-level metal on the CMOS chip and do not interfere with the operation or layout of the underlying circuits. The flip-chip bonding is accomplished using 10×10 µm bumps. The VCSELs have submilliamp threshold currents and thresholds below 1.5 V after being attached to the CMOS circuits.

3.4 Performance Comparison of Electrical and Free-Space Optical Links

The performance of VLSI systems is increasingly limited by long electrical interconnections, which involve increased wire resistance from the smaller feature size, the residual wire capacitance, the aspect ratio of the interconnection wires, and the interwire cross talk. When electrical interconnections are replaced by FSOI, a quantitative performance comparison between them is necessary to identify the trade-offs and offer guidelines for applications. A performance comparison between electrical and free-space optical interconnects based on power and speed considerations was done by Feldman et al. [3] in 1988. Recently a more complete study has been done by Yayla et al. [4], which includes more comprehensive interconnection models and evaluates on-chip and off-chip electrical interconnects and FSOI based on MWQ modulators and VCSELs. In the optical interconnection model, the transmitter is modeled by a current source and a capacitor. In most cases, the capacitance to be driven is small enough (< 100 pF) so that a single inverter is sufficient. However, if the transmitter is too large to be driven by by the minimum logic, a superbuffer can be used to amplify the minimum logic before it is sent to the transmitter driver. Larger superbuffers can increase the switching speed but consume more electrical power. Each detector is assumed to receive the signal from only one transmitter, but each transmitter may have a fan out. The routing optics is modeled by a time-of-flight delay and a power transfer efficiency. The detected photocurrent is modeled by a current source, and the photodiode output voltage is amplified and thresholded by the associated electronics. For MQW modulators, the important parameters are the inversion loss (IL), the contrast ratio (CR), and the capacitance. For VCSELs, the main characteristics are the threshold current and voltage, the L-I slope efficiency, and the series resistance. VCSEL based FSOI requires less overall system energy than MQW modulator based FSOI, and the optical system is greatly simplified by the elimination of an external light source. Hence it may offer better optical link efficiency and be easier to implement. With these assumptions, analysis shows that FSOI based on MQW modulators or VCSELs as a transmitters has a significant speed advantage over both off-chip and on-chip electrical interconnections. In comparison with the fastest off-chip VLSI interconnects, FSOI provides a speed advantage of 200%, 50%, and 20–40% for short (a few centimeters), medium-length (5–15 cm), and long off-chip interconnections, respectively. An MQW modulator offers the best one-to-one speed performance due to its smaller transmitter current requirement. In comparison with the on-chip electrical interconnects, FSOI offers even better speed performance when the interconnection length increases. For one-to-one on-chip interconnects, FSOI generally requires less energy than electrical interconnects. The break-even length between them for equal energy is on the order of a few centimeters. For large array applications, FSOI can offer a speed-energy product advantage greater than 20 over electrical interconnects.

The transmitter technologies for FSOI have been evaluated in terms of data throughput and energy requirements [61,62], and optimization methods for optical receiver design have been developed [63]. In addition, Woodward et al. [64] presented four different receiver designs and three operating modes, including 1 Gb/s high impedance, two-beam diode-clamped FET-SEED receivers, single and dual-beam transimpedence receivers realized with a hybrid attachment of MQW devices to 0.8 μm CMOS operating to 1 Gb/s, and synchronous sense amplifier based optical receivers with low (~1 mW) power consumption. Recently a design methodology has been presented to minimize the power dissipation of a FSOI link [65]. The method optimizes all the components of a link (transmitter driver, transmitter, photodetector, and amplifier) so that the total power dissipation of the link is at a minimum at a given operating bit rate, which is determined by the rise time of the transmitter and that of the receiver circuits. For a given bit rate, the design variables are optimized in an iterative way. The results show that at high bit rates VCSELs offer comparable interconnect densities with MQW modulators. At low bit rates, the VCSEL threshold power dominates, whereas at high bit rates or high fan outs a high slope efficiency is more important than a low threshold current. As VLSI technology is scaled down, the maximum bit rate capacity can be improved.

4   IMAGING OF DILUTE ARRAYS

To connect the outputs of the transmitters or logic gates in one array to the subsequent array of photodetectors or logic gates, many computing and switching systems proposed so far have used free-space optical imaging [66,67]. If the neighboring transmitters or logic gates have the same interconnection pattern, the system is said to be space invariant. If different transmitters undergo separate interconnection operations, the system is said to be space variant. The general requirements on such imaging systems are high resolution (small spot size), low loss, low cross talk, large image size (field of view), low cost, and small volume. Several approaches have been implemented, including macro-optics with conventional compound lenses [68–71], micro-optical lens arrays [72–64], hybrid optics consisting of microlens arrays and simple large aperture lenses [75–79], and min-ilens arrays or minihybrid lenses [80–85]. If MQW modulators are used, an array illuminator is necessary. Before presenting the trade-offs among the different optical designs, we first introduce the scaling laws for the imaging lens [86].

For a conventional bulk lens with focal length f and diameter D, the image quality is affected by two factors: diffraction due to the limited size of the lens and aberration due to the lens structure. Assuming that a collimated beam illuminates a perfect lens, the field distribution in the image plane is an Airy pattern caused by diffraction. The diameter of the central Airy disc, which contains 84% of the power entering the lens, is given by $d = 2.44\lambda f$/number, where λ is the wavelength and f/number is the effective focal length divided by the illuminated aperture. A perfect lens with an f/number of 3 generates a 6.2 μm diameter Airy spot at a wavelength of 850 nm. To avoid energy loss, the image spot size must be smaller than the size of the detector. Therefore a low f/number is desirable for high-resolution imaging, and it allows the detector to collect the light efficiently. Further, the driver current of a transmitter is minimized if the area is small and the speed of the system can be increased. The interconnectivity or channel capacity is given by the space bandwidth product (SBWP) of the imaging system, which is defined as the number of resolvable points in the image plane. SBWP can be calculated by the ratio of the field in the image plane with an area A and the area of one spot δx^2, which is determined by lateral aberration and diffraction. Assume that the lens is scaled uniformly by a factor M; the focal length f, the diameter D, the field extent, and the radii and separations are all scaled by the same factor. The lateral aberration and thus wave aberration that describes the deviation of a real wavefront from a perfect ideal wave will scale with M, and the Gaussian moment with M^2. The travel time of the light scales with M, the area of image field with M^2, and the volume and weight with M^3. However, the resolution does not change, since the f/number is left unchanged. If the lens is aberration free, then SBWP scales with M^2. If the diffraction blur is negligible in comparison with the lateral aberration, SBWP stays the same.

Imaging mapping needs to be considered when designing any FSOI systems. Image inversion may occur in many optical systems. The image mapping will affect the physical layout of the optoelectronic device arrays, which will further influence the architecture embedded in the optical system.

4.1 Macro-Optics

The conventional technique for imaging discrete 2D arrays of spots uses a telecentric $4f$ imaging configuration as shown in Fig. 7 [66], where f is the focal length of the two lenses. The distance between the two lenses is large enough to insert a beam splitter for beam in/out-coupling purposes when an external source is needed to read the signals from the device array. Since the optical setup

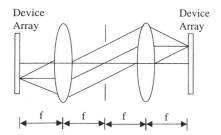

Figure 7 Macrolens optical imaging system.

is symmetric with respect to the central line, one can fold the system by use of beam steering elements and a reflection mirror. An aperture stop with a diameter D is placed between the two lenses. In this case, the resolution should be calculated according to the effective f/number, which is given by the ratio of the focal length and the stop diameter. This system can offer a SBWP as high as 10^5–10^6. The VCSEL window is typically a few μm, and the detector window is a few tens of μm. Consider a 32×32 device array with a pitch of 250 μm. The corresponding field of view is about 8×8 mm². Because the devices are sparsely distributed, the crosstalk is negligible even for significant aberrations. Imaging errors will only cause energy loss. Unfortunately, due to the dilute arrangement of the device array, only a small portion of the SBWP is used in imaging the optical windows of the devices, and a large portion of the SBWP is wasted in resolving the space between them. When the field extent is increased, the field angle and thus the field-dependent aberrations such as coma, astigmatism, and field curvature are also increased. A complex compound lens is needed to correct the aberrations at the expense of cost. A longer focal length decreases the angle subtended by the device array. However, to keep the lower f/number, the effective aperture of the lens must be increased, too. As discussed above, the scaling of the lens will increase the lateral aberrations. Similarly, a new lens design may be found to reduce the magnitude of aberrations at the cost of complexity. A compromise has to be made between the field of view and the magnitude of aberrations. This trade-off leads us to consider the use of other simpler designs.

A lot of optical interconnection systems based on macro-optical imaging have been demonstrated [68–71]. As an example, an FSOI system that was built recently at UCSD for intra-MCM-board interconnects is shown here (see Fig. 8)

Figure 8 (a) System schematics. (b) Optical system configuration using only commercially available devices (right) and Code V view of the optics (left). The point spread function shows that 90% of the energy will be encircled in a 17 μm spot. (From Ref. 70.)

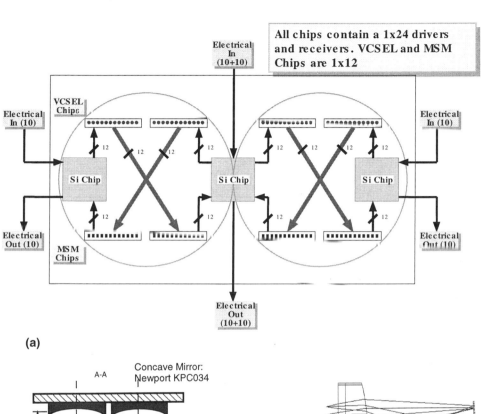

All chips contain a 1x24 drivers and receivers. VCSEL and MSM Chips are 1x12

(a)

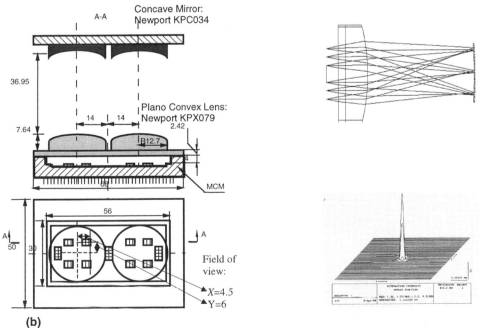

(b)

[69]. In this system, four 1D VCSELs (1 × 12 elements each) and four 1D MSM detector arrays (1 × 12 elements each) are used as light source and photodetectors, respectively. The lasers and the detectors are on a 250 µm pitch. The VCSELs operate at 850 nm with a 15° divergence angle (full angle at $1/e^2$), and the detector aperture is 80 × 80 µm. Laser drivers, receivers, and router circuits are integrated on three Si chips. An optical layer that consists of a modified folded 4-f imaging system based on off-the-shelf macro-optical components was selected and implemented. The optical system is simple and has only a planoconvex lens and a concave mirror. The concave mirrors are used for folding the system as well as for aberration compensation. Simulations in CODE V shows that 90% of the energy emitted from the VCSEL will be encircled in a spot no bigger than 17 µm in diameter. With all possible misalignments caused by both the fabrication and the assembly taken into account, the simulated worst-case condition shows that the spot size increases to approximately 20 µm with a 25 µm maximum lateral shift at the detector plane. The optical links were demonstrated at a speed of 200 MHz.

4.2   Micro-Optics

The micro-optics approach [72–74], as shown in Fig. 9b, has multiple microchannels. Each communication channel has its own optical system consisting of two microlenses. The diameter of the microlens is equal to or smaller than the pitch of the optoelectronic device array. Since the transmitter and detector are located on the axis of its optical system, field-dependent aberration contributions are negligible, and only spherical aberration needs to be considered. A small f/number is still required for high resolution. Currently there are many different technologies available for mass production of refractive and diffractive microlens arrays. The small diameter and focal length of the microlens allows a more compact system. Additionally, a single optical system for each transmitter/detector pair offers the maximum flexibility in interconnection topology, since space-variant

Device Microlens   Microlens Device
Array   Array   Array   Array

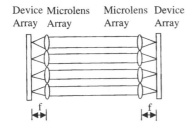

Figure 9   Microlens optical imaging system.

interconnection can be achieved by using microprisms or diffractive elements for beam deflection.

The problem with microchannel imaging is the significant divergence of the beam between the two microlens arrays due to diffraction at the small lens aperture. This may have two effects: cross talk by light falling into the neighboring channels and limited interconnection distance between the two microlens arrays. The cross talk can be controlled by constraining the longitudinal separation. Analysis through Gaussian beam propagation shows that the maximum transmission distance is of the order of D^2/λ [72], where D is the diameter of the microlens. For short distance communications, this is sufficient. If longer interconnection distance is required, additional relay optics must be used. A possible choice is to use a hybrid optical imaging system. Tolerance calculations can be performed using a variety of models such as Gaussian beam theory, ray tracing by use of a Gaussian apodization of the source, and wave propagation based on the Fresnel–Kirchhoff scalar diffraction theory.

4.3 Hybrid Optics

A hybrid optical imaging system is a combination of a pair of small-field low-f/number microlenses and a pair of large-field high-f/number macrolenses. Figure 10 shows such a system as suggested by Lohmann [66]. In this system, a $4f$ imaging setup is added to the above microlens relay structure to increase the interconnection distance. Both the microlens and the macrolens are used at infinite conjugates, and the beams are focused at the common focal plane of the two macrolenses. A high resolution is provided by the microlenses, and the macrolenses need only resolve the aperture of the microlenses. Hence the performance requirements on the macrolenses are greatly reduced. This system combines most of the strengths of the previous two configurations. It is very suitable for imaging of large dilute arrays with a larger field than conventional imaging

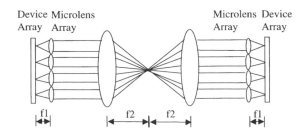

Device Microlens   Microlens Device
Array Array   Array   Array

f1   f2   f2   f1

Figure 10 Hybrid optical imaging system with both the microlens and the macrolens used at infinite conjugates and the beams focused at the central line.

and without the cross talk limitations of micro-optical imaging. Infinite conjugate systems are easier to align and increase the tolerance to mechanical misalignment. This system has been employed for practical implementations [75,76]. It has been demonstrated that a significant reduction in the diameter and numerical aperture of the macrolens can be achieved in the hybrid system. Since the aberrations of an imaging system increase with increasing numerical aperture, it is helpful to improve the imaging quality. More design flexibility can be achieved if the microlens arrays not only collimate the beam but also deflect it [77]. If at the same time the focal length of each microlens is adjusted individually, the hybrid imaging system can be optimized. Multilevel diffractive optical elements can offer the necessary design freedom without increasing the complexity of the fabrication process. A modified hybrid $2f/2f$ setup is shown in Fig. 11 [77], where a single imaging lens with a focal length f is used as a field lens that images one microlens array to another. The diffractive gratings are used for beam deflection. The advantage of this configuration is that we are able to optimize the imaging properties for each individual channel by the individual design of each microlens. The possible aberrations are smaller than those in the hybrid $4f$ imaging system.

However, in the above configuration (Fig. 10), the small f/number microlenses have severe mechanical tolerance constraints. An alternative approach is that the microlens be used at finite conjugates and the macrolens at infinite conjugates [78]. This allows the collimated beams to transmit between the two macrolenses with minimal angular deflection. This eliminates the aberration when prisms, beam splitters, and waveplates are inserted between them. Figure 12 shows an architecture based on this consideration.

If the optoelectronic chips are located in the same plane, the hybrid optical system can be folded at the central plane with a reflection mirror. A novel design that allows a large alignment tolerance by using a concave reflection mirror has been implemented [79]. The system is presented in Section 7.

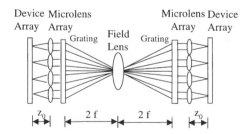

Figure 11   Hybrid $2f/2f$ optical imaging system with deflecting microlens arrays A1 and A2 and one field lens L. (From Ref. 77.)

Device Microlens   Microlens Device
Array Array   Array Array

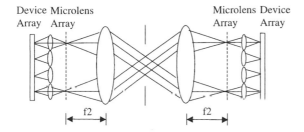

$$f2 \qquad\qquad f2$$

Figure 12 Hybrid optical imaging system with both the microlenses used at finite conjugates and the macrolenses used at infinite conjugates and the beams collimated at the central line. (From Ref. 78.)

4.4 Minilens Optics

Previous studies of microchannel optical interconnection geometries have typically concentrated on interconnects in which a single optical beam is relayed along each microchannel. The minilens optical system described here is used for communication between two clustered SPAs [80–84]. In this scheme, multiple transmitter/receiver pairs in a clustered window are relayed by a single minilens channel with a small aperture, and 2D arrays of lens relays are used to support large SPAs. As an example, Fig. 13 shows an optical interconnection system

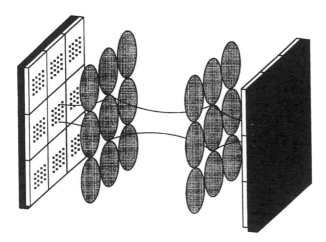

Figure 13 Minilens 4-f optical imaging system with clustered windows below each lenslet. (From Ref. 83.)

consisting of two 2×2 lenslet arrays with each lenslet supporting 4×4 clusters of windows. The selffoc microlens array [80] and refractive [84] and diffractive [81,82] lens arrays have been considered for applications in such systems. Analysis with Gaussian beam propagation reveals that by use of a 4×4 clustered-window scheme, a connection density of 4000 connections/cm^2 may be obtained with a microlens aperture of 632 μm, device spacing of 30 μm, device size of 35 μm, and an angular field of view of 2.5°. This concept can be extended to a minihybrid lens array system as suggested in [85]. Figure 14 demonstrates an SPA based optical interconnection module currently being developed under the Free-Space Accelerator for Switching Terabit Networks (FAST-Net) project [84]. All SPA devices are distributed across a single multichip plane. The interleaved 2D arrays of VCSELs and photodetectors are arranged in clusters. The clusters are arranged in a pattern that is self-similar to the 4×4 lens array pattern. The SPA chip contains 16 clusters. As shown in the magnified inset, each cluster consists of a square array of two VCSELs and two photodetectors. The 16 lenses are arrayed on a center-to-center spacing of 1.7 cm. The lenses used are compound $f/1.12$ microprojection lenses with a focal length of 13 mm. The images of the two VCSELs of a cluster are made to fall on the two detectors of the associated cluster on one of the chips. Conversely, the two VCSELs of the associated cluster are imaged onto the two detectors of the first cluster. All 16 chip

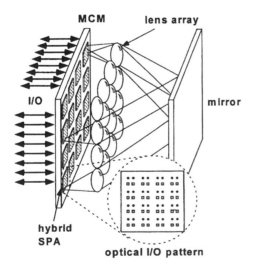

Figure 14   Schematic view of the optical interconnection module being developed under the FAST-Net project. (From Ref. 84.) The inset depicts a magnified view of the I/O pattern for one of the SPA chips, partitioned into 16 clusters of VCSEL's (dots) and photodetectors (squares).

locations are connected in this manner. Thus this system functions as a fully connected network, in which each chip location is connected to every other with a bidirectional data path. An overall efficiency of greater than 80% is expected for optimized transmission of the optical system at the VCSEL wavelength.

4.5 Performance Scaling Comparison

Previous results have indicated that SPA-based 3D FSOI has advantages over planar metallic interconnection for high-bisection bandwidth (BSSW) link patterns in multichip interconnection topologies. It is necessary to explore and compare the scaling rules for different 3D FSOI and 2D planar metallic interconnection architectures. The overall results may provide guidance in determining whether and how strongly an FSOI approach can be applied to a given multiprocessor problem. Scalability of an interconnection system refers to the increase of the size of the system with nominal changes in the existing configuration but with a comparable increase in performance. Recently a performance scaling comparison has been done for macro- and micro-FSOI and electronic interconnections in terms of area, path length, volume, and power consumption [87]. The results show that a 3D macro-optical approach scales uniformly and with a significantly lower slope at high BSBW, while 2D electronic interconnection scales poorly for high BSBW requirements. Furthermore, a micro-optical approach also scales poorly compared with macro-optical systems in the 1–10 Tb/s BSBW regime. The reason for the latter case is the diffraction effect of the microapertures, which limits the interconnection length. For global interconnections this requires the use of repeaters to overcome the losses and cross talk. If the product of the volume and the power consumption is used as a parameter, the scaling advantages of macro-optical FSOI architectures are amplified. However, micro-optical FSOI possesses the flexibility in space-variant operations. As described above, the combination of macro- and micro-optical elements in a hybrid system [88] can exploit the best potential of both, compensate for the drawback of micro-optical systems, and simplify the macro-optical elements design.

5 SPOT ARRAY GENERATOR

A spot array generator splits an incoming beam into a large number of beamlets to form a 2D array of equal-intensity light spots [89]. It can be an important optical element in FSOI. It can be used for the illumination of MQW modulators and fan out for one-to-many broadcast connection. The spot array may be focused or collimated, depending on different applications. The performance parameters for spot array generators include splitting ratio, light efficiency, uniformity, signal-to-noise ratio, compression ratio, modal shape of the beams, cost, volume, and manufacturability. Splitting ratio is the number of generated beams. Light efficiency is the ratio of the total intensity of the spot array and the intensity of the input beam. Uniformity means the deviation of the actual intensity of each spot from the desired value. Compression ratio is defined as the ratio of the area of a spot and the whole area of an elementary cell. Various optical techniques have been proposed to generate spot arrays. These techniques can be classified as geometric imaging approaches, interferometric approaches, and diffractive approaches. For different applications, the common goal is to maximize the light efficiency and to minimize the nonuniformity of the spot array.

In the geometrical imaging approach, four architectures have been implemented. One uses a lenslet array (refractive, diffractive, or holographic) to focus different parts of the beam into a spot array (Fig. 15a) [90]. Array sizes of 10,000 or more are possible. The second method (Fig. 15b) is to use a telescope to

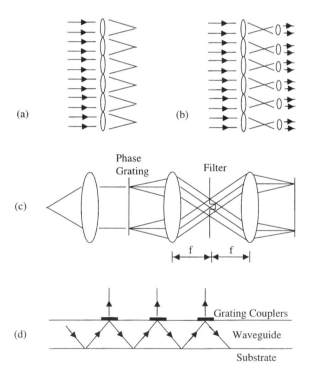

Figure 15 Image plane spot array generators. (a) Lenslet array. (From Ref. 90.) (b) Telescope array. (From Ref. 91.) (c) Phase contrast system. (From Ref. 92.) (d) Tapped waveguide system. (From Ref. 93.)

compress a large beam into small areas [91]. The third idea is based on a phase-contrast imaging technique (Fig. 15c [92]), in which a pure phase distribution is converted into an intensity distribution by putting a phase spatial filter in the Fourier plane. A splitting ratio of 32×32 has been demonstrated. A further approach is based on a waveguide [93], which has grating couplers on its surface that couple out light at the desired spot positions on the waveguide. A spitting ratio of 16×16 was demonstrated.

In the interferometric approach, there exist two different configurations. As shown in Fig. 16a [94], lights coming from four coherent point sources, namely, $\delta(x' - a, y' - b)$, $\delta(x' + a, y' - b)$, $\delta(x' + a, y' + b)$, $\delta(x' - a, y' + b)$, pass through a Fourier transform lens. The intensity distribution of an interference pattern in the x–y plane behind the lens is an array of diffraction free beams. Another method is to use two Michelson interferometers arranged in tandem with polarizing elements (Fig. 16b) [95]. A collimated beam from a laser launched into the first Michelson setup splits into two beams before it enters the second Michelson setup. Each of the two beams will be further split into two beams, resulting in four beams at the output plane. Five quarter-wave plates are placed in front of the four reflection mirrors and between the two polarized beam splitters, respectively. All five quarter-wave plates are oriented so that their eigenvectors enclose an angle of 45° with respect to the polarization plane of the incident light. The only element that causes amplitude loss is the polarizer at the last stage of the interferometer, since it picks up the component of the electric field vibra-

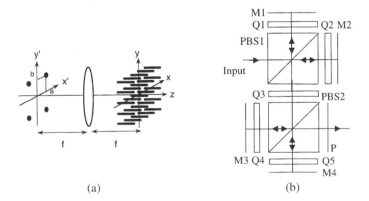

(a)   (b)

Figure 16 Interferometric spot array generators. (a) Setup with four coherent point sources and a Fourier lens. (From Ref. 94.) (b) Michelson interferometer in tandem setup. (From Ref. 95.) M, mirror; PBS, polarizing beam splitter; Q, quarterwave plate; P, polarizer.

tions of the light parallel to its pass plane. The complex amplitudes of the four light beams at the output plane are given by $a/(4\sqrt{2})$, $a/(4\sqrt{2})\exp(-j2\pi ux)$, $a/(4\sqrt{2})\exp(-j2\pi vy)$, $a/(4\sqrt{2})\exp[-j2\pi(ux + vy)]$, where a is the amplitude of the input beam, u and v are the spatial frequencies related to the tilts θ_1 and θ_2 given to mirrors M_2 and M_4, respectively, $u = \sin(\theta_1)/\lambda$, $v = \sin(\theta_2)/\lambda$, where λ is the wavelength of the light. The intensity distribution resulting from the interference of the four beams is given by $I(x, y) = (a^2/8)[1 + \cos(2\pi ux)][1 + \cos(2\pi vy)]$, which represents a 2D array of equal-intensity light spots.

The diffractive approach includes two subdivisions, Fresnel plane and Fourier plane array generators. The Fresnel plane spot array generator is based on the Talbot effect (Fig. 17a) [96–99]. When a periodic amplitude object is illuminated by a collimated beam, identical images will appear at integer multiples of the Talbot distance, $Z_T = 2d^2/\lambda$, where d is the grating period. However, at a distance of $Z_T/2$, a pure phase image is formed. Thus if this phase image is fabricated, at a certain distance from the grating the diffracted light is concentrated

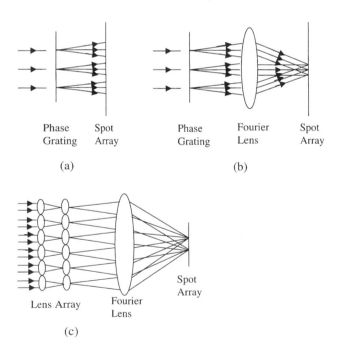

Figure 17 Diffractive spot array generators. (a) Fresnel plane based on the Talbot effect. (b) Fourier plane using a binary phase grating or kinform. (c) Fourier plane using a lenslet array. (From Ref. 109.)

in bright spots. The splitting ratio is given by the number of the grating period and could be very large (100 × 100 has been demonstrated experimentally [89]). Compared with the geometric imaging approaches, the inhomogeneity due to the input laser beam profile is alleviated. Recently a planar-integrated Talbot array illuminator designed to generate one-dimensional spot arrays has been demonstrated [99]. Some modified algorithms for the design of the Talbot array illuminator have also been proposed.

The Fourier plane spot array generator is most widely used in FSOI. A special diffractive component (usually a phase grating) is Fourier transformed (Fig. 17b), generating the desired spot array. The intensity distribution in the Fourier plane is determined by the pattern in one period of the phase component. The periodic pattern may be binary (Dammann grating [100–103]), multilevel (kinform [104–108]), or continuous (lenslet array (Fig. 17c [109]). A Dammann grating is a kind of special binary phase grating with phase values of 0 and π. One period contains a set of transition positions where the phase values vary from 0 to π or vice versa. The location of transitions determines the intensity distribution in the Fourier plane. There is no unique solution for the problem of designing Dammann gratings with high efficiency and ease of manufacturability, and optimization procedures are used [102]. For large splitting ratios, the spatial resolution may limit the performance of the spot array such as the uniformity [103]. Both kinoform and lenslet arrays can offer higher light efficiency. For kinoforms, each period has a number of pixels, and each pixel has a specific phase value. Large spot arrays have been demonstrated. In general, the larger the splitting ratio, the more difficult it is to produce uniform arrays by Fourier methods.

Recently a kinematically aligned, cascaded spot array generator for a modulator based FSOI has been proposed as shown in Fig. 18 [110], where two Fourier plane array generators are cascaded to produce the required array of 512 beams. The first stage produces an array of 4 × 8 beams on a 1600 × 800 μm pitch, which corresponds to the layout of modulator clusters on the SPA chip. Each of these beams is then fanned out into an array of 4 × 4 spots on a 90 μm pitch. The Fourier plane array generators are computer-generated holograms implemented as a binary-level (4 × 4 fan out) and a multilevel (4 × 4 fan out) phase grating. There are four lens transformations in the array generation system: (1) the collimation of the beam emitted from the fiber, (2) the focusing of the collimated beam at the output of the Fourier lens, (3) the collimation of the beam at the output of the optical power supply module, and (4) the focusing of the collimated beam onto the modulators. The system is designed to be scaleable (to an arbitrary even number of boards), modular (to allow easy assembly and maintenance), compact, robust, and misalignment tolerant. Experimental results have been obtained with high uniformity in the spot size, the array pitch, and the array power.

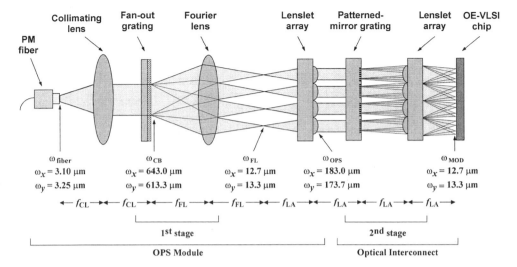

Figure 18 Schematic drawing of a spot array-generation system in which two Fourier plane array generators are cascaded to produce the required spot patterns. (From Ref. 110.) (For simplicity, polarization and beam-steering components are not shown.)

6 OPTICAL INTERCONNECTION NETWORKS

Communication among processing elements or between processing elements and memory is the limiting factor that determines the performance of computer systems, especially multiprocessor and multicomputer systems [111]. Optical interconnection networks are expected to improve the communication performance for these massively parallel computers. Many optical interconnection networks have been proposed and implemented by free-space optics. Depending on the computational problems to be solved by the system, different interconnection networks are preferred. Usually there is no interconnection network that is optimum for all applications, so the design is a compromise among several different architectures. Sometimes an interconnection network is a combination of different networks. The performance of an interconnection network can be evaluated in terms of communication bandwidth, latency, connectivity, control, hardware cost, reliability, modularity, etc. Connectivity refers to the degree of a node, i.e., the number of interconnections per node. The connectivity may affect other performance parameters. As the connectivity increases, the system is more fault tolerant, and the diameter (the maximum number of interconnections that a message has to travel between any source and any destination along the shortest path) can be reduced, but the cost may increase. A network with high bandwidth, low latency, minimum hardware, and simple control is desired. Reliability includes the number of different ways available to reach a node and the bit error rate of an interconnection. Modularity indicates the ability to construct a larger network from several smaller networks.

Interconnection networks can be classified into two categories in terms of topology, static and dynamic. Static networks are often used for communications between neighboring processing elements. These architectures include pipeline processing and systolic array processing. Some examples of the static network structures are stars, binary trees, completely connected nodes hypercubes, meshes, and rings. The general requirement for a static network include small diameter, good symmetry, and reasonable addressing for routing algorithms. Dynamically reconfigurable interconnections among processors are necessary for a general-purpose parallel computing system [112,113]. The reconfiguration is accomplished by switching elements in the network. A dynamic network includes switching elements, routing optics, and a control system. The control system can be either centralized or distributed. Centralized systems use a special unit to control the switching states. The signal transmission between the individual switch elements and the control unit may limit the bandwidth of the network, and failure of the central unit will break down the entire network. In the distributed mode, the switching elements themselves determine the data path. A dynamic network can be blocking or nonblocking. A nonblocking network (e.g., Clos, Cantor) is characterized by the feature that any permutation between senders and receivers is possible at all times, whereas in a blocking network (e.g., Banyan, n-cube, Omega), an arbitrary connection between a sender and a free receiver pair cannot always be fulfilled, because the necessary paths may be occupied by another connection. Finally, a network may have only a single stage or multiple stages. A single-stage network like the crossbar takes a single step to transmit the message from the sender to the receiver. On the other hand, a multistage interconnection network (MIN, e.g., perfect shuffle, Omega, Banyan, Crossover, Clos, Benes) processes parallel communication lines by a pipeline of stages. Each stage comprises an interconnection pattern where the lines are permuted in a certain way, followed by a layer of switching elements that operate on adjacent data lines.

A design issue related to routing optics is the degree of spatial variance, which is a measure of the complexity of the interconnections. If each source has a totally different interconnection operation from the other, the system is called fully space variant. In between the two extremes of space invariant and fully space variant lies the space semi-invariant system, in which some of the sources have the same interconnection pattern. Generally speaking, the SBWP reduces with increasing complexity of the interconnection scheme. A regular interconnection can be specified by means of a set of mathematical relations. Examples of regular interconnections are shuffle, Banyan, and hypercube. In contrast, an irregular interconnection exhibits a random component. The use of regular networks

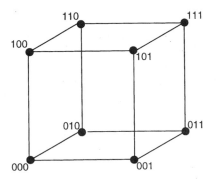

Figure 19 Hypercube connection with $n = 3$. The nodes with the binary addresses differing in one position are interconnected.

seems to limit the flexibility of designing a digital general-purpose computer, but it has been shown that regular networks can be used efficiently in terms of gate count and throughput [114]. A regular interconnection can be implemented either by a space invariant or space semivariant optics design, while an irregular interconnection cannot be done in this way. However, a space-variant optics system can be used to perform both regular and irregular interconnections. A space-invariant interconnect is desirable for optics because it is easier to implement by use of simple space-invariant optical elements. Such space-invariant operations would better use the SBWP of the optical systems and take full advantage of the parallelism of free-space optics. Space-variant operations can be decomposed into a set of space-invariant ones. However, the use of space-variant optical elements to implement space-variant interconnections can reduce the system complexity, volume, and energy consumption. Such space-variant operations can be achieved by a patterned mask or mirror [158], controllable polarization elements, faceted diffractive elements [20,156], and volume holograms [21] in combination with microprisms and micro-optical lenses.

   In this section, some typical interconnection networks and their optical implementations are reviewed, including the optical multimesh hypercube network (OMMH), the crossbar, and various MINs. The optical implementation of a ring-array network can be found in Ref. 115.

6.1 Optical Multimesh Hypercube Network

The binary n-cube, also called the hypercube, is one of the most popular networks. A binary n-cube (Fig. 17, $n = 3$) has $N = 2^n$ processing elements, which are placed at the corners of an n-dimensional cube. If the addresses of the processors are numbered from 0 to $2^n - 1$, the processors are arranged so that there is only one binary digit difference between any processor. The diameter, dimension, and degree are all equal to $n = \log_2 N$. The arrangement of the binary addresses greatly simplifies the routing of messages through the network. The routing algorithm can be implemented by digit-by-digit Exclusive OR operation of the addresses of the source node and the destination node, and only along the dimensions corresponding to the bit positions which contain a 1 does a message have to be passed from the source to the destination. Hence a message needs at most $n = \log_2 N$ steps if every digit of the exclusive OR operation is 1. Because of the regular and symmetric nature of this network, there are many separate paths between a source and a destination, and thus the hypercube network is highly fault tolerant. It is difficult to implement the hypercube electronically for large N due to the large fan out and the large number of interconnections. Optical implementations based on multiple Imaging with titled mirror arrays [116] and computer generated holograms [117] have been proposed. The limitation of the hypercube is its lack of size scalability. As the dimension of the hypercube is increased by one, one additional link needs to be added to every node and the existing nodes are doubled.

A mesh (Fig. 20) is formed by arranging all the nodes in a uniform 2D array. It has a simple and regular connection, and the degree of a node is fixed to be 4 except for the boundary nodes. If the boundary nodes are also connected together, the mesh is referred to as a torus. Because of its constant node degree, a mesh is easily size scalable. However, it suffers from a large diameter, which is $\sqrt{N}$ for an N-node network. A connection between the source and the destination is achieved by transmitting data along the direction determined by the relative positions of the source and destination nodes. To transmit a message from a node at (i, j) to another node at (k, l), it takes $(k - i)$ steps horizontally and $(l - j)$ steps vertically, so the latency is proportional to the distance between the two

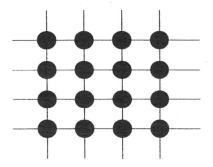

Figure 20 Mesh connection.

nodes. According to the routing scheme, the mesh is a kind of blocking network. The mesh/torus topology has been used in many massively parallel computers. A free-space optical mesh-connected bus network architecture has been proposed [118] using wavelength-division multiple access.

The hypercube has advantages such as small diameter, high connectivity, symmetry, simple control and routing, and fault tolerance, but it has poor scalability, whereas the mesh has good size scalability but a large diameter. On this basis, a scalable network called the optical multimesh hypercube (OMMH) has been proposed and demonstrated [119] by combining the advantages of the hypercube and the mesh while avoiding their disadvantages. As shown in Fig. 21 [119], an OMMH is characterized by a triplet (l, m, n), where l and m represent the row and column dimensions, respectively, and n is the dimension of the hypercube. In

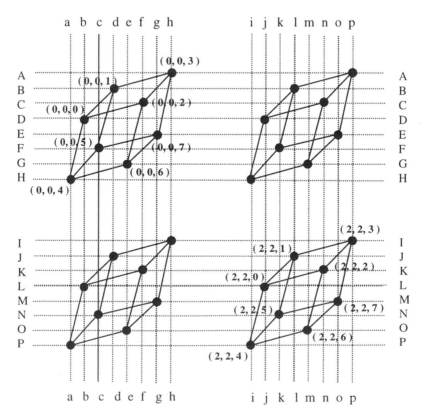

Figure 21 Sample (2, 2, 3)-multimesh hypercube network with four hypercubes, eight meshes, and 32 nodes. (From Ref. 119.)

this example, $l = 2$, $m = 2$, and $n = 3$. An (l, m, n)-OMMH contains $l \times m \times 2^n$ nodes. For clarity, some addresses of the nodes are shown in the figure. For two nodes (l_1, m_1, n_1) and (l_2, m_2, n_2), there is (1) a torus link between two nodes if (a) $n_1 = n_2$ and (b) the components l_1 and l_2, and m_1 and m_2, differ by 1 in one component while the other component is identical. (2) A torus link for the wraparound connection in the row if (a) $n_1 = n_2$ and (b) $l_1 = l_2$; $m_1 = 0$; $m_2 = m-1$, or, for the wraparound connection in the column, if (a) $n_1 - n_2$ and (b) $m_1 = m_2$, $l_1 = 0$, $l_2 = l - 1$. (3) A hypercube link between two nodes if and only if (a) $l_1 = l_2$, (b) $m_1 = m_2$, and (c) $n_1 = n_2$ differ by one bit position in their binary representation. With reference to Fig. 21, an OMMH interconnection consists of two layers: a local connection layer representing a set of hypercube modules (solid lines) and a global connection layer representing the torus network (dashed lines) that connects the hypercube modules. For optical implementation, space-invariant free-space optical implementation of the local level interconnection leads to a compact and modular architecture, while the long-distance high-level interconnections favor fiber-optic arrays. In addition, wavelength-division multiplexing techniques can be used to reduce the number of fibers. A conceptual view of the model is shown in Fig. 22 [119], where free-space and fiber links are shown separately. An imaging system with space-invariant hologram is used to provide connections between planes L and R, and the nodes of plane L (R) are interconnected by fibers to achieve the mesh connectivity. For example, modules n24, n27, n29, and n30 from plane L and modules n25, n26, n28, and n31 from plane R belong to the same hypercube. The connection patterns for the free-space links are all identical in both directions, as shown by the solid and dashed arrows in Fig. 22a.

6.2 Dynamic Interconnection Networks

The simplest dynamic network is based on the idea of a bus. A bus has a common communication line to which all the processors are connected. During a time interval, only one node can occupy the line and transmit its data, while the other nodes either receive the signal or wait, depending on the receiver address. The bus is simple, cheap, and easy to implement. An optics bus can be simply built with a series of beam splitters [120,121]. The architecture is a blocking network, since no parallel data transmission is possible. It can provide good performance when the number of processors is small. The bandwidth of a bus is determined by the system clock frequency. To increase the bandwidth and reliability, a multibus network can be used. For large systems with hundreds of processors or more, an interconnection network that can achieve higher bandwidth is necessary. Such a network may be a crossbar or a multistage interconnection network (MIN), which accepts an array of inputs and generates an array of outputs that are the permutation of the inputs.

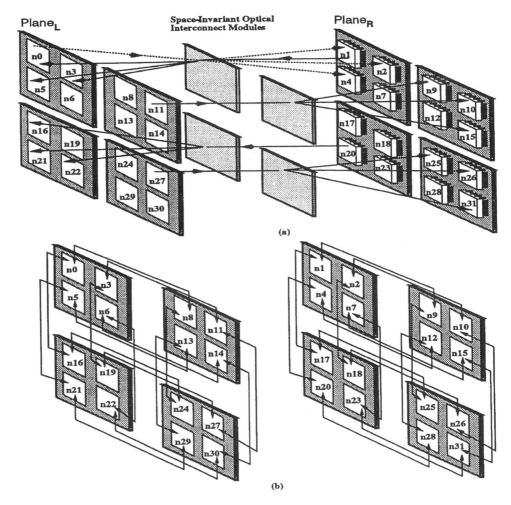

Figure 22 3D view of a (2, 2, 3)-optical multimesh hypercube network showing plane L and plane R interconnected by four space-invariant optical interconnection modules. (From Ref. 119.) (a) The free-space hypercube links are shown separated from (b) the mesh links.

6.2.1 Crossbar

The crossbar is one of the most general reconfigurable networks. As shown in Fig. 23, an $N \times N$ crossbar has N^2 switches. This network can generate a path from each of the nodes in the input stage to every node in the output stage without blocking. Any path includes only one switching element, so the diameter is 1 and

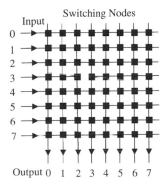

Figure 23 Crossbar network for 1D input.

thus the latency is minimal. In addition, the crossbar can realize full or selective broadcast interconnections. However, since the number of switching elements increases with N^2, the cost is high when N is large. A crossbar network with large nodes and large bandwidth is difficult to construct in electronics.

The most straightforward optical implementation of the crossbar is to use the architecture of a matrix-vector multiplier (Fig. 24). The 2D spatial light modulator (SLM) is used as the interconnection matrix, which has an $N \times N$ array of controllable elements. The input vector (a row of N sources) is fanned out by using, e.g., a cylindrical lens so that the light from each source is broadcast over a corresponding column of the SLM. After transmitting through the SLM, similar optics is used to collect the light from each row of the SLM to each corresponding output element. This architecture takes full advantage of optics in parallel information processing. The interconnection pattern can be configured by changing the transmittance (ON or OFF) of each pixel on the SLM. To make more efficient use of the SBWP of the optical system, input and output channels can be arranged

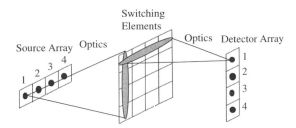

Figure 24 Matrix-vector multiplier architecture for 1D crossbar switching using a spatial light modulator.

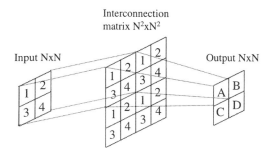

Figure 25 Crossbar network for 2D input. The center plane represents the interconnection matrix.

in 2D fashion, extending to a matrix–matrix structure (Fig. 25). Then diffractive optical elements can be employed to replicate the input pattern across the SLM. The light passing through the SLM is redirected so that all signals from a single replication set are focused to one output channel [122]. Thus each output channel can receive messages from any one of the inputs. Based on this concept, two 64 × 64 free-space optical matrix–matrix crossbar switches have been demonstrated [122,123]. One system [122] used a bundle of 64 single-mode polarization-preserving fibers configured in the form of a regular 8 × 8 square array. By using a 4-f imaging system composed of a 2D binary phase grating and a relay lens, a close-packed 8 × 8 array of replica images of the input fiber bundle was formed in the Fourier plane. This created 64 × 64 optical channels on the SLM. The SLM was a polarization-rotating reflective device (ferroelectric liquid crystal over a CMOS backplane), and thus the signals returning from the SLM could be discriminated by a polarized beam splitter (PBS), depending on the ON and OFF states of the pixels. The desired signals were reflected by the PBS and imaged in front of the fan in optics of the output fibers. Finally the 8 × 8 groups of optical signals were focused into each output fiber. In another system [123], the signals were input with an electrically addressed 8 × 8 VCSEL array, and a MQW modulator based SPA was used for crossbar switching. The system aimed to demonstrate an aggregate bandwidth of 1 Tb/s. In addition, different types of optical image crossbar switch based on multiple imaging and an optical shutter array have been suggested [124].

The disadvantage of the above systems is the low efficiency in energy utilization. For each source, at most only $1/N$ of the light reaches the destination while $(N - 1)/N$ of the light does not pass through the SLM. The energy loss increases with the size of the input array. For example, the loss can be as high as 99.9% when the input has 1000 elements. Considering the insertion loss of the SLM, the loss will be even higher. To overcome this problem, point-to-point beam-steering optics can be used, instead of the fan out optics. In practice, an acousto-optic Bragg cell array is often used for beam deflection. As shown in Fig. 26 [125], light from each input source is collimated along the z-axis and incident at the Bragg angle on one channel of a multichannel acousto-optic cell. By applying acoustic waves to each of the piezoelectric transducers along the x-axis in each channel, the Bragg cell diffracts a large portion of the incident light at an angle in the x–z plane. The deflection angle is proportional to the addressing frequency of the monotone radio frequency signals, which is used to generate acoustic waves. Any interconnection between the input and output array can be realized by tuning the rf signal to provide the proper deflection angle. In another angle-multiplexed beam-steering crossbar architecture (Fig. 27) [126], a $\sqrt{N} \times \sqrt{N}$ matrix of VCSELs is used, corresponding to each input source. An electronic N-to-1 selection switch selects a particular laser pixel to transfer optical data from a data line. The optical power from an activated VCSEL will be directed through an imaging system to the designated receiver. Because there is no fan out power loss, the interconnect capacity of this system is determined by the diffraction limited receiver power cutoff, and therefore a crossbar network of more than

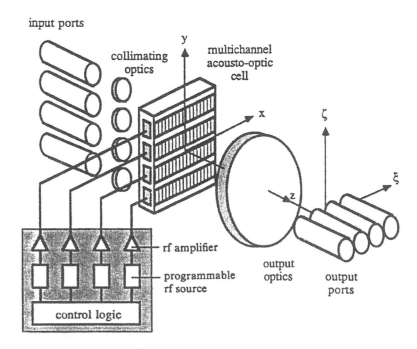

Figure 26 Configuration of an acousto-optic photonic crossbar switch. (From Ref. 125.)

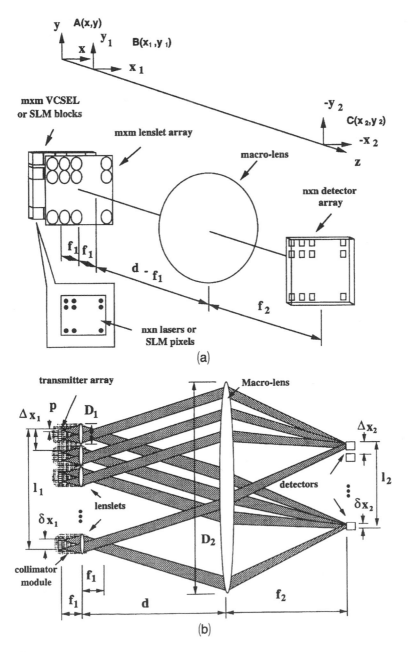

Figure 27 Schematic of the angle multiplexed beam steering-type crossbar. (From Ref. 126.) (a) A general 3D view. (b) A top or side view of (a). SLM, spatial light modulator.

1000 nodes with a per-node bandwidth of 1 GHz is possible with state-of-the-art technology.

Dynamic photorefractive holograms are capable of performing reconfigurable optical interconnects with high energy efficiency. A schematic diagram based on the matrix-vector architecture is shown in Fig. 28 [127]. A small portion of the incident laser beam is coupled out of the beam using a beam splitter. This small portion of beam is expanded and then transmitted through the SLM. The transmitted beam (called the probe beam) meets the main beam in the photorefractive crystal, where a volume hologram is recorded. Due to the nonreciprocal energy transfer in the two-beam coupling, almost all the energy in the main beam is transferred to the probe beam, resulting in high energy efficiency [128]. An architecture using integrated photorefractive waveguides has also been suggested [129].

With the advances in wavelength-tunable VCSEL technology, it is feasible to build an all-optical crossbar switch using wavelength division multiplexing (WDM). The conceptual overview is shown in Fig. 29 [130]. Each processor contains a single tunable VCSEL and a single fixed-frequency receiver. The VCSEL is integrated with the processor. A continuous-wavelength tuning range of 31.6 nm has been reported for the VCSEL with a base wavelength of ~960 nm [131]. The transmit waveguides from each processor are coupled into a single polymer waveguide by use of a passive optical combiner, and the combined optical signal in the polymer waveguide contains multiple individual wavelength signals to be routed to different processors. The multiple wavelength signals

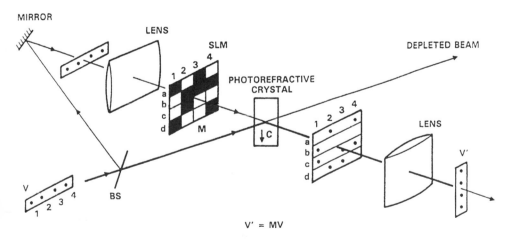

Figure 28 Schematic diagram of a photorefractive reconfigurable crossbar interconnect. (From Ref. 127.)

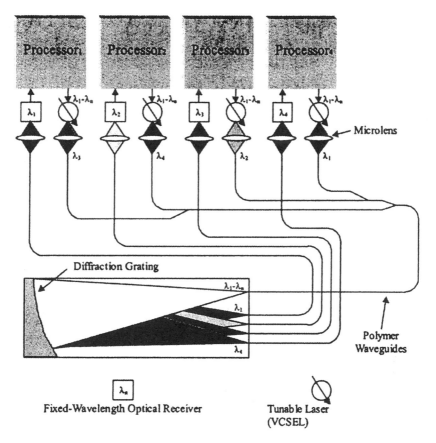

Figure 29 Conceptual view of a free-space optical crossbar with WDM for interprocessor interconnects. (From Ref. 130.)

are guided to a concave diffraction grating based free-space optical crossbar-demultiplexer, which routes the optical signals to the appropriate output waveguide. Each processor is assigned a fixed wavelength with which to receive signals, and other processors can transmit to that processor by sending signals on the wavelength assigned to the receiving processor. This configuration has several features. The system is relatively simple. It fully exploits the high bandwidth and cost advantages of WDM and the benefits of FSOI. There is no fan out energy loss. The use of integrated polymer waveguides permits routing of spatially separated processors to a centralized optical crossbar.

Other optical crossbar implementations include integrated optical form [132,133], a free-space optical implementation with several 2D arrays of devices

[134], another version that uses variable grating mode liquid-crystal devices [135], a crossbar for fiber networks based on planar arrays of either PLZT or magneto-optical material [136], a holographic implementation with folded mirrors [137], and free-space implementation based on directional switches [138].

6.2.2 Multistage Interconnection Networks

The structure of a dynamic MIN usually consists of alternating layers of static interconnection patterns followed by an array of switching elements. The switching element array is called an exchange stage. Such a dynamic exchange stage and a static permutation stage form the basis of an exchange network. A MIN for N inputs typically has $O(\log_2 N)$ stages with $N/2$ switches per stage. In comparison with the crossbar, MINs are less expensive but have larger latencies. MINs provide less interconnection capability than a crossbar in broadcast connections. However, as noted previously, the crossbar quickly becomes prohibitive in cost as the number of processors increases, while MINs are more scaleable. Different optical MIN networks have been studied extensively, including perfect shuffle/Omega, Banyan, Crossover, Clos, baseline, Banes, and gamma. It has been shown that these networks are topologically equivalent and have equivalent functional performance. The 1D interconnection pattern can be extended to a 2D one in which switch elements are arranged in a 2D array, and hence a volume and dynamic 3D network can be established. 3D omega and Clos networks have been analyzed [139–141]. In this section, we are mainly focused on the MINs with 1D interconnection patterns. The performance characteristics of optoelectronic and VLSI MINs have been compared in terms of speed, bandwidth, power consumption, and footprint area [20,22]. Based on that comparison, optoelectronics scales better than VLSI. For large networks, optoelectronics offers higher speed and lower area than VLSI.

Optical Switch Elements. The basic building block of a MIN is a $K \times K$ self-contained switching element. The simplest case is a 2×2 element with two input ports and two output ports, as shown in Fig. 30. For packet-switching mode,

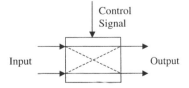

Figure 30 A 2×2 bypass-exchange switch. Solid lines, bypass mode; dashed lines, exchange mode.

the function of the switch is to route input data packets to one or both outputs. A data packet is a unit of information containing the packet header with self-identifying instructions that pertain to the unit's source address, destination address, the intended treatment, and the data message. The 2×2 switch element has four allowed states: bypass, exchange, broadcast, and combine. Here broadcast and combine are not considered, and the element is called a bypass-exchange switch. In the bypass state, the two input signals are connected directly to their respective output ports. In the exchange state, the input signals are crossed between the output ports. Guided-wave switches based on electro-optic $LiNbO_3$, polymers, or MQW heterostructures are commonly used. Another possible optical implementation of the 2×2 switch is to use the polarization state for switching. Two possible orthogonal polarization states of the input beams can be controlled by a rotator or an electrooptic modulator at each channel while polarization-sensitive optical elements such as birefringent crystals, polarized beam splitters, Wollaston prisms, and polarization selective diffractive elements can be used to direct the beams. Alternatively, one can employ electroholography, which permits the governing of the reconstruction process of volume holograms by means of an externally applied electric field. Here we show three examples. There are many other optical switches based on nonlinear optical bistable devices [142], S-SEEDs [8], EARS [38], microelectromechanical systems, liquid crystals, tunable lasers, or ''bubble technology.''

Switch Based on Waveguide.   Lithium niobate ($LiNbO_3$) is often used to fabricate a bypass-exchange switch [143]. As shown in Fig. 31a, two waveguides are arranged with a small separation so that the two beams can interfere with each other in the interaction region. By applying a voltage across the waveguides, the refractive index of the material will be changed due to the electro-optic effect. The index change causes a phase mismatch between the two beams. By controlling the relative phase difference, the two beams can interfere constructively or destructively, and switching of the beams can be achieved. The disadvantage of this switch is the long interaction length required.

Nonlinear optical switches using orthogonally intersecting waveguides have been proposed [144] (Fig. 31b). The switch is based on the all-optical interaction at the intersection region of the waveguides that are composed of MQW materials. A control beam can change the index and absorption, and thus the transmission of the waveguide. The control beam may come from the crossing waveguide or an optical fiber above the intersection region. An input beam propagating in one of the waveguides can be modulated or switched by the control beam. Since the nonlinear region can be confined to the intersection, the interaction length is only about 5 μm.

Switch Based on a Birefringent Crystal.   A birefringent crystal, e.g., calcite, has the nature of double refraction. If an ordinary beam is incident normally on the crystal, the beam will pass straight through the crystal without deviation. How-

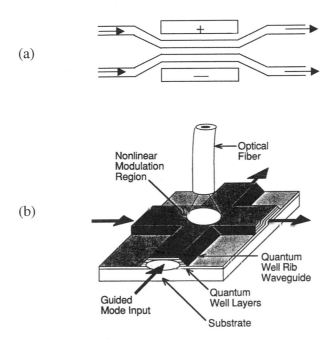

Figure 31 Waveguide switches. (a) A LiNbO3 switch. (b) An MQW switch. (From Ref. 144.)

ever, if an extraordinary beam is incident in the same direction the beam will be deflected at a certain angle inside the crystal, and at the exit plane the beam comes out parallel to but deviated from the input beam, in the principal section. On this basis a calcite plate can split a normally incident beam into two parallel separate beams with orthogonal polarization components. On the other hand, it can also combine two separated normally incident beams with orthogonal polarization components into a single beam. Figure 32a shows the basic birefringence-customized building block for a beam bypass or a deflection operation [145,146]. It consists of a $\lambda/2$ retarder and a calcite plate. The arrow represents the optic axis of the crystal. A dark cell in the retarder represents a hole etched through the retarder. An incident 2D array of beams with an identical linear polarization can be programmed to achieve the desired beam displacement if the through hole pattern on the retarder is tailored. Therefore, by using a polarization controller and two calcite plates, a 2×2 switch can be fabricated [145]. As shown in Fig. 32b, two $\lambda/2$ plates are placed at the entrance and the exit in the second channel of each switch, and this guarantees that all the input and output signals are in the same polarization state, making cascadability possible. A controllable $\lambda/2$ device is put in between the two calcite plates. Two electro-optic modulator

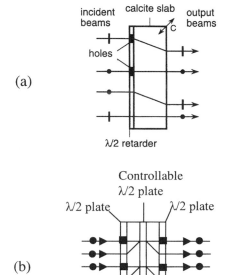

Figure 32 Switch based on birefrigent crystal. (From Ref. 146.) (a) Basic birefrigent building block for a beam bypass or a deflection operation. (b) Bypass-exchange switch.

arrays are used, each in front of the calcite plate. The polarization state of each channel can be controlled by applying a half-wave voltage to the corresponding pixel or not. If the device is in the ON state, the orthogonal polarizations of the two incident beams undergo a 90° rotation, and after transmitting through the second calcite, the two beams are exchanged. If the device is in the OFF state, the two beams will remain in their respective polarizations. The beam in the first channel of each switch goes straight through the second calcite, and the beam in the second channel of each switch is deflected to its original position at the exit plane. This corresponds to the bypass state of the switch.

Switch Based on a Birefringent Computer-Generated Hologram. A birefringent computer-generated hologram (BCGH) [147], which is fabricated by etching the surface-relief profile into a birefringent substrate, has a different phase profile and hence an independent impulse response for each of the two orthogonal linear polarizations that illuminate the hologram. A 2 × 2 switch constructed by using two BCGHs and an electro-optic polarization rotator sandwiched between them

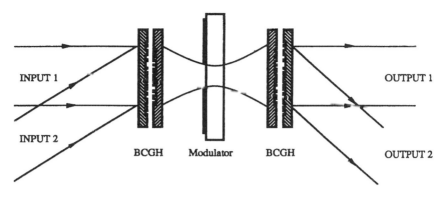

Figure 33 A switch using BCGH. (From Ref. 147.)

is shown in Fig. 33 [140]. The first BCGH combines and focuses the two inputs into the modulator. After being combined, the two beams propagate in the same direction through the modulator, which either exchanges their polarizations or not. The second BCGH deflects the beams into different directions, depending on their polarization states. Note that this switch also has the same functionality for beams propagating through the optics in reverse, making a reverse network and a folded architecture possible. To improve the cross talk performance, a dilated bypass-exchange switch was proposed, which comprises four 1 × 2 switches coupled together [148].

Switch Based on Electroholography. The electroholographic switch was demonstrated recently [149]. Electroholography (EH) is based on the use of the voltage-controlled photorefractive effect in special crystals, e.g., potassium lithium tantalate niobate (KLTN) doped with copper and vanadium. The photorefractive index change is caused by the quadratic electro-optic effect in the presence of an externally applied electric field. In the reconstruction stage, the reading beam is diffracted only when an additional electric field is applied to the crystal. Assume that a refractive-index grating that is spatially correlated with the space-charge field is induced in the crystal when an electric field is applied. In principle, a 2 × 2 bypass-exchange switch can be implemented by using a single EH grating (Fig. 34a). Consider an input beam incident upon the grating at the Bragg angle. If no field is applied, both the reading beams transmit through the crystal in their respective original directions, corresponding to the bypass state of the switch. If an external field is applied, both beams are diffracted, with one beam in the direction of the other beam and vice versa. This corresponds to the exchange state of the switch. In comparison with the above two switches, the advantage of using the EH switch is that the switch is controlled directly by applying the

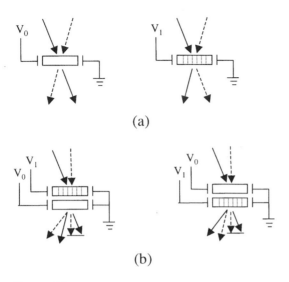

(a)

(b)

Figure 34 Implementation of a switch by use of the electroholography effect. (From Ref. 149.) Each implementation shows one of the two possible states of the switch. (a) A switch using a single crystal. (b) A switch using two crystals.

electrical field to the crystal instead of using a modulator to change the polarization of the beam. The switching speed is fast (limited to about 5 ns). This switch also contains the spatial routing information, which may eliminate the need for additional optics between stages. The device can be operated in the near infrared wavelengths (in particular, 1.55 and 1.31 μm) making it compatible with standard communication systems. The insensitivity of the photorefractive effect to the 1.55 μm wavelength also eliminates the need for hologram fixing.

The simple EH switch suffers from cross talk problems if the diffraction efficiency is not 100%. Fortunately, this problem can be avoided completely by using two crystals, each containing two independent EH gratings, as presented in Fig. 34b [149]. In this case, the switching state is changed by flipping the applied voltages between the two crystals. Both the output beams are diffracted beams. Thus the transmitted beam would not cause cross talk but only a small power loss. The two configurations in Figs. 34a and 34b are termed analog and digital switches, respectively. By adding more crystals and by recording more EH gratings on each crystal, one can increase the number of routes that can be switched. A 4 × 4 cross-connect switch that connects high-speed fiber-optic communication channels has been demonstrated.

Perfect Shuffle Networks. A perfect shuffle (shuffle-exchange) MIN is shown in Fig. 35. The interconnection pattern is identical for each stage, leading to a modularity of interchangeable optical stages. A 1D perfect shuffle performs a

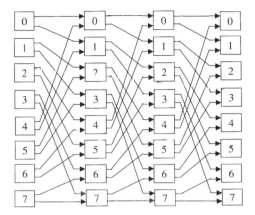

Figure 35 Perfect shuffle network.

specific permutation of $N = 2^k$ channels. The N channels are divided into two halves; these two halves are then interleaved perfectly. At each stage, neighboring pairs of 1D channels are connected through switch elements. If the addresses of the N channels are represented by binary numbers, ranging from 0 to $N - 1$, after the perfect shuffle operation the binary addresses of the output lines represent a right shift operation. Mathematically the perfect shuffle can be expressed by i' $= 2i + [2i/N] \bmod N$ $(i = 0, 1, \ldots, N)$, where i is the source channel index, i' is the destination channel index, and the brackets [ ] indicate the largest integer less than or equal to the arguments. The perfect shuffle interconnection is of great versatility. It can be used for sorting, routing, and performing fast Fourier transforms.

Perfect shuffle is a highly space-variant operation. It can be implemented by space-variant optical elements or by the combination of two space-invariant operations in one system. Several optical implementations of 1D perfect shuffle interconnects have been proposed. They can be classified into three different categories: imaging systems [150–152], Michelson-type systems [153], and spatial filtering systems [154]. Each of these systems has its own advantages and disadvantages. In imaging systems, both refractive and diffractive elements can be used. Figure 36 shows a system using conventional optical elements [150]. The first pair of prisms separates the two halves of the channels. In the Fourier plane, the second pair of prisms introduces different shifts to the upper and lower halves. In the output plane the shuffled version of the input appears in an overall reversed order. Two more compact optical perfect shuffle systems were later proposed [152]. One used two identical negative cylindrical lenses, and the other used a single cylindrical lens together with two identical prism wedges. The perfect shuffle can be achieved by producing two copies of the input and then in-

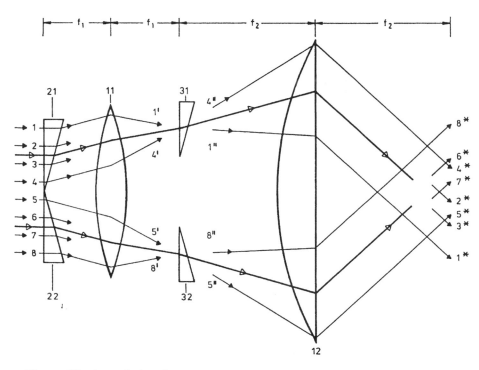

Figure 36 An optical perfect shuffle setup. (From Ref. 150.)

terleaving them. Different designs based on a Michelson, a Mach–Zehnder, or a Sagnac setup have been presented [153]. In this approach, beamsplitters divide the input into two equal images, and mirrors deflect these images to produce a relative shift in the output plane. Only the central interleaved region is used, thus one-half of the optical power is lost. Inplementations using facetted Fresnel mirrors [155] or computer-generated holograms [156] and a lenslet array [157] have been demonstrated. To avoid loss of power, a space-variant patterned mirror may be used in combination with polarizing elements such as waveplates and polarizing beam splitters [158]. Alternatively, one can use partitioned space-variant polarization elements to separate and recombine the input signals [150]. Extension of the 1D perfect shuffle to a 2D version has been discussed and implemented [157]. A 1024-channel free-space sorting demonstrator with a 2D perfect shuffle was designed using diffractive gratings and conventional refractive lenses [159].

Banyan Networks. The flow diagram for a 16-input Banyan network is shown in Fig. 37, where three interconnection stages are interleaved with four switching stages. In contrast to the perfect shuffle, the Banyan interconnects vary from stage

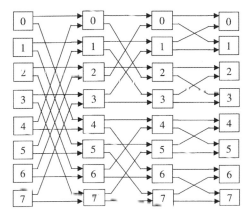

Figure 37 Banyan network.

to stage. Each stage m $(m = 0, 1, \ldots, \log_2 N - 1)$ of the network can be described mathematically by two permutations. The first permutation is the straight connection, and the second permutation is a cross connection, or a cyclic shift of the network. The total space-variant operation can be decomposed into three space-invariant operations: straight connection, shift up, and shift down.

Optical implementations based on Michelson-type and Mach–Zehnder-type systems have been presented [160,158]. Figure 38 shows a Michelson setup using polarization optics [160]. Suppose the light entering the system in plane P1 is circularly polarized. The p polarized light is transmitted downward to plane

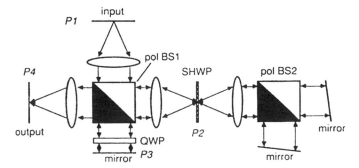

Figure 38 Implementation of a Banyan network based on a Michelson setup. (From Ref. 160, with permission from Elsevier Science). Pol BS, polarizing beam splitter; SHWP, array of halfwave plates; QWP, quarterwave plate.

P3, and this branch implements the straight connections. The s polarized light is reflected to plane P2, and this branch implements the cross connection. In the straight-connection branch, the beam passes through the quarterwave plate and is reflected from the mirror P3. Then the beam passes through the quarterwave plate again and is reflected by PBS1 to the output plane P4. In the cross connection branch, the beams will be separated into two parts: some beams will shift upward and others downward. An individually addressable halfwave plate is placed at plane P2 so that the two classes of beams are separated when they enter the polarized beam splitter PBS2, and are all kept p polarized after they are reflected and transmitted through the waveplate again. The shifts of the beams are controlled by the two mirrors M2 and M3. Finally these beams undergo PBS1 to the output plane.

Crossover Networks. There are two versions of the crossover network, the half-crossover network and the full-crossover network. Figure 39 shows the 16-input half-crossover network with three interconnection stages. Similar to the Banyan, each interconnection stage is different from the others, and each stage contains one or more smaller crossovers. The first stage has a single 16-input crossover, the second stage has two 8-input crossovers, and the third stage includes four 4-input crossovers.

An optical implementation based on a Michelson setup is depicted in Fig. 40a [161]. The beam splitter BS splits the input signal into two different paths. The path P1 → P2 → P4 implements the straight-through connections, and the path P1 → P3 → P4 implements the crossover connections. The crossover is achieved by placing a reflective 90° prism in P3. Figure 40b shows the prism to be used in the first stage, and Fig. 40c depicts the use of the prism by showing

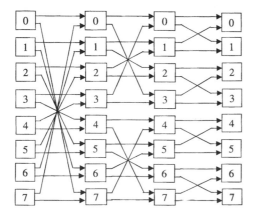

Figure 39 Half-crossover network.

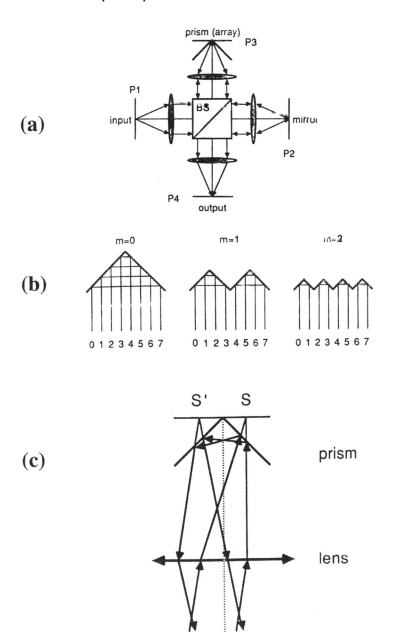

Figure 40 Optical implementation of the crossover network. (From Ref. 161.) (a) Setup. (b) Partitioned prisms to be used in the three stages ($m = 0$, 1, and 2) of the crossover network for flipping of the input pixels. (c) Operation of the reflective 90° prism.

the light paths. Light to be focused to a position S in the back focal plane of the lens is reflected twice and is imaged back to the system as if it would emerge from S'. S and S' are symmetric with respect to the center position of the prism. Note that for light traveling to different positions the path lengths are exactly the same. Therefore no relative delays are introduced between different signals. A double prism and a quadruple prism are required for the second stage and for the third stage, respectively. Other implementations include using logic gates and polarizing elements [162].

Benes Networks.   The Benes network is a regular network that achieves arbitrary rearrangeable nonblocking interconnections with the minimum number of switching sources. Figure 41 shows one possible form of the Benes networks, which is formed by combining $\log_2 N$ stages of a MIN with $\log_2 N - 1$ stages of a reversed MIN. The resulting network consists of $2[\log_2 N] - 1$ switching stages and $2[\log_2 N] - 2$ interconnection stages. There are many ways to get from the source node to the destination node. The increased number of stages leads to an increase in costs, size, and latency. The Benes network is useful for networks that can use out-of-band reconfigurations or that can store a precompiled set of interconnection patterns. 2D fast Fourier transform is an application example. Different types of Benes networks can be formed using different interconnection patterns. In Ref. 163, two isomorphic Benes networks based on Banyan and perfect shuffle interconnects are discussed. The identical plane-to-plane interconnection patterns make possible another topological transformation in which the interconnection module is interleaved and folded back onto itself. A reflective optical interconnection module that interleaved multiple k shuffles on a 4×4 simulated SPA was demonstrated [163].

Transpose Network.   The transpose interconnection is a one-to-one interconnection that is useful for shuffle-based MINs, mesh matrix processors, and hypercube interconnections [164]. Assume there are L input signals $\{l_i\}$ and L output signals $\{l_o\}$, where $l_i, l_o = 0, 1, \ldots, L - 1$ and $L = MN$ (M and N are integers). Figure 42a shows the physical layout of the input and output planes for $M = 16$ and

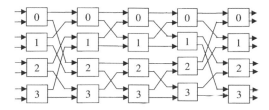

Figure 41   Benes network.

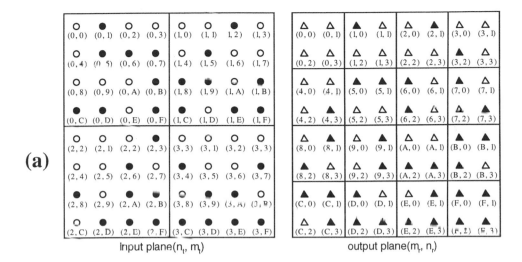

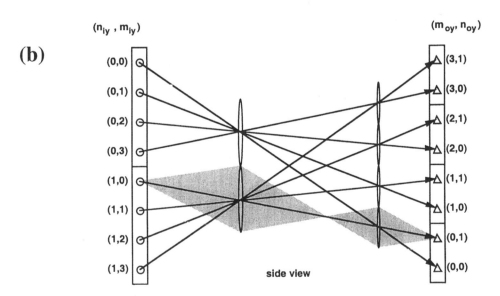

Figure 42 Optical transpose interconnection system (OTIS). (From Ref. 164.) (a) Input and output planes of a 16 × 4 OTIS. (b) Side view of the OTIS.

$N = 4$. The indices l_i and l_o can be divided into ordered pairs (n_i, m_i) and (m_o, n_o), respectively, whereas m_i, $m_o = 0, 1, \ldots, M - 1$, and n_i, $n_o = 0, 1, \ldots, N - 1$, such that $l_i = Mn_i + m_i$ and $l_o = Nm_o + n_o$. The indices n_i and m_o refer to the major indices of the inputs and outputs, respectively. The minor indices are m_i and n_o and l_i and l_o are termed the scalar indices. The major index indicates the region where an input or output signal is located, i.e., the signals in the same regions are assigned the same major index, while the minor index indicates the location of an input or output signal within a region. In the transpose interconnection, (n_i, m_i) is connected to (m_o, n_o) if and only if $m_i = m_o$ and $n_i = n_o$. The major pattern (2×2 in this case) of the input plane is the same as the minor pattern of the output plane, except for a scaling factor $\sqrt{M}$. On the other hand, the minor pattern (4×4 in this case) of the input plane is the same as the major pattern of the output plane but is smaller by a factor of $\sqrt{N}$. Such an interconnection is defined as an $M \times N$ transpose and is functionally equivalent to a k ($k = N$) shuffle network.

The optical transpose interconnection system (OTIS) uses only a pair of lenslet arrays. Figure 42b shows the side view of the basic system [164]. The top view is similar. Interconnections between the input and output planes are realized by employing a $\sqrt{N} \times \sqrt{N}$ array of lenslets and a $\sqrt{M} \times \sqrt{M}$ array of lenslets. Note that the output plane is rotated 180° relative to the input plane. An optical implementation by using BCGH has been demonstrated [165]. The OTIS interconnection can be used to implement a k shuffle MIN. The k shuffle MIN provides full connectivity between L input channels and L output channels in $\log_k L$ stages of switch arrays and $\log_k L - 1$ stages of optical k shuffles. Especially when $k = \sqrt{L}$, only two stages of optoelectronic switches and one stage of optics are required for full routing between the input and output channels. This matches the implementation of the OTIS network. Another useful application of OTIS is in support of the 2D mesh architecture, which is useful for matrix computations. Details can be found in Ref. 164.

7 A 3D OPTOELECTRONIC PROCESSOR ARCHITECTURE

Recent advances in 3D VLSI packaging technologies have made it possible to integrate many chips into a smaller volume by stacking electronic chip dies on top of each other. This packaging approach has the potential to reduce the length of interconnections between one chip and its neighbors. However, this approach cannot support the density and globality of interconnects, which are required by many operations when a stack contains many chips or when the overall system contains several stacks. In this case, free-space optics provides the much needed global, lower power, high-density communication capability. By combining the strengths of 3D chip packaging and optoelectronic array interconnection technologies, it is possible to build an efficient compact system allowing fast processing and handling of large data arrays [23]. To this end, a 3D optoelectronic stacked processor system has been investigated at UCSD.

Figure 43a shows a schematic diagram of the demonstration system in which three 3D VLSI stacks are assembled with optoelectronic modules for global communications [79,166]. Each stack consists of 16 VLSI chips, and a single 16 × 16 VCSEL/MSM detector array is flip-chip bonded on top of the chip stack. Each chip in the stack supports 16 optical I/Os at 1 Gb/s, which is realized by 64 electronic I/Os at 250 Mb/s. Furthermore, the electronic chip implements a 16 × 16 crossbar switch that can arbitrarily route data packets between its inputs and its outputs. The electronic switch, the transmitter driver, and the detector amplifier are built on the same layer. In each stack, the 16 chips are interleaved in the stack with diamond layers for heat removal. A new corner-bonding technique was developed to attach the stack onto a support board. A flexible cable attached to the side of the stack provides the system electrical address, control, clock, and power and ground lines to the stack. An optical interconnection layer is designed for global communications between neighboring stacks. The central stack is rotated 90° so that a signal in one end stack can be routed to any chip on the opposite end stack. To realize communication with both the left and the right neighbor, the optoelectronic array is separated into two halves, and the beams in the two halves are deflected differently. The bandwidth is designed to be 128 Gb/s with a 16 Gb/s throughput per chip. Figure 43b depicts the conceptual view of the three-stack system assembly with each stack corner bonded to the flexible cable.

Figure 44a shows a prototype cube stack with a VCSEL/MSM detector array attached on top of it. The size of the cube is $14 \times 14 \times 8$ cm^3. The VCSEL and detector arrays are offset by 250 μm in the transverse directions. The size of the optoelectronic device array is 8.8×8.8 mm^2. Figure 44b shows a top view of the cube stack with five VCSELs in the upper left corner switched on. The L-I curve of one VCSEL working at 850 nm is depicted in Fig. 44c. The threshold current and the threshold voltage are about 1.3 mA and 2.0 V, respectively. The VCSEL has an aperture of 5 μm with a full divergent angle of 30°, and the detector is a 50 μm octagon. Figure 44d is one of the final cubes to be assembled in the system. The threshold current of the VCSELs is lower than the previous value.

A novel hybrid optical imaging system that permits large misalignment tolerance has been designed for interconnecting two neighboring stacks. The system is a combination of a pair of small field, low f/number microlenses, a pair of large field, high f/number macrolenses, a pair of deflecting elements (prisms or gratings), and a concave reflection mirror. Due to the small feature sizes of the VCSELs and detectors and the relatively large device array, stringent requirements are placed on the optical imaging system. The system must guarantee high

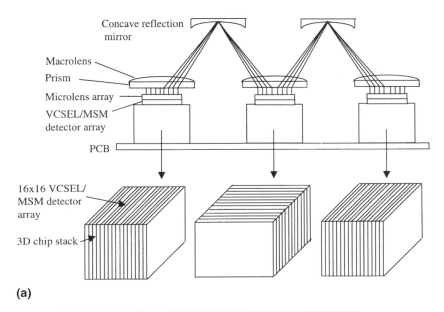

Concave reflection mirror

Macrolens

Prism

Microlens array

VCSEL/MSM detector array

PCB

16x16 VCSEL/MSM detector array

3D chip stack

(a)

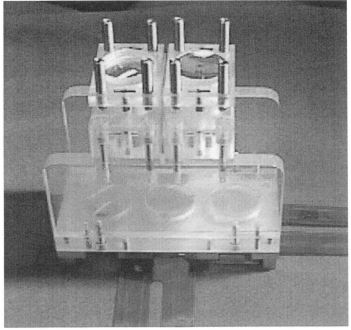

(b)

Figure 43 (a) Schematic diagram of the 3D optoelectronic processor. (b) Conceptual view of the three 3D stack assembly. Each stack is corner-bonded to the flexible cable.

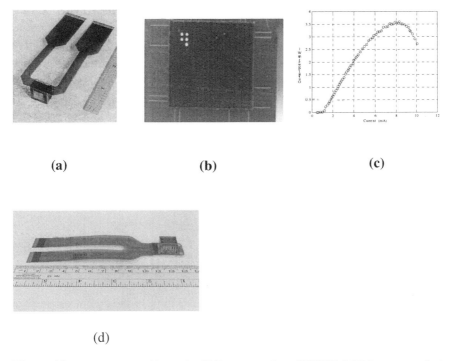

Figure 44 (a) Prototype chip stack. (b) Demonstration of VCSEL/MSM array attached on top of a cube stack with five VCSELs on. (c) L-I curve of one VCSEL. (d) Final chip stack.

image quality (sometimes diffraction limited) over a large image field. Furthermore, it must be compact and cost-effective with simple alignment. This system combines most of the strengths of the macro- and micro-optical configurations. The functions of the microlenses and macrolenses are the same as in the common hybrid system, i.e., the microlenses offer a high resolution and the macrolenses provide a large interconnection distance. The microlenses are used to collimate the beams coming from the VCSELs and refocus the beams onto the detectors. Infinite conjugate systems are easier to align and tend to increase the tolerance to mechanical misalignment. The deflecting elements are inserted between the micro- and macrolenses so that they do not generate additional aberrations. Since the aberrations of an imaging system increase with increasing numerical aperture, it is helpful to use a large f/number macrolens to improve the imaging quality. However, in this configuration the small f/number microlenses have severe mechanical tolerance constraints. A small alignment error of the microlens may greatly deteriorate the images. To compensate the aberrations caused by the misalignment of the microlens, a concave reflection mirror whose radius is close to the focal length of the macrolens is employed. By adjusting the vertical position of the concave mirror, the system can tolerate a significant misalignment of the microlens. Moreover, the transverse displacement of the cubes can also be compensated by adjusting the top mirror. These features permit the cost of packaging to be reduced. The diameter of the microlens is 300 μm, the radius of curvature 250 μm, the lens sag 50 μm, and the refractive index 1.52. The macrolens is an off-the-shelf device from Newport (KPX088) with an $f/2.9$, and so is the concave mirror (KPC031, with reflection coating on the concave side). In contrast, if a plane reflection mirror is used in the top, specially designed macrolenses are required and the alignment tolerance is quite small (~10 μm). For the above Newport macrolens, a plane mirror cannot generate good images at all. Thus the use of a concave mirror has significant advantages. The optics module is made of plastics and can be removed and replaced.

8   ALIGNMENT AND PACKAGING

Alignment of optical systems involves the positioning of the optoelectronic devices and optical elements to maximize the coupling of the light through the optical system from a transmitter to a receiver. Two issues are of concern in the alignment, tolerance and reliability. Alignment tolerance represents the accuracy with which the system should be aligned. It determines the cost and the degree of difficulty of alignment of the optoelectronic devices and optical elements. The alignment also affects other system performance measures such as link efficiency, cross talk, bit-error rate, and system bandwidth. Reliability determines how well the alignment can be maintained over time and environmental changes (e.g., mechanical vibration and temperature change). Three packaging technologies have been investigated: optomechanics, planar optics, and stacked optics.

8.1   Optomechanics

In an optical interconnection system packaged by optomechanics, there are three interfaces: the optical interface, the mechanical interface, and the electrical interface. Compared with an all-electronic system, the addition of optomechanics into the optical interconnection system alters the nature of electrical packaging design methodologies and increases the packaging complexity. This includes the alignment between the optical elements and the optoelectronic chip, and the optical elements themselves. Alignment of optical components is usually a critical issue, since each component has three positional and three rotational degrees of freedom. In assembly, each component has to be positioned according to these degrees of freedom. There are three main sources of misalignment: mechanical effects, thermal effects, and aberration of the optical elements. Errors due to mechanical effects are caused by manufacturing tolerance and assembly tolerance. Thermal effects may result in the wavelength shift of a transmitter and thus a change in focal length. Tolerance analysis can be performed by commercially available optical design programs such as CODE V©. Other advanced computer-aided-design tools such as Chatoyant [167] have recently been developed. Numerical models include Gaussian beam propagation, ray tracing, and Fresnel–Kirchhoff scalar diffraction theory or Rayleigh–Sommerfeld diffractive formulations [168]. Some specific computer-aided-design tools for free-space optoelectronic systems are also being developed. Such tools are capable of modeling optical signal propagation with mechanical tolerancing at the system level. It should be noted that for micro-optical systems diffractive theory may generate better results. Typically the window size of the optoelectronic device is on the order of a few micrometers or tens of micrometers. Precise alignment in the micrometer or submicrometer range is required. Such a requirement is usually satisfied at a high cost. Recently an optomechanics design based on the use of LEGO was presented [169]. With this method, most of the mechanical parts of an optical setup can be built with little effort and at an extremely reduced cost. For optomechanical FSOI systems, active alignment is usually required [170]. Adaptive optics has been used for dynamic alignment [171]. To make the use of optical interconnects more practical, it is necessary to consider designs that alleviate or even remove the need for adjustments [172]. Recently some cost-effective and self-alignment methods of packaging optomechanical and optoelectronic devices have been reported [173,174]. For example, a self-alignment can be achieved by use of microconnectors [173], each of which consists of an optical plug and a socket. The optical plugs are made of a photosensitive resin exposed to light through the optical system, and the sockets are fabricated by etching of the substrate. Another packaging approach uses transparent optical molded plastic [174], which can realize sufficient alignment precision that further adjustment is not required during system assembly. These methods satisfy the requirements for free space to become more widely used.

8.2 Planar Optics

One method that reduces the alignment problem and produces a better package for optical interconnection systems involves the use of micro-optical integration technologies. In comparison with bulk optics, integrated micro-optical systems permit small VLSI compatible dimensions with built-in alignment. Two distinct approaches have been developed, planar optics and stacked optics.

The basic idea of planar optics [18] is to fold 3D optical systems into a more compact geometry and integrate all elements on two sides of a single substrate with the light traveling inside the substrate along a zigzag path. It was proposed to make passive free-space optics compatible with the manufacture of integrated circuits. The passive micro-optical elements include microlenses, beam splitters, and computer-generated holograms, and they can be refractive, reflective, and diffractive. The substrate may be either glass or GaAs that is transparent at the working wavelength. The micro-optical elements can be etched into the substrate at predetermined positions with submicron precision, and all the elements on one side can be fabricated simultaneously using photolithography. Optical elements on the two sides of the substrate can also be aligned with submicron precision by using the through-wafer alignment technique. As described before, optoelectronic chips can be mounted on the substrate by using hybrid integration techniques such as flip-chip solder bump bonding, which allows one to achieve a precision of a few micrometers or less. The system is compact and requires a smaller volume than systems packaged with optomechanics. Furthermore, the system is insensitive to environmental influences; it is both mechanically and thermally stable.

If large active arrays are integrated in the planar architecture, heat removal is an important issue, since the power dissipated on a single chip may be several watts. Generally speaking, one solution is to spread the dissipated heat over a large area so that the heat can be removed by a convenient sink such as Si or diamond. Therefore an appropriate wafer can be sandwiched between the substrate and the optoelectronic chip plane [175]. The wafer has two functions. One is to act as a spacer to provide the desired separation between the optoelectronic devices and the microlenses for collimation and focusing. The other function is to spread the heat generated by the chip.

Planar optics have been applied to optical clock distribution [176], optical imaging [72,168], optical interconnections for MCMs [178,179], optical backplanes and buses [180], compact optical correlator [181], etc. In the planar optical clock distribution schemes, a cascade of N 1×2 beam splitters is used to split up a signal into 2^N output beams, and the beam splitters are implemented as binary diffraction gratings designed to generate only two diffraction orders. Two integrated optical imaging systems have been demonstrated. One system [177] is a hybrid $4f$ system with 32×32 channels in an area of 1.6×1.6 mm. All optical elements are implemented as diffractive optical elements and monolithically integrated on a quartz glass substrate of 50 mm diameter and 6 mm thickness. Very low optical cross talk was measured between the channels. Another system [77] is a hybrid $2f/2f$ imaging system. The setup is schematically shown in Fig. 45. An array of 50×50 microlenses with an interconnection density of 400 mm^{-2} is used to couple the light into the glass substrate. The light propagates within the fused silica glass substrate of 6 mm thickness. The microlenses of arrays A1 and A2 are individually designed to focus the light in plane L, where the imaging lens is located. All these lenses are fabricated lithographically as four-level diffractive optical elements.

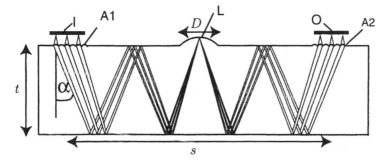

Figure 45 Planar optical implementation of the imaging system shown in Fig. 11 with 2500 parallel data channels. (From Ref. 77.) A1, A2, optimized microlens arrays; L, imaging lens; substrate thickness $t = 6$ mm, interconnection length $s = 8.6$ mm.

Tolerance studies for planar optical systems have been performed [179,180]. In Ref. 179, multiple misalignments and active-element variations are considered. A computer-aided analysis procedure was designed that can determine both active and passive element tolerances to achieve some system-level specifications such as cost or yield (the percentage of systems that have acceptable performance). The procedure includes a detailed design flow chart that can be studied in a standard optical design program, such as CODE V. The optical energy transfer efficiency can be determined by performing Gaussian propagation or general Fresnel propagation. On this basis, system-level performance such as bit-error rate can also be determined. To evaluate the cost and yield, a tolerance analysis by Monte Carlo techniques is used. The calculated design variables can be used for a subsequent tolerancing design to optimize performance.

8.3 Stacked Optics

The concept of stacked micro-optics was first proposed by Iga et al. [15] in 1982. The idea is to stack various planar 2D array components to build a real 3D integrated optical system. The arrayed elements include microlenses, mirrors, prisms, gratings, holograms, filters, etc, which can be achieved from planar technology with the help of lithographic fabrication of the integrated circuits [182]. This packaging approach allows mass production of standardized micro-optical components, and optical components of different materials can be connected in tandem. For packaging of a stacked micro-optical system, one advantage is the reduction of lateral degrees of freedom by use of prearranged arrays. The lateral alignment of the individual elements in an array is eliminated since the lateral position is determined by the lithographic fabrication. Only the lateral position

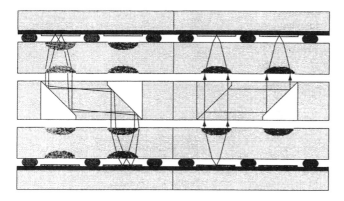

Figure 46   Schematic of a stacked micro-optical imaging system. (From Ref. 182.)

of different layers is important. Several schemes have been used. For mechanical alignment, one can either align the layers with alignment structures onto separate alignment grooves fabricated by wet chemical etching in silicon, or directly stack all the layers together by fabricating mechanical alignment structures (naps and grooves) for plug-and-play applications. For higher accuracy alignment, active alignment can be used. This can be done with the existing mask alignment technologies. The third choice is to use flip-chip bonding by making metallic bonding pads at the side of each element array. This technique is faster and cheaper than the previous methods. Stacked optics can be used for various applications. In FSOI, it has been used for board-to-board optical bus interconnection by cascading Selfoc lens arrays [80]. In another application case [183], microlens arrays were stacked as a beam coupling module between the laser diode array and a fiber array. Optical imaging systems between the optoelectronic arrays can be established using a similar architecture. Figure 46 shows a stacked micro-optical system with two different imaging schemes [182]. The left-hand side uses a two-lens system (light pipe) for Fourier transforms while the right-hand side uses a one-lens system ($2f$ system) for Fourier transforms. Each lens is able to image a whole data array. Implementations for an optical interconnection network [145,146] and weighted interconnection patterns [17,184] using stacked optics have been demonstrated.

9 SUMMARY

The role of optics in digital computing is being considered as an interconnection tool. Significant progress has been made in FSOI in the past decade. This is driven by the advances in device technologies. The performance of SLMs, MQW modulators, VCSELs, and SPAs, the integration of optoelectronic devices and electronic processing elements, micro-optical fabrication, and optical system packaging technologies have improved significantly. With the increased research interest in both the optical and electrical communities, integration of some form of FSOI in commercial digital systems will likely soon become reality.

Numerous studies in the literature of optical interconnection have been based on technological arguments. The global operation of the targeted architecture has not yet been fully analyzed. This analysis should consider the trade-offs to be made in optically interconnected digital systems. A most critical issue in computer architectures, from the monoprocessor to large multiprocessor systems, is access time to the main memory [185]. In monoprocessors the memory access time is dominated by the electronic latency of the memory itself. Thus implementing FSOIs inside the memory hierarchy without changing the memory architecture cannot dramatically improve the global performance. In strongly compiled multiprocessors, node-bypass latency dominates. Therefore the higher connectivity possible with optics will shorten the node-to-node path but increase the expense and complexity of the network. This trade-off leaves open the choice of the best network in terms of simplicity and latency reduction.

REFERENCES

1. National Technology Roadmap for Semiconductors. The Semiconductor Industry Association, 1994, 1997.
2. J. W. Goodman, F. I. Leonberger, S.-Y. Kung, and R. A. Athale. Optical interconnections for VLSI systems. Proc. IEEE 72, 850–866 (1984).
3. M. R. Feldman, S. C. Esener, C. C. Guest, and S. H. Lee. Comparison between optical and electrical interconnects based on power and speed considerations. Appl. Opt. 27, 1742–1751 (1988).
4. G. I. Yayla, P. J. Marchand, and S. C. Esener. Speed and energy analysis of digital interconnections: comparison of on-chip, off-chip, and free-space technologies. Appl. Opt. 205–227 (1998).
5. D. A. B. Miller. Physical reasons for optical interconnection. Inter. J. Optoelectronics 11, 155–168 (1997).
6. V. Krishnamoorthy and D. A. B. Miller. Scaling optoelectronic-VLSI circuits into the 21st century: a technology roadmap. IEEE J. Sel. Top. Quantum Electron. 2, 55–76 (1996).
7. D. A. B. Miller and H. M. Ozaktas. Limit to the bit-rate capacity of electrical interconnects from the aspect ratio of the system architecture. J. Parallel Distrib. Comput. 41, 42–52 (1997).
8. H. S. Hinton, T. J. Cloonan, F. B. McCormick, A. L. Lentine, and F. A. P. Tooley. Free-space digital optical systems. Proc. IEEE 82, 1632–1649 (1994).
9. B. Mansoorian, V. Ozguz, C. Fan, and S. Esener. Design and implementation of flip-chip bonded Si/PLZT smart pixels. Proc. IEEE LEOS Summer Top. Meet. on Smart Pixels, Santa Barbara, CA, Aug. 1992, Paper MC2.

10. K. W. Goossen, J. A. Walker, L. A. D'Asaro, S. P. Hui, B. Tseng, R. Leibenguth, D. Kossives, D. D. Bacon, D. Dahringer, L. M. F. Chirovsky, A. L. Lentine, and D. A. B. Miller. GaAs MQW modulators integrated with silicon CMOS. IEEE Photon Technol. Lett. 7, 360–362 (1995).

11. A. V. Krishnamoorthy, L. M. F. Chirovsky, W. S. Hobson, R. E. Leibenguth, S. P. Hui, G. J. Zydzik, K. W. Goosen, J. D. Wynn, B. J. Tseng, J. Lopata, J. A. Walker, J. E. Cunningham, and L. A. D'Asaro. Vertical-cavity surface-emmiting lasers flip-chip bonded to gigabit-per-second CMOS circuits. IEEE Photon. Technol. Lett. 11, 128–130 (1999) and references therein.

12. A. L. Lentine, K. W. Goosen, J. A. Walker, L. M. F. Chirovsky, L. A. D'Asaro, S. P. Hui, B. J. Tseng, R. E. Leibenguth, J. E. Cunningham, W. Y. Jan, J.-M. Kuo, D. W. Dahringer, D. P. Kossives, D. D. Bacon, G. Livescu, R. L. Morrison, R. A. Novotny, and D. B. Buchholz. High-speed optoelectronic VLSI switching chip with >4000 optical I/O based on flip-chip bonding of MQW modulators and detectors to silicon CMOS. IEEE J. Sel. Top. Quantum Electron. 2, 77–84 (1996).

13. N. McArdle, M. Naruse, and M. Ishikawa. Optoelectronic parallel computing using optically interconnected pipeline processing arrays. IEEE J. Sel. Top. Quantum Electron. 5, 250–260 (1999).

14. G. A. Betzos and P. A. Mitkas. Performance evaluation of massively parallel processing architectures with three-dimensional optical interconnections. Appl. Opt. 37, 315–325 (1998).

15. K. Iga, M. Oikawa, S. Misawa, J. Banno, and Y. Kokubun. Stacked planar optics: an application of the planar mircolens. Appl. Opt. 21, 3456–3460 (1982).

16. K. H. Brenner, M. Kufner, S. Kunfer, J. Moisel, A. Müller, S. Sinzinger, M. Testorf, J. Göttert, and J. Mohr. Application of three-dimensional microoptical components formed by lithography, electroforming and plastic molding. Appl. Opt. 32, 6464–6469 (1993).

17. L. Liu, G. Li, L. Shao, and Y. Yin. Optical 3-D stacked morphological processor module with variable thresholding. Optical Review 3, 406–409 (1996).

18. J. Jahns. Planar packaging of free-space optical interconnections. Proc. IEEE 82, 1623–1631 (1994).

19. A. V. Krishnamoorthy and D. A. B. Miller. Firehose architectures for free-space optically interconnected VLSI circuits. J. Parallel Distrib. Comput. 41, 109–114 (1997).

20. F. E. Kiamilev, P. Marchand, A. V. Krishnamoorthy, S. C. Esener, and S. H. Lee. Performance comparison between optoelectronic and VLSI multistage interconnection networks. J. Lightwave Technol. 9, 1674–1692 (1991).

21. F. Wang, L. Liu, Z. Wang, G. Li, and L. Xu. Reconfigurable interconnection using fixed regional angular volume holograms in photorefractive materials. Journal of Modern Optics 45, 153–165 (1997).

22. A. V. Krishnamoorthy, P. J. Marchand, F. E. Kiamilev, and S. C. Esener. Grain-size considerations for optoelectronic multistage interconnection networks. Appl. Opt. 31, 5480–5507 (1992).

23. S. Esener and P. Marchand. 3D optoelectronic stacked processors: design and analysis. Proc. SPIE 3490, 541–545 (1998).

24. J. Rorie, P. J. Marchand, J. Ekman, F. E. Kiamilev, and S. Esener. Application of

networking concepts to optoelectronic multiprocessor architectures. IEEE J. Sel. Top. Quantum Electron. Vol. 5, 353–359 (1999).

25. F. B. McCormick, F. A. B. Tooley, J. M. Sasian, J. L. Brubaker, A. L. Lentine, T. J. Cloonan, R. L. Morrison, S. L. Walker, and R. J. Crisci. Parallel interconnection of two 64 × 32 symmetric self electrooptic effect device arrays. Electron. Lett. 27, 1869–1871 (1991).

26. H. M. Ozaktas and M. F. Erden. Comparison of fully three-dimensional optical, normally conducting, and superconducting interconnections. Appl. Opt. 7264–7275 (1999).

27. H. S. Hinton. Progress in the smart pixel technologies. IEEE J. Sel. Top. Quantum Electron. 14–23 (1996).

28. D. R. Rolston, B. Robertson, H. S. Hinton, and D. V. Plant. Analysis of a microchannel interconnect based on clustering of smart-pixel-device windows. Appl. Opt. 35, 1220–1233 (1996).

29. D. A. B. Miller, D. S. Chemla, and S. Schmitt-Rink. Electric field dependence of optical properties of semiconductor quantum wells: physis and applications. In Optical Nonlinearities and Instabilities in Semiconductors. Academic Press, New York, 1988.

30. R. Morgan, L. Chirovsky, M. Focht, and R. Leibenguth. High-power quantum well modulators exploiting resonant tunneling. Appl. Phys. Lett. 59, 3524–3526 (1991).

31. M. W. Hansen. High intensity phase modulation using absorption saturated multiple quantum wells. Ph.D diss., University of California, San Diego, 1998.

32. G. D. Boyd, D. A. B. Miller, S. L. McCall, A. C. Gossard, and J. H. English. Multiple quantum well reflection modulators. Appl. Phys. Lett. 50, 1119–1121 (1987).

33. A. L. Lentine, L. M. F. Chirovsky, L. A. D'Asaro, R. F. Kopf, and J. M. Kuo. High speed 2 × 4 array of differential quantum well modulators. IEEE Photon. Technol. Lett. 2, 477–480 (1990).

34. D. A. B. Miller, D. S. Chemla, T. C. Damen, A. C. Gossard, W. Weigmann, T. H. Wood, and C. A. Burrus. Novel hybrid optically bistable switch: the quantum well self-electro-optic effect device. Appl. Phys. Lett. 45, 13–15 (1984).

35. G. Livescu, D. A. B. Miller, J. E. Henry, A. C. Gossard, and J. H. English. Spatial light modulator and optical dynamic memory using a 6 × 6 array of self-electro-optic-effect devices. Opt. Lett. 13, 297–299 (1988).

36. A. L. Lentine, H. S. Hinton, D. A. B. Miller, J. E. Henry, J. E. Cunningham, and L. M. F. Chirovsky. Symmetric self-electrooptic effect device: optical set-reset latch, differential logic gate, and differential modulator, detector. IEEE J. Quantum Electron. 25, 1928–1936 (1989).

37. D. A. B. Miller, M. D. Feuer, T. Y. Chang, S. C. Shunk, J. E. Henry, D. J. Burrows, and D. S. Chemla. Field-effect transistor self-electrooptic effect device: integrated photodiode, quantum well modulator and transistor. IEEE Photon. Technol. Lett. 1, 62–64 (1989).

38. M. Yamaguchi, T. Yamamoto, K. Hirabayashi, S. Matsuo, and K. Koyabu. High-density digital free-space photonic-switching fabrics using exciton absorption reflection-switch (EARS) arrays and microbeam optical interconnections. IEEE J. Sel. Top. Quantum Electron. 2, 47–54 (1996).

39. D. W. Shih, C. Fan, M. W. Hansen, S. C. Esener, and H. H. Wieder. Integration of InAlGaAs/InGaAs MODFETs on MQW modulators on GaAs substrates. Electron. Lett. 30, 1713–1714 (1994).

40. A. L. Lentine and D. A. B. Miller. Evolution of the SEED technology: bistable logic gates to optoelectronic smart pixels. IEEE J. Quantum Electron. 29, 655–669 (1993).

41. F. B. McCormick, T. J. Cloonan, A. L. Lentine, J. M. Sasian, R. L. Morrison, M. G. Beckman, S. L. Walker, M. J. Wojcik, S. J. Hinterlong, R. J. Crisci, R. A. Novotny, and H. S. Hinton. A 5-stage free-space optical switching network with field-effect transistor self-electro-optic-effect smart pixel arrays. Appl. Opt. 33, 1601–1618 (1994).

42. T. K. Woodward, A. V. Krishnamoorthy, A. L. Lentine, K. W. Goossen, J. A. Walker, J. E. Cunningham, W. Y. Jan, L. A. D'Asaro, L. M. F. Chirovsky, S. P. Hui, B. Tseng, D. Kossives, D. Dahringer, and R. E. Leibenguth. 1-Gb/s two beam transimpedance smart-pixel optical receivers made from hybrid GaAs MQW modulators bonded to 0.8-µm silicon CMOS. IEEE Photon. Technol. Lett. 8, 422–424 (1996).

43. D. V. Plant, A. Z. Shang, M. R. Otazo, D. R. Rolston, B. Roberston, and H. S. Hinton. Design, modeling, and characterization of FET-SEED smart pixel transceiver arrays for optical backplanes. IEEE J. Quantum Electron. 32, 1391–1398 (1996).

44. K. W. Goossen, G. D. Boyd, J. E. Cunningham, W. Y. Jan, D. A. B. Miller, D. S. Chemla, and R. M. Lum, GaAs-AlGaAs multiple quantum well reflection modulators grown on GaAs and silicon substrates. IEEE Photon. Technol. Lett. 1, 304–306 (1989).

45. K. W. Goossen, J. A. Walker, L. A. D'Asaro, S. P. Hui, B. Tseng, R. Leibenguth, D. D. Bacon, D. Dahringer, L. M. F. Chirovsky, A. L. Lentine, and D. A. B. Miller. GaAs MQW modulators integrated with silicon CMOS. IEEE Photon. Technol. Lett. 7, 360–362 (1995).

46. H. Soda, K. Iga, C. Kitahara, and Y. Suematsu. GaInAsP/InP surface emitting injection lasers. Jpn. J. Appl. Phys. 18, 2329–2330 (1979).

47. K. Choquette and H. Q. Hou. Vertical-cavity surface emitting lasers: moving from research to manufacturing. Proc. IEEE 85, 1730–1739 (1997).

48. J. L. Jewell, J. P. Harbison, A. Scherer, Y. H. Lee, and L. T. Florez. Vertical-cavity surface-emitting lasers: design, growth, fabrication, characterization. IEEE J. Quantum Electron. 27, 1332–1346 (1991).

49. J. L. Jewell, Y. H. Lee, S. L. McCall, J. P. Harbison, and L. T. Florez. High-finesse (Al,Ga)As interference filters grown by molecular beam epitaxy. Appl. Phys. Lett. 53, 640–642 (1988).

50. M. K. Hibbs-Brenner, R. A. Morgan, R. A. Walterson, J. A. Lehman, E. L. Kalweit, S. Bounnak, T. Marta, and R. Gieske. Performance, uniformity, and yield of 850-nm VCSEL's deposited by MOVPE. IEEE Photon. Technol. Lett. 8, 7–9 (1996).

51. K. D. Choquette, R. P. Schneider, K. L. Lear, and K. M. Geib. Low threshold voltage vertical-cavity lasers fabricated by selective oxidation. Electron. Lett. 30, 2043–2044 (1995).

52. G. M. Yang, M. H. MacDougal, and P. D. Dapkus. Ultralow threshold current vertical-cavity surface-emitting lasers obtained with selective oxidation. Electron. Lett. 31, 886–888 (1995).

53 K. L. Lear, K. D. Choquette, R. P. Schneider, S. P. Kilcoyne, and K. M. Geib. Selectively oxidized vertical cavity surface emitting lasers with 50% power conversion efficiency, Electron. Lett. 31, 208–209 (1995).

54. B. J. Thibeault, K. Bertilsson, E. R. Hegblom, E. Strzelecka, P. D. Floyd, and L. A. Coldren. High-speed characteristics of low-optical loss oxide-apertured vertical-cavity lasers. IEEE Photon. Technol. Lett. 9, 11–13 (1997).

55. H. Q. Hou, H. C. Chui, K. D. Choquette, B. E. Hammons, W. G. Breiland, and K. M. Geib. Highly uniform and reproducible vertical-cavity surface-emitting lasers grown by metalorganic vapor phase epitaxy with in situ refelctometry. IEEE Photon. Technol. Lett 8, 1285–1287 (1996).

56. C. L. Chua, R. L. Thornton, and D. W. Treat. Planar laterally oxidized vertical-cavity lasers for low-threshold high-density top-surface-emitting arrays. IEEE Photon. Technol. Lett. 9, 1060–1062 (1997).

57. E. M. Strzelecka, R. A. Morgan, Y. Liu, B. Walterson, J. Skogen, E. Kalweit, S. Bounak, H. Chanhvongsak, T. Marta, D. Skogman, J. Nohava, J. Gieske, J. Lehman, and M. K. Hibbs-Brenner. VCSEL based modules for optical interconnects. Proc. SPIE 3627, 2–13 (1999).

58. T. Kurokawa, S. Matso, T. Nakahara, K. Tateno, Y. Ohiso, A. Wakatsuki, and H. Tsuda. Design approaches for VCSEL's and VCSEL-based smart pixels toward parallel optoelectronic processing system. Appl. Opt. 37, 194–204 (1998).

59. D. A. Louderback, O. Sjölund, E. R. Hegblom, S. Nakagawa, J. Ko, and L. A. Coldren. Modulation and free-space link characteristics of monolithically integarted vertical-cavity lasers and photodetectors with microlenses. IEEE J. Sel. Top. Quantum Electron. 5, 157–165 (1999).

60. R. Pu, C. Duan, and C. W. Wilmsen. Hybrid integration of VCSEL's to CMOS integrated circuits. IEEE J. Sel. Top. Quantum Electron. 5, 201–208 (1999).

61. C. Fan, B. Mansoorian, D. A. V. Blerkom, M. W. Hansen, V. H. Ozguz, S. C. Esener, and G. C. Marsden. Digital free-space optical interconnections: a comparison of transmitter technologies. Appl. Opt. 34, 3103–3115 (1995).

62. T. Nakahara, S. Matsuo, S. Fukushima, and T. Kurokawa. Performance comparison between multiple-quantum-well modulator-based and vertical-cavity-surface-emitting laser-based smart pixels. Appl. Opt. 35, 860–871 (1996).

63. D. A. Van Blerkom, C. Fan, M. Blume, and S. C. Esener. Transiimpedance receiver design optimization for smart pixel arrays. Proc. IEEE LOES Summer Topical Meeting on Smart Pixels, Keystone, PA, Aug. 1996.

64. T. K. Woodward, A. V. Krishnamoorthy, A. L. Lentine, and L. M. F. Chirovsky. Optical receivers for optoelectronic VLSI. IEEE J. Sel. Top. Quantum Electron. 106–116 (1996).

65. O. Kibar, D. A. V. Blerkom, C. Fan, and S. Esener. Power minimization and technology comparisons for digital free-space optoelectronic interconnections. J. Lightwave Technol. 17, 546–555 (1999).

66. A. W. Lohmann. Image formation of dilute arrays for optical information processing. Opt. Commun. 86, 365–370 (1991).

67. F. A. P. Tooley. Challenges in optically interconnecting electronics. IEEE J. Sel. Top. Quantum Electron. 2, 3–13 (1996).

68. R. K. Kostuk, J. W. Goodman, and L. Hesselink. Optical imaging applied to microelectronic chip-to-chip interconnections. Appl. Opt. 24, 2851–2858 (1985).

69. X. Zheng, P. J. Marchand, D. Huang, O. Kibar, N. S. E. Ozkan, and S. C. Esener. Optomechanical design and characterization of a printed-circuit-board-based free-space optical interconnect package. Appl. Opt. 38, 5631–5640 (1999).

70. X. Zheng, P. J. Marchand, D. Huang, and S. Esener. Free-space parallel multichip interconnection system. Appl. Opt. 39, 3516–3524 (2000).

71. C. V. Plant, B. Roberston, H. S. Hinton, M. H. Ayliffe, G. C. Boisset, W. Hsiao, D. Kabal, N. H. Kim, Y. S. Liu, M. R. Otazo, D. Pavlasek, A. Z. Shang, J. Simmons, K. Song, D. A. Thompson, and W. M. Robertson. 4 × 4 vertical-cavity surface-emitting laser (VCSEL) and metal-semiconductor-metal (MSM) optical backplane demonstrator system. Appl. Opt. Vol. 35, 6365–6368 (1996).

72. F. B. McCormic, F. A. P. Tooley, T. J. Cloon, J. M. Sasian, and H. S. Hinton. Optical interconnects using microlens arrays. Opt. Quantum Electron. 24, 465 (1992).

73. C. V. Plant, B. Robertson, H. S. Hinton, W. M. Robertson, G. C. Boisset, N. H. Kim, Y. S. Liu, M. R. Otazo, D. R. Rolston, and A. Z. Shang. An optical backplane demonstrator system based on FET-SEED smart pixel arrays and diffractive lenslet arrays. IEEE Photon. Technol. Lett. 7, 1057–1059 (1995).

74. E. M. Strzelecka, D. A. Louderback, B. J. Thibeault, G. B. Thompson, K. Bertilsson, and L. A. Coldren. Parallel free-space optical interconnect based on arrays of vertical-cavity lasers and detectors with monolithic microlens. Appl. Opt. 37, 2811–2821 (1998).

75. J. Jahns, F. Sauer, B. Tell, K. F. Brown-Goebeler, A. Y. Feldblum, C. R. Nijander, and W. P. Townsend. Parallel optical interconnections using surface-emitting microlasers and a hybrid imaging system. Opt. Commun. 109, 328–337 (1994).

76. C. Berger, J. Ekman, X. Wang, P. Marchand, H. Spaanenburg, F. Kiamilev, S. Esener. Parallel distributed free-space optoelectronic computer engine using flat "plug-on-top" optics package. In Optics in Computing 2000. R. A. Lessard, T. Galstian, eds. SPIE 4089, 1037–1045 (2000).

77. S. Sinzinger and J. Jahns. Integrated micro-optical imaging system with a high interconnection capacity fabricated in planar optics. Appl. Opt. 36, 4729–4735 (1997).

78. F. A. P. Tooley, S. M. Prince, M. R. Taghizadeh, F. B. McCormick, M. W. Derstine, and S. Wakelin. Implementation of a hybrid lens. Appl. Opt. 34, 6471–6470 (1995).

79. G. Li, D. Huang, E. Yuceturk, et al. Three-dimensional optoelectronic stacked processor (3DOESP) by use of free-space optical interconnection and 3D VLSI chip stacks. Appl. Opt. 41, 348–360 (2002).

80. K. Hamanaka. Optical bus interconnection system using Selfoc lenses. Opt. Lett. 16, 1222–1224 (1991).

81. Y. S. Liu, B. Robertson, D. V. Plant, H. S. Hinton, and W. M. Robertson. Design and characterization of a microchannel optical interconnect for optical backplanes. Appl. Opt. 36, 3127–3141 (1997).

82. B. Robertson. Design of an optical interconnect for photonic backplane applications. Appl. Opt. 37, 2974–2984 (1998).

83. D. R. Rolston, B. Robertson, H. S. Hinton, and D. V. Plant. Analysis of a microchannel interconnect based on clustering of smart-pixel-device windows. Appl. Opt. 35, 1220–1233 (1996).

84. M. W. Haney, M. P. Christensen, P. Milojkovic, J. Ekman, P. Chandramani, R. Rozier, F. Kiamilev, Y. Liu, and M. H. Brenner. Multichip free-space global optical interconnection demonstration with integrated arrays vertical-cavity surface-emitting lasers and photodetectors. Appl. Opt. 38, 6190–6200 (1999).

85. S. Araki, M. Kajita, K. Kasahara, K. Kobota, K. Kurihara, I. Redmond, E. Schenfeld, and T. Suzaki. Experimental free-space optical network for massively parallel computers. Appl. Opt. 35, 1269–1281 (1996).

86. A. W. Lohmann. Scaling laws for lens systems. Appl. Opt. 28, 4996–4998 (1989).

87. M. W. Haney and M. P. Christensen. Performance scaling comparison for free-space optical and electrical interconnection approaches. Appl. Opt. 37, 2886–2894 (1998).

88. M. P. Christensen, P. Milojkovic, and M. W. Haney. Low-distortion hybrid optical shuffle concept. Opt. Lett. 24, 169–171 (1999).

89. N. Streibl. Beam shaping with optical array generators. J. Modern Opt. 36, 1559–1573 (1989).

90. A. C. Walker, M. R. Taghizadeh, J. G. H. Mathew, I. Redmond, R. J. Campbell, S. D. Smith, J. Dempsey, and G. Lebreton. Optical biastable thin-film interference devices and holographic techniques for experiments in digital optics. Opt. Eng. 27, 38–44 (1988).

91. A. W. Lohmann and F. Sauer. Holographic telescope arrays. Appl. Opt. 27, 3003–3007 (1988).

92. A. W. Lohmann, J. Schwider, N. Streibl, and J. Thomas. Array illuminator based on phase contrast. Appl. Opt. 27, 2915–2921 (1988).

93. M. Takeda and T. Kubota. Integrated optic array illuminator: a design for efficient and uniform power distribution. Appl. Opt. 30, 1090–1095 (1991).

94. O. Matoba, K. Itoh, and Y. Ichioka. Array of photorefractive waveguides for massively parallel optical interconnections in lithium niobate. Opt. Lett. 21, 122–124 (1996).

95. P. Senthikumaran. Interferometric array illuminator with analysis of nonobservable fringes. Appl. Opt. 38, 1311–1316 (1999).

96. A. W. Lohmann. An array illuminator based on the Talbot-effect. Optik 79, 41–45 (1988).

97. A. W. Lohmann and J. A. Thomas. Making an array illuminator based on the Talbot effect. Appl. Opt. 29, 4337–4340 (1990).

98. T. J. Suleski. Generation of Lohmann images from binary-phase Talbot array illuminators. Appl. Opt. 36, 4686–4691 (1997).

99. M. Testorf and J. Jahns. Planar-integrated Talbot array illuminators. Appl. Opt. 37, 5399–5407 (1998).

100. H. Dammann and K. Görtler. High-efficiency in-line multiple imaging by means of multiple phase holograms. Opt. Commun. 3, 312–315 (1971).

101. J. Turunen, A. Vasara, J. Westerholm, G. Jin, and A. Salin. Optimisation and fabrication of grating beamsplitters. J. Phys. D 21, 102 (1988).

102. U. Krackhardt and N. Streibl. Design of Dammann-gratings for array generation. Opt. Commun. 74, 31 (1989).

103. J. Jahns, M. M. Downs, M. E. Prise, N. Streibl, and S. J. Walker. Dammann gratings for laser beam shaping. Opt. Eng. 28, 1267–1275 (1989).

104. M. R. Feldman and C. C. Guest. Iterative encoding of high-efficiency holograms for generation of spot arrays. Opt. Lett. 14, 479–481 (1989).

105. F. Wryoski. Diffraction optics: design, fabrication and applications. Technical Digest (Optical Society of America, Washington D.C. 1992), Vol. 9, 147–150.

106. V. Arrizón and M. Testorf. Performance of discrete Fourier array illuminators. Opt. Lett. 22, 197–199 (1997).

107. P. Blair, H. Lüpken, M. R. Taghizadeh, and F. Wyrowski. Multilevel phase-only array generators with a trapezoidal phase topology. Appl. Opt. 36, 4713–4721 (1997).

108. R. Iyer, Y. Liu, G. C. Boisset, D. J. Goodwill, M. H. Ayliffe, B. Robertson, W. M. Robertson, D. Kabal, F. Lacroix, and D. V. Plant. Design, implementation, and characterization of an optical power supply spot-array generator for a four-stage free-space optical backplane, Appl. Opt. 36, 9230–9242 (1997).

109. N. Streibl, J. Jahns, U. Nölscher, and S. J. Walker. Array generation with lenslet arrays. Appl. Opt. 30, 2739–2742 (1991).

110. D. F. Brosseau, F. Lacroix, M. H. Ayliffe, E. Bernier, B. Robertson, F. A. P. Tooley, D. V. Plant, and A. G. Kirk. Design, implementation, and characterization of a kinematically aligned, cascaded spot-array generator for a modulator-based free-space optical interconnect. Appl. Opt. 39, 733–745 (2000).

111. J. Duato, S. Yalamanchili, and L. Ni. Interconnection Networks: An Engineering Approach. IEEE Computer Society, Los Alomitos, California, 1997.

112. J. Ford, Y. Fainman, and S. H. Lee. Reconfigurable array interconnection by photorefractive correlation. Appl. Opt. 33, 5363–5377 (1994).

113. T. M. Pinkston and J. W. Goodman. Design of an optical reconfigurable shared-bus-hypercube interconnect. Appl. Opt. 33, 1434–1443 (1994).

114. M. J. Murdocca, A. Huang, J. Jahns, and N. Streibl. Optical design of programmable logic arrays. Appl. Opt. 27, 1651 (1988).

115. B. Ha and Y. Li. Reflective optical ring-array interconnects: an optical system design study. Appl. Opt. 32, 5727–5740 (1993).

116. Y. Sheng. Space invariant multiple imaging for hypercube interconnections, Appl. Opt. 29, 1101–1105 (1990).

117. A. Louri and H. Sung. Efficient implementation methodology for three-dimensional space-invariant hypercube-based optical interconnection networks. Appl. Opt. 32, 7200–7209 (1993).

118. Y. Li, A. W. Lohmann, and S. B. Rao. Free-space optical mesh-connected bus networks using wavelength-division multiple access. Appl. Opt. 32, 6425–6437 (1993).

119. A. Louri, S. Furlonge, and C. Neocleous. Experimental demonstration of the optical multi-mesh hypercube: scaleable interconnection network for multiprocessors and multicomputers. Appl. Opt. 6909–6919 (1996), and the references therein.

120. J.-H. Yeh, R. K. Kostuk, and K.-Y. Tu. Hybrid free-space optical bus system for board-to-board interconnections. Appl. Opt. 35, 6354–6364 (1996).

121. M. Kajita, K. Kasahara, T. J. Kim, D. T. Neilson, I. Ogura, I. Redmond, and E. Schenfeld. Wavelength-division multiplexing free-space optical interconnect networks for massively parallel processing systems. Appl. Opt. 37, 3746–3755 (1998).

122. B. P. Barrett, P. Blair, G. S. Buller, D. T. Neilson, B. Robertson, E. C. Smith, M. R. Taghizadeh, and A. C. Walker. Components for the implementation of free-space optical crossbars. Appl. Opt. 35, 6934–6944 (1996).

123. A. C. Walker, M. P. Y. Desmulliez, M. G. Forbes, S. J. Fancey, G. S. Buller, M. R. Taghizadeh, J. A. B. Dines, C. R. Stanley, G. Pennelli, A. R. Boyd, P. Horan, D. Byrne, J. Hegarty, S. Eitel, H.-P. Gauggel, K.-H. Gulden, A. Gauthier, P. Benabes, J.-Louis Gutzwiller, and M. Goetz. Design and construction of an optoelectronic crossbar switch containing a terabit per second free-space optical interconnect. IEEE J. Sel. Top. Quantum Electron. 5, 236–247 (1999).

124. M. Fukui and K.-I. Kitayama. Design considerations of the optical image crossbar switch. Appl. Opt. 31, 5542–5547 (1992).

125. A. O. Harris. Multichannel acousto-optic crossbar switch. Appl. Opt. 30, 4245–4256 (1991).

126. Y. Li, T. Wang, and R. A. Linke. VCSEL-array-based angle-multiplexed optoelectronic crossbar interconnects. Appl. Opt. 35, 1282–1295 (1996). Y. Li, T. Wang, and S. Kawai. Distributed crossbar interconnects with vertical-cavity surface-emitting laser-angle multiplexing and fiber image guides. Appl. Opt. 37, 254–263 (1998).

127. E. Chiou and P. Yeh. 2 × 8 photorefractive reconfigurable interconnect with laser diodes. Appl. Opt. 31, 5536–5541 (1992).

128. G. Li, L. Liu, B. Liu, and Z. Xu. High-efficiency volume hologram recording with a pulsed signal beam. Optics Letters 23, 1307–1309 (1998).

129. D. J. Brady and D. Psaltis. Holographic interconnections in photorefractive waveguides, Appl. Opt. 30, 2324–2333 (1991).

130. B. Webb amd A. Louri. All-optical crossbar switch using wavelength division multiplexing and vertical-cavity surface-emitting lasers. Appl. Opt. 38, 6176–6183 (1999).

131. M. Y. Li, W. Yuen, and C. J. Chang-Hasnain. Top-emitting micromechanical VCSEL with a 31.6 nm tuning range. IEEE Photon. Technol. Lett. 10, 18–20 (1998).

132. L. McCaughan and G. A. Bogert. 4 × 4 Ti:LiNbO3 integrated-optical crossbar switch array. Appl. Phys. Lett. 47, 348–350 (1985).

133. L. B. Aronson and L. Hesselink. Photorefractive integrated-optical switch arrays in LiNbOs. Opt. Lett. 15, 30–32 (1990).

134. T. J. Cloon. Free-space optical implementation of a feed-forward crossbar network, Appl. Opt. 29, 2006–2012 (1990).

135. H.-I. Jeon and A. A. Sawchuck. Optical crossbar interconnections using variable grating mode devices, Appl. Opt. 26, 261–269 (1987).

136. H. M. Ozaktas, Y. Amitai, and J. W. Goodman. Comparison of system size for some optical interconnection architectures and the folded multi-facet architecture, Opt. Commun. 82, 225–228 (1991).

137. M. E. Marhic and S. G. Lee. Mirror-folded free-space crossbars with holographic implementation, Appl. Opt. 32, 6438–6444 (1993).
138. K.-H. Brenner and T. M. Merklein. Implementation of an optical crossbar network based on directional switches, Appl. Opt. 31, 2446–2451 (1992).
139. L. Cheng and A. A. Sawchuk. Three-dimensional Omega networks for optical implementation, Appl. Opt. 31, 5468–5479 (1992).
140. Y. Wu, L. Liu, and Z. Wang. Characteristics, routing algorithm, and optical implementation of two-dimensional perfect-shuffle networks, Appl. Opt. 32, 7210–7216 (1993).
141. T. M. Slagle and K. H. Wagner. Optical smart-pixel-based Clos crossbar switch, Appl. Opt. 36, 8336–8351 (1997).
142. M. Guizani. Picosecond multistage interconnection network architecture for optical computing, Appl. Opt. 33, 1587–1599 (1994).
143. L. Thylén. Integrated optics in LiNbO3: recent developments in devices for telecommunications. J. Lightwave Technol. 6, 847–861 (1988).
144. M. J. Brinkman and G. J. Sonek. Low power response of all-optical crossbar networks in quantum well heterostructures. Appl. Opt. 31, 338–349 (1992).
145. L. Liu, Y. Yin, H. Peng, W. Xiong, G. Li, B. Wang, N. Wang, L. Shao, F. Liang, and L. Zhao. Design and fabrication of parallel processor modules by polarization-optical stacked integration. International Journal of Optoelectronics 10, 39–49 (1995). L. Shao, L. Liu, H. Peng, and G. Li. Routing scheme, permutation perproties, and optical implementation of a four-shuffle-exchange-based Omega network. Opt. Eng. 36, 1542–1547 (1997).
146. L. Liu and Y. Li. Free-space optical shuffle implementations by use of birefrigent-customized modular optics. Appl. Opt. 36, 3854–3865 (1997).
147. F. Xu, J. E. Ford, and Y. Fainman. Polarization-selective computer-generated holograms: design, fabrication, and applications. Appl. Opt. 34, 256–266 (1995).
148. D. Marom, P. E. Shames, F. Xu, and Y. Fainman. Folded free-space polarization-controlled multistage interconnection network. Appl. Opt. 37, 6884–6891 (1998).
149. B. Pesach, G. Bartal, E. Refaeli, A. J. Agranat, J. Krupnik, and D. Sadot. Free-space optical cross-connect switch by use of electroholography. Appl. Opt. 39, 746–758 (2000).
150. A. W. Lohmann, W. Stork, and G. Stucke. Optical perfect shuffle. Appl. Opt. 25, 1530–1531 (1986).
151. A. W. Lohmann. What classical optics can do for digital optical computer, Appl. Opt. 25, 1543–1549 (1986).
152. G. Eichmann and Y. Li. Compact optical generalized perfect shuffle, Appl. Opt. 26, 1167–1169 (1987).
153. K.-H. Brenner and A. Huang. Optical implementation of the perfect shuffle interconnection, Appl. Opt. 27, 135–137 (1988).
154. Q. W. Song and F. T. S. Yu. Generalized perfect shuffle using optical spatial filtering, Appl. Opt. 27, 1222–1223 (1988).
155. Y. Sheng. Light effective 2-D optical perfect shuffle using Fresnel mirrors, Appl. Opt. 28, 3290–3292 (1989).
156. K. S. Urquhart, P. Marchand, Y. Fainman, and S. H. Lee. Diffractive optics applied to free-space optical interconnects, Appl. Opt. 33, 3670–3682 (1994).

157. B. W. Stirk, R. A. Athale, and M. W. Haney. Folded perfect shuffle optical processor, Appl. Opt. 27, 202–203 (1988).

158. F. B. McCormick and M. E. Prise. Optical circuitry for free-space interconnections, Appl. Opt. 29, 2013–2018 (1990).

159. D. T. Neilson, S. M. Prince, D. A. Baillie, and F. A. P. Tooley. Optical design of a 1024-channel free-space sorting demonstrator, Appl. Opt. 36, 9243–9252 (1997).

160. J. Jahns. Optical implementation of the banyan network, Opt. Commun, 76, 321–324 (1990).

161. J. Jahns and M. J. Murdocca. Crossover networks and their optical implementation, Appl. Opt. 3155–3160 (1988).

162. T. J. Cloon and F. B. McCormick. Photonic switching applications of 2-D and 3-D crossover networks based on 2-input, 2-output switching nodes, Appl. Opt. 30, 2309 2323 (1991).

163. M. P. Christensen and M. W. Haney. Two-bounce optical arbitrary permutation network, Appl. Opt. 37, 2879–2885 (1998).

164. G. C. Marsden, P. J. Machand, P. Harvey, and S. C. Esener. Optical transpose interconnection system architectures, Opt. Lett. 18, 1083–1085 (1993).

165. W. L. Hendrick, P. J. Marchand, F. Xu, and S. Esener. A compact, packaged free-space optical interconnect, IN Optics in Computing, R. A. Lessard, T. Galstian, Editors, SPIE 4089, 520–529 (2000).

166. D. Huang, G. Li, E. Yuceturk, M. M. Wang, C. Berger, X. Zheng, P. J. Marchand, and S. C. Esener. 3D Optical interconnect distributed crossbar switching architecture. Submitted to Optics in Computing 2001.

167. S. P. Levitan, T. P. Kurzweg, P. J. Marchand, M. A. Rempel, D. M. Chiarulli, J. A. Martinez, J. M. Bridgen, C. Fan, and F. B. McCormick. Chatoyant: a computer-aided-design tool for free-space optoelectronic systems. Appl. Opt. 37, 6078–6092 (1998).

168. F. Lacroix, M. Chateauneuf, X. Xue, and A. G. Kirk. Experimental and numerical analyses of misalignment tolerances in free-space optical interconnects. Appl. Opt. 39, 704–713 (2000).

169. F. Quercioli, B. Tiribilli, A. Mannoni, and S. Acciai. Optomechanics with LEGO. Appl. Opt. 37, 3408–3416 (1998).

170. B. Robertson, Y. Liu, G. C. Boisset, M. R. Tagizadeh, and D. V. Plant. In situ interferometric alignment systems for the assembly of microchannel relay systems. Appl. Opt. 36, 9253–9260 (1997).

171. J. Gourlay, T.-Y. Yang, M. Ishikawa, and A. C. Walker. Lower-order adaptive optics for free-space optoelectronic interconnects. Appl. Opt. 39, 714–720 (2000).

172. C. T. Neilson. Tolerance of optical interconnections to misalignment. Appl. Opt. 38, 2282–2290 (1999).

173. D. Miyazaki, S. Masuda, and K. Matsushita. Self-alignment with optical microconnectors for free-space optical interconnections. Appl. Opt. 37, 228–232 (1998).

174. D. T. Neilson and E. Schenfeld. Plastic modules for free-space optical interconnects. Appl. Opt. 37, 2944–2952 (1998).

175. C. Acklin and J. Jahns. Packaging considerations for planar optical interconnection systems. Appl. Opt. 33, 1391–1397 (1994).

176. S. J. Walker and J. Jahns. Optical clock distribution using integrated free-space optics. Opt. Commun. 86, 359–371 (1992).

177. J. Jahns and B. Acklin. Integrated planar optical imaging system with high interconnection density. Opt. Lett. 18, 1594–1596 (1993).

178. M. R. Feldman, J. E. Morris, I. Turlik, P. Magill, G. Adema, and M. Y. A. Raja. Holographic optical interconnects for VLSI multichip modules. IEEE Trans. Components, Packing, and Manufacturing Technol. (part B) 17, 223–227 (1994).

179. I. Zaleta, S. Patra, V. Ozguz, J. Ma, and S. H. Lee. Tolerancing of board-level-free-space optical interconnects. Appl. Opt. 35, 1317–1327 (1996).

180. R. S. Beech and A. K. Ghosh. Optimization if alignability in integrated planar-optical interconnect packages. Appl. Opt. 32, 5741–5749 (1993).

181. W. Eckert, V. Arrizon, S. Sinzinger, and J. Jahns. Compact planar-integrated optical correlator for spatially incoherent signals. Appl. Opt. 39, 759–765 (2000).

182. K. H. Brenner, M. Kufner, S. Kunfer, J. Moisel, A. Müller, S. Sinzinger, M. Testorf, J. Göttert, and J. Mohr. Application of three-dimensional microoptical components formed by lithography, electroforming and plastic molding. Appl. Opt. 32, 6464–6469 (1993).

183. A. Sasaki, T. Baba, and K. Iga. Focusing characteristics of convex-shaped distributed-index microlens. Jpn. J. Appl. Phys. 31, 1611–1617 (1992).

184. B. Wang, L. Liu, Y. Yin, and G. Li. A various structuring element morphological image processor with a compact weighted interconnection module. Microwave & Optical Technology Letters 13, 38–42 (1996).

185. J. H. Collet, D. Litaize, J. V. Campenhout, C. Jesshope, M. Desmulliez, H. Thienpont, J. Goodman, and A. Louri. Architecture approach to the role of optics in monoprocessor and multiprocessor machines. Appl. Opt. 39, 671–682 (2000).

11

Optical Computing Using Trinary Signed-Digit Arithmetic

Abdallah K. Cherri
Kuwait University, Safat, Kuwait

Mohammad S. Alam
The University of South Alabama, Mobile, Alabama

1   INTRODUCTION

The inherent parallelism of optics offers an attractive approach of attaining high-speed computation (almost at the speed of light), high temporal/spatial bandwidth, and noninterfering communications. The most primitive operation in arithmetic computation is addition since other arithmetic operations (subtraction, multiplication, and division) can be realized through addition.

The primary limitations of the computing speed based on the binary number system are the long carry/borrow propagation paths extending from the least-significant bit to the most-significant bit position [1]. This motivates researchers to look for alternative approaches for designing high-speed arithmetic units. Using nonconventional number representations (nonbinary) [2] for designing fast arithmetic units has gained much attention in recent years. Several nonbinary number representation schemes, such as multiple-valued fixed radix-number [3,4], residue number [5–8], redundant number [9], and signed-digit [9–25] were reported in the past decade to implement efficient arithmetic operations.

Signed-digit number system (proposed by Aviziens [26]) is being extensively explored, in both electronics [27–32] and optics [9–25], by many researchers for use in reliable designs and high speed addition, subtraction, multiplication, division, and square root algorithms. The redundancy provided in signed-digit representation allows for fast addition and subtraction because the sum or differ-

377

ence digit is a function of only the digits in two adjacent digit positions for radix greater than 2, and three adjacent digit positions for a radix of 2. Thus the add time for two redundant signed-digit numbers is constant independent of the word length of the operands, which is the key to high-speed computation. Further, signed-digit number systems allow higher information storage density, less complexity, fewer system components, and fewer cascaded gates and operations.

In this chapter, we report efficient trinary signed-digit (TSD) arithmetic operations such as addition/subtraction, multiplication and division. Also, we report three different optical implemetations for TSD arithmetic.

2 SIGNED-DIGIT NUMBERS

Signed-digit (SD) numbers are formally defined as follows: given a radix r, each digit of an SD number assumes $(2\alpha + 1)$ values of the digit set $\{-\alpha, \ldots, -1, 0, 1, \ldots, \alpha\}$ where $\alpha \le r - 1$. In general, a decimal number D may be represented in terms of an n-digit signed-digit number as

$$D = \sum_{j=0}^{n-1} b_i r^j \tag{1}$$

where b_i digit is selected from the set $\{-\alpha, \ldots, -1, 0, 1, \ldots, \alpha\}$ to produce the appropriate decimal representation. For trinary signed-digit (TSD) numbers, $r = 3$, the digit set is $\{\bar{2}, \bar{1}, 0, 1, 2\}$. Here the digits $\bar{2}$, and $\bar{1}$ denote -2, and -1, respectively.

An SD negative number can be obtained directly by complementing the corresponding SD positive number. For example, using primes to denote complementation, we have $\bar{2}' = 2$, $2' = \bar{2}$, $1' = \bar{1}$, $\bar{1}' = 1$, and $0' = 0$, and therefore $(14)_{10} = [2\,\bar{1}\,\bar{1}]_{TSD}$ or, equivalently, $(14)_{10} = [2\,\bar{2}\,2]_{TSD} = [12\,\bar{1}]_{TSD} = [112]_{TSD}$.

To illustrate the higher information storage density of SD representation, consider the following TSD numbers and their corresponding binary and decimal representation:

$$(22)_3 = (1000)_2 = (8)_{10} \tag{2a}$$

$$(222222222222222)_3 = (1101\ 1010\ 1111\ 0010\ 0110\ 1010)_2 \tag{2b}$$
$$= (143489906)_{10}$$

In the first example, we observe that at most we need four binary bits to represent a two-digit TSD number. In the second case we need a maximum of 24 binary bits to represent a 15-digit TSD number. In the two examples cited, the TSD number system requires 50% and 37.5% fewer digits than the binary number system, which may result in substantial saving of memory space.

2.1 TSD Addition/Subtraction

When studying the TSD representation for the addition of two single digits, it is found that the combination digits $(1, 2)$, or $(2, 2)$, or $(\bar{2}, \bar{1})$, or $(\bar{2}, \bar{2})$ cause a carry propagation to the next higher order digit. The generation of a carry can be avoided by mapping the two digits in question into an intermediate sum and intermediate carry (also known as weight and transfer digits) such that the ith intermediate sum and the $(i - 1)$th intermediate carry never form any one of the aforementioned four combinations. The addition operation is performed in two successive steps which are governed by the following two equations [10]:

$$\text{Step one} \rightarrow x_i + y_i = 3c_{i+1} + s_i \tag{3}$$

$$\text{Step two} \rightarrow S_i = c_i + s_i \tag{4}$$

where x_i, y_i are the operand TSD digits, c_i, s_i represent the intermediate carry and sum, respectively, and S_i is the final carry-free sum. The totally parallel adder for signed-digit representation is shown in Fig. 1, where the functional blocks A and B represent the computing rules generated from Eqs. (3) and (4), respectively. All possible two TSDs that satisfy Eqs. (3) and (4) are listed in Tables 1a and 1b, respectively.

The pairs of to-be-added trinary digits are first divided into nine groups as shown in the second column of Table 1a, whereas the third column shows the necessary intermediate carry and intermediate sum digits for carry-free addition. Thus Tables 1a and 1b represent the 25 and the 9 computational rules used in

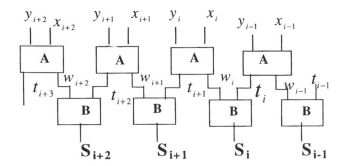

Figure 1 A diagram for a two-step TSD adder/subtracter where the functional blocks A and B employ the computation rules shown in Table 1.

Table 1a　Truth Table for First Step TSD Addition

Group number	Addend/augend $x_i y_i$	Intermediate carry/sum $t_{i+1} w_i$
G1	2 2	1 1
G2	2 1	
	1 2	1 0
G3	1 1	
	0 2	1 $\bar{1}$
	2 0	
G4	0 1	
	1 0	0 1
	2 $\bar{1}$	
	$\bar{1}$ 2	
G5	0 0	
	1 $\bar{1}$	
	$\bar{1}$ 1	0 0
	2 $\bar{2}$	
	$\bar{2}$ 2	
G6	0 $\bar{1}$	
	$\bar{1}$ 0	
	$\bar{2}$ 1	·0 $\bar{1}$
	1 $\bar{2}$	
G7	$\bar{1}$ $\bar{1}$	
	0 $\bar{2}$	$\bar{1}$ 1
	$\bar{2}$ 0	
G8	$\bar{1}$ $\bar{2}$	$\bar{1}$ 0
	$\bar{2}$ $\bar{1}$	
G9	$\bar{2}$ $\bar{2}$	$\bar{1}$ $\bar{1}$

blocks A and B of Fig. 1. The following example shows that Table 1 results in carry-free addition:

$$
\begin{array}{rcccccccccc}
\textit{Addend} & = & 1961_{10} & = & 2 & 2 & 1 & \bar{2} & \bar{1} & 0 & \bar{1} \\
\textit{Augend} & = & -315_{10} & = & \bar{1} & 1 & 2 & 1 & \bar{2} & 0 & 0 \\
\textit{Intermediate carry} & = & & & 0 & 1 & 1 & 0 & \bar{1} & 0 & 0 & \varnothing \\
\textit{Intermediate sum} & = & & & \varnothing & 1 & 0 & 0 & \bar{1} & 0 & 0 & \bar{1} \\
\textit{Final sum} & = & 1646_{10} & = & 0 & 2 & 1 & 0 & \bar{2} & 0 & 0 & \bar{1}
\end{array}
$$

where $\varnothing$ denotes a padded zero.

Table 1b Truth Table for the Second
TSD Addition

Group	Intermediate carry/sum $t_{i-1} w_i$	Final sum s_i
1	1 1	2
2	0 1	
	1 0	1
3	0 0	
	1 $\bar{1}$	0
	$\bar{1}$ 1	
4	0 $\bar{1}$	$\bar{1}$
	$\bar{1}$ 0	
5	$\bar{1}$ $\bar{1}$	$\bar{2}$

Once the computing rules for addition are developed, the computing rules
for subtraction follow directly from them. Note that TSD subtraction is realized
by adding the complement of the subtrahend. The following equation describes
the process:

$$x_i - y_i = x_i + \bar{y}_i \tag{5}$$

The y_i entries of Table 1a are thus complemented (the first step subtraction rules)
to obtain the second step subtraction rules shown in Table 2. The third step for
subtraction operation, however, is the same as the second step for addition,
namely, the computation rules shown in Table 1b.

2.2 Two-Step Recoded TSD Addition/Subtraction

An alternative method to eliminate the carry propagation is to employ a simple
recoding algorithm, which was proposed recently [10]. Table 3 illustrates the
recoding truth table for generating carry-free addition. Therefore an n-digit TSD
number $D = D_{n-1} D_{n-2} \ldots D_1 D_0$ when recoded with Table 3 results in an $(n +
1)$-digit recoded TSD number $Z = Z_n Z_{n-1} Z_{n-2} \ldots Z_1 Z_0$ such that both D and Z
are numerically equal. The recoded TSD output Z_i of Table 3 does not include
the $\bar{2}$ and 2 literals. Thus the recoding operation maps the TSD set $\{\bar{2}, \bar{1}, 0, 1, 2\}$
into a smaller set $\{\bar{1}, 0, 1\}$. The addition truth table corresponding to the recoded
TSD output of Table 3 is shown in Table 4, where A_i and B_i, and S_i represent
addend, augend, and sum operands, respectively. Note that it is necessary for us
to pad three zeros trailing the least significant digit and one zero preceding the
most significant digit in order to apply Table 3 to recode the TSD numbers. Also,

Table 2 Truth Table for the Second Step TSD Subtraction (Intermediate Difference and Borrow)

Group number	Minuend/subtrahend $X_i Y_i$	Intermediate borrow/difference $B_{i+1} D_i$
1	0 0	
	1 1	
	$\bar{1}$ $\bar{1}$	0 0
	2 2	
	$\bar{2}$ $\bar{2}$	
2	0 $\bar{1}$	
	1 0	
	2 1	0 1
	$\bar{1}$ $\bar{2}$	
3	0 1	
	$\bar{1}$ 0	
	$\bar{2}$ $\bar{1}$	0 $\bar{1}$
	1 2	
4	1 $\bar{1}$	
	0 $\bar{2}$	
	2 0	1 $\bar{1}$
5	$\bar{1}$ 1	
	0 2	
	$\bar{2}$ 0	$\bar{1}$ 1
6	2 $\bar{1}$	
	1 $\bar{2}$	1 0
7	$\bar{1}$ 2	
	$\bar{2}$ 1	$\bar{1}$ 0
8	2 $\bar{2}$	1 1
9	$\bar{2}$ 2	$\bar{1}$ $\bar{1}$

note that the recoding algorithm of the TSD number representation is totally parallel. Therefore a fast electronics hardware unit can be designed to process the recoding algorithm. The application of Table 3 and Table 4 on the previous addition example yields

$$
\begin{array}{rcrccccccccccccc}
Addend & = & 1961_{10} & = & \varnothing & 2 & 2 & 1 & \bar{2} & \bar{1} & 0 & \bar{1} & \varnothing & \varnothing & \varnothing \\
Augend & = & -315_{10} & = & \varnothing & \bar{1} & 1 & 2 & 1 & \bar{2} & 0 & 0 & \varnothing & \varnothing & \varnothing \\
Recoding & = & & & 1 & 0 & \bar{1} & 0 & 1 & \bar{1} & 0 & \bar{1} & & & \\
Recoding & = & & & 0 & 0 & \bar{1} & \bar{1} & 0 & 1 & 0 & 0 & & & \\
Final\ sum & = & 1646_{10} & = & 1 & 0 & \bar{2} & \bar{1} & 1 & 0 & 0 & \bar{1} & & &
\end{array}
$$

Table 3   Recoding Truth Table for TSD Numbers

Groups for $D_{i-2}D_{i-3}$ digits	Groups for D_iD_{i-1} digits	$N_1 = 00, 10, 01,$ $0\bar{1}, \bar{1}0, 02, 0\bar{2},$ $1\bar{2}, \bar{1}1, \bar{1}\bar{1}, 1\bar{1},$ $11, 2\bar{2}, \bar{2}2, \bar{1}2$	$N_2 = 2\bar{1}, 20,$ $1\bar{2}, 21, 22$	$N_3 = \bar{2}1, \bar{2}0,$ $\bar{1}2, \bar{2}\bar{1}, \bar{2}\bar{2}$
		OUTPUT		Z_i
$C_1 =$	$\bar{2}\bar{2}, 01, 1\bar{2}$	0	1	0
$C_2 =$	$1\bar{1}, 02, \bar{2}\bar{1}$	1	1	0
$C_3 =$	$\bar{2}1, \bar{1}\bar{2}, 11, 2\bar{2}$	1	$\bar{1}$	1
$C_4 =$	$\bar{2}0, 10$	1	1	1
$C_5 =$	00	0	0	0
$C_6 =$	$20, \bar{1}0$	$\bar{1}$	$\bar{1}$	$\bar{1}$
$C_7 =$	$2\bar{1}, 12, \bar{1}\bar{1}, \bar{2}2$	$\bar{1}$	$\bar{1}$	1
$C_8 =$	$\bar{1}1, 0\bar{2}, 21$	$\bar{1}$	0	$\bar{1}$
$C_9 =$	$22, 0\bar{1}, \bar{1}2$	0	0	$\bar{1}$

Source: Ref. 10.

The subtraction truth table corresponding to the recoded TSD output of Table 3 is shown in Table 5, which is obtained from Table 4 by complementing the augend (B_i).

2.3   One-Step Nonrecoded TSD Addition

In TSD operation, a carry propagation is restricted to one digit position to the left. The final sum digit depends on the next lower order position (see Fig. 1 and

Table 4   Truth Table for the Recoded TSD Addition

Group number	Addend/augend A_iB_i	Final sum S_i
1	1 1	2
2	0 1	1
	1 0	
3	0 0	0
	1 $\bar{1}$	
	$\bar{1}$ 1	
4	0 $\bar{1}$	$\bar{1}$
	$\bar{1}$ 0	
5	$\bar{1}$ $\bar{1}$	$\bar{2}$

Table 5 Truth Table for Recoded TSD Subtraction

Group number	Addend/augend A_iB_i	Final sum S_i
1	1 $\bar{1}$	2
2	0 $\bar{1}$ 1 0	1
3	0 0	0
4	1 1 $\bar{1}$ $\bar{1}$	0
5	0 1 $\bar{1}$ 0	$\bar{1}$
6	$\bar{1}$ 1	$\bar{2}$

Table 1). In order to achieve a one-step scheme, two digit pairs have to be considered at the same time. Therefore, there are $5^4 = 625$ possible combinations that need to be considered for a fully parallel TSD adder. Table 6 arranges the 625 computation rules in a simple truth table, which is obtained by combining the two truth tables shown in Table 1. The one-step TSD addition (using Table 6) of the previous example is

Table 6 Truth Table for the One-Step TSD Addition

	$x_{i-1}y_{i-1}x_iy_i$	$N_1 = G_4 + G_5 + G_6$	$N_2 = G_1 + G_2 + G_3$	$N_3 = G_7 + G_8 + G_9$
		10, 01, $\bar{1}$2, $2\bar{1}$, 00, $1\bar{1}$, $\bar{1}$1, $2\bar{2}$, $\bar{2}$2, $0\bar{1}$, $\bar{1}$0, $\bar{2}$1, $1\bar{2}$	22, 12, 21, 11, 02, 20	$\bar{1}\,\bar{1}$, $0\bar{2}$, $\bar{2}0$, $\bar{1}\,2$, $\bar{2}\,\bar{1}$, $2\bar{2}$
$C_1 = G_2 + G_5 + G_8$	12, 21, 00, $1\bar{1}$, $\bar{1}$1, $2\bar{2}$, $\bar{2}$2, $\bar{1}\,2$, $\bar{2}\,\bar{1}$	0	1	$\bar{1}$
$C_2 = G_1 + G_4 + G_7$	22, 10, 01, $\bar{1}$2, $2\bar{1}$, $\bar{1}\,\bar{1}$, $0\bar{2}$, $\bar{2}0$	1	2	0
$C_3 = G_3 + G_6 + G_9$	11, 02, 20, $0\bar{1}$, $\bar{1}$0, $\bar{2}$1, $1\bar{2}$, $\bar{2}\,\bar{2}$	$\bar{1}$	0	$\bar{2}$

$$\textit{Addend} \;=\; 1961_{10} \;=\; \varnothing \;\; 2 \;\; 2 \;\; 1 \;\; \bar{2} \;\; \bar{1} \;\; 0 \;\; \bar{1} \;\; \varnothing$$

$$\textit{Augend} \;=\; -315_{10} \;=\; \varnothing \;\; \bar{1} \;\; 1 \;\; 2 \;\; 1 \;\; \bar{2} \;\; 0 \;\; 0 \;\; \varnothing$$

$$\textit{Final sum} \;\vdash\; 1646_{10} \;=\; 0 \;\; 2 \;\; 1 \;\; 0 \;\; \bar{2} \;\; 0 \;\; 0 \;\; \bar{1}$$

3 TSD MULTIPLICATION

Multiplication involves two basic operations: the generation of multiplication partial products and their accumulation. Consequently, there are two ways to speed up multiplication: reduce the number of partial products or accelerate their accumulation. Generating a small number of multiplication partial products reduces the complexity, and as a result it reduces the time needed to accumulate the partial products. In this section, we will consider generating all partial products in parallel, and then we will use the fast TSD adders that we have developed earlier for their accumulation.

The multiplication of two n-bit binary numbers A and B will create a $2n$-bit product

$$P = AB = \sum_{i=0}^{2n-1} P_i = \sum_{i=0}^{n-1} pp_i$$

where pp_i is called the partial product, A is called the multiplicand, and B is the multiplier as shown in Fig. 2. Generating the partial products can be achieved using vector outer product matrix of the multiplicand and the multiplier and using optical perfect shuffle [33–35]. Appropriate shifting of partial products can be performed using reflectors, prisms, fibers, or holograms [36]. For TSD numbers, the accumulation process of the multiplication partial products is achieved through a binary-tree architecture where $\log_2(n+1)$ steps and $(n+1)/2$ adders

					a_4	a_3	a_2	a_1	a_0		$= A =$ Multiplicand
					b_4	b_3	b_2	b_1	b_0		$= B =$ Multiplier
$\varnothing$	$\varnothing$	$\varnothing$	$\varnothing$	$\varnothing$	b_0a_4	b_0a_3	b_0a_2	b_0a_1	b_0a_0	pp_0	
$\varnothing$	$\varnothing$	$\varnothing$	$\varnothing$	b_1a_4	b_1a_3	b_1a_2	b_1a_1	b_1a_0	$\varnothing$	pp_1	
$\varnothing$	$\varnothing$	$\varnothing$	b_2a_4	b_2a_3	b_2a_2	b_2a_1	b_2a_0	$\varnothing$	$\varnothing$	pp_2	$\Leftarrow$ Partial Products
$\varnothing$	$\varnothing$	b_3a_4	b_3a_3	b_3a_2	b_3a_1	b_3a_0	$\varnothing$	$\varnothing$	$\varnothing$	pp_3	
$\varnothing$	b_4a_4	b_4a_3	b_4a_2	b_4a_1	b_4a_0	$\varnothing$	$\varnothing$	$\varnothing$	$\varnothing$	pp_4	
P_9	P_8	P_7	P_6	P_5	P_4	P_3	P_2	P_1	P_0	$=$ Final Product	

Figure 2 A two five-digit multiplication.

are required to obtain the final sum (the product of two n-digit numbers). Our main concern in this section is to obtain the TSD computation rules (truth tables) to generate the partial products that will be used in the TSD adders. First, we will consider the multiplication operation when both the multiplicand and the multiplier are recoded and then sequences of addition-recoding operations are performed to obtain the final product. Second, only the multiplier will be recoded and the generated partial products are accumulated using the nonrecoding TSD adder. Third, nonrecoded multiplicand and multiplier are used directly to produce multiplication partial products and multiplication partial carries, and consequently the nonrecoded TSD adder is employed to obtain the final product. Finally, nonrecoded multiplicand and multiplier are used directly to produce partial products without carries, and the nonrecoded TSD adder is used to obtain the final product.

3.1 Recoded TSD Multiplier and Multiplicand

When both the TSD multiplicand and the multiplier are recoded using Table 3, one obtains digits from the set $\{\bar{1}, 0, 1\}$. Consequently, simple computation rules are obtained for generating the partial products: a multiplication partial product is the multiplicand itself, 0, or the complement of the multiplicand, depending on whether the multiplier digit is 1, 0, or $\bar{1}$, respectively. Table 7 shows the truth table for generating the partial products for the recoded TSD multiplication. The recoded TSD multiplication is illustrated by the example shown in Fig. 3 for two three-digit TSD numbers. Note that recoding and generation of the partial products are executed in constant time independent of operand length. However, the addition algorithm takes $\log_2(n + 1)$ iterations.

Table 7 Truth Table for TSD Multiplication When Both the Multiplicand and Multiplier Are Recoded

Group	Multiplicand/ multiplier $a_i b_i$	Product P_i
1	1 1	1
2	$\bar{1}$ $\bar{1}$	1
3	0 1	0
	1 0	
4	0 0	0
5	0 $\bar{1}$	0
	$\bar{1}$ 0	
6	1 $\bar{1}$	
	$\bar{1}$ 1	$\bar{1}$

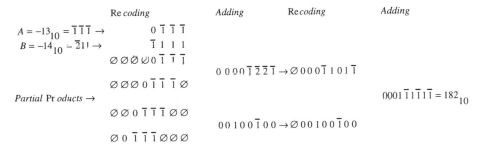

3.2 Nonrecoded TSD Multiplication Using Only Recoded Multiplier

The required recoding steps in the previous subsection can be eliminated if one uses the nonrecoding TSD addition algorithm instead of the recoded one. This has the effect of speeding up the multiplication process. However, in order to obtain the partial products in parallel, we need to recode only the multiplier. This guarantees that the product of two digits in the partial products is carry free.

Table 8 Truth Table for TSD Multiplication When Only the Multiplier is Recoded

Group	Multiplicand/ multiplier $a_i b_i$	Product P_i
1	2 1	2
2	$\bar{2}$ $\bar{1}$	2
3	1 1	1
4	$\bar{1}$ $\bar{1}$	1
5	2 0	0
6	$\bar{2}$ 0	0
7	0 1	0
	1 0	
8	0 0	0
9	0 $\bar{1}$	0
	$\bar{1}$ 0	
10	1 $\bar{1}$	
	$\bar{1}$ 1	$\bar{1}$
11	2 $\bar{1}$	$\bar{2}$
12	$\bar{2}$ 1	$\bar{2}$

This can be seen by examining the truth table shown in Table 8. Again, simple computation rules are obtained for generating the partial products: a partial product is the multiplicand itself, 0, or the complement of the multiplicand, depending on whether the multiplier digit is 1, 0, or $\bar{1}$, respectively.

3.3 Nonrecoded TSD Multiplication with Carries

When performing nonrecoded TSD multiplication, carries may be transmitted to the next higher positions when the a_i or b_i digit is either 2 or $\bar{2}$ as demonstrated in Table 9. Figure 4 shows the process of generating the partial products and partial carries. Note that, once generated, the partial products as well as the partial

Table 9 Truth Table for the Nonrecoded TSD Multiplication with Carries

Group	Multiplicand/ multiplier $a_i b_i$	Product $c_{i+1} P_i$
1	2 2	1 1
2	$\bar{2}\,\bar{2}$	1 1
3	1 1	0 1
4	$\bar{1}\,\bar{1}$	0 1
5	$2\,\bar{1}$ $\bar{1}\,2$	$\bar{1}$ 1
6	$\bar{2}\,1$ $1\,\bar{2}$	$\bar{1}$ 1
7	0 2 2 0	0 0
8	0 1 1 0	0 0
9	0 0	0 0
10	$0\,\bar{1}$ $\bar{1}\,0$	0 0
11	$0\,\bar{2}$ $\bar{2}\,0$	0 0
12	2 1 1 2	1 $\bar{1}$
13	$\bar{1}\,\bar{2}$ $\bar{2}\,\bar{1}$	1 $\bar{1}$
14	$1\,\bar{1}$ $\bar{1}\,1$	0 $\bar{1}$
15	$2\,\bar{2}$ $\bar{2}\,2$	$\bar{1}$ $\bar{1}$

			a_3	a_2	a_1	a_0		$= A =$ Multiplicand
			b_3	b_2	b_1	b_0		$= B =$ Multiplier
∅	∅	∅	c_{03}	c_{02}	c_{01}	c_{00}	∅	
∅	∅	∅	∅	p_{03}	p_{02}	p_{01}	p_{00}	
∅	∅	c_{13}	c_{12}	c_{11}	c_{10}	∅	∅	Partial Products
∅	∅	∅	p_{13}	p_{12}	p_{11}	p_{10}	∅	and
∅	c_{23}	c_{22}	c_{21}	c_{20}	∅	∅	∅	Partial Carries
∅	∅	p_{23}	p_{22}	p_{21}	p_{20}	∅	∅	
c_{33}	c_{32}	c_{31}	c_{30}	∅	∅	∅	∅	
∅	p_{33}	p_{32}	p_{31}	p_{30}	∅	∅	∅	

Figure 4 A four-digit nonrecoded TSD multiplication: p and c are partial product and carry, respectively.

carries can all be processed in parallel. However, the accumulation process is doubled in size, i.e., instead of only n partial product words to be added, we have extra n-partial carry words to be processed. Consequently the speed of the multiplication process is decreased by a factor of 2.

3.4 Nonrecoded TSD Multiplication Without Carries

In order to increase the speed of multiplication, the partial carry generation of the multiplication process of the previous subsection can be avoided by considering the product of two consecutive digits of the multiplicand A ($a_i a_{i-1}$) and one digit of the multiplier B (b_j) as demonstrated in Table 10. The results in Table 10 are obtained directly from the results of Table 9.

4 TSD DIVISION

For TSD division, we use the multiplicative normalization algorithm, which consists of computing two sequences such that as the values of one sequence converges to some constant such as 1, the values of the other sequence approach the function to be computed [37]. Consider a dividend N and a divisor D, both TSD fractions in normalized form. Our objective is to obtain the quotient

$$Q = \frac{N}{D} \tag{7}$$

such that there is no remainder. To compute Q, a set of multiplication factors $\{m_0, m_1, m_2, \ldots, m_n\}$ are obtained such that

Table 10 Truth Table for the Nonrecoded TSD Multiplication Without Carries

$a_i a_{i-1}$	b_j $\bar{2}$	$\bar{1}$	2	1	0
$2\bar{2},\ \bar{1}\ \bar{2}$	0	2	0	$\bar{2}$	0
$1\bar{2},\ \bar{2}2$	0	$\bar{2}$	0	2	0
$11,\ \bar{2}1$	0	$\bar{1}$	0	1	0
$2\bar{1},\bar{1}\ \bar{1}$	0	1	0	$\bar{1}$	0
$20,\ \bar{1}0$	$\bar{1}$	1	1	$\bar{1}$	0
$21,\ \bar{1}1$	$\bar{2}$	1	2	$\bar{1}$	0
$22,\ \bar{1}2$	$\bar{2}$	0	2	0	0
$10,\ \bar{2}0$	1	$\bar{1}$	$\bar{1}$	1	0
$1\bar{1},\ 2\ \bar{1}$	2	$\bar{1}$	$\bar{2}$	1	0
$1\bar{2},\ \bar{2}\ 2$	2	0	$\bar{2}$	0	0
$0\bar{2}$	1	1	$\bar{1}$	$\bar{1}$	0
$0\bar{1}$	1	0	$\bar{1}$	0	0
01	$\bar{1}$	0	1	0	0
02	$\bar{1}$	$\bar{1}$	1	1	0
00	0	0	0	0	0

$$D \prod_{i=0}^{n} m_i \rightarrow 1 \tag{8}$$

within an acceptable error margin. Thus Eq. (7) can be written in the form

$$Q = \frac{N \prod_{i=0}^{n} m_i}{D \prod_{i=0}^{n} m_i} \tag{9}$$

Applying Eq. (8) in the denominator of Eq. (9), we get

$$Q = N \prod_{i=0}^{n} m_i \tag{10}$$

Assuming $D_0 = D$, $N_0 = N$, $N_{n+1} = N \prod_{i=0}^{n} m_i$, and $D_{n+1} = D \prod_{i=0}^{n} m_i$, the numerator and denominator of Eq. (10) can be represented by the two recurrence equations

$$N_{n+1} = N_n m_n \tag{11}$$

$$D_{n+1} = D_n m_n \tag{12}$$

where for small n, $D_n \to 1$ and $Q = N \prod_{i=0}^{n} m_i$. The key to the successful implementation of this algorithm is to obtain the multiplication factors in such a way that computing with Eqs. (11) and (12) is a simple task. This can be achieved by obtaining the roots of Eq. (12) using a suitable algorithm such as the Newton–Raphson iterative algorithm [37]. Now Eq. (12) can be written as $D_{n+1} = f(D_n)$, where the function $f(D_n)$ is expected to converge to 1 starting from an initial value $D_0 = D$. Therefore Eq. (12) can be rewritten in the form of a polynomial as

$$D_{n+1} - f(D_n) = 0 \tag{13}$$

Earlier research [38] has shown that quadratic convergence is more suitable for optical implementation. For the function $f(D_n)$ to converge quadratically to 1, the multiplication factors m_i must be selected such that [39]

$$m_i = 2 - D_i = 2 + (-D_i) = 2 + \overline{D}_i \tag{14}$$

The multiplication factors expressed by Eq. (14) can be readily computed by digit by digit complementing D_i and then applying the TSD addition rules presented in Sec. 3. Since the convergence is quadratic, the accumulated denominator length is doubled after each iteration. Therefore for a quotient length of n, at most $\log_2 n$ iterations are needed. The following example shows an illustration of the TSD division algorithm using $N = (0.\overline{1}2)_3 = (0.5555)_{10}$ and $D = (0.20)_3 = (0.6666)_{10}$.

Iteration number	Multiplication factor	Accumulated numerator	Accumulated denominator
0	$m_0 = 2 - D_0$ $= (1.1000)_3$ $= (1.3333 \ldots)_{10}$	$N_1 = N_0 m_0$ $= (0.\overline{2}0\overline{1}00)_3$ $= (-0.7407 \ldots)_{10}$	$D_1 = D_0 m_0$ $= (0.2200)_3$ $= (0.8888 \ldots)_{10}$
1	$m_1 = 2 - D_1$ $= (1.0\overline{1}00)_3$ $= (1.1111 \ldots)_{10}$	$N_2 = N_1 m_1$ $= (0.\overline{2}\overline{1}\overline{1}02)_3$ $= (-0.8230 \ldots)_{10}$	$D_2 = D_1 m_1$ $= (0.2222)_3$ $= (0.9876 \ldots)_{10}$
2	$m_2 = 2 - D_2$ $= (1.000\overline{1})_3$ $= (1.0124 \ldots)_{10}$	$N_3 = N_2 m_2$ $= (0.\overline{2}\overline{1}\overline{1}\overline{1}\overline{1} \ldots)_3$ $= (-0.8333 \ldots)_{10}$	$D_3 = D_2 m_2$ $= (0.2222222)_3$ $= (0.9998 \ldots)_{10}$

From the above example, it is evident that the proposed technique yields the quotient just after three iterations, i.e., $Q = N_3 = (0.\overline{2}\overline{1}\overline{1}\overline{1}\overline{1} \ldots)_3 = (-0.8333 \ldots)_{10}$. Each iteration of the TSD division algorithm involves three operations—a subtraction operation (or a complement and addition operation) for calculating the multiplication factors m_i and two TSD multiplication operations for calculating the numerator and the denominator. The subtraction (or complement plus

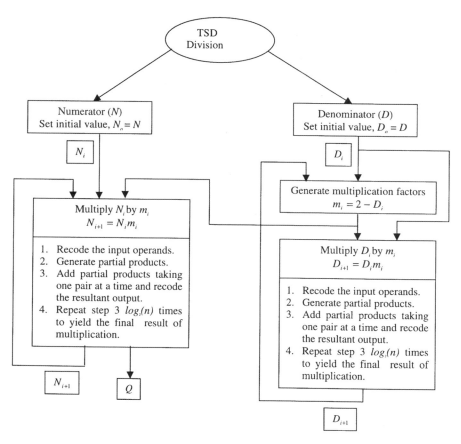

Figure 5   A block diagram showing the different steps of the division process.

addition) and multiplication operations required for each step of the division algorithm can be performed in constant time using TSD subtraction (addition) and multiplication algorithms developed in the previous sections. A block diagram for the division process is shown in Fig. 5.

5   TSD ADDERS AND MULTIPLIERS USING REDUNDANT BIT REPRESENTATIONS

Recently, optical two-step TSD addition/subtraction [10,13] and two-step higher order TSD schemes have been proposed [9]. These schemes perform TSD operations based on the method of content-addressable memory (CAM). However, the

CAM method suffers from the disadvantage that the reference pattern is proportional to the length of the input operands. Therefore the number of reference patterns becomes large when high-accuracy calculation is performed [13,19,40,41]. Further, two new more efficient optical implementation schemes—correlation and matrix multiplication—have recently been proposed for modified signed-digit (MSD) numbers using fixed reference pattern masks for arbitrary-length operands [20,22]. One of these schemes introduces a new optical encoding (redundant bit representation) for the digit operands [22].

On the other hand, many signal and image processing applications require a large number of multiplication operations to be performed at very high speed. Various digital multiplication/division algorithms have been proposed to meet this demand [27–32]. To utilize the full advantage of the parallelism offered by optics, iterative-tree architecture is proposed to perform multiplication operations where multiplication partial products are first generated in parallel and then summed in parallel in a treelike fashion. The parallel generation of all partial products is done in a constant time, independent of the word length. It is the adder tree that usually requires $\log_2 n$ time for the multiplication of two n-digit numbers.

In this section, one-step nonrecoded and two-step recoded trinary signed-digit addition operations based on redundant bit representation (RBR) of the digits are presented. We use TSD adders as the basic building blocks for highly parallel multiplication. We derive the necessary reduced minterms for both the adders and the multipliers for which the optical implementation is carried out using correlation or matrix multiplication based schemes.

5.1 Redundant Bit Representation (RBR)

Redundant bit representation (RBR) [22] uses a group of binary bits to represent each digit of the signed-digit number. Table 11 shows the RBR for the prime TSD numbers $\{\bar{2}, \bar{1}, 0, 1, 2\}$ where five bits are used to represent each of the five TSD digits. To encode optically the TSD digits using RBR coding, it is necessary to use five spatial channels (pixels) to represent the five possible input signals $0, 1, \bar{1}, 2$, and $\bar{2}$. The encoding scheme for the five input TSD digits may be expressed mathematically as

$$I_{\bar{2}}^i = \bar{p}_1 \bar{p}_2 \bar{p}_3 \bar{p}_4 p_5$$
$$I_{\bar{1}}^i = \bar{p}_1 \bar{p}_2 \bar{p}_3 p_4 \bar{p}_5$$
$$I_0^i = \bar{p}_1 \bar{p}_2 p_3 \bar{p}_4 \bar{p}_5 \qquad (15)$$
$$I_1^i = \bar{p}_1 p_2 \bar{p}_3 \bar{p}_4 \bar{p}_5$$
$$I_2^i = p_1 \bar{p}_2 \bar{p}_3 \bar{p}_4 \bar{p}_5$$

Table 11 Five-for-One Redundant Bit
Representation (RBR) Coding for the Prime
and the Fuzzy TSD Digits

TSD digit	RBR code	Denotation
$\bar{2}$	10000	[16]
$\bar{1}$	01000	[8]
0	00100	[4]
1	00010	[2]
2	00001	[1]
either 1 or 2	00011	[3]
either 2 or 0	00101	[5]
either 1 or 0 or 2	00111	[7]
either 2 or $\bar{1}$	01001	[9]
either $\bar{1}$ or 1	01010	[10]
either $\bar{1}$ or 0	01100	[12]
either $\bar{1}$ or 0 or 1	01110	[14]
either $\bar{2}$ or 1	10010	[18]
either $\bar{2}$ or 0	10100	[20]
either $\bar{2}$ or $\bar{1}$	11000	[24]
either 0 or $\bar{1}$ or $\bar{2}$	11100	[28]

where I^i refers to the ith input and $p_j (1 \le j \le 5)$ represents a pixel associated with each spatial pattern. A pixel p_j with an overhead bar represents a dark pixel (transmittance $= 0$), while a pixel p_j without an overhead bar represents a transparent pixel (transmittance $= 1$). For example, the encoding expression for the digit 1 indicates that the second pixel (p_2) is transparent while the remaining pixels are dark.

Also shown in Table 11 are some of the fuzzy digits, which can be used in the RBR codes. Note that a fuzzy digit can be obtained by performing a bitwise OR operation between the correspondent prime TSD digits. For example, the fuzzy digit 01110 for either (0 or $\bar{1}$ or 1) is the OR operation of the prime TSD codes 00100, 01000, and 00010, respectively, which is denoted as decimal [14]. Figures 6a and 6b show the RBR spatial encoded patterns used in the encoding of the prime and the fuzzy TSD numbers. Further, in RBR coding a pair of TSD digits (x_i, y_i) can be denoted by a bracketed string of two digits. For instance, combination digits (1, 1) and (either $\bar{1}$ or 0 or 1, $\bar{1}$) are coded as [2 2] and [14 8], respectively. Furthermore, the RBR code [4 14] represents the combination (0, either 0 or $\bar{1}$ or 1) which is equivalent to the digit combinations (0, 0), (0, 1), and (0, $\bar{1}$) and, equivalently, to the RBR code combinations [4 4], [4 2], and [4 8], respectively.

5.2 RBR Minimization for the Recoded TSD Adder

By studying Table 3, one can recognize that $D_i D_{i-1}$ ($D_{i-2} D_{i-3}$) digits are classified into nine (three) different groups; namely, C_1, C_2, . . ., C_9 (N_1, N_2, and N_3). Group C_1, for instance, consists of combinations $(\bar{2}, \bar{2})$, $(0, 1)$, and $(1, \bar{2})$, which can be denoted by RBR codes [16 16], [4 2], and [2 16], respectively. Therefore the combinations of digits in group C_1 may be minimized to the two minterms [18 16] and [4 2]. Similarly, groups C_2, C_3, . . ., C_9 digit combinations are minimized to the minterms $\{[18\ 8], [4\ 1]\}$, $\{[18\ 2], [9\ 16]\}$, $\{[18\ 4]\}$, $\{[4\ 4]\}$, $\{[9\ 4]\}$, $\{[9\ 8], [18\ 1]\}$, $\{[9\ 2], [4\ 16]\}$, and $\{[9\ 1], [4\ 8]\}$, respectively. Using similar minimization procedures for the $D_{i-2} D_{i-3}$ digits, it is found that groups N_1, N_2, and N_3 could be minimized to minterms $\{[2\ 1], [1\ 15]\}$, $\{[2\ 1], [1\ 15]\}$, and $\{[8\ 16], [16\ 30]\}$, respectively.

Now minimizing the rows and the columns of Table 3, using the RBR coding, a computation rule for the recoding algorithm for TSD numbers can be expressed as $[D_i D_{i-1} D_{i-2} D_{i-3}] \rightarrow Z_i$. Consequently, the recoded digits of Table 3 can be minimized into the following 15 computation rules: $C_1 N_2 \rightarrow 1$; $C_2 (N_1 + N_2) \rightarrow 1$; $C_3 (N_1 + N_3) \rightarrow 1$; $C_4 (N_1 + N_2 + N_3) \rightarrow 1$; $C_7 N_3 \rightarrow 1$; $C_3 N_2 \rightarrow \bar{1}$; $C_6 (N_1 + N_2 + N_3) \rightarrow \bar{1}$; $C_7 (N_1 + N_2) \rightarrow \bar{1}$; $C_8 (N_1 + N_3) \rightarrow \bar{1}$; $C_9 N_3 \rightarrow \bar{1}$; $C_1 (N_1 + N_3) \rightarrow 0$; $C_2 N_3 \rightarrow 0$; $C_5 (N_1 + N_2 + N_3) \rightarrow 0$; $C_8 N_2 \rightarrow 0$; $C_9 (N_1 + N_2) \rightarrow 0$. Here, for example, the $A(B + C)$ rule means A AND (B OR C).

Table 12a summarizes the reduced minterms of the TSD recoding algorithm obtained by using the above 15 computation rules. Table 12a includes 63 minterms (21, 21, and 21 minterms for outputs 1, $\bar{1}$, and 0, respectively). However, in the optical implementation, the output 0 can be regarded as neither 1 nor $\bar{1}$, and consequently only 42 minterms are enough for the TSD algorithm implementation. Table 12b summarizes the reduced minterms of the recoded TSD addition truth tables obtained from entries of Table 4.

5.3 RBR Minimization for the Nonrecoded One-Step TSD Adder

The digits $x_i y_i$ ($x_{i-1} y_{i-1}$) of Table 6 are classified into three different groups, namely, C_1, C_2, and C_3 (N_1, N_2, and N_3). Now similar to the previous minimization procedure of rows and columns of Table 3, using the RBR coding, a computation rule for the one-step nonrecoding trinary adder (Table 6) may also be expressed as $[x_i y_i x_{i-1} y_{i-1}] \rightarrow S_i$, where x_i and y_i are the RBR codes of the ith digits of the operands X and Y, respectively. Consequently, the sum digits of Table 6 can be minimized into the following nine computation rules: $C_1 N_2 \rightarrow 1$; $C_2 N_1 \rightarrow 1$; $C_1 N_3 \rightarrow \bar{1}$; $C_3 N_1 \rightarrow \bar{1}$; $C_2 N_2 \rightarrow \bar{2}$; $C_3 N_3 \rightarrow \bar{2}$; $C_1 N_1 \rightarrow 0$; $C_2 N_3 \rightarrow 0$; and $C_3 N_2 \rightarrow 0$. Table 13 summarizes the reduced minterms for the one-step nonrecoded TSD adder. Table 13 includes 90 minterms (21, 21, 30, 9, and 9 minterms

Table 12 Reduced Minterms for (a) the Recoding TSD Algorithm and (b) the Addition of Recoded TSD Numbers, Based on RBR Coding

1	$\bar{1}$	0	1	$\bar{1}$	0	2	$\bar{2}$
[18 16 2 1]	[18 2 8 16]	[9 2 2 1]	[4 2]	[4 8]	[4 4]	[2 2]	[8 8]
[18 16 1 15]	[18 2 16 30]	[9 2 1 15]	[2 4]	[8 4]	[2 8]		
[4 2 2 1]	[9 16 8 16]	[4 16 2 1]			[8 2]		
[4 2 1 15]	[9 16 16 30]	[4 16 1 15]					
[18 8 7 31]	[9 8 7 31]	[9 1 7 31]			(b)		
[18 8 8 15]	[9 8 8 15]	[9 1 8 15]					
[18 8 14 14]	[9 8 14 14]	[9 1 16 1]					
[4 1 7 31]	[18 1 7 31]	[4 8 7 31]					
[4 1 8 15]	[18 1 8 15]	[4 8 8 15]					
[4 1 16 1]	[18 1 16 1]	[4 8 16 1]					
[18 2 28 1]	[9 2 28 11]	[18 16 28 11]					
[18 2 7 16]	[9 2 7 16]	[18 16 7 16]					
[18 2 14 14]	[9 2 14 14]	[18 16 14 14]					
[9 16 28 1]	[4 16 28 1]	[4 2 28 1]					
[9 16 7 16]	[4 16 7 16]	[4 2 7 16]					
[9 16 14 14]	[4 16 14 14]	[4 2 14 14]					
[18 4 31 31]	[9 4 31 31]	[4 4 31 31]					
[9 8 8 16]	[9 1 8 16]	[18 8 8 16]					
[9 8 16 30]	[9 1 16 30]	[18 8 16 30]					
[18 1 8 16]	[4 8 8 16]	[4 1 8 16]					
[18 1 16 30]	[4 8 16 30]	[4 1 16 30]					
	(a)						

for outputs 1, $\bar{1}$, 0, 2, and $\bar{2}$, respectively). Again, in practice, however, the output 0 can be regarded as neither 1, nor $\bar{1}$, nor 2, nor $\bar{2}$; and consequently, only 60 minterms are enough for the one-step TSD addition implementation.

5.4 RBR-Based Optical Implementations of TSD Addition

Recently, two optical implementations for MSD arithmetic using RBR coded digits were proposed [27]. These two optical schemes are based on correlation and matrix multiplication. To implement the computation rules mentioned in the previous section, we adopt the matrix multiplication scheme for illustration purposes. In this scheme, an input matrix (formed by the RBR codes of the operands) is multiplied by another matrix (formed by the RBR codes of the minterms) to perform the TSD addition. Note that the codes of Fig. 6a and 6b can be used to encode each TSD as well as the fuzzy TSD digits of the minterms.

Table 13 Reduced Minterms for the One-Step Nonrecoded TSD Addition Based on RBR Coding

1	$\bar{1}$	0	2	$\bar{2}$
[4 18 4 14]	[4 9 4 14]	[4 4 4 14]	[4 18 7 1]	[4 9 28 16]
[4 18 14 4]	[4 9 14 4]	[4 4 14 4]	[4 18 1 7]	[4 9 16 28]
[4 18 3 24]	[4 9 3 24]	[4 4 3 24]	[4 18 2 2]	[4 9 8 8]
[4 18 24 3]	[4 9 24 3]	[4 4 24 3]		
[18 4 4 14]	[9 4 4 14]	[9 18 4 14]	[18 4 7 1]	[9 4 28 16]
[18 4 14 4]	[9 4 14 4]	[9 18 14 4]	[18 4 1 7]	[9 4 16 28]
[18 4 3 24]	[9 4 3 24]	[9 18 3 24]	[18 4 2 2]	[9 4 8 8]
[18 4 24 3]	[9 4 24 3]	[9 18 24 3]		
[9 9 4 14]	[18 18 4 14]	[18 9 4 14]	[9 9 7 1]	[18 18 28 16]
[9 9 14 4]	[18 18 14 4]	[18 9 14 4]	[9 9 1 7]	[18 18 16 28]
[9 9 3 24]	[18 18 3 24]	[18 9 3 24]	[9 9 2 2]	[18 18 8 8]
[9 9 24 3]	[18 18 24 3]	[18 9 24 3]		
[4 4 7 1]	[4 4 28 16]	[4 4 28 16]		
[4 4 1 7]	[4 4 16 28]	[4 4 16 28]		
[4 4 2 2]	[4 4 8 8]	[4 4 8 8]		
[9 18 7 1]	[9 18 28 16]	[18 4 28 16]		
[9 18 1 7]	[9 18 16 28]	[18 4 16 28]		
[9 18 2 2]	[9 18 8 8]	[18 4 8 8]		
[18 9 7 1]	[18 9 28 16]	[9 9 28 16]		
[18 9 1 7]	[18 9 16 28]	[9 9 16 28]		
[18 9 2 2]	[18 9 8 8]	[9 9 8 8]		
		[4 9 7 1]		
		[4 9 1 7]		
		[4 9 2 2]		
		[9 4 7 1]		
		[9 4 1 7]		
		[9 4 2 2]		
		[18 18 7 1]		
		[18 18 1 7]		
		[18 18 2 2]		

The 60 minimized minterms of Table 13 can be arranged in a 20×60 matrix where a single minterm consists of 20×1 elements. Therefore, the 60 minterms are encoded as 20×60 matrix where the first and the second 21 columns and the third and the fourth 9 columns encode minterms for the output digits 1, $\bar{1}$, and 2, $\bar{2}$, respectively. Also, note that the input digit matrix needs to include the operands as well as a shifted version of them. Thus for two n digit addition, the input matrix is an $(n + 1) \times 20$. Therefore the multiplication of the input matrix $(n + 1) \times 20$ with the minterms matrix 20×60 results in a matrix of $(n + 1) \times 60$ elements.

In our RBR encoding, the code of the digit 1 or 2 is a 180° rotated version of the code of the digit $\bar{1}$ or $\bar{2}$ (see Fig. 6). This rotational property of the encoding permits us to reduce the number of minterms to be used in the minterms matrix by 50%. Thus only thirty minterms (21 and 9 for the output digits 1 and 2, respectively) in the matrix minterms are needed for the addition algorithm. The optical setup shown in Fig. 7 can be used to implement the proposed TSD adder. In this figure, the input operands are displayed in the input plane SLM1 and the minterms corresponding to the output digits 1 and 2 are stored in the second SLM2. The light beam originating from the SLM1 is split using a 50% beam splitter. The beam propagating straight through and reflected by mirror $\bar{M}_1$ is used to detect the occurrences of all the input patterns that produce output digits 1 and 2. The other beam is rotated 180° using the same beam splitter and mirror $\bar{M}_2$ to identify the presence of all the input patterns that generate $\bar{1}$ and $\bar{2}$ digits. Figure 8 shows the spatial encoding patterns of the minterms' matrix for both the adder and the subtracter.

The details of the operation of the matrix processor can be found in Ref. 20 and 42. However, for illustration purposes, we consider the addition process of two three-digit TSD numbers $X = 0\bar{2}\bar{2} = (-8)_{10}$ and $Y = \bar{1}01 = (-8)_{10}$. In

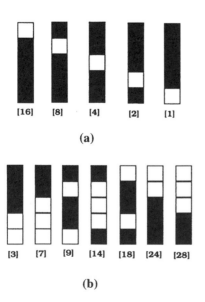

(a)

(b)

Figure 6 Optical spatial encoding for the prime (a) and the fuzzy (b) TSD numbers of Table 11.

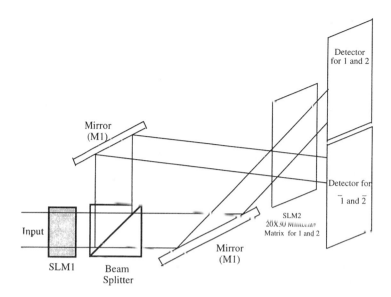

Figure 7 A matrix multiplication optoelectronics implementation of the TSD adder.

general, four parallel digital registers are used to download the n-digit TSD (and one shifted version of it) input number into the optical processor of Fig. 7. The registers' contents are used to control the individual pixels of an electronically addressed $(n + 1) \times 20$ pixel SLM1 forming the input matrix A. Each column of the SLM1 contains four TSD $(x_i y_i x_{i-1} y_{i-1})$ digits. Also, the four-variable min-terms of Table 13 corresponding to the TSD digits 1 and 2 are stored in the second SLM2 forming 20×30 matrix B. The optical processor of Fig. 7, which performs an optical analog multiplication of matrix A (and a rotated version of matrix A denoted as A') with matrix B, produces an $(n + 1) \times 30$ pixel output matrix C_U (C_L) on the output detector plane. C_U and C_L represent respectively the upper half and the lower half of the optical detector array. Notice that matrix C_U is used to identify the presence of the TSD digit $\bar{1}$ and $\bar{2}$ while matrix C_L detects the presence of the digits 1 and 2. By adjusting the bias of the optical detector array, the intensity of each element of matrices C_U and C_L is thresholded to only two levels: 0 and 1 where the threshold value is set to 4. After thresholding, the electronic signals are then grouped and passed through two $(n + 1) \times 30$ OR gate arrays, to generate the outputs corresponding to 1, 2, and $\bar{1}$, $\bar{2}$. The OR gate array includes $2(n + 1)$ twenty one-input OR gates for outputs 1 and $\bar{1}$ and $2(n + 1)$ nine-input OR gates for outputs 2 and $\bar{2}$.

For our example, the 3-digit inputs are regrouped to produce a 4×4 input matrix:

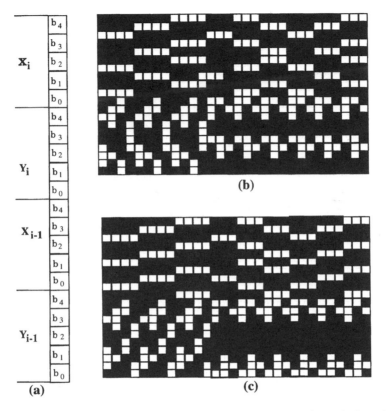

Figure 8 Input TSDs and the encoding of the 30 minterms for optical matrix multiplication schemes; (a) column-order arrangement for encoding a minterm; (b) and (c) encoded minterms mask which produces output digits $\bar{1}$ and $\bar{2}$ for the TSD addition and subtraction, respectively.

$$
\begin{array}{cccc}
\varnothing & \varnothing & x_2 & y_2 \\
x_2 & y_2 & x_1 & y_1 \\
x_1 & y_1 & x_0 & y_0 \\
x_0 & y_0 & \varnothing & \varnothing
\end{array}
=
\begin{array}{cccc}
\varnothing & \varnothing & 0 & \bar{1} \\
0 & \bar{1} & 0 & 1 \\
\bar{2} & 0 & \bar{2} & 1 \\
\bar{2} & 1 & \varnothing & \varnothing
\end{array}
\tag{16}
$$

Using the RBR coding rules of Fig. 6a, we can generate the following 4×20 A and A^r matrices:

$$
A =
\begin{array}{llll}
0\,0\,1\,0\,0 & 0\,0\,1\,0\,0 & 0\,0\,1\,0\,0 & 0\,0\,0\,1\,0 \\
0\,0\,1\,0\,0 & 0\,0\,0\,1\,0 & 0\,0\,0\,0\,1 & 0\,0\,1\,0\,0 \\
0\,0\,0\,0\,1 & 0\,0\,1\,0\,0 & 0\,0\,0\,0\,1 & 0\,1\,0\,0\,0 \\
0\,0\,0\,0\,1 & 0\,1\,0\,0\,0 & 0\,0\,1\,0\,0 & 0\,0\,1\,0\,0
\end{array}
\tag{17}
$$

and

$$
A^r = \begin{array}{llll}
0\,0\,1\,0\,0 & 0\,0\,1\,0\,0 & 0\,0\,1\,0\,0 & 0\,1\,0\,0\,0 \\
0\,0\,1\,0\,0 & 0\,1\,0\,0\,0 & 1\,0\,0\,0\,0 & 0\,0\,1\,0\,0 \\
1\,0\,0\,0\,0 & 0\,0\,1\,0\,0 & 1\,0\,0\,0\,0 & 0\,0\,0\,1\,0 \\
1\,0\,0\,0\,0 & 0\,0\,0\,1\,0 & 0\,0\,1\,0\,0 & 0\,0\,1\,0\,0
\end{array}
\qquad (18)
$$

Similarly, using the symmetric coding of Figs. 6a and 6b, the 20×30 B matrix corresponding to the reduced minterms that output digits 1 and 2 is given by

$$
B = \quad (19)
$$

```
0 0 0 0  0 0 0 0  1 1 1 1  0 0 0  1 1 1  0 0 0   0 0 0  0 0 0  1 1 1

0 0 0 0  1 1 1 1  0 0 0 0  0 0 0  0 0 0  1 1 1   0 0 0  1 1 1  0 0 0

1 1 1 1  0 0 0 0  0 0 0 0  1 1 1  0 0 0  0 0 0   1 1 1  0 0 0  0 0 0

0 0 0 0  0 0 0 0  1 1 1 1  0 0 0  1 1 1  0 0 0   0 0 0  0 0 0  1 1 1

0 0 0 0  1 1 1 1  0 0 0 0  0 0 0  0 0 0  1 1 1   0 0 0  1 1 1  0 0 0


0 0 0 0  0 0 0 0  1 1 1 1  0 0 0  0 0 0  1 1 1   0 0 0  0 0 0  1 1 1

1 1 1 1  0 0 0 0  0 0 0 0  0 0 0  1 1 1  0 0 0   1 1 1  0 0 0  0 0 0

0 0 0 0  1 1 1 1  0 0 0 0  1 1 1  0 0 0  0 0 0   0 0 0  1 1 1  0 0 0

0 0 0 0  0 0 0 0  1 1 1 1  0 0 0  0 0 0  1 1 1   0 0 0  0 0 0  1 1 1

1 1 1 1  0 0 0 0  0 0 0 0  0 0 0  1 1 1  0 0 0   1 1 1  0 0 0  0 0 0


0 0 1 0  0 0 1 0  0 0 1 0  1 1 0  1 1 0  1 1 0   1 1 0  1 1 0  1 1 0

0 1 1 0  0 1 1 0  0 1 1 0  1 0 1  1 0 1  1 0 1   1 0 1  1 0 1  1 0 1

1 1 0 0  1 1 0 0  1 1 0 0  1 0 0  1 0 0  1 0 0   1 0 0  1 0 0  1 0 0

0 1 0 1  0 1 0 1  0 1 0 1  0 0 0  0 0 0  0 0 0   0 0 0  0 0 0  0 0 0

0 0 0 1  0 0 0 1  0 0 0 1  0 0 0  0 0 0  0 0 0   0 0 0  0 0 0  0 0 0


0 0 0 1  0 0 0 1  0 0 0 1  1 1 0  1 1 0  1 1 0   1 1 0  1 1 0  1 1 0

1 0 0 1  1 0 0 1  1 0 0 1  0 1 1  0 1 1  0 1 1   0 1 1  0 1 1  0 1 1

1 1 0 0  1 1 0 0  1 1 0 0  0 1 0  0 1 0  0 1 0   0 1 0  0 1 0  0 1 0

1 0 1 0  1 0 1 0  1 0 1 0  0 0 0  0 0 0  0 0 0   0 0 0  0 0 0  0 0 0

0 0 1 0  0 0 1 0  0 0 1 0  0 0 0  0 0 0  0 0 0   0 0 0  0 0 0  0 0 0
```

↑   ↑

Producing   Producing
Output 1   Output 2

When both A and A^r are multiplied by matrix B, they generate the following 4×30 output matrices for C_U and C_L:

$$C_U = A \times B = \begin{matrix} 3\,2\,2\,1 & 3\,2\,2\,1 & 2\,1\,1\,0 & 3\,2\,2 & 1\,0\,0 & 1\,0\,0 & 2\,0\,1 & 2\,1\,1 & 1\,0\,0 \\ 2\,1\,1\,2 & 1\,0\,0\,1 & 2\,1\,1\,2 & 1\,2\,1 & 0\,1\,0 & 1\,2\,1 & 1\,2\,2 & 1\,2\,0 & 1\,2\,0 \\ 1\,0\,0\,2 & 3\,2\,2\,4 & 1\,0\,0\,2 & 1\,2\,2 & 0\,1\,1 & 1\,2\,2 & 0\,1\,1 & 2\,3\,3 & 0\,1\,1 \\ 3\,3\,1\,1 & 2\,3\,1\,1 & 2\,2\,0\,0 & 1\,1\,0 & 2\,2\,1 & 2\,2\,1 & 2\,2\,1 & 2\,2\,1 & 1\,1\,0 \end{matrix} \qquad (20)$$

$$\begin{matrix} \qquad\qquad\qquad \uparrow \qquad\qquad\qquad\qquad\qquad\qquad \uparrow \\ \qquad\qquad\quad \text{Output 1} \qquad\qquad\qquad\qquad\qquad \text{Output 2} \end{matrix}$$

and

$$C_L = A^r \times B = \begin{matrix} 3\,2\,1\,2 & 3\,2\,1\,2 & 2\,1\,0\,1 & 3\,3\,3 & 1\,1\,1 & 1\,1\,1 & 2\,2\,2 & 2\,2\,2 & 1\,1\,1 \\ 3\,3\,3\,2 & 1\,1\,1\,0 & 1\,1\,0\,0 & 3\,2\,1 & 3\,2\,1 & 1\,2\,0 & 3\,4\,2 & 1\,2\,0 & 1\,2\,0 \\ 1\,0\,2\,0 & 2\,1\,3\,1 & 2\,1\,3\,1 & 2\,2\,1 & 2\,2\,1 & 1\,1\,0 & 1\,1\,0 & 2\,2\,1 & 2\,2\,1 \\ 2\,2\,0\,0 & 2\,2\,0\,0 & 4\,4\,2\,2 & 1\,1\,0 & 2\,2\,1 & 2\,2\,1 & 1\,1\,0 & 1\,1\,0 & 3\,3\,2 \end{matrix} \qquad (21)$$

$$\begin{matrix} \qquad\qquad\qquad \uparrow \qquad\qquad\qquad\qquad\qquad\qquad \uparrow \\ \qquad\qquad\quad \text{Output } \overline{1} \qquad\qquad\qquad\qquad\qquad \text{Output } \overline{2} \end{matrix}$$

From the matrix C_U, we observe the value of 4 as an output in the third row for the 1 output region. Thus we expect an output of 1 in this position in the result matrix. From the matrix C_L, we observe 4 in the second and the fourth rows for the $\overline{2}$ and $\overline{1}$ output regions, respectively; and consequently, $\overline{2}$ and $\overline{1}$ outputs are expected in these positions. Electronic postprocessing of C_U and C_L (i.e., thresholding, ORing operations) yields the output vectors: $O_U = [0\ 0\ 1\ 0]^T$ and $O_L = [0\ \overline{2}\ 0\ \overline{1}]^T$. By combining these two output vectors, we get $S = [0\ \overline{2}\ 1\ \overline{1}]^T$ as final output vector (the addition result), which corresponds to $(-16)_{10}$.

5.5 RBR Based TSD Multiplication

5.5.1 RBR Based Recoded TSD Multiplier and Multiplicand

Table 7 shows the truth table for generating the partial products for the recoded TSD multiplication when both the multiplicand and the multiplier are recoded using Table 3. Table 14 illustrates the necessary reduced minterms of Table 7 that may be used in the optical system of Fig. 7 to generate the partial products simultaneously. The system of Fig. 7 can also be used to accumulate the partial products once they are generated. One needs only to process the generated partial products using the minterms presented in Table 12. It is worth mentioning that the minterms for generating the partial products, for output digit 1, as shown in Table 14, are not digit-by-digit complements to the ones that generate output

Table 14 Reduced Minterms for
Generating Partial Products When
Both the Multiplicand and the
Multiplier Are Recoded

Output digits	Reduced minterms
1	[2 2]
	[8 8]
$\overline{1}$	[2 8]
	[8 2]
0	[4 14]
	[14 4]

digit $\overline{1}$. Therefore one cannot use minterms for only output 1 or only $\overline{1}$, for SLM2, as for TSD adders. However, we can still save 50% of the SLM2 area by using only half the minterms that produce output digit 1 and half the minterms that produce output digit $\overline{1}$ SLM2. In this case, the upper and the lower detectors in Fig. 7 are both used to detect the presence of output digits 1 and $\overline{1}$. As a result, the multiplication design uses basically the same optical system shown in Fig. 7 and it requires only two minterms to be stored in SLM2. The recoding, the addition, and the multiplication operations require 21, 3, and 2 reduced minterms, respectively.

5.5.2 RBR Based Nonrecoded TSD Multiplication Using Only Recoded Multiplier

In order to obtain the partial products in parallel, one may recode only the multiplier. This guarantees that the product of two digits in the partial products is carry free. This can be seen by examining the truth table shown in Table 8. Table 15 shows the necessary reduced minterms for generating the partial products in this case. This design requires only four minterms for generating the partial products and 30 minterms for the nonrecoded TSD addition. Note that the multiplier can be recoded digitally. Otherwise, we need additional 21 minterms for its optical implementation.

5.5.3 RBR Based Nonrecoded TSD Multiplication with Carries

Table 9 shows the truth table for performing nonrecoded TSD multiplication. Table 16 shows the required reduced minterms for generating the partial products and carries. Note that, once generated, the partial products as well as the partial carries can all be processed in parallel using the optical system of Fig. 7 where

Table 15 Reduced Minterms
for Generating Partial Products
When Only the Multiplier Is Recoded

Output digits	Reduced minterms
1	[2 2]
	[8 8]
$\bar{1}$	[2 8]
	[8 2]
0	[4 14]
	[31 4]
2	[1 2]
	[16 8]
$\bar{2}$	[1 8]
	[16 2]

the reduced minterms of Table 13 are used in SLM2. This multiplication design requires 30 and 6 reduced minterms for addition and multiplication, respectively.

5.5.4 RBR Based Nonrecoded TSD Multiplication Without Carries

Table 10 shows the partial carry generation of the multiplication process by considering the product of two consecutive digits of the multiplicand A $(a_i a_{i-1})$ and one digit of the multiplier B (b_j). The RBR minimization of the entries of Table

Table 16 Reduced Minterms for Generating Partial
Products and Partial Carries Digits for Nonrecoded
TSD Multiplication

Partial carries		Partial products	
1	$\bar{1}$	1	$\bar{1}$
[1 3]	[1 24]	[18 18]	[18 9]
[3 1]	[24 1]	[9 9]	[9 18]
[16 24]	[16 3]		
[24 16]	[3 16]		

Table 17 Reduced Minterms for Generating Partial
Products Without Carries for the Nonrecoded TSD
Multiplication

1	$\bar{1}$	2	$\bar{2}$
[18 14 2]	[18 14 16]	[9 16 8]	[9 16 2]
[18 4 16]	[18 4 1]	[9 3 1]	[9 24 16]
[9 14 8]	[9 14 2]	[18 1 2]	[18 1 8]
[9 4 1]	[9 4 16]	[18 24 16]	[18 24 1]
[4 8 16]	[4 2 16]		
[4 2 1]	[4 8 1]		
[4 16 24]	[4 1 24]		
[4 1 3]	[4 16 3]		

10 produces the reduced minterms shown in Table 17 for the generation of the
partial products. This multiplication design requires 30 and 12 (three variables)
minterms for addition and multiplication, respectively. Table 18 is a comparison
of the four multiplication designs in terms of the number of reduced minterms
required for the recoding, the addition, and the multiplication operations, the
number of accumulation steps, and the number of adders required to accumulate
the partial products to obtain the final product.

5.6 Summary

Highly efficient two-step recoded and one-step nonrecoded trinary signed-digit
carry-free adder/subtracters are presented based on redundant bit representation (RBR) for the operand digits. It has been shown that only 24 (30) minterms
are needed to implement the two-step recoded TSD (the one-step nonrecoded)
addition for any operand length. The proposed TSD adder/subtracter unit is
much simpler and uses fewer minterms than any previously reported techniques
[10,13,19]. Optical implementation of the proposed arithmetic can be carried using correlation or matrix multiplication based schemes. An efficient matrix multiplication based optical implementation that employs a fixed number of minterms
for any operand length has been developed. Further, the RBR encoding along
with the proposed optical matrix–matrix processor save 50% of the system memory. Four different multiplication designs based on the proposed recoded and
nonrecoded TSD adders have been considered. Our multiplication designs require
a small number of reduced minterms to generate the partial products. We have
shown that the TSD adder optical implementation can also be used to process
the sum of the multiplication partial products. Finally, the pipelined iterative-tree

Table 18 Comparison of the Four Multiplication Designs

Type of multiplicand and multiplier	Number of Reduced Minterms needed for			Number of accumulation steps	Number of adders
	Recoding	Addition	Partial products generation		
Nonrecoded multiplicand and recoded multiplier	21	3	2	$\log_2(n+1) +$ $(n+1)/2$ recoding steps	$(n+1)/2$
Only recoded multiplier		30	4	$\log_2(n+1)$	$(n+1)/2$
Nonrecoded multiplicand and multiplier with carries		30	6	$\log_2(2n+1)$	$(2n+1)/2$
Nonrecoded multiplicand and multiplier without carries		30	12	$\log_2(n+1)$	$(n+1)/2$

algorithm, which was proposed recently [33], can be used in these proposed adders/multipliers.

6 TSD ADDERS AND MULTIPLIERS USING OPTICAL CASCADED CORRELATION

In this section, we propose a symbolic substitution (SS [43]) based optical numeric processor using the recoded TSD [10] and nonrecoded TSD number representations [26]. The proposed techniques perform parallel carry-free addition, borrow-free subtraction, and multiplication in constant time, independent of the operand length. Also, we propose new joint spatial encodings for the TSD numbers. The new spatial encodings reduce the number of the SS computation rules of the TSD algorithm. Optoelectronic implementation of the proposed recoded adder is feasible because the recoding algorithm of the TSD number representation is totally parallel. Therefore a fast electronics hardware unit can be designed to process the recoding algorithm, and the addition step can be performed optically in one step. On the other hand, the nonrecoded TSD adder can also be performed optically in one or two steps. Both the proposed recoded and nonrecoded TSD optical adders are more compact than the MSD counterpart and use fewer correlators and spatial light modulators (SLM).

6.1 SS Cascaded Correlator Architecture

The SS [43] is an algorithm for performing digital optical computing on information present in a 2D input image. In this algorithm, a specific group of spatially oriented 2D patterns is first recognized and then another set of 2D patterns is substituted for it in that location. Thus the SS consists of two pattern processing steps: (1) a recognition phase, where all the occurrences of a search pattern are simultaneously searched in the input image, and (2) a substitution phase, where a scribe pattern is substituted in all the locations where the search pattern is found. Recently, Casasent et al. [33,44–48] proposed the basic correlator architecture shown in Fig. 9 to implement: (1) a production system neural net, (2) hit-or-miss morphological transforms for automatic target recognition (ATR); (3) morphological inspection; and (4) an optical numeric coprocessor based on the SS algorithm. In Fig. 9, the first correlator (P1–P3) performs the recognition phase and the second correlator (P3–P5) performs the substitution phase of the SS algorithm. In Fig. 9, the parallel wave passing through plane P1 is Fourier transformed by lens L1 and focused on Plane P2, which contains a matched spatial filter (MSF) representing the Fourier transform of the recognized encoded symbols. Thus at P2 we have the product of the Fourier transform of the input image and the Fourier transform of the encoded recognition symbols. This product image

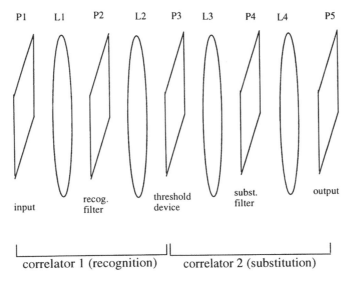

P1   L1   P2   L2   P3   L3   P4   L4   P5

input   recog.   threshold   subst.   output
          filter   device   filter

| correlator 1 (recognition) | correlator 2 (substitution) |

Figure 9 Basic SS cascaded correlator architecture. (From Ref. 48.)

is Fourier transformed by lens L2 to produce the correlation peak intensities at Plane P3. These peaks' intensities indicate the presence of the required recognized symbols in the input image at the P1 plane. The correlation peaks can be thresholded using a nonlinear optically addressed SLM or a photorefractive crystal. However, many proposed optical demonstrations use an image grabber to execute the thresholding and the decoding processing [24,27,48]. Now the second correlator (P3–P5) performs the substitution phase in the same manner as the first correlator (P1–P3). Here the binarized peak intensities appearing at P3 that correspond to the recognized symbols at the input P1 are substituted by the required substitution symbols using the MSF at P4. Thus P5 has a substitution symbol that corresponds to each occurrence of the recognition symbol in the input P1. The description of the cascaded correlation in the previous paragraphs represents the execution of one SS computation rule. In general, there are several SS rules for optical computation. Consequently, one may use multichannel correlator architecture based on multiple space- and frequency-multiplexed filters [33,44–48] at P2.

Recently, the above-mentioned SS based correlator optical processor was used to perform modified signed digit (MSD) arithmetic [33,48]. The MSD addition is a three-stage process, and thus six cascaded correlators are necessary for fully parallel operations. However, it was shown [48] that the number of correlators can be reduced to three when each pair of correlator stages (for recognition

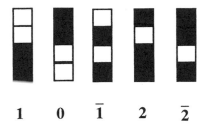

Figure 10 Four-pixel TSD encodings that can be used in both X and Y matrices.

and substitution phases) are combined into a single correlator. This can be achieved if each SS stage is formulated as a matrix-vector (M-V) multiplication ($\mathbf{y} = M\mathbf{x}$) for each input of the recognition digits (the input vector $\mathbf{x}$) and the associated pair of the output substitution digits (the output vector $\mathbf{y}$). The unknown matrix M is the solution of this M-V multiplication equation and it is used as the MSF filter at P2 of Fig. 9.

6.2 Design of Recoded TSD Adder

To illustrate the design methodology of the above SS based cascaded correlator scheme, consider the design of the recoded TSD adder represented by the truth table shown in Table 4. In this table, the second column represents the input recognition digits (the x-values) and the third column represents the output substitution digits (the y-values). By a fast comparison of the MSD symbolic substitution rules (SS rules for each MSD stage) to the recoded TSD ones (Table 4), one can recognize that the TSD truth table has also nine computation rules that are very similar to the MSD truth table ones; however, here we have only one stage to perform the addition instead of three stages. Therefore provided that the TSD numbers are recoded and downloaded to the SLM at plane P1, our proposed recoded TSD adder uses a single SS stage with one correlator to carry the addition operation for the nine computation rules shown in Table 4.

Note that an exact solution for the M-V equation is possible through the use of six-pixel encoding for each TSD. However, this will decrease the SBWP of an SLM. Therefore as with the MSD optical processor [48], four pixels per digit can be used to encode the TSDs as shown in Fig. 10. To save space, we use the symbols o, z, ob, t, and tb to represent the input pixel patterns for the TSD encoding for 1, 0, $\bar{1}$, 2, and $\bar{2}$, respectively. The nine possible inputs TSD of Table 4 can be written as the columns of the input matrix X:

$$X = \begin{bmatrix} o & o & o & z & z & z & ob & ob & ob \\ o & z & ob & o & z & ob & o & z & ob \end{bmatrix}$$

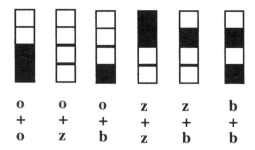

Figure 11   New superimposed four-pixel TSD encodings that are used in the input X matrix.

The nine corresponding output substitution patterns TSD of Table 4 can be written as the columns of the output matrix Y:

$$Y = [\text{t} \quad \text{o} \quad \text{z} \quad \text{o} \quad \text{z} \quad \text{ob} \quad \text{z} \quad \text{ob} \quad \text{tb}]$$

In order to reduce the spatial bandwidth product (SBWP) and hence increase the throughput for a given SLM, it is preferable to reduce the number of columns in the X matrix [48]. This can be achieved by noticing that the order of the operands does not affect the resultant sum (i.e., 1 plus 0 yield the same sum as 0 plus 1, $\bar{1}$ plus 0 yield the same sum as 0 plus $\bar{1}$, and $\bar{1}$ plus 1 yield the same sum as 1 plus $\bar{1}$). Therefore three columns from the X and Y matrices can be eliminated. Now it is necessary to introduce new spatial encodings to replace those of Fig. 10 by superimposing [49] (OR) all the encoded digits for each operands. The new encoding is shown in Fig. 11, and the new input matrix X is given as

$$X = [\text{o} + \text{o} \quad \text{o} + \text{z} \quad \text{o} + \text{ob} \quad \text{z} + \text{z} \quad \text{z} + \text{ob} \quad \text{ob} + \text{ob}]$$

where the symbols '+' denotes a logical OR and not a sum. The corresponding output matrix Y is

$$Y = [\text{t} \quad \text{o} \quad \text{z} \quad \text{z} \quad \text{ob} \quad \text{tb}]$$

Using the encodings of Figs. 11 and 10 in the corresponding X and Y input and output matrices, respectively, we obtain

$$X = \begin{bmatrix} 1 & 1 & 1 & 0 & 1 & 1 \\ 1 & 1 & 1 & 0 & 0 & 0 \\ 0 & 1 & 1 & 1 & 1 & 1 \\ 0 & 1 & 0 & 1 & 1 & 0 \end{bmatrix} \quad Y = \begin{bmatrix} 0 & 1 & 0 & 0 & 1 & 0 \\ 1 & 1 & 0 & 0 & 0 & 0 \\ 0 & 0 & 1 & 1 & 1 & 1 \\ 0 & 0 & 1 & 1 & 0 & 0 \end{bmatrix}$$

The solution matrix M of $Y = MX$ (which is the required MSF) is

$$
M = \begin{bmatrix}
0.54 & -0.18 & -0.45 & 0.82 \\
0.36 & 0.54 & -0.64 & 0.54 \\
0.09 & -0.36 & 1.09 & -0.36 \\
-0.91 & 0.64 & 1.09 & -0.36
\end{bmatrix}
$$

Note that because the X matrix is not a square, a nonexact pseudoinverse M matrix solution is obtained. Therefore a thresholding at the P3 output correlation plane is needed to yield the exact results. Note that the output P3 plane would be thresholded in practice in any optical architecture. Cross talk noise may affect the thresholding decision required to compute the exact output for Y. However, noise sensitivity issues related to digital optical computing can be carried out in a very similar manner to the ones demonstrated in Refs. 20 and 50–52. Further, a thresholding SLM (ferroelectric liquid crystal [53]) with a good sensitivity is a good candidate as an output device. Using the M matrix in P2, we obtain the following Y matrix:

$$
Y = MX = \begin{bmatrix}
0.36 & 0.73 & -0.09 & 0.36 & 0.91 & 0.09 \\
0.91 & 0.82 & 0.27 & -0.09 & 0.27 & -0.27 \\
-0.27 & 0.45 & 0.82 & 0.73 & 0.82 & 1.18 \\
-0.27 & 0.45 & 0.82 & 0.73 & -0.18 & 0.18
\end{bmatrix}
$$

After thresholding at 0.5 we obtain the exact corrected output:

$$
Y = \mathrm{thresh}(MX) = \begin{bmatrix}
0 & 1 & 0 & 0 & 1 & 0 \\
1 & 1 & 0 & 0 & 0 & 0 \\
0 & 0 & 1 & 1 & 1 & 1 \\
0 & 0 & 1 & 1 & 0 & 0
\end{bmatrix}
$$

It is worth mentioning that the encoding of the TSDs is not unique. For instance, the encodings of Fig. 12 can be used in the X and Y matrices, and the following results are obtained:

$$
X = \begin{bmatrix}
0 & 1 & 1 & 1 & 1 & 1 \\
0 & 0 & 1 & 0 & 1 & 1 \\
1 & 1 & 1 & 1 & 1 & 0 \\
1 & 1 & 1 & 0 & 0 & 0
\end{bmatrix}
\qquad
Y = \begin{bmatrix}
1 & 0 & 1 & 1 & 1 & 0 \\
0 & 0 & 0 & 0 & 1 & 1 \\
0 & 1 & 1 & 1 & 0 & 1 \\
1 & 1 & 0 & 0 & 0 & 0
\end{bmatrix}
$$

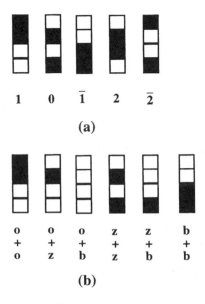

(a)

(b)

Figure 12 Alternative four-pixel encodings for the recoded TSD adder: (a) encodings for the digits in the Y matrix and (b) encodings for the digits in the X matrix.

$$M = \begin{bmatrix} -0.64 & 0.54 & 1.36 & -0.45 \\ 0.1 & 0.64 & 0.09 & -0.36 \\ 1.36 & -0.45 & -0.64 & 0.54 \\ 0.09 & -0.36 & 0.09 & 0.64 \end{bmatrix}$$

$$Y = MX = \begin{bmatrix} 0.91 & 0.27 & 0.82 & 0.73 & 1.27 & -0.09 \\ -0.27 & -0.81 & 0.45 & 0.18 & 0.82 & 0.73 \\ -0.09 & 1.27 & 0.82 & 0.73 & 0.27 & 0.91 \\ 0.73 & 0.82 & 0.45 & 0.18 & -0.18 & -0.27 \end{bmatrix}$$

$$Y = \text{thresh}(MX) = \begin{bmatrix} 1 & 0 & 1 & 1 & 1 & 0 \\ 0 & 0 & 0 & 0 & 1 & 1 \\ 0 & 1 & 1 & 1 & 0 & 1 \\ 1 & 1 & 0 & 0 & 0 & 0 \end{bmatrix}$$

Further improvement in the SBWP of the system is possible by considering the following important observation. At the input plane P1, one can download the

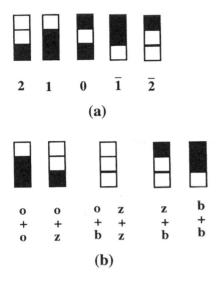

(a)

2	1	0	$\bar{1}$	$\bar{2}$

(b)

o	o	o	z	z	b
+	+	+	+	+	+
o	z	b	z	b	b

Figure 13 Alternative three pixel encodings for the recoded TSD adder: (a) encodings for the digits in the Y matrix and (b) encodings for the digits in the X matrix.

encoding of the three pairs of digits 00, $1\bar{1}$, and $\bar{1}1$ (the three pairs produce a sum of 0, see Table 4) instead of the pair 00 by itself and the pairs $1\bar{1}$ and $\bar{1}1$ by themselves. This will eliminate an extra column from the X and Y matrices as well as reduce the number of pixels in the encoding, as illustrated in Fig. 13. The corresponding system matrices for this case are

$$Y = \begin{bmatrix} 1 & 1 & 0 & 0 & 0 \\ 1 & 0 & 1 & 0 & 1 \\ 0 & 0 & 0 & 1 & 1 \end{bmatrix} \quad X = \begin{bmatrix} 1 & 1 & 1 & 0 & 0 \\ 0 & 1 & 1 & 1 & 0 \\ 0 & 0 & 1 & 1 & 1 \end{bmatrix}$$

$$M = \begin{bmatrix} 0.75 & 0 & -0.25 \\ 1 & -1 & 1 \\ -0.25 & 0 & 0.75 \end{bmatrix}$$

$$Y = MX = \begin{bmatrix} 0.75 & 0.75 & 0.5 & -0.25 & -0.25 \\ 1 & 0 & 1 & 0 & 1 \\ -0.25 & -0.25 & 0.5 & 0.75 & 0.75 \end{bmatrix}$$

After thresholding at 0.75 the output is

$$Y = \text{thresh}(MX) = \begin{bmatrix} 1 & 1 & 0 & 0 & 0 \\ 1 & 0 & 1 & 0 & 1 \\ 0 & 0 & 0 & 1 & 1 \end{bmatrix}$$

6.3 Design of Nonrecoded Two-Step TSD Adder

The optical processor of Fig. 9 can also be used to process the two-step nonre-coded TSD addition algorithm of Table 1. In the first step rules (Table 1a), we have 25 different TSD combinations, which can be reduced to 15 TSD combina-tions using the superposition of only one pair of digits, as was demonstrated in the previous subsection. However, further reduction to nine or ten input patterns is possible if one uses the process of superposition of more than one pair of TSD digits, as demonstrated in the last case of the previous subsection. In order to obtain an exact solution for the M-V equation, the 10-pixel encoding of Fig. 14a and 14b is used for the input matrix X, whereas Fig. 14c shows a simple encoding for the output matrix Y. The Y, X, and M matrices for this case are

$$Y = \begin{bmatrix} 1 & 1 & 1 & 0 & 0 & 0 & 0 & 0 & 0 & 0 \\ 0 & 0 & 0 & 1 & 1 & 1 & 1 & 0 & 0 & 0 \\ 0 & 0 & 0 & 0 & 0 & 0 & 0 & 1 & 1 & 1 \\ 1 & 0 & 0 & 1 & 0 & 0 & 0 & 1 & 0 & 0 \\ 0 & 1 & 0 & 0 & 1 & 1 & 0 & 0 & 1 & 0 \\ 0 & 0 & 1 & 0 & 0 & 0 & 1 & 0 & 0 & 1 \end{bmatrix}$$

$$X = \begin{bmatrix} 1 & 1 & 1 & 1 & 0 & 1 & 0 & 0 & 0 & 0 \\ 0 & 1 & 1 & 1 & 0 & 1 & 1 & 0 & 0 & 0 \\ 0 & 0 & 1 & 1 & 1 & 0 & 1 & 1 & 0 & 0 \\ 0 & 0 & 0 & 1 & 0 & 1 & 1 & 1 & 1 & 0 \\ 0 & 0 & 0 & 0 & 0 & 1 & 1 & 1 & 1 & 1 \\ 1 & 1 & 1 & 1 & 1 & 1 & 1 & 1 & 0 & 0 \\ 1 & 1 & 1 & 1 & 0 & 1 & 1 & 1 & 1 & 1 \\ 1 & 1 & 1 & 1 & 0 & 1 & 1 & 1 & 1 & 0 \\ 0 & 0 & 1 & 1 & 1 & 1 & 1 & 1 & 1 & 1 \\ 0 & 1 & 1 & 1 & 1 & 1 & 1 & 1 & 1 & 1 \end{bmatrix}$$

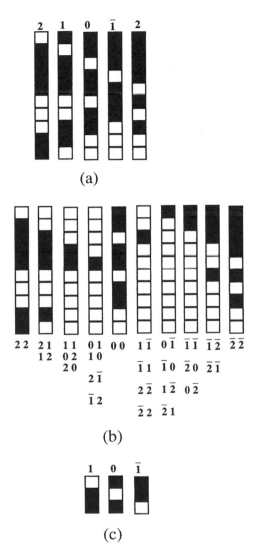

Figure 14 Ten-pixel encodings for the first step nonrecoded TSD adder: (a) the input TSD digits, (b) the superimposed inputs for the X matrix, and (c) the encodings for the Y matrix.

$$M = \begin{bmatrix} 0 & 0 & 0 & -1 & 0 & 0 & 0 & 1 & 0 & 0 \\ -2 & 1 & -2 & 1 & -2 & 2 & 1 & -1 & 2 & -1 \\ 1 & -1 & 1 & 0 & 1 & -1 & 0 & 0 & -1 & 1 \\ 1 & -1 & 1 & 1 & 0 & 0 & 1 & -1 & -1 & 0 \\ -1 & 0 & -2 & 0 & -1 & 1 & -1 & 1 & 1 & 1 \\ -1 & 1 & 0 & -1 & 0 & 0 & 1 & 0 & 1 & -1 \end{bmatrix}$$

Also, an exact 9 × 9 solution for the M-V equation is possible by using the encodings of Fig. 15 in the following matrices:

$$Y = \begin{bmatrix} 1 & 1 & 1 & 0 & 0 & 0 & 0 & 0 & 0 \\ 0 & 0 & 0 & 1 & 1 & 1 & 0 & 0 & 0 \\ 0 & 0 & 0 & 0 & 0 & 0 & 1 & 1 & 1 \\ 1 & 0 & 0 & 1 & 0 & 0 & 1 & 0 & 0 \\ 0 & 1 & 0 & 0 & 1 & 0 & 0 & 1 & 0 \\ 0 & 0 & 1 & 0 & 0 & 1 & 0 & 0 & 1 \end{bmatrix}$$

$$X = \begin{bmatrix} 1 & 1 & 1 & 1 & 1 & 0 & 0 & 0 & 0 \\ 0 & 1 & 1 & 1 & 1 & 1 & 0 & 0 & 0 \\ 0 & 0 & 1 & 1 & 1 & 1 & 1 & 0 & 0 \\ 0 & 0 & 0 & 1 & 1 & 1 & 1 & 1 & 0 \\ 0 & 0 & 0 & 0 & 1 & 1 & 1 & 1 & 1 \\ 1 & 1 & 1 & 1 & 1 & 1 & 1 & 0 & 0 \\ 1 & 1 & 1 & 1 & 1 & 1 & 1 & 1 & 0 \\ 0 & 0 & 1 & 1 & 1 & 1 & 1 & 1 & 1 \\ 0 & 1 & 1 & 1 & 1 & 1 & 1 & 1 & 1 \end{bmatrix}$$

$$M = \begin{bmatrix} 0 & 0 & 0 & -1 & 0 & 0 & 1 & 0 & 0 \\ 0 & 1 & -1 & 1 & 0 & 1 & -1 & 1 & -1 \\ 0 & -1 & 0 & 0 & 0 & 0 & 0 & 0 & 1 \\ 0 & -1 & -1 & 1 & -1 & 2 & -1 & 1 & 0 \\ 1 & 0 & 1 & 0 & 1 & -2 & 1 & -2 & 1 \\ -1 & 1 & -1 & -1 & 0 & 1 & 0 & 2 & -1 \end{bmatrix}$$

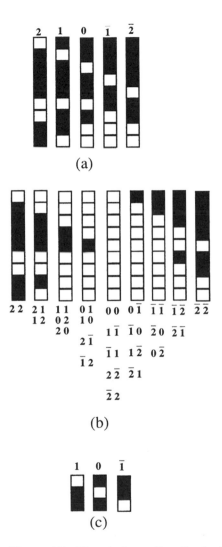

Figure 15 Nine-pixel encodings for the first step nonrecoded TSD adder: (a) the input TSD digits, (b) the superimposed inputs for the X matrix, and (c) the encodings for the Y matrix.

Finally, an approximate five-pixel solution using the encodings of Fig. 16 can be obtained as

$$Y = \begin{bmatrix} 1 & 1 & 1 & 0 & 0 & 0 & 0 & 0 & 0 \\ 0 & 0 & 0 & 1 & 1 & 1 & 0 & 0 & 0 \\ 0 & 0 & 0 & 0 & 0 & 0 & 1 & 1 & 1 \\ 1 & 0 & 0 & 1 & 0 & 0 & 1 & 0 & 0 \\ 0 & 1 & 0 & 0 & 1 & 0 & 0 & 1 & 0 \\ 0 & 0 & 1 & 0 & 0 & 1 & 0 & 0 & 1 \end{bmatrix}$$

$$X = \begin{bmatrix} 1 & 1 & 1 & 1 & 1 & 0 & 0 & 0 & 0 \\ 0 & 1 & 1 & 1 & 1 & 1 & 0 & 0 & 0 \\ 0 & 0 & 1 & 1 & 1 & 1 & 1 & 0 & 0 \\ 0 & 0 & 0 & 1 & 1 & 1 & 1 & 1 & 0 \\ 0 & 0 & 0 & 0 & 1 & 1 & 1 & 1 & 1 \end{bmatrix}$$

$$M = \begin{bmatrix} 0.75 & 0 & 0 & -0.5 & 0.25 \\ -0.17 & 0.5 & 0 & 0.5 & -0.17 \\ 0.25 & -0.5 & 0 & 0 & 0.75 \\ 0.75 & -1 & 0.5 & 0.5 & -0.25 \\ 0.33 & 0.5 & -1 & 0.5 & 0.33 \\ -0.25 & 0.5 & 0.5 & -1 & 0.75 \end{bmatrix}$$

$$Y = MX = \begin{bmatrix} 0.75 & 0.75 & 0.75 & 0.25 & 0.5 & -0.25 & -0.25 & -0.25 & 0.25 \\ -0.17 & 0.33 & 0.33 & 0.83 & 0.67 & 0.83 & 0.33 & 0.33 & -0.17 \\ 0.25 & -0.25 & -0.25 & -0.25 & 0.5 & 0.25 & 0.75 & 0.75 & 0.75 \\ 0.75 & -0.25 & 0.25 & 0.75 & 0.5 & -0.25 & 0.75 & 0.25 & -0.25 \\ 0.33 & 0.83 & -0.17 & 0.33 & 0.67 & 0.33 & -0.17 & 0.83 & 0.33 \\ -0.25 & 0.25 & 0.75 & -0.25 & 0.5 & 0.75 & 0.25 & -0.25 & 0.75 \end{bmatrix}$$

After thresholding at 0.75,

$$Y = \text{thresh}(MX) = \begin{bmatrix} 1 & 1 & 1 & 0 & 0 & 0 & 0 & 0 & 0 \\ 0 & 0 & 0 & 1 & 1 & 1 & 0 & 0 & 0 \\ 0 & 0 & 0 & 0 & 0 & 0 & 1 & 1 & 1 \\ 1 & 0 & 0 & 1 & 0 & 0 & 1 & 0 & 0 \\ 0 & 1 & 0 & 0 & 1 & 0 & 0 & 1 & 0 \\ 0 & 0 & 1 & 0 & 0 & 1 & 0 & 0 & 1 \end{bmatrix}$$

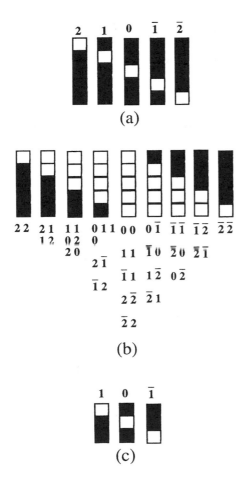

Figure 16 Five-pixel encodings for the first step nonrecoded TSD adder: (a) the input TSD digits, (b) the superimposed inputs for the X matrix, and (c) the encodings for the Y matrix.

Note that the second step rules of Table 1b have the same format as the recoded TSD of Table 4. Therefore the second step process can use any encodings presented in Sec. 6.2 as a solution for the M-V equation.

6.4 Design of Nonrecoded One-Step TSD Adder

Consider Table 6, where the digit classes C_i and N_i are each composed of three different groups of digit pairs. For instance, the C_1 class is made from the digit groups G_2, G_5, and G_8. Similar observations apply to classes C_2, C_3, N_1, N_2, and N_3. To differentiate between these classes, one may use three pixels per group in the spatial encoding assignment; this leads to a nine-pixel per class spatial encoding as shown in Fig. 17a. Note that in Fig. 17a each pixel denotes a position for one of the nine digit groups G_1 through G_9.

Now, to implement the TSD computation rules of Table 6, we superimpose the spatial encoding of the C_i and N_i classes, and consequently we can form the X matrix of the M-V equation. Figures 17b and 17c represent the spatial encod-

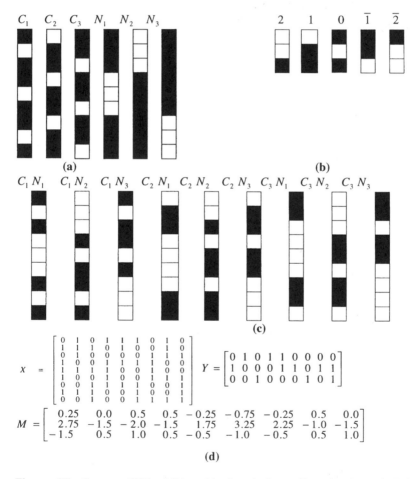

Figure 17 One-step TSD addition: (a) nine-pixel encodings for CP and NLOP digit classes; (b) spatial encodings that are used in the input Y matrix; (c) the superimposed spatial encodings that are used in the X matrix; and (d) solution of the M-V equation.

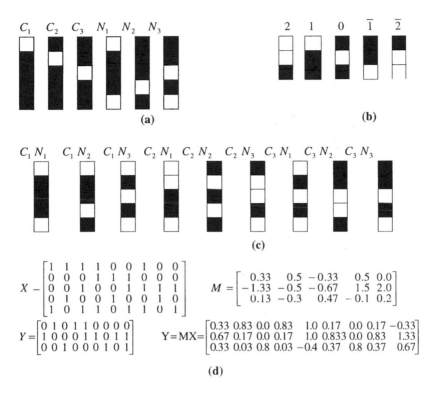

Figure 18 One-step TSD addition: (a) five-pixel encodings for CP and NLOP digit classes; (b) spatial encodings that are used in the input Y matrix; (c) the superimposed spatial encodings that are used in the X matrix; and (d) solution of the M-V equation.

ings of the Y matrix and the superimposed X matrix, respectively. Figure 17d shows the solution (the M matrix) of the M-V equation. It is worth mentioning that the spatial encodings of the C_i and N_i classes are not unique. Further, as with the TSD adder design example of the previous subsection, a nonexact solution that requires a thresholding operation is possible, as demonstrated in Fig. 18, where five-pixel per class spatial encoding is used. This represents an improvement in the SBWP of the SLMs and the size of the MSFs used in the optical system.

6.5 Design of Recoded TSD Multiplication

From Table 7, we can present three different designs for generating the partial products. By studying Table 7, one can recognize 6 different groups of input

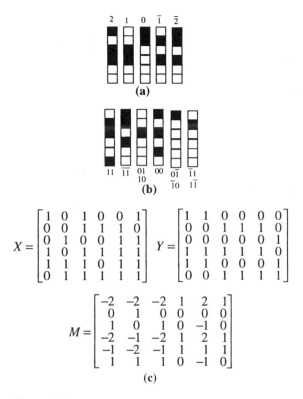

Figure 19 Recoded TSD multiplication design I: (a) six-pixel TSD encodings that can be used in both X and Y matrices; (b) joint spatial encodings that are used in the input X matrix; and (c) solution of the M-V equation.

TSD numbers. If six pixels per TSD encoding for the digits $\{1, 0, \bar{1}\}$ (Fig. 19a) and the superimposed digits shown in Fig. 19b are used in the X and Y matrices, then an exact solution for the MV equation is obtained as shown in Fig. 19c. However, a reduction in the number of pixel encoding is possible by the use of the encoding of Fig. 20 in the M-V equation. In this case, the solution is

$$
M = \begin{bmatrix}
0.45 & 0.18 & 0.45 & -0.82 \\
0.82 & -0.27 & -0.18 & -0.27 \\
-0.36 & 0.45 & 0.64 & 0.45 \\
0 & 0 & 0 & 1
\end{bmatrix}
$$

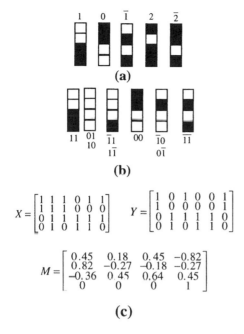

Figure 20 Recoded TSD multiplication design II: (a) alternative four-pixel TSD encodings that can be used in both X and Y matrices; (b) joint spatial encodings that are used in the input X matrix; and (c) solution of the M-V equation.

$$Y = MX = \begin{bmatrix} 0.64 & 0.27 & 1.09 & -0.36 & 0.09 & 0.91 \\ 0.54 & 0.09 & 0.36 & -0.45 & 0.36 & 0.64 \\ 0.09 & 1.18 & 0.73 & 1.09 & 0.73 & 0.27 \\ 0 & 1 & 0 & 1 & 1 & 0 \end{bmatrix}$$

$$thresh(MX) = \begin{bmatrix} 1 & 0 & 1 & 0 & 0 & 1 \\ 1 & 0 & 0 & 0 & 0 & 1 \\ 0 & 1 & 1 & 1 & 1 & 0 \\ 0 & 1 & 0 & 1 & 1 & 0 \end{bmatrix}$$

Now, if superposition (ORing) of more than two TSD pairs is allowed in the encoding process, then one can obtain an even simpler solution. Combining the digit groups 3, 4, and 5 of Table 7, then only four entries are necessary in the

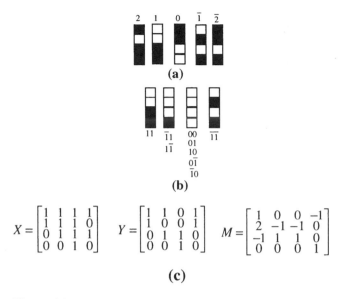

Figure 21 Recoded TSD multiplication design III: (a) an alternative four-pixel TSD encoding that can be used in both X and Y matrices; (b) joint spatial encodings that are used in the input X matrix; and (c) solution of the M-V equation.

X and Y matrices and four pixels per digit is enough for TSD encoding. Figure 21 shows the resulting design for this case.

6.6 Design of Nonrecoded TSD Multiplication Using Only Recoded Multiplier

This can be seen by examining the truth table shown in Table 8 where a 12 pixel per digit encoding guarantees an exact M-V solution as demonstrated in Fig. 22. The number of pixels in the TSD encoding can be lowered at the expense of a nonexact M-V solution and a thresholding operation of the output matrix. Figure 23 represents a six pixel per digit design with a thresholding operation at 0.45. Note that in this design, at most two pairs of input TSD digits are superimposed. However, if we assign one single superimposed code for the digits 5, 6, 7, 8, and 9 of Table 8, then an eight pixel per digit encoding and 8×8 pixels for matrix X is achieved as illustrated in Fig. 24. On the other hand, Fig. 25 shows a better six pixel per digit encoding design and 6×8 pixels for matrix X with a nonexact solution and a required threshold value of 0.5. Further, if we combine digits 11 and 12 as well, then a seven pixel per digit encoding and 7×7 pixels for matrix X is possible, as shown in Fig. 26. Also, we can still obtain a six pixel

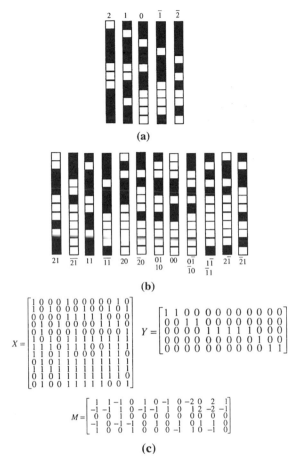

(a)

(b)

(c)

Figure 22 Nonrecoded TSD multiplication when only multiplier is recoded design I: (a) 12-pixel encodings which can be used in both X and Y matrices; (b) joint spatial encodings that are used in the input X matrix; and (c) solution of the M-V equation.

per digit encoding design with 6×7 pixel size matrix X but with a nonexact solution and a thresholding value of 0 at the output device (see Fig. 27).

6.7 Design of Nonrecoded TSD Multiplication with Carries

This design is based on Table 9, which generates partial products and partial carries. Note that, once generated, the partial products as well as the partial carries can all be processed in parallel using the optical system of Fig. 9.

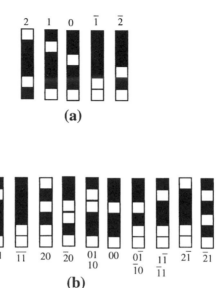

$$X = \begin{bmatrix} 1 & 0 & 0 & 0 & 1 & 0 & 0 & 0 & 0 & 0 & 1 & 0 \\ 1 & 0 & 1 & 0 & 0 & 0 & 1 & 0 & 0 & 1 & 0 & 1 \\ 0 & 0 & 0 & 0 & 1 & 1 & 1 & 1 & 1 & 0 & 0 & 0 \\ 0 & 1 & 0 & 0 & 0 & 01 & 0 & 0 & 0 & 0 & 0 & 1 \\ 1 & 1 & 0 & 1 & 1 & 0 & 0 & 1 & 1 & 1 & 0 \\ 1 & 1 & 1 & 1 & 1 & 1 & 1 & 1 & 1 & 1 & 1 \end{bmatrix} \quad Y = \begin{bmatrix} 1 & 1 & 0 & 0 & 0 & 0 & 0 & 0 & 0 & 0 & 0 & 0 \\ 0 & 0 & 1 & 1 & 0 & 0 & 0 & 0 & 0 & 0 & 0 & 0 \\ 0 & 0 & 0 & 0 & 1 & 1 & 1 & 1 & 1 & 0 & 0 & 0 \\ 0 & 0 & 0 & 0 & 0 & 0 & 0 & 0 & 0 & 1 & 0 & 0 \\ 0 & 0 & 0 & 0 & 0 & 0 & 0 & 0 & 0 & 0 & 1 & 1 \end{bmatrix}$$

$$M = \begin{bmatrix} 0.17 & 0.21 & -0.03 & 0.42 & 0.37 & -0.27 \\ -0.31 & -0.36 & -0.61 & -0.56 & -0.37 & 1 \\ 0 & 0 & 1 & 0 & 0 & 0 \\ -0.29 & 0.31 & 0.06 & -0.04 & 0.37 & -0.21 \\ 0.42 & -0.16 & -0.41 & 0.17 & -0.37 & 0.47 \end{bmatrix}$$

$$Y = MX = \begin{bmatrix} 0.49 & 0.52 & -0.06 & 0.1 & 0.24 & 0.11 & -0.1 & -0.31 & 0.06 & 0.31 & 0.27 & 0.36 \\ -0.04 & 0.07 & 0.64 & 0.63 & -0.29 & -0.16 & 0.03 & 0.39 & 0.02 & 0.27 & 0.32 & 0.09 \\ 0 & 0 & 0 & 0 & 1 & 1 & 1 & 1 & 1 & 0 & 0 & 0 \\ 0.19 & 0.12 & 0.10 & 0.16 & -0.06 & -0.19 & 0.16 & -0.14 & 0.23 & 0.48 & -0.12 & 0.06 \\ 0.36 & 0.27 & 0.31 & 0.1 & 0.11 & 0.23 & -0.1 & 0.06 & -0.31 & -0.06 & 0.52 & 0.49 \end{bmatrix}$$

(c)

Figure 23 Nonrecoded TSD multiplication when only multiplier is recoded design II: (a) six-pixel encodings which can be used in both X and Y matrices; (b) joint spatial encodings that are used in the input X matrix; and (c) solution of the M-V equation.

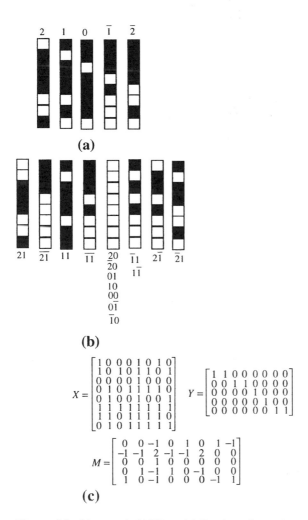

Figure 24 Nonrecoded TSD multiplication when only multiplier is recoded design III: (a) an alternative eight-pixel encodings which can be used in both X and Y matrices; (b) joint spatial encodings that are used in the input X matrix; and (c) solution of the M-V equation.

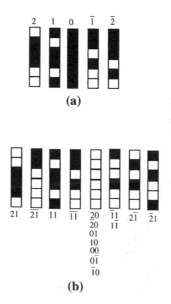

$$X=\begin{bmatrix}1&0&0&0&1&0&1&0\\1&0&1&0&1&1&0&1\\0&1&0&1&1&1&1&0\\0&1&0&0&1&0&0&1\\1&1&1&1&1&1&1&1\\1&1&0&1&1&1&1&0\end{bmatrix}\quad M=\begin{bmatrix}-0.35&-0.35&-1.39&0.43&0.13&1.5\\-0.22&-0.22&0.13&-0.48&0.96&-0.52\\0.39&0.39&0.56&0.26&-0.52&-0.26\\-0.29&0.61&0.43&-0.26&-0.48&0.26\\0.56&-0.43&0.26&0.04&0.91&-1.0\end{bmatrix}$$

$$Y=\begin{bmatrix}1&1&0&0&0&0&0&0\\0&0&1&1&0&0&0&0\\0&0&0&0&1&0&0&0\\0&0&0&0&0&1&0&0\\0&0&0&0&0&0&1&1\end{bmatrix}\quad Y=MX=\begin{bmatrix}1.0&0.73&-0.21&0.30&0.04&-0.04&-0.04&0.21\\0.0&0.08&0.73&0.56&-0.34&0.34&0.34&0.26\\0.0&0.04&-0.13&-0.21&0.82&0.17&0.17&0.13\\0.0&-0.04&0.13&0.21&0.17&0.82&-0.17&-0.13\\0.0&0.17&0.43&0.13&0.30&-0.30&0.79&0.54\end{bmatrix}$$

(c)

Figure 25 Nonrecoded TSD multiplication when only multiplier is recoded design IV: (a) a fourth alternative six-pixel encoding which can be used in both X and Y matrices; (b) joint spatial encodings that are used in the input X matrix; and (c) solution of the M-V equation.

A direct application of Table 9 in our design process leads to a 15 pixel per digit encoding 15×15 pixels for matrix X design as illustrated in Fig. 28. A reduction to 11 pixel per digit design 11×15 pixels for matrix X is possible as demonstrated in Fig. 29. Finally, by superimposing more than two pairs of digits of Table 9 (groups 5 and 6; groups 7, 8, 9, 10, and 11), we can achieve a ten pixel per digit encoding design and 10×10 pixels for matrix X as illustrated in Fig. 30.

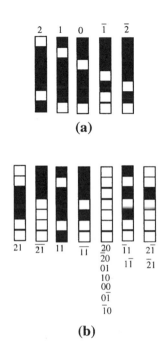

$$X = \begin{bmatrix} 1 & 0 & 0 & 0 & 1 & 0 & 1 \\ 1 & 0 & 1 & 0 & 1 & 1 & 1 \\ 0 & 0 & 0 & 0 & 1 & 0 & 0 \\ 0 & 1 & 0 & 1 & 1 & 1 & 1 \\ 0 & 1 & 0 & 0 & 1 & 0 & 1 \\ 1 & 1 & 0 & 1 & 1 & 1 & 1 \\ 1 & 1 & 1 & 1 & 1 & 1 & 1 \end{bmatrix} \quad Y = \begin{bmatrix} 1 & 1 & 0 & 0 & 0 & 0 & 0 \\ 0 & 0 & 1 & 1 & 0 & 0 & 0 \\ 0 & 0 & 0 & 0 & 1 & 0 & 0 \\ 0 & 0 & 0 & 0 & 0 & 1 & 0 \\ 0 & 0 & 0 & 0 & 0 & 0 & 1 \end{bmatrix}$$

$$M = \begin{bmatrix} -1 & 0 & 0 & -2 & 1 & 2 & 0 \\ 1 & -1 & 0 & 1 & -1 & -2 & 2 \\ 0 & 0 & 1 & 0 & 0 & 0 & 0 \\ -1 & 1 & 0 & 0 & 0 & 1 & -1 \\ 1 & 0 & -1 & 1 & 0 & -1 & 0 \end{bmatrix}$$

(c)

Figure 26 Nonrecoded TSD multiplication when only multiplier is recoded design V: (a) a fifth alternative seven-pixel encoding which can be used in both X and Y matrices; (b) joint spatial encodings that are used in the input X matrix; and (c) solution of the M-V equation.

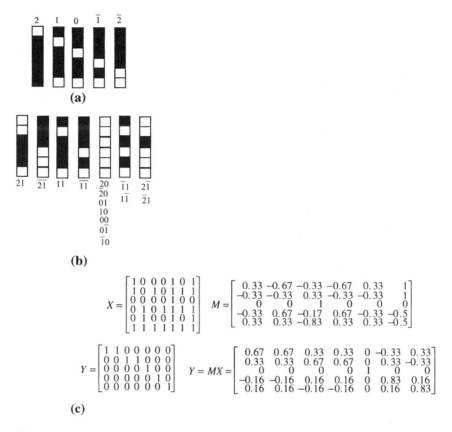

Figure 27 Nonrecoded TSD multiplication when only multiplier is recoded design VI: (a) a sixth alternative six-pixel encoding which can be used in both X and Y matrices; (b) joint spatial encodings that are used in the input X matrix; and (c) solution of the M-V equation.

6.8 Summary

We have presented compact optoelectronic TSD addition and multiplication. A detailed design methodology for the trinary input encodings and the required optical matched spatial filters for the optical correlators were presented. The size of our proposed TSD adders is less than that of the MSD counterparts, which were reported in Ref. 48. The TSD adders employ only one or two correlators instead of the three of the MSD adder; and they can handle a much larger dynamic range, since the TSD representation uses many fewer digits than the MSD to represent the same decimal numbers. As we reduce the hardware requirements and increase the data dynamic range, our proposed trinary based adders become attractive for practical implementations.

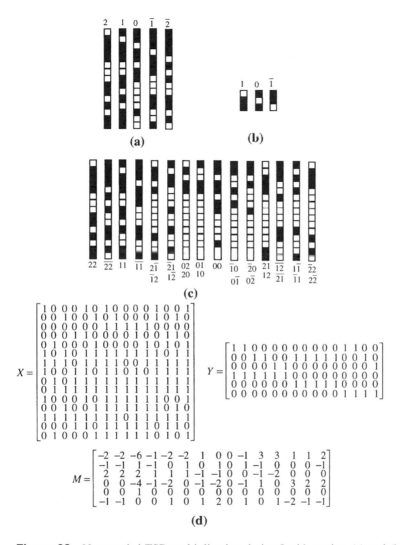

Figure 28 Nonrecoded TSD multiplication design I with carries: (a) and (b) 15-pixel and three-pixel encodings which can be used in X and Y matrices, respectively; (c) joint spatial encodings that are used in the input X matrix; and (d) solution of the M-V equation.

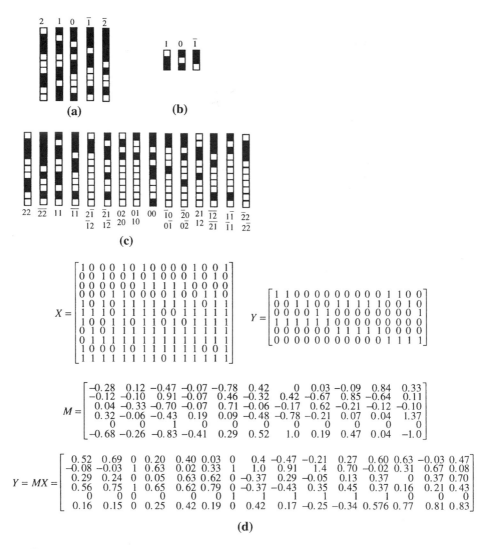

Figure 29 Nonrecoded TSD multiplication design II with carries: (a) and (b) 11-pixel and three-pixel encodings which can be used in X and Y matrices, respectively; (c) joint spatial encodings that are used in the input X matrix; and (d) solution of the M-V equation.

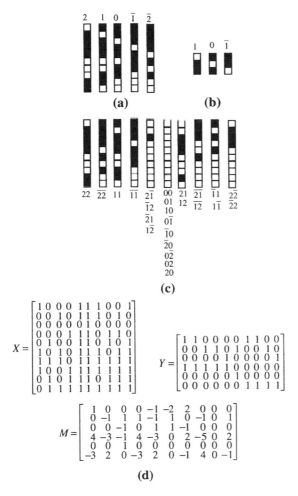

Figure 30   Nonrecoded TSD multiplication design III with carries: (a) and (b) 10-pixel and three-pixel encodings which can be used in X and Y matrices, respectively; (c) joint spatial encodings that are used in the input X matrix; and (d) solution of the M-V equation.

7   ONE-STEP CLASSIFIED TSD ADDITION AND ITS INCOHERENT IMPLEMENTATION

In this section, a simple one-step TSD arithmetic is proposed for parallel optical computing based on the joint spatial encoding technique. This technique performs multidigit carry-free addition and borrow-free subtraction in constant time. The

TSD addition is based on classifying the neighboring digit pairs into various groups in order to reduce the computation rules. A new joint spatial encoding technique is developed to present both the operands and the computation rules. This technique increases the efficiency usage of the spatial bandwidth product (SBWP) of the spatial light modulators (SLM) of the system. An optical implementation of TSD arithmetic operations is also presented.

7.1 Digit Classification and Joint Spatial Encoding

Table 6 represents all the computation rules needed to perform TSD addition. For example, if the CP digits belong to group C_3 and the NLOP digits belong to group N_3, then the output digit is $\overline{2}$. Thus Table 6 shows that we can perform TSD addition by classifying the two neighboring digit pairs into six groups and by making a decision according to these groups as illustrated in Fig. 31. Note that, in practice, the output 0 can be regarded as neither 1, $\overline{1}$, 2, nor $\overline{2}$; and consequently, only six computing rules are enough for the one-step TSD addition implementation.

The sum digits of TSD numbers depend only on the match of CP and NLOP digit pairs with the entries of the truth tables of nine elements. To encode spatially the TSD digits, one needs $5^2 = 25$ pixels to represent all possible input digit combinations of $x_i y_i$. These pixels can be arranged into 5×5 cells. The large number of pixels needed for representing the TSD numbers reduces the space bandwidth product (SBWP) of a given spatial light modulator (SLM) and decreases the speed of addition operation (the throughput of the SLM). To compensate for this problem, one can notice that the encoding of the TSD on the SLM

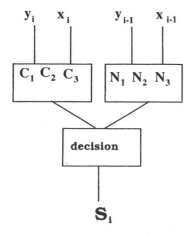

Figure 31   Function decision blocks for one-step TSD addition.

can be done electrically, and at the same time, for addition, the order of operands does not affect the resultant sum [44] (i.e., 1 plus 0 yields the same sum as 0 plus 1, $\bar{1}$ plus 0 yields the same sum as 0 plus $\bar{1}$, 2 plus 0 yields the same sum as 0 plus 2, etc.). Therefore, using simple and small lookup tables (combinational circuits) and parallel digital registers, one can download to an SLM the joint spatial encoding shown in Fig. 32. Thus one can omit some of the required pixels in the original $5 \times 5 = 25$ to only $3 \times 5 = 15$ pixels. This saves 40% of the SLM pixels area. The corresponding classified groups for the CP and NLOP digit pair encoding are shown in Fig. 33.

A further increase of the SBWP and an increase in the speed throughput of an SLM can be achieved if one considers a joint spatial encoding for more than two pairs of digits as was illustrated in Sec. 6. For instance, by examining Table 1a, five or fewer input digit pairs are grouped together. If one pixel is assigned for each group, then only nine pixels (instead of 15) are necessary to classify the input digit pairs $x_i y_i$. The TSD joint spatial encodings can be arranged in 9×1 (as shown in Fig. 34); this represents a savings of 64% of an SLM pixel area. Figure 35 shows the new corresponding CP and NLOP digit group classifications based on the new joint spatial encodings of Fig. 34.

$\begin{array}{c}\mathbf{x}\ \mathbf{y}\\ \mathbf{i}\ \ \mathbf{i}\end{array}$				
$\bar{2}\,\bar{2}$	$\bar{1}\,\bar{1}$	$0\,0$	$1\,1$	$2\,2$
$\bar{2}\,\bar{1}$ $\bar{1}\,\bar{2}$	$\bar{2}\,1$ $\bar{1}\,2$	$0\,\bar{1}$ $\bar{1}\,0$	$\bar{2}\,1$ $\bar{1}\,2$	$2\,1$ $1\,2$
$\bar{2}\,0$ $0\,\bar{2}$	$1\,\bar{1}$ $\bar{1}\,1$	$0\,1$ $1\,0$	$2\,\bar{2}$ $\bar{2}\,2$	$2\,0$ $0\,2$

Figure 32 A new joint spatial encoding scheme for the TSD based on combining one pair of TSD digits.

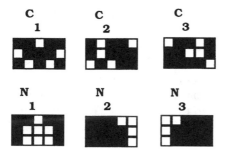

Figure 33   Encoded patterns for the six classified groups of digits for the TSD adder.

G1 G2 G3 G4 G5 G6 G7 G8 G9

Figure 34   Nine-pixel joint spatial encodings for TSD. G1, G2, G3, . . . represent the digit group numbers of the operands as shown in Table 1a.

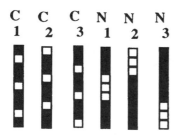

Figure 35   The new CP and NLOP group classifications of input digits representing the nine-pixel encodings for the TSD adder.

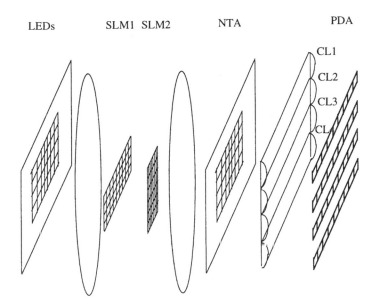

Figure 36 Incoherent optical correlator processor to implement signed-digit arithmetic. LED, light emitting diode; SLM, spatial light modulator; NTA, nonlinear thresholding array; CL, cylindrical lens array; PDA, photodetector array. (From Ref. 23.)

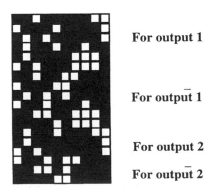

For output 1

For output $\bar{1}$

For output 2

For output $\bar{2}$

Figure 37 Computation masks for the the 15-pixel encoding of the TSD adder.

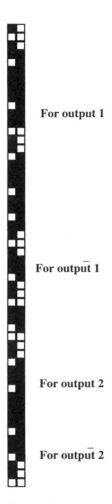

For output 1

For output $\overline{1}$

For output 2

For output $\overline{2}$

Figure 38 Computation masks for the the nine-pixel encoding of the TSD adder.

7.2 Optical Implementations

To implement the computation rules mentioned in the previous sections, we adopt the optical incoherent correlator with thresholding that was proposed recently [23]. The incoherent optical correlator that performs the one-step SD addition is shown in Fig. 36, where the encoded input digit patterns are introduced in SLM1, and a rule mask representing the computation rules (CP and NLOP) is stored in SLM2. The system in Fig. 36 simply performs a discrete correlation between the encoded input digit matrix and the stored computation rules matrix of the mask. The results of the correlation are then thresholded (threshold value is set to 2 units) using a nonlinear thresholding array (NTA) device. Finally, the cylindrical lenses perform summation operations in one dimension to collect the light at photodetector arrays to produce the corresponding TSD outputs. Figure 37 shows the mask rules for the 15-pixel joint spatial encoding for TSD adders, whereas Fig. 38 illustrates the mask rules for the nine-pixel joint spatial encoding.

More details of the operation of the incoherent correlator processor of Fig. 11 can be found in Ref. 23. However, for illustration purposes, we will consider the addition process of two three-digit TSD numbers $X = 1\bar{2}1 = (4)_{10}$ and $Y = 202 = (20)_{10}$ using the nine-pixel joint spatial encoding for the input digits of Fig. 34 and the corresponding computation mask rules of Fig. 38. In general, two parallel digital registers are used to download the n-digit TSD input number into the optical processor of Fig. 36. The register contents are used to control the individual pixels of an electronically addressed $9 \times (n + 2)$ pixels SLM1 forming the input matrix A. Note that the addition of two n-digit SD numbers will generally be an $(n + 1)$-digit number. To obtain correctly the $(n+1)^{th}$ output digit, one zero pair must be padded at the left of the most significant digit of the input operands. Similarly, to obtain correctly the least significant digit of the result, we should pad another pair of zeros at the right of the least significant digit of the input operands.

Each column of the SLM1 contains the joint spatial encoding of two TSD $(x_i y_i)$ digits. Also, the computation rules mask of Fig. 38 corresponding to the TSD digits 1, $\bar{1}$, 2, and $\bar{2}$ is stored in the second SLM2, forming the 54×2 matrix B. The optical processor of Fig. 36, which performs an optical correlation of matrix A with matrix B, produces a $62 \times (n + 3)$-pixel output matrix C on the NTA plane. A matrix of $54 \times (n + 1)$ elements can be obtained by sampling the $(9i, 1 + j)^{th}$ $(i = 1, 6; j = 1, 4)$ elements of the correlation matrix. After thresholding the sampled matrix, the cylindrical lenses (CLs) collect light transmitted from the NTA device. CL1, CL2, CL3, and CL4 collect light transmitted from the 18, 18, 9, and 9 rows of the NTA device, which correspond, respectively, to outputs 1, $\bar{1}$, 2, and $\bar{2}$.

For our example, using the spatial encoding of Fig. 34, the three-digit TSD inputs are arranged

$$
A = \begin{bmatrix} \varnothing & 1 & \bar{2} & 1 & \varnothing \\ \varnothing & 2 & 0 & 2 & \varnothing \end{bmatrix} = \begin{bmatrix} 0 & 0 & 0 & 0 & 0 \\ 0 & 1 & 0 & 1 & 0 \\ 0 & 0 & 0 & 0 & 0 \\ 0 & 0 & 0 & 0 & 0 \\ 1 & 0 & 0 & 0 & 1 \\ 0 & 0 & 0 & 0 & 0 \\ 0 & 0 & 1 & 0 & 0 \\ 0 & 0 & 0 & 0 & 0 \\ 0 & 0 & 0 & 0 & 0 \end{bmatrix}
$$

where $\varnothing$ represents a padded zero. Similarly, using the encoding mask rules of Fig. 38, the 54×2 B matrix is given by

$$
B =
\begin{bmatrix}
0 & 1 \\
1 & 1 \\
0 & 1 \\
0 & 0 \\
1 & 0 \\
0 & 0 \\
0 & 0 \\
1 & 0 \\
0 & 0 \\
1 & 0 \\
0 & 0 \\
0 & 0 \\
1 & 1 \\
0 & 1 \\
0 & 1 \\
1 & 0 \\
0 & 0 \\
0 & 0 \\
0 & 0 \\
1 & 0 \\
0 & 0 \\
0 & 0 \\
1 & 0 \\
0 & 0 \\
0 & 1 \\
1 & 1 \\
0 & 1 \\
0 & 0 \\
0 & 0 \\
1 & 0 \\
0 & 1 \\
0 & 1 \\
1 & 1 \\
0 & 0 \\
0 & 0 \\
1 & 0 \\
1 & 1 \\
0 & 1 \\
0 & 1 \\
1 & 0 \\
0 & 0 \\
0 & 0 \\
1 & 0 \\
0 & 0 \\
0 & 0 \\
0 & 0 \\
0 & 0 \\
1 & 0 \\
0 & 0 \\
0 & 0 \\
1 & 0 \\
0 & 1 \\
0 & 1 \\
1 & 1
\end{bmatrix}
$$

The resulting correlation 62×6 matrix C is

$$
C = \begin{bmatrix}
0 & 0 & 0 & 0 & 0 & 0 \\
0 & 0 & 0 & 0 & 0 & 0 \\
0 & 0 & 1 & 0 & 0 & 0 \\
0 & 0 & 1 & 1 & 0 & 0 \\
1 & 0 & 1 & 0 & 1 & 0 \\
1 & 1 & 0 & 0 & 1 & 1 \\
1 & 0 & 0 & 1 & 1 & 0 \\
0 & 1 & 0 & 1 & 0 & 0 \\
0 & 2 & 1 & 1 & 1 & 1 \\
0 & 1 & 0 & 2 & 0 & 0 \\
0 & 0 & 0 & 0 & 0 & 0 \\
0 & 1 & 1 & 1 & 1 & 1 \\
0 & 0 & 0 & 0 & 0 & 0 \\
0 & 1 & 0 & 0 & 0 & 1 \\
0 & 0 & 2 & 1 & 1 & 0 \\
0 & 0 & 1 & 0 & 0 & 0 \\
1 & 1 & 2 & 0 & 2 & 1 \\
1 & 0 & 0 & 1 & 1 & 0 \\
1 & 0 & 0 & 0 & 1 & 0 \\
0 & 2 & 1 & 1 & 1 & 1 \\
0 & 1 & 0 & 1 & 0 & 0 \\
0 & 1 & 0 & 2 & 0 & 0 \\
0 & 0 & 1 & 0 & 1 & 0 \\
0 & 1 & 0 & 0 & 0 & 1 \\
0 & 0 & 0 & 1 & 0 & 0 \\
0 & 0 & 0 & 0 & 0 & 0 \\
0 & 1 & 2 & 0 & 1 & 1 \\
0 & 0 & 1 & 1 & 0 & 0 \\
1 & 0 & 1 & 0 & 1 & 0 \\
1 & 0 & 0 & 0 & 1 & 0 \\
0 & 1 & 0 & 2 & 0 & 0 \\
0 & 1 & 2 & 1 & 1 & 0 \\
0 & 2 & 1 & 1 & 0 & 1 \\
1 & 0 & 1 & 1 & 1 & 0 \\
1 & 0 & 0 & 0 & 1 & 0 \\
1 & 1 & 1 & 0 & 2 & 1 \\
0 & 1 & 0 & 2 & 0 & 0 \\
0 & 1 & 1 & 2 & 0 & 0 \\
0 & 2 & 2 & 1 & 1 & 1 \\
1 & 1 & 1 & 0 & 1 & 1 \\
1 & 0 & 0 & 1 & 1 & 0 \\
1 & 0 & 1 & 0 & 2 & 0 \\
0 & 2 & 1 & 1 & 1 & 1 \\
0 & 1 & 0 & 2 & 0 & 0 \\
0 & 1 & 0 & 1 & 0 & 0 \\
0 & 1 & 1 & 0 & 1 & 1 \\
0 & 0 & 0 & 0 & 0 & 0 \\
0 & 0 & 0 & 0 & 0 & 0 \\
0 & 0 & 1 & 1 & 1 & 0 \\
0 & 0 & 0 & 0 & 0 & 0 \\
0 & 1 & 0 & 0 & 0 & 1 \\
0 & 0 & 0 & 1 & 0 & 0 \\
0 & 0 & 1 & 0 & 0 & 0 \\
0 & 1 & 2 & 0 & 1 & 1 \\
1 & 0 & 1 & 1 & 1 & 0 \\
1 & 0 & 0 & 0 & 1 & 0 \\
1 & 1 & 1 & 0 & 2 & 1 \\
0 & 1 & 0 & 1 & 0 & 0 \\
0 & 1 & 0 & 1 & 0 & 0 \\
0 & 1 & 1 & 1 & 1 & 0 \\
0 & 0 & 0 & 0 & 0 & 0 \\
\end{bmatrix}
$$

A matrix of $54 \times (n + 1)$ elements can be obtained by sampling and thresholding (at a value of 2) the above correlation C matrix:

$$
\begin{bmatrix}
1 & 0 & 0 & 0 \\
0 & 0 & 0 & 0 \\
0 & 0 & 0 & 0 \\
0 & 0 & 0 & 0 \\
0 & 0 & 0 & 0 \\
0 & 0 & 0 & 0 \\
0 & 0 & 0 & 0 \\
0 & 0 & 0 & 0 \\
0 & 0 & 0 & 0 \\
0 & 0 & 0 & 0 \\
0 & 0 & 0 & 0 \\
0 & 0 & 0 & 0 \\
0 & 0 & 0 & 0 \\
0 & 0 & 0 & 0 \\
0 & 0 & 0 & 0 \\
0 & 0 & 0 & 0 \\
0 & 0 & 0 & 0 \\
0 & 0 & 0 & 0 \\
0 & 1 & 0 & 0 \\
0 & 0 & 0 & 0 \\
0 & 0 & 0 & 0 \\
0 & 0 & 0 & 0 \\
0 & 0 & 0 & 0 \\
0 & 0 & 0 & 0 \\
0 & 0 & 0 & 0 \\
0 & 0 & 0 & 0 \\
0 & 0 & 0 & 0 \\
0 & 0 & 0 & 0 \\
0 & 0 & 0 & 0 \\
0 & 0 & 0 & 0 \\
0 & 0 & 0 & 0 \\
0 & 0 & 0 & 0 \\
0 & 0 & 0 & 0 \\
0 & 0 & 0 & 0 \\
0 & 0 & 0 & 0 \\
0 & 0 & 0 & 0 \\
0 & 0 & 1 & 0 \\
0 & 0 & 0 & 0 \\
0 & 0 & 0 & 0 \\
0 & 0 & 0 & 0 \\
0 & 0 & 0 & 0 \\
0 & 0 & 0 & 0 \\
0 & 0 & 0 & 0 \\
0 & 0 & 0 & 0 \\
0 & 0 & 0 & 0 \\
0 & 0 & 0 & 0 \\
0 & 0 & 0 & 0 \\
0 & 0 & 0 & 0 \\
0 & 0 & 0 & 0 \\
0 & 0 & 0 & 0 \\
0 & 0 & 0 & 0 \\
0 & 0 & 0 & 0 \\
0 & 0 & 0 & 0 \\
0 & 0 & 0 & 0
\end{bmatrix}
$$

and the four cylindrical lenses produce the four output rows

$$
output = \begin{bmatrix} 1 & 0 & 0 & 0 & \leftarrow & \text{for output} & 1 \\ 0 & 1 & 0 & 0 & \leftarrow & \text{for output} & \bar{1} \\ 0 & 0 & 1 & 0 & \leftarrow & \text{for output} & 2 \\ 0 & 0 & 0 & 0 & \leftarrow & \text{for output} & \bar{2} \end{bmatrix}
$$

From the above output matrix, we observe a value of 1 in row 1 and column 1. Thus we expect an output of 1 in this position in the resulting sum. Similarly, we expect outputs of $\bar{1}$ and 2 in the second and third positions of the sum. Finally, a 0 output digit is obtained in the output sum for the fourth digit position, since the fourth row has all zero elements. Therefore the output sum is $[1\bar{1}20]_{TSD} = (24)_{10}$ as expected.

Fully parallel one-step TSD adder/subtracter units are presented based on new joint spatial encodings of the TSD operands. This technique increases the spatial bandwidth product of the SLM of the system. It has been shown that a one-step scheme and only six computation rules are needed to implement the TSD addition instead of a two-step scheme and 24, 44, and tens of computations rules as reported in Refs. 10, 13, and 19, respectively. Thus our proposed TSD adder/subtracter units are much simpler and use fewer computation rules than any previously reported techniques [10,13,17,18,19,24]. Optical implementation of the proposed arithmetic can be carried out using an incoherent correlation scheme. Further, It is worth mentioning that our proposed TSD adder/subtracter can also be implemented using matrix multiplication techniques [18,20,22,24].

REFERENCES

1. E. L. Johnson and M. A. Karim. Digital Design: A Pragmatic Approach. PWS-Kent, Boston, 1987.
2. K. Hwang. Computer Arithmetic Principles: Architecture and Design. John Wiley, New York, 1979.
3. S. L. Hurst. Multiple-valued threshold logic: its status and its realization. Opt. Eng. 25(1), 44–53 (1986).
4. M. M. Mirsalehi and T. K. Gaylord. Logical minimization of multilevel coded functions. Appl. Opt. 25, 3078–3088 (1986).
5. N. S. Szabo and R. T. Tanaka. Residue Arithmetic and its Applications to Computer Technology. McGraw-Hill, New York, 1967.
6. A. P. Gautzouilis, E. C. Malarkey, D. K. Davies, J. C. Bradley, and P. R. Beaudet. Optical processing with residue LED/LD lookup tables. Appl. Opt 27, 1674–1681 (1988).

7. D. Psaltis and D. Casasent. Optical residue arithmetic: a correlator approach. Appl. Opt. 18, 163–1171 (1979).
8. A. Hwang, Y. Tsunida, J. W. Goodman, and S. Ishihara. Optical computation using residue arithmetic. Appl. Opt. 18, 149–162 (1979).
9. G. A. De Biase and A. Massini. High efficiency redundant binary number representations for parallel arithmetic on optical computers. Optics and Laser Technology 26, 219–224 (1994).
10. M. S. Alam. Parallel optical computing using recoded trinary signed-digit numbers. Appl. Opt. 33(20), 4392–4397 (1994).
11. A. K. Cherri and M. A. Karim. Modified signed digit arithmetic using an efficient symbolic substitution. Appl. Opt. 27(18), 3824–3827 (1988).
12. R. P. Bocker, B. L. Drake, M. E. Lasher, and T. B. Henderson. Modified Signed-digit addition and subtraction using optical symbolic substitution. Appl. Opt. 25(15), 2456–2457 (1986).
13. A. A. S. Awwal, M. N. Islam, and M. A. Karim. Modified signed-digit trinary arithmetic by using symbolic substitution. Appl. Opt. 31(11), 1687–1694 (1994).
14. A. A. S. Awwal. Recorded signed-digit binary addition-subtraction using opto-electronic symbolic substitution. Appl. Opt. 31, 3205–3208 (1992).
15. A. K. Cherri. Symmetrically recoded modified signed-digit optical addition and subtraction. Appl. Opt. 33(20), 4378–4382 (1994).
16. Y. Li and G. Eichmann. Conditional symbolic modified-signed-digit arithmetic using optical content-addressable memory logic elements. Appl. Opt. 26(12), 2328–2333 (1987).
17. K. Hwang and D. K. Panda. High-radix symbolic substitution and superposition techniques for optical matrix algebraic computations. Opt. Eng. 31(11), 2422–2433 (1992).
18. A. K. Cherri and N. I. Khachab. Canonical quaternary signed-digit arithmetic using optoelectronics symbolic substitution. Optics and Laser Technology 28(5), 397–403 (1996).
19. M. S. Alam, M. A. Karim, A. A. S. Awwal, and J. J. Westerkamp. Optical processing based on conditional higher-order trinary modified signed-digit symbolic substitution. Appl. Opt. 31(26), 5614–5621 (1992).
20. B. Ha and Y. Li. Parallel modified signed-digit arithmetic using an optoelectronic shared content-addressable-memory processor. Appl. Opt. 33(17), 3647–3662 (1994).
21. Y. Li, D. H. Kim, A. Kostrzewski, and G. Eichmann. Content-addressable-memory-based single stage optical modified-signed-digit arithmetic. Opt. Let. 14, 1254–1256 (1989).
22. H. Huang, M. Itoh, and T. Yatagai. Modified signed-digit arithmetic based on redundant bit representation. Appl. Opt. 33(26), 6146–6156 (1994).
23. H. Huang, M. Itoh, T. Yatagai, and L. Liu. Classified one-step modified signed-digit arithmetic and its optical implementation. Opt. Eng. 35(4), 1134–1140 (1996).
24. A. K. Cherri. High-radix arithmetic using symbolic substitution computation. Optics Communications 128, 108–122 (1996).
25. A. K. Cherri, M. A. Habib, and M. S. Alam. Efficient implementation of arithmetic units based on polarization-encoded optical shadow-casting. Opt. Eng. 36(1), 94–101 (1997).

26. A. Avizienis. Signed-digit number representations for fast parallel arithmetic. IRE Trans. Electronic Computers EC-10(3), 389–400 (1961).

27. F. Li and M. Morisue. A novel Josephson adder without carry propagation delay. IEEE Transactions on Applied Superconductivity 3(1), 2683–2686 (1993).

28. T. Stouraitis and C. Chen. Hybrid signed-digit logarithmic number system processor. IEE Proceedings-E 140(11), 205–210 (1993).

29. W. Balakrishnan and N. Burgess. Very-high-speed VLSI 2s-complement multiplier using signed binary digits. IEE Proc.-Comput. Digit. Tech. 71(139), 29–34 (1994)

30. N. Takagi and S. Yajima. Modular multiplication hardware algorithms with redundant representation and their application to RSA cryptosystem. IEEE Transactions on Computers 41(7), 887–891 (1992).

31. P. Srinivasan and F. E. Petry. Constant-division algorithms. IEE Proc.-Comput. Digit. Tech. 141(6), 334–340 (1994).

32. N. Burgess. Radix-2 SRT division algorithm with simple quotient digit selection. Electronics Letters 27(21), 1910–1911 (1991).

33. K. Al-Ghoneim and D. P. Casasent. High-accuracy pipelined iterative-tree optical multiplication. Appl. Opt. 33, 1517–1527 (1994).

34. K. H. Brenner and A. Huang. Optical implementations of the perfect shuffle interconnections. Appl. Opt. 27, 135–137 (1988).

35. M. S. Alam and M. A. Karim. Programmable optical perfect shuffle interconnection network using Fredkin gates. Microwave Opt. Technol. Lett. 5, 330–333 (1992).

36. K. Hwang and A. Louri. Optical multiplication and division using modified signed-digit symbolic substitution. Opt. Eng. 28, 364–372 (1989).

37. A. R. Omondi. *Computer Arithmetic Systems*. Prentice-Hall, 1994.

38. K. Hwang and A. Louri. Optical arithmetic using symbolic signed-digit substitution. Proceedings of the 17th International Conference on Parallel Processing. Pennsylvania State University Press, 1988, pp. 55–64.

39. E. V. Krishnamurthy. On optimal iterative schemes for high speed division. IEEE Transactions on Computers, C-19, 227–231 (1970).

40. M. S. Alam. Parallel optical computing using recoded trinary signed-digit numbers. Appl. Opt. 33, 4392–4397 (1994).

41. T. K. Gaylord and M. M. Mirsalehi. Truth-table look-up processing: number representation, multilevel coding, and logical minimization. Opt. Eng. 25, 22–33 (1986).

42. M. S. Alam, Y. Ahuja, A. K. Cherri, and A. Chatterjea. Symmetrically recoded quaternary signed-digit arithmetic using a shared content-addressable memory. Opt. Eng. 35, 1141–1149 (1996).

42. J. U. Ahmed and A. A. S. Awwal. Multiplier design using RBSD number system. Proc. IEEE 1993 National Aerospace & Electronics Conf., 1993, pp. 180–184.

43. A. Huang. Parallel algorithms for optical digital computers. In proceedings, Tenth International Optical Computing Conference (S. Horvitz, ed.). *IEEE Computer Society and ICO, New York*, 1983, pp. 13–17.

44. D. P. Casasent and E. C. Botha. Multifunctional optical processor based on symbolic substitution. Opt. Eng. 28, 425–433 (1989).

45. E. Botha, J. Richards, and D. P. Casasent. Optical laboratory morphological inspection processor. Appl. Opt. 28, 5342–5350 (1989).

46. D. P. Casasent and R. Sturgill. Optical hit-or-miss morphological transforms for

ATR. SPIE Proc. Applications of Digital Image Processing XII 1153, 500–510 (1989).

47. D. P. Casasent and E. C. Botha. Optical correlator production system neural net. Apl. Opt. 31, 1030–1040 (1992).
48. D. P. Casasent and P. Woodford. Symbolic substitution modified signed-digit optical adder. Appl. Opt. 33, 1498–1506 (1994).
49. A. Louri. Throughput enhancement for optical symbolic substitution systems. Appl Opt. 29, 2979–2980 (1990).
50. H. E. Michel and A. A. S. Awwal. Noise and cross talk study in an optical neural network. In Proceedings, National and Aerospace Electronics Conference, IEEE 2, 662–669 (1996).
51. R. P. Webb. Performance of an optoelectronic neural network in the presence of noise. Appl. Opt. 34, 5230–5240 (1995).
52. S. Yamamoto, R. Sekura, J. Yamanaka, T. Ebihara, N. Kato, and H. Hosi. Optical recognition with LAPS-SLM. In Computer and Optically Formed Holographic Optics (I. Cindrich and S. H. Lee, eds.). Proc. SPIE 1211, 273–283 (1990).
53. M. Roe and K. Schnehrer. High-speed and high-contrast operation of ferroelectric liquid crystal optically addressed spatial light modulators. Opt. Eng. 32, 1662–1667 (1993).
54. A. K. Cherri, M. K. Habib, and M. S. Alam. Optoelectronic recoded and non-recoded trinary signed-digit adder using optical correlation. Applied Optics 37(4), 2153–2163 (1998).
55. A. K. Cherri and M. S. Alam. Recoded and nonrecoded trinary signed-digit adders and multipliers using redundant bit representations. Applied Optics 37(7), 4405–4418 (1998).
56. A. K. Cherri. Classified one-step high-radix signed-digit arithmetic units. Optical Engineering 37(8), 2324–2333 (1998).
57. A. K. Cherri. Signed-digit arithmetic for optical computing: digit grouping and pixel assignment for spatial encoding. Optical Engineering 38(3), 422–431 (1999).
58. A. K. Cherri. Designs of optoelectronic trinary signed-digit multiplication using joint spatial encodings and optical correlation. Applied Optics 38, 828–837 (1999).
59. M. S. Alam. Parallel optoelectronic trinary signed-digit division. Optical Engineering 38(3), 441–448 (1999).

12

Ultrafast Arithmetic, Logical, and Image Processing Operations Using Polarization-Encoded Optical Shadow-Casting

Abdallah K. Cherri
Kuwait University, Safat, Kuwait

Mohammad S. Alam
The University of South Alabama, Mobile, Alabama

1   INTRODUCTION

The main performance constraint toward realizing a massively parallel electronic digital computer is its interconnection bottleneck. An attractive solution to this problem would be to employ optical digital systems that utilize the high speed and parallelism of optics at all computation levels [1,2]. To achieve this goal, various optical computing techniques have been proposed and investigated [3–7]. Among these methods, the lensless optical shadow-casting (OSC) technique [5] has been found to exploit completely the advantages of optics in performing combinatorial logic operations. In the OSC scheme, since all of the operations are linear, the computation speed is limited only by the speed with which light travels from the input plane to the output plane. Initially, the scope of OSC was somewhat limited because only sixteen output operations (as input pixel elements are binary) were possible, which in turn limited its application toward the implementation of complex logic elements (e.g., the trinary multiplier). The OSC system was later extended to incorporate polarized codes [7] for encoding the input pixels and the output mask transparency. The corresponding encoding conditions are transparent (T), opaque (F), horizontally polarized (H), and vertically polarized (V). The modified OSC system increases the degrees of freedom for the choice of coding, thereby making it possible to design complicated logic circuits. Also, this extended scheme, referred to as polarization-encoded optical shadowcasting (POSC), is more efficient, since memory saving is achieved by storing multiple minterms in the same pixel subcell location using orthogonal codes.

Using a generalized algorithm for the design of POSC combinational logic circuits [2], a number of binary and nonbinary multioutput combinational circuits [2,8] have been designed. Later on, by modifying this algorithm, a number of trinary combinational circuits were implemented [9,10]. More recently, Awwal et al. designed a programmable logic array [11], a J-K flip-flop [12], a grey level serial/parallel image encoder [13] and a multiprocessor [14]. For multioutput systems, the POSC algorithm [2] requires switching of the source pattern to perform particular logic operations after the input mask has been encoded [15] or switching of the decoding mask for a fixed input pattern. Very recently, Rizvi et al. [15,16] proposed a further extension of this algorithm in order to avoid the aforementioned limitation. A careful observation of this algorithm [15] reveals that it may not always lead to a minimized design. However, design minimization is especially important for POSC based circuits, since the size of the input overlap pattern, the LED source pattern, and the output overlap pattern is directly dependent on the number of minterms and their proper spatial allocation.

In this chapter, we report efficient POSC designs for combinational logic [17], two-dimensional register [18], multiple-valued logic multiprocessor [19], modified signed-digit arithmetic units [20]. Also, the POSC algorithm is applied for performing image processing operations such as edge detection [21] and morphological transformations [22,23].

2 THE POSC SYSTEM

A POSC system consists of a source pattern, an input overlap pattern, an output overlap pattern, and a detector mask as shown in Fig. 1. The spatially encoded two-dimensional input pixel patterns are introduced in perfect contact at the input plane and illuminated by a set of light emitting diodes (LEDs) from the source plane. The overlap of the projected shadows results in an output overlap pattern at the output plane. A decoding mask placed at the output plane is used to filter spatially and detect the output in the output plane. The logical inputs are encoded by dividing the input pixel into a number of subpixels and then assigning each subpixel a particular code from the POSC code set $\{V, H, T, F\}$. In Fig. 1, the 2×2 LED source pattern causes each overlapped pixel to produce a 3×3 output pixel. The combination of corresponding input pixel patterns determines the minterms corresponding to a particular output while the source plane LEDs control the minterms to produce the desired output. A POSC overlap equation

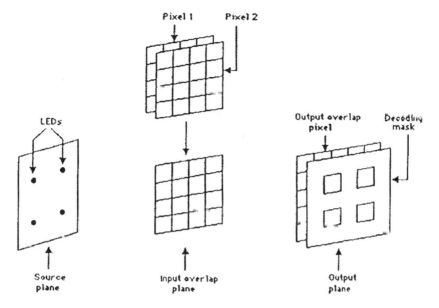

Figure 1   A lensless POSC system.

[2], for each of the input overlap pixel subcells, may be developed to relate the input condition with the POSC code of the corresponding output. Such POSC overlap equations are solved to identify the input coding patterns and the LED source pattern.

2.1   The POSC Algorithm

In POSC systems, the source, the input, and the output variables are represented in the form of spatial matrices. For illustration, consider a POSC system consisting of two input variables A and B (of spatial dimension $m \times n$) and a source plane L (of spatial dimension $p \times q$) spatial matrices:

$$L = \begin{bmatrix} L_{11} & L_{12} & L_{13} & \cdots & L_{1q} \\ L_{21} & Q_{22} & L_{23} & \cdots & L_{2q} \\ L_{31} & Q_{32} & L_{33} & \cdots & L_{3q} \\ \cdots & \cdots & \cdots & \cdots & \cdots \\ L_{p1} & L_{p2} & L_{p3} & \cdots & L_{pq} \end{bmatrix} \quad A = \begin{bmatrix} a_{11} & a_{12} & a_{13} & \cdots & a_{1n} \\ a_{21} & a_{22} & a_{23} & \cdots & a_{2n} \\ a_{31} & a_{32} & a_{33} & \cdots & a_{3n} \\ \cdots & \cdots & \cdots & \cdots & \cdots \\ a_{m1} & a_{m2} & a_{m3} & \cdots & a_{mn} \end{bmatrix}$$

$$B = \begin{bmatrix} b_{11} & b_{12} & b_{13} & \cdots & b_{1n} \\ b_{21} & b_{22} & b_{23} & \cdots & b_{2n} \\ b_{31} & b_{32} & b_{33} & \cdots & b_{3n} \\ \cdots & \cdots & \cdots & \cdots & \cdots \\ b_{m1} & b_{m2} & b_{m3} & \cdots & b_{mn} \end{bmatrix}$$

The input overlap spatial matrix I can be expressed in terms of the input variables A and B as

$$I = F_1(A, B)$$

Where F_1 represents an overlap function and I has the same dimension as A or B. Therefore the output overlap matrix can be derived in terms of the L and I matrices:

$$D = F_2(L, I)$$

where F_2 represents another overlap function. The dimension $u \times v$ of the output overlap pattern (D), i.e., the decoding mask, is determined by the dimensions of the source and input overlap pixel patterns:

$$u = m + p - 1$$
$$v = n + q - 1$$

$$D = \begin{bmatrix} D_{11} & D_{12} & D_{13} & \cdots & D_{1v} \\ D_{21} & D_{22} & D_{23} & \cdots & D_{2v} \\ D_{31} & D_{32} & D_{33} & \cdots & D_{3v} \\ \cdots & \cdots & \cdots & \cdots & \cdots \\ D_{u1} & D_{u2} & D_{u3} & \cdots & D_{uv} \end{bmatrix}$$

To augment the performance of the existing design algorithms [2,15] even more, we present the following efficient algorithm for the design of arbitrary POSC logic circuits:

Step 1: Establish the logic truth table of the desired circuit with the input combinations corresponding to a common combination of the output listed into separate groups.

Step 2: Starting with the output having the maximum number of minterms, group the minterms in pairs so that they may have complementary [2] outputs and so that the total number of minterms required to generate all the outputs is a minimum. The number of minterm pairs is a minimum only when the number of common minterm pairs among different outputs is a maximum. One way to achieve this is to search exhaustively all possible minterm pairings and then to choose the minimum pair condition. For circuits with large numbers of inputs and outputs, a computer program may be used for this purpose. Alternatively, the input combinations of the truth table can be divided into two categories: input combinations generating a single output (ISO) and input combinations generating multiple outputs (IMO). The ISO conditions are paired first and then the IMO conditions are paired. The total number of pixel subcells required to encode the input conditions is equal to the sum of the paired minterms and the unpaired minterms (which may result in a case of an odd number of minterms).

Step 3: Based on the total number of minterms determined in Step 2, represent the input pixel pattern by an $m \times n$ spatial matrix and the source overlap pattern by a $p \times q$ matrix. The dimension of the decoding mask will be determined by the dimensions of the source and the input overlap pixel pattern. The minterms corresponding to different outputs are mapped beginning with the output corresponding to the maximum number of minterms. In general, the ISO minterms are placed near the boundary of the input overlap pixel, while the IMO minterms are placed toward the inside of the input overlap pixel so that they can be shared with the suitable minterms of other outputs. Next, the output having the second maximum number of input conditions is mapped, and the same procedure is followed for other outputs, so that the resulting mapping block has the minimum perimeter with a minimum number of sides.

Step 4: Determine the projection (input–output) equations [15] for each pixel subcell D_{ij} of the output detector mask as a function of the LED source pixel subcells and the input overlap pixel subcells.

Step 5: Select a suitable output pixel subcell for a particular output and assign a POSC code to it. Then, develop the POSC overlap equations [2] from the projection equations corresponding to the paired minterms.

Step 6: Find the input coding pattern and LED source pattern by projecting the output overlap pixel backward and solving the corresponding POSC overlap equations simultaneously.

Note that the POSC overlap equations corresponding to input conditions with nonzero outputs will automatically satisfy the input conditions with zero outputs only if either pair of equations is solved simultaneously or a single POSC equation is solved using nonpolarized codes. Accordingly, it is not necessary to determine the hardware corresponding to zero outputs. Also, for larger POSC circuits, the bit-slice approach may be used, where the entire logic circuit is first split into a number of similar smaller and less complex logic units and then the final circuit is implemented using these smaller units.

3 ENCODER DESIGN

To illustrate the efficiency of the POSC algorithm, consider the design of an 8-to-3 line encoder. The truth table of the 8-to-3 line encoder is shown in Table 1

Table 1 Truth Table for the 8-to-3 Encoder

Condition number	Inputs								Outputs		
	a	b	c	d	e	f	g	h	x	y	z
1	1	0	0	0	0	0	0	0	0	0	0
2	0	1	0	0	0	0	0	0	0	0	1
3	0	0	1	0	0	0	0	0	0	1	0
4	0	0	0	1	0	0	0	0	0	1	1
5	0	0	0	0	1	0	0	0	1	0	0
6	0	0	0	0	0	1	0	0	1	0	1
7	0	0	0	0	0	0	1	0	1	1	0
8	0	0	0	0	0	0	0	1	1	1	1

where a through h represent the eight input conditions of the encoder while x, y, and z represent the three outputs. In Sec. 3.1, we design this encoder using the algorithm proposed in Sec. 2, while in Sec. 3.2, we design the same logic circuit following the procedure mentioned in Ref. 15.

3.1 Design I

In an 8-to-3 line encoder, each of the inputs is encoded into three output lines. Since there are eight input conditions, eight input patterns need to be introduced in the input overlap plane. From Table 1, we observe that all of the outputs have the same number of minterms. As mentioned before, the minimal number of minterm pairs can be obtained by either exhaustive search or ISO/IMO technique. Applying the exhaustive search technique, the outputs with their corresponding input conditions may be grouped as follows:

   x: {5, 6, 7, 8}
   y: {3, 4, 7, 8}
   z: {2, 4, 6, 8}

To achieve optimum pairing, all possible pairs of the minterms are identified for the individual outputs. For each output, since there are four minterms, a total of six minterm pairs are possible and each output can be obtained in three different ways:

   x: [{5, 6}, {7, 8}], [{5, 7}, {6, 8}], [{5, 8}, {6, 7}]
   y: [{3, 4}, {7, 8}], [{3, 7}, {4, 8}], [{3, 8}, {4, 7}]
   z: [{2, 4}, {6, 8}], [{2, 6}, {4, 8}], [{2, 8}, {4, 6}]

Table 2 Possible Minterm Combinations for the 8-to-3 Encoder

1	[{56}{78}]	[{34}{78}]	[{24}{68}]	{78}
2	[{56}{78}]	[{34}{78}]	[{26}{48}]	{78}
3	[{56}{78}]	[{34}{78}]	[{28}{46}]	{78}
4	[{56}{78}]	[{37}{48}]	[{24}{68}]	none
5	[{56}{78}]	[{37}{48}]	[{26}{48}]	{48}
6	[{56}{78}]	[{37}{48}]	[{28}{46}]	none
7	[{56}{78}]	[{38}{47}]	[{24}{68}]	none
8	[{56}{78}]	[{38}{47}]	[{26}{48}]	none
9	[{56}{78}]	[{38}{47}]	[{28}{46}]	none
10	[{56}{68}]	[{37}{48}]	[{24}{68}]	{68}
11	[{56}{68}]	[{37}{48}]	[{26}{48}]	{48}
12	[{56}{68}]	[{37}{48}]	[{28}{46}]	none
13	[{56}{68}]	[{34}{78}]	[{24}{68}]	{68}
14	[{56}{68}]	[{34}{78}]	[{26}{48}]	none
15	[{56}{68}]	[{34}{78}]	[{28}{46}]	none
16	[{56}{68}]	[{38}{47}]	[{24}{68}]	{68}
17	[{56}{68}]	[{38}{47}]	[{26}{48}]	none
18	[{56}{68}]	[{38}{47}]	[{28}{46}]	none
19	[{58}{67}]	[{38}{47}]	[{24}{68}]	none
20	[{58}{67}]	[{38}{47}]	[{28}{46}]	none
21	[{58}{67}]	[{38}{47}]	[{26}{48}]	none
22	[{58}{67}]	[{37}{48}]	[{24}{68}]	none
23	[{58}{67}]	[{37}{48}]	[{26}{48}]	{48}
24	[{58}{67}]	[{37}{48}]	[{28}{46}]	none
25	[{58}{67}]	[{38}{47}]	[{24}{68}]	none
26	[{58}{67}]	[{38}{47}]	[{26}{48}]	none
27	[{58}{67}]	[{38}{47}]	[{28}{46}]	none

Thus a total of $27 = 3^3$ conditions need to be considered as shown in Table 2. We observe that at best only one minterm can be shared among the different combinations of the minterm pairs. For example, we can consider the input combinations [{5, 6}, {7, 8}], [{3, 4}, {7, 8}] and [{2, 4}, {6, 8}] for the proposed system. Since {7, 8} is common between x and y, the total number of required minterm pairs is five. Now the ISO/IMO technique may be employed to obtain the combination of minimal minterm pairs. The input combinations for the three outputs corresponding to the ISO category are

x: {5}

y: {3}

z: {2}

while those for the IMO category are

x: {6, 7, 8}

y: {4, 7, 8}

z: {4, 6, 8}

Since in the ISO category, each output has only one minterm, no minterm pair can be formed for any output solely from the ISO category. From the IMO category, however, we observe that two input conditions are common between any two outputs, while the input condition 8 is common among all these outputs. Consider the pair {7, 8} to generate the output x. Thus the only pair that is left to generate x is {5, 6}. Since the pair {7, 8} is also common with y, the other necessary minterm pair to produce y is {3, 4}.

After this minterm assignment, the output z will not have any minterm pair common with either x or y. Therefore z can be generated by a combination like {6, 8} and {2, 4} or {6, 2} and {8, 4} or {6, 4} and {8, 2}. Considering the minterm pairs {6, 8} and {2, 4} for z, for example, the minimal minterm pairs required for the encoder become {5, 6}, {7, 8}, {3, 4}, {2, 4} and {6, 8}. Thus five minterm pairs are sufficient to realize the encoder, and accordingly a 2 × 3 spatial matrix given by

$$I = \begin{bmatrix} I_{11} & I_{12} & I_{13} \\ I_{21} & I_{22} & I_{23} \end{bmatrix}$$

will be required for encoding the input overlap pixel. The five input conditions are mapped into the input overlap matrix in such a way that the size of the mapping block is rectangular with minimum perimeter. The address maps for the individual inputs and the overall address map are shown in Fig. 2, where pixel subcells (11), (12), (13), (21), and (23) house the minterm pairs {7, 8}, {5, 6}, {2, 4}, {3, 4}, and {6, 8}, respectively. For the simultaneous generation of the outputs and the necessary overlap of the inputs in the output overlap plane, a 2 × 2 pixel matrix given by

$$L = \begin{bmatrix} L_{11} & L_{12} \\ L_{21} & L_{22} \end{bmatrix}$$

is required for the LED source pattern. Thus the output overlap pattern will consist of a 3 × 4 (since L is a 2 × 2 and I is a 2 × 3 matrix) spatial pattern:

$$D = \begin{bmatrix} D_{11} & D_{12} & D_{12} \\ D_{21} & D_{22} & D_{23} \\ D_{31} & D_{32} & D_{33} \end{bmatrix}$$

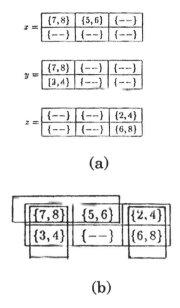

$$x = \boxed{\begin{array}{|c|c|c|} \hline \{7,8\} & \{5,6\} & \{--\} \\ \hline \{--\} & \{--\} & \{--\} \\ \hline \end{array}}$$

$$y = \boxed{\begin{array}{|c|c|c|} \hline \{7,8\} & \{--\} & \{--\} \\ \hline \{3,4\} & \{--\} & \{--\} \\ \hline \end{array}}$$

$$z = \boxed{\begin{array}{|c|c|c|} \hline \{--\} & \{--\} & \{2,4\} \\ \hline \{--\} & \{--\} & \{6,8\} \\ \hline \end{array}}$$

(a)

$$\begin{array}{|c|c|c|} \hline \{7,8\} & \{5,6\} & \{2,4\} \\ \hline \{3,4\} & \{--\} & \{6,8\} \\ \hline \end{array}$$

(b)

Figure 2 Overall address map for the input overlap pixel.

Assume that the outputs x, y, and z be detected at subcells D_{12}, D_{21}, and D_{23}, respectively. The corresponding projection equations are

$$L_{21} \wedge I_{11} + L_{21} \wedge I_{12} = D_{12} \tag{1}$$

$$L_{22} \wedge I_{21} + L_{12} \wedge I_{11} = D_{21} \tag{2}$$

$$L_{11} \wedge I_{13} + L_{21} \wedge I_{23} = D_{13} \tag{3}$$

where $\wedge$ is the overlap operation [2]. Using Eq. (1), for example, the POSC logic equations corresponding to minterms 7 and 8 are obtained as

$$L_{21} \wedge \{\bar{a}_{11} \wedge \bar{b}_{11} \wedge \bar{c}_{11} \wedge \bar{d}_{11} \wedge \bar{e}_{11} \wedge \bar{f}_{11} \wedge g_{11} \wedge \bar{h}_{11}\} = H \tag{4}$$

$$L_{21} \wedge \{\bar{a}_{11} \wedge \bar{b}_{11} \wedge \bar{c}_{11} \wedge \bar{d}_{11} \wedge \bar{e}_{11} \wedge \bar{f}_{11} \wedge g_{11} \wedge h_{11}\} = V \tag{5}$$

where minterms 7 and 8 are assigned orthogonal codes (H and V), since they share the same input overlap pixel subcell. Solving these two equations simultaneously, the POSC codes for the different elements are found be $L_{21} = T$, $a_{11} = b_{11} = c_{11} = d_{11} = e_{11} = f_{11} = F$ and $g_{11} = H$, $h_{11} = V$. Following the identical procedure, the POSC equations corresponding to the other minterm pairs can be obtained and solved. Consequently, the final coding for the LED source pattern and input overlap pattern becomes

$$L = \begin{bmatrix} T & T \\ T & T \end{bmatrix}$$

$$a = \begin{bmatrix} F & F & F \\ F & F & F \end{bmatrix} \quad b = \begin{bmatrix} F & F & H \\ F & F & F \end{bmatrix} \quad c = \begin{bmatrix} F & F & F \\ H & F & F \end{bmatrix} \quad d = \begin{bmatrix} F & F & V \\ V & F & F \end{bmatrix}$$

$$e = \begin{bmatrix} F & H & F \\ F & F & F \end{bmatrix} \quad f = \begin{bmatrix} F & V & F \\ F & F & F \end{bmatrix} \quad g = \begin{bmatrix} H & F & F \\ F & F & H \end{bmatrix} \quad h = \begin{bmatrix} V & F & F \\ F & F & V \end{bmatrix}$$

3.2 Design II

The 8-to-3 line encoder is designed next according to the algorithm of Ref. 15. For simultaneous generation of the outputs at the decoding plane and the necessary overlap of the inputs at the output overlap plane, each of the inputs is represented by a 3×3 pixel matrix, and the LED source pattern is represented by a 3×2 pixel matrix. Thus the output overlap pattern will consist of a 5×4 spatial pattern. Assume that the outputs x, y, and z be detected at subcells D_{22}, D_{33}, and D_{44}, respectively. The projection equations corresponding to these pixel subcells are given by

$$L_{21} \wedge I_{11} + L_{22} \wedge I_{12} + L_{31} \wedge I_{21} + L_{32} \wedge I_{22} = D_{22} \tag{6}$$

$$L_{11} \wedge I_{12} + L_{12} \wedge I_{13} + L_{21} \wedge I_{22} + L_{22} \wedge I_{23} = D_{33} \tag{7}$$

$$L_{11} \wedge I_{23} + L_{21} \wedge I_{33} = D_{44} \tag{8}$$

where the minterm pairs {7, 8}, {5, 6}, {3, 7}, {4, 8}, {2, 4}, and {6, 8} have been housed at the pixel subcells (11), (12), (13), (22), (23), and (33) of the input overlap pattern I. The POSC equations corresponding to this pixel assignment can be obtained from Eqs. (6–8). Upon solving these POSC equations, the LED source pattern and input encoding schemes are found to be

$$L = \begin{bmatrix} T & T \\ T & T \\ F & T \end{bmatrix}$$

$$a = \begin{bmatrix} F & F & F \\ F & F & F \\ F & F & F \end{bmatrix} \quad b = \begin{bmatrix} F & F & F \\ F & F & H \\ F & F & F \end{bmatrix}$$

$$c = \begin{bmatrix} F & F & H \\ F & F & F \\ F & F & F \end{bmatrix} \quad d = \begin{bmatrix} F & F & F \\ F & H & V \\ F & F & F \end{bmatrix}$$

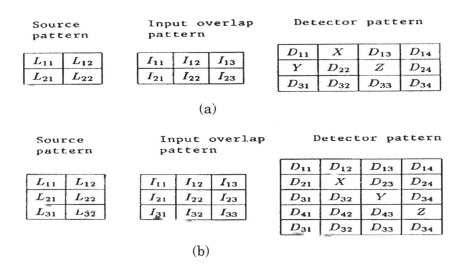

Figure 3 shown at top.

Source pattern

L_{11}	L_{12}
L_{21}	L_{22}

Input overlap pattern

I_{11}	I_{12}	I_{13}
I_{21}	I_{22}	I_{23}

Detector pattern

D_{11}	X	D_{13}	D_{14}
Y	D_{22}	Z	D_{24}
D_{31}	D_{32}	D_{33}	D_{34}

(a)

Source pattern

L_{11}	L_{12}
L_{21}	L_{22}
L_{31}	L_{32}

Input overlap pattern

I_{11}	I_{12}	I_{13}
I_{21}	I_{22}	I_{23}
I_{31}	I_{32}	I_{33}

Detector pattern

D_{11}	D_{12}	D_{13}	D_{14}
D_{21}	X	D_{23}	D_{24}
D_{31}	D_{32}	Y	D_{34}
D_{41}	D_{42}	D_{43}	Z
D_{31}	D_{32}	D_{33}	D_{34}

(b)

Figure 3 POSC implementation of the source, the input, and the detector pattern using (a) design I and (b) design II.

$$e = \begin{bmatrix} F & H & F \\ F & F & F \\ F & F & F \end{bmatrix} \quad f = \begin{bmatrix} H & F & V \\ F & F & F \\ F & F & H \end{bmatrix}$$

$$g = \begin{bmatrix} H & F & V \\ F & F & F \\ F & F & F \end{bmatrix} \quad h = \begin{bmatrix} V & F & F \\ F & V & F \\ F & F & V \end{bmatrix}$$

The source, the input overlap, and the detector patterns corresponding to design I and design II are shown in Fig. 3a and Fig. 3b, respectively. From these two designs, we observe that the modified algorithm is significantly more efficient when compared to the algorithm presented in Ref. 15. For example, an 8-to-3 line encoder, using the modified algorithm, requires 33% fewer LEDs in the source pattern, 33% less space for spatial encoding of the inputs, and 40% less space for the detector pattern.

3.3 Conclusion

A modified algorithm (an extension of the algorithms presented in Refs. 2 and 15) for the design of POSC based logic circuits is presented. Applying this algorithm, we designed an 8-to-3 line encoder. The same circuit is also designed using the most recently published design technique [15]. The modified design algorithm is found to be more efficient than the other schemes in terms of both the number of LEDs and the spatial encoding size for the input overlap and detector patterns. The saving of space in pixel subcell area is very important in POSC based sequential circuits, since relatively expensive spatial light modulators are expected to be used in encoding the inputs. Since the proposed algorithm requires less space for encoding the inputs and detector pattern, the resultant logic design is cost-effective. Also, the modified algorithm is especially suitable for logic circuits having large numbers of inputs and outputs.

4 POSC: TWO-DIMENSIONAL REGISTER DESIGN

Although the POSC technique has been widely used for the design of combinational logic circuits, its application for the design of sequential logic circuits has been minimal. To date, the design of only one POSC based sequential logic unit, a J-K flip-flop (FF) [12], has been reported in the literature. Since the POSC technique possesses some attractive advantages when compared to the other optical implementation techniques, it is important to investigate the application of this technique for the design of sequential logic circuits. In this section, we investigate the application of the POSC technique in the design of a multibit optical register. Applying the POSC algorithm of Sec. 2, we design a one-input/one-output toggle FF, two-input/two-output S-R and J-K FFs, and then a two-dimensional n^2-bit optical register is designed using toggle FFs. Notice that the application of the proposed algorithm becomes especially important when designing multiinput/multioutput sequential logic units such as registers.

4.1 Flip-Flop Design

The building block of digital sequential circuits is a FF. A FF may be considered as a 1-bit memory device that remembers its current state even after its inputs have been withdrawn [24]. An internal path connects the FF output to its input, thus making the FF next state a function of its present state. Thus, a FF may also be considered as a two-input (as in toggle FF or D FF) or a three-input (J-K FF) device. In this section we first consider the design of a toggle FF, then an S-R FF, and finally a J-K FF.

4.1.1 Toggle Flip-Flop

The two particular inputs of the toggle FF are the control input (T') and the feedback input (Q). The Q output of the FF represents the current state. Therefore

Table 3 Truth Table for the Toggle Flip-Flop

Condition number	Input T'	Present state Q'	Next state $Q^{t+\Delta t}$
1	0	0	0
2	0	1	1
3	1	0	1
6	1	1	0

the truth table of the toggle FF as shown in Table 3 has two inputs, T' and Q', and one output, $Q^{t+\Delta t}$. In Table 3, Q' corresponds to the current state and $Q^{t+\Delta t}$ corresponds to the next state of the FF. There are only two nonzero outputs corresponding to the input conditions 2 and 3, and therefore a 1×2 pixel matrix can be used for the inputs I. Since there are two inputs, two LEDs (represented by $L = [L_{11} \, L_{12}]$) are sufficient to illuminate the input overlap pattern (I) and generate the necessary overlap at the output plane. The output overlap pattern D will consist of three pixel subcells. Assume that the output Q will be detected at pixel subcell D_{12} of the detector. The projection equation for D_{12} becomes

$$L_{11} \wedge I_{11} + L_{12} \wedge I_{12} = D_{12} \tag{9}$$

Let the minterms 2 and 3 be represented by the input overlap pixel subcells $\{11\}$ and $\{12\}$, respectively. Using unpolarized codes (T for 1 and F for 0) for the input overlap pixels and a T code for D_{12}, the POSC overlap equations for the aforementioned outputs become

$$L_{11} \wedge (\overline{T}'_{11} \wedge Q_{11}) = T \tag{10}$$

$$L_{12} \wedge (T'_{12} \wedge \overline{Q}_{12}) = T \tag{11}$$

Solving these two equations, the codings for the source and input overlap patterns are given by

$$L = [T \, T], \, T' = [F \, T], \, Q = [T \, F] \tag{12}$$

The pixel subcell requirement can be further reduced by assigning the minterms corresponding to the two nonzero outputs to a single pixel. This can be achieved by using the POSC codes H and V for the input conditions two and three, respectively. Thus only one unpolarized LED in the source pattern would be sufficient for this system. Therefore only one output overlap pixel will be generated, and the corresponding projection equation becomes

$$L \wedge I = D \tag{13}$$

Assigning T code for the output detector pixel, the POSC equations corresponding to the aforementioned input conditions reduce to

$$\bar{T}' \wedge Q = H \tag{14}$$

$$T' \wedge \bar{Q} = V \tag{15}$$

Thus the overall coding for the source and input overlap patterns becomes

$$L = [T] \qquad T' = [V] \qquad Q = [H] \tag{16}$$

Following an identical design procedure, we observe that a delay FF can be realized by only one input overlap pixel and one source LED.

4.1.2   S-R/J-K Flip-Flop

The truth table of the S-R FF is shown in Table 4, which shows that there are three nonzero outputs corresponding to input conditions 2, 5, and 6, respectively. Since three minterms are necessary for realizing an S-R FF, at least two pixel subcells are required in the input overlap pixel pattern. Assigning the minterms 2 and 6 to the input overlap pixel subcell {11} and minterm 5 to input overlap pixel subcell {12}, we observe that two LEDs are required in the source plane to generate the necessary overlap of input pattern shadows at the output plane. Assuming that the output Q will be detected at pixel subcell D_{12} of the detector, the projection equation becomes similar to Eq. (9). Assigning a V code to minterm 2 and an H code to minterm 6, the POSC overlap equations for the minterm pair (2, 6) becomes

$$L_{11} \wedge (\bar{S}_{11} \wedge \bar{R}_{11} \wedge Q_{11}) = V \tag{17}$$

$$L_{11} \wedge (S_{11} \wedge \bar{R}_{11} \wedge Q_{11}) = H \tag{18}$$

Table 4   Truth Table for the S-R Flip-Flop

Condition number	Inputs		Present state	Next state
	S	R	Q^t	$Q^{t+\Delta t}$
1	0	0	0	0
2	0	0	1	1
3	0	1	0	0
4	0	1	1	0
5	1	0	0	1
6	1	0	1	1
7	1	1	x	x
8	1	1	x	x

Similarly, the POSC equation corresponding to the remaining minterm 5 can be written as

$$L_{12} \wedge (S_{12} \wedge \overline{R}_{12} \wedge \overline{Q}_{12}) = T \tag{19}$$

Solving Eqs. (17–19), the overall coding for the source and input overlap patterns becomes

$$L = [T\ T] \qquad S = [H\ T] \qquad R = [F\ F] \qquad Q = [T\ F] \tag{20}$$

Following similar design procedure, we may design a J-K FF with two 1×2 input overlap pixels and two source LEDs. The overall coding (using polarized codes) for the source and input overlap patterns of the J-K FF using the proposed algorithm is given by

$$L = [T\ T] \qquad J = [H\ T] \qquad K = [F\ H] \qquad Q = [V\ V] \tag{21}$$

Using such POSC based toggle, delay, S-R, and J-K FFs as basic building blocks, complex sequential circuits such as POSC registers and counters can be designed. Thus the proposed POSC algorithm can be used for the implementation of sequential POSC architectures.

4.1.3   Optical Implementation

An optical implementation of the toggle FF is shown in Fig. 4. The encoding of the feedback input Q needs to be changed to assure that the output of the FF changes. Again, the T' input needs to be changed synchronously to generate the next state of the FF. The encoder changes the coding of T' and Q input patterns. The feedback connection from the detector to the encoder can be established optically (or electronically) in the form of light (1) or no light (0). A new output value can be transmitted in nearly real time from the detector to the input overlap plane. To maintain a steady output, the LED is always left turned on to generate the output from the present value of Q. However, when $T' = 0$, the coding of T' and Q inputs remains unchanged, thus keeping the output the same. Notice that the optical implementation will be the same for a delay FF. However, for an S-R or J-K FF, the source and the input overlap patterns will have to be replaced by 1×2 patterns.

4.2   Optical Register Design

Registers are relatively complex sequential circuits (built using FFs) capable of storing several bits of information having a wide range of applications. For example, we may consider the design of an optical register using the toggle FF developed in the previous section. The registers can be parallel-in parallel-out (PIPO), serial-in parallel-out (SIPO), parallel-in serial-out (PISO), or serial-in serial-out

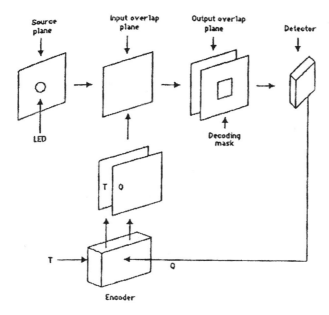

Figure 4 Optical implementation of a toggle flip-flop.

(SISO) types. Among these four categories, the PIPO registers, which can be used to store or retrieve all bits of a word simultaneously, are especially suitable for optical implementation since such circuits can utilize fully the massive parallelism of light.

4.2.1 Design I

The architecture of a two-dimensional optical register is shown in Fig. 5, where the input overlap pixel pattern consists of $n \times n$ pixel subcells. Since only one pixel subcell is required for a toggle FF, such a register may consist of n^2 FFs given by the following matrix:

$$
FF_r = \begin{bmatrix}
FF_{11} & FF_{12} & FF_{13} & \ldots & FF_{1n} \\
FF_{21} & FF_{22} & FF_{23} & \ldots & FF_{2n} \\
FF_{31} & FF_{32} & FF_{33} & \ldots & FF_{3n} \\
\ldots & \ldots & \ldots & \ldots & \ldots \\
FF_{n1} & FF_{n2} & FF_{n3} & \ldots & FF_{nn}
\end{bmatrix}
$$

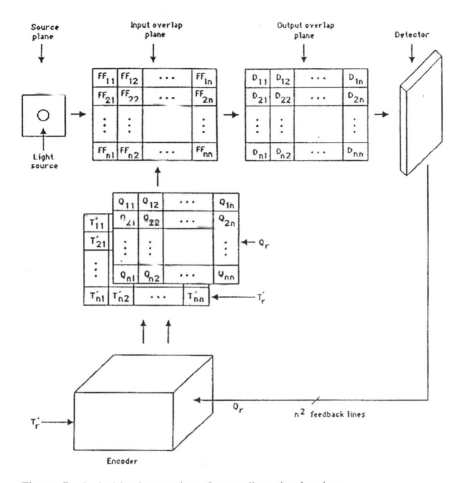

Figure 5 Optical implementation of a two-dimensional register.

Accordingly, n^2 bits of information can be processed in parallel. Now, each of the two inputs (T'_r and Q_r) of this register consists of n^2 input lines, which can be represented by the following matrices:

$$T'_r = \begin{bmatrix} T'_{11} & T'_{12} & T'_{13} & \cdots & T'_{1n} \\ T'_{21} & T'_{22} & T'_{23} & \cdots & T'_{2n} \\ T'_{31} & T'_{32} & T'_{33} & \cdots & T'_{3n} \\ \cdots & \cdots & \cdots & \cdots & \cdots \\ T'_{n1} & T'_{n2} & T'_{n3} & \cdots & T'_{nn} \end{bmatrix}$$

and

$$
Q_r = \begin{bmatrix}
Q_{11} & Q_{12} & Q_{13} & \cdots & Q_{1n} \\
Q_{21} & Q_{22} & Q_{23} & \cdots & Q_{2n} \\
Q_{31} & Q_{32} & Q_{33} & \cdots & Q_{3n} \\
\cdots & \cdots & \cdots & \cdots & \cdots \\
Q_{n1} & Q_{n2} & Q_{n3} & \cdots & Q_{nn}
\end{bmatrix}
$$

Each element of the FF_r matrix is related to the corresponding elements of the T'_r and Q_r matrices by the following equation:

$$FF_r = T'_r \circ Q_r$$

where $\circ$ represents an AND operation. The spatial encoding of the inputs T'_r and Q_r are performed in the encoder by means of two polarization-sensitive spatial light modulators (SLMs). The procedure for encoding each pixel subcell of the inputs is the same as developed for the toggle FF in Sec. 4.1. The coded input patterns are kept in perfect contact at the input overlap plane, which is illuminated by an LED, and the outputs are detected in the output plane by an $n \times n$ detector array (e.g., CCD detector array). Since the next state $Q_r^{t+\Delta t}$ of the register is a function of its present state Q_r^t, the present state inputs (i.e., the feedback inputs) are fed back to the encoder from the detector array either optically or electronically.

4.2.2 Design II

In the previous design, the following problems are encountered:

While it is very important to keep the coded input patterns in intimate contact in the input overlap plane, it may become especially difficult since the input encoding needs to be changed frequently.

Since the capacity of a detector for simultaneous detection of outputs is limited (typically 1 byte or 1 word at a time), it may necessitate the feedback process to become partially sequential in nature. Use of multiple detector arrays may solve the problem, but this will increase cost, system size, and synchronization problems among the different detector arrays.

To alleviate these problems, we propose an alternative optical implementation for the register as shown in Fig. 6 that uses an additional SLM such as the liquid crystal light value (LCLV) [16] in the feedback path. The FF outputs (Q_r) are passed through a beam splitter, and the output beam passing straight through the beam splitter is detected by single/multiple detector arrays. The other light beam available from the beam splitter is fed directly to the LCLV, which func-

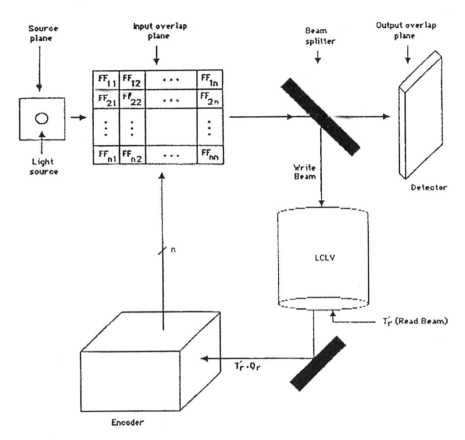

Figure 6   Optical implementation of a two-dimensional register using LCLV.

tions as the write beam. The T'_r input beam is applied to the LCLV in the form of a readout beam. The LCLV performs an AND operation between the T'_r and Q_r inputs and produces the FF_r output, which is then sent to the encoder using a mirror. An SLM in the encoder encodes the FF_r input, and the coded pattern is placed in the input overlap plane to produce the next state of the register. Since there is only one input pattern in the input overlap plane, the problem of keeping inputs in perfect contact does not arise in this design. Also, the problem of detector inadequacy is entirely eliminated, since the feedback inputs are taken from the beam splitter instead of the detector. This design may yield better results with an intense spatially coherent light source. Since LEDs have limited intensity and are incoherent, a He-Ne laser may be used for this purpose.

Notice that in this design, the output light is fed back to the encoder using free-space communication by means of a mirror. However, accurate alignment of the mirror will be essential for successful operation of the system. Alternatively, an optical fiber cable may be used. With optical fibers remote operation is possible [26]. However, a microscopic objective lens will then be needed to launch the light into the fiber. The main problem of this design is the relatively slow speed of the LCLV. However, with continual advancement of technology, this speed limitation of LCLV is expected to be overcome. Alternatively, to speed up the feedback process an optically addressed SLM (OASLM) [27] may be used instead of the LCLV. In this case, the T_r signal is used to control each pixel of the OASLM optically. As soon as the Q_r beam coming from the beam splitter is available, the FF_r signal is generated at the output of the OASLM, thereby avoiding the processing speed bottleneck of the LCLV.

4.3 Conclusion

Using the generalized algorithm for the design of POSC, sequential circuits are developed. We presented two designs for a toggle FF. The second design using polarization codes is more efficient than the first using unpolarized codes in terms of the number of LEDs and the size of the input patterns. Using polarization codes, a multi-input/multioutput S-R/J-K FF is also designed. Based on an array of toggle flip-flops (using the second design), two different optical implementations of two-dimensional optical register are proposed. Of the two designs, the second design using an LCLV is more efficient and cost-effective. Since LCLVs are relatively slow, the system speed may be much improved by replacing the LCLV with an OASLM.

The main problems associated with electronic digital circuits are the fan-in/fan-out limitation, nonimmmunity to noise, interconnection (especially with large systems), and clock dependency. In comparison, the POSC based schemes provide fast, synchronous, and massively parallel system. In this paper, we designed a two-dimensional optical register using only toggle FFs. Optical registers using other kinds of FFs such as the S-R FF or the J-K FF can also be implemented using a similar scheme. For example, if S-R (J-K) FFs were used for the register developed in Sec. 4.2, the resulting design would contain $(n^2/2)$ FFs having $(n^2/2)$ S (J), $(n^2/2)$ R (K), and $(n^2/2)$ Q inputs. This is because a 1×2 pixel matrix will be required for the implementation of each S-R (J-K) FF. Notice that the POSC realization of the two-dimensional optical register employs an all-optical implementation. The operational speed of the two-dimensional register is the same as the operational speed of a single FF. It may be mentioned that this is the first optical register design reported using the POSC technique.

5 POSC: DESIGN OF MULTIPLE-VALUED LOGIC BASED MULTIPROCESSOR

Multiple instruction and multiple data (MIMD) type multiprocessor computing architectures are desirable in digital computing systems to achieve massive parallelism and high speed. Binary logic based systems have reached their present level of complexity and physical limit. There is an ongoing search for new optical architectures and algorithms to employ the parallelism, speed, and noninterferring communication of optics. The performance of an MIMD multiprocessor can be enhanced many times by using multiple-valued logic [28], which is capable of increasing both effective data rates and storage density as well as reducing interconnection requirements and circuit complexity. However, if the number of logic levels are increased, more elementary values will have to be distinguished, and so it may become more sensitive to noise. Among the multiple valued logic systems, however, trinary logic appears to be the most promising in terms of the number of logic levels and storage complexity [29].

To implement multiple-valued logic, using POSC, for a trinary logic system, the three logical values P, Q, and R must be orthogonal to each other. Thus the possible overlap of the trinary logical values are given by $P \wedge Q \wedge R = F$, $P \wedge Q = F$, $Q \wedge R = F$, and $R \wedge P = F$, where $\wedge$ represents an overlap operation [2] and P, Q, and R are members of the POSC code set $\{T, F, V, H\}$. Accordingly, for a trinary logic system, the orthogonal variable set is allowed to have any one of the three code combinations $\{T, F, F\}$, $\{V, H, F\}$, and $\{F, F, F\}$. Karim et al. [2] presented a generalized algorithm for the design of POSC logic circuits. Applying this algorithm, a number of multioutput binary/trinary combinational circuits [7–9,29] had been already designed. By extending this algorithm, it was possible to design a J-K flip-flop [12] as well as a binary logic based multiprocessor [14]. This multiprocessor is capable of performing concurrent binary addition and subtraction. To exploit the potential of multiple-valued logic in optical computing, in this section, we extend the scope of Ref. 14 by applying the POSC algorithm of Sec. 2 for the design of a multiple-valued logic based optical multiprocessor. As an illustration, we furnish the design of a trinary logic based multiprocessor that can perform parallel trinary full addition and full subtraction. The POSC algorithm presented in this paper uses fixed source and detector patterns (which was a limitation of the algorithm of Ref. 2) and yields a minimized design, an improvement over the POSC algorithm presented in Ref. 14.

5.1 Multiprocessor Design

This section deals with the design of a trinary multiprocessor using the algorithm of Sec. 2. Consider that the to-be-designed multiprocessor, for example, is capa-

Table 5 Truth Table for the Trinary POSC Logic Based Multiprocessor

Condition number	Inputs			Adder outputs		Subtracter outputs	
	X	Y	Z	S	C	D	S
1	0	0	0	0	0	0	0
2	0	0	1	1	0	2	1
3	0	1	0	1	0	2	1
4	0	1	1	2	0	1	1
5	1	0	0	1	0	1	0
6	1	0	1	2	0	0	0
7	1	1	0	2	0	0	0
8	1	1	1	0	1	2	1
9	0	2	0	2	0	1	1
10	0	2	1	0	1	0	1
11	2	0	0	2	0	2	0
12	2	0	1	0	1	1	0
13	2	2	0	1	1	0	0
14	2	2	1	2	1	2	1
15	1	2	0	0	1	2	1
16	2	1	0	0	1	1	0
17	1	2	1	1	1	1	1
18	2	1	1	1	1	0	0

ble of performing parallel trinary addition and subtraction. The corresponding truth table is shown in Table 5. The inputs to the multiprocessor are X, Y, and Z, representing the addend (minuend), augend (subtrahend) and carry-in (borrow-in), respectively, and the outputs are S, C, D, and B, representing the sum, carry-out, difference, and borrow-out respectively. Since there are three input conditions, three input patterns need to be introduced in the input overlap plane. From Table 5, we observe that the outputs S and D correspond each to 12 input conditions, i.e., minterms, while the outputs C and D correspond each to 9 minterms, as shown in the following list:

$$S = \{2^1, 3^1, 4^2, 5^1, 6^2, 7^2, 9^2, 11^2, 13^2, 14^2, 17^1, 18^2\}$$
$$C = \{8^1, 10^1, 12^1, 13^1, 14^1, 15^1, 16^1, 17^1, 18^1\}$$
$$D = \{2^2, 3^2, 4^1, 5^1, 8^2, 9^1, 11^2, 12^1, 14^2, 15^2, 16^1, 18^1\}$$
$$B = \{2^1, 3^1, 4^1, 8^1, 9^1, 10^1, 14^1, 15^1, 17^1\}$$

where an entry m^n indicates that the output of the condition number (i.e., minterm) m ($1 \leq m \leq 18$) in Table 5 is n ($1 \leq n \leq 2$). For example, the output corresponding to minterm 2 of Table 5 for the D output is trinary 2, which is represented by the first entry 2^2 of D in the above list.

Assume the trinary numbers 0, 1, and 2 to be represented by the POSC codes F, H, and V, respectively. The minimal number of minterm pairs can be obtained either by exhaustive search or by ISO/IMO technique. The minterms are paired in such a way that they have complementary POSC codes. Since there are 12 (9) minterms for the S and D (C and B) outputs, a total of 66 (36) minterm pairs are possible, and each output can be obtained in 11 (9) different ways. Applying the ISO/IMO technique, the output having ISO entries (in this case, only one) is

$$S = \{6^2, 7^2\}$$

while the rest of the outputs do not have any minterm in this category. The outputs having IMO entries are

$$S = \{2^1, 3^1, 4^2, 5^1, 9^2, 11^2, 13^2, 14^2, 17^1, 18^1\}$$
$$C = \{8^1, 10^1, 12^1, 13^1, 14^1, 15^1, 16^1, 17^1, 18^1\}$$
$$D = \{2^2, 3^2, 4^1, 5^1, 8^2, 9^1, 11^2, 12^1, 14^2, 15^2, 16^1, 17^1\}$$
$$B = \{2^1, 3^1, 4^1, 8^1, 9^1, 10^1, 14^1, 15^1, 17^1\}$$

In the ISO category, each of the minterms 6^2 and 7^2 belonging to S has an output value of trinary 2, so no minterm pair can be formed of these two input conditions. From the IMO category, however, we observe that the minterm pairs $\{5^1, 11^2\}$ and $\{14^2, 17^1\}$ are common between the outputs S and D. Note that the minterm pairs $\{2, 4\}$ and $\{3, 9\}$, though common between S and D, have complementary outputs, so S and D outputs cannot share the minterm pairs $\{2, 4\}$ and $\{3, 9\}$. The remaining minterm pairs of S are different from those of D. Considering the B and C outputs, we see from Table 5 that these two outputs have either 0 or 1 as their output values. This implies that if we assign F code for 0, then either H or V can be assigned for 1. Accordingly, by assigning appropriate POSC codes, the minterm pairs $\{2^1, 4^1\}$, $\{3^1, 9^1\}$, and $\{14^1, 17^1\}$ of B can be shared with either S or D. Note that if the minterm pairs of B are taken as $\{2^1, 3^1\}$ and $\{4^1, 9^1\}$, they cannot be shared with either S or D, because the input conditions in each minterm pair of S or D will have the same POSC output code. Again, minterm pairs $\{8^1, 10^1\}$ and $\{14^2, 15^2\}$ can be shared between B and C. The remaining minterms are paired arbitrarily for each output, since no more suitable sharing of minterms is possible among the desired outputs. Thus the minterm assignment for the outputs may become

$$S = [\{2^1, 4^2\}, \{3^1, 9^2\}, \{5^1, 11^2\}, \{14^2, 17^1\}, \{7^2, 13^1\}, \{6^2, 18^1\}]$$
$$C = [\{8^1, 10^1\}, \{12^1, 13^1\}, \{14^1, 15^1\}, \{16^1, 18^1\}, \{12^1, 17^1\}]$$
$$D = [\{2^2, 4^1\}, \{3^2, 9^1\}, \{5^2, 11^1\}, \{14^2, 17^1\}, \{8^2, 12^1\}, \{15^1, 16^1\}]$$
$$B = [\{2^1, 4^1\}, \{3^1, 9^1\}, \{8^1, 10^1\}, \{14^1, 15^1\}, \{14^1, 17^1\}]$$

which shows that 22 minterms are required to generate the required outputs. Since three minterms are shared between B and S, two minterms are shared between B and C, and S and D and one minterm among B, D, and S, the total number of subpixels required to generate the four outputs is 15. Notice that all the minterms of B are shared with outputs S and C. From this minterm assignment, we observe that to produce S or D (B or C) output in the output overlap plane, an overlap of six (five) minterm pairs is required. This means that the input overlap pattern should be illuminated so that it produces the overlap of the appropriate input overlap pattern subpixels into the same subpixel of the output overlap pattern to produce the desired output. Since the input conditions must be mapped into the input overlap matrix using a rectangular mapping block with minimum perimeter, a 3×2 or 2×3 pixel submatrix in the input overlap pattern is required for each output as shown in Fig. 7a by the pixel matrices (of minterm pairs) for each output.

The overall input overlap pixel matrix I is formed by mapping the pixel

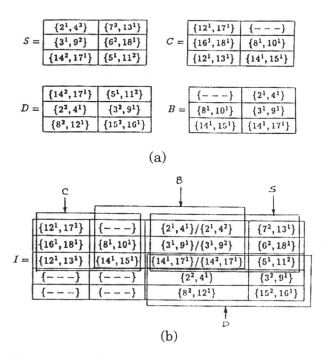

(a)

(b)

Figure 7 (a) Pixel matrix for each output depicting the spatial allocation of the minterm pairs required to generate the corresponding output and (b) overall pixel matrix for the input overlap pixel required to generate all the output simultaneously at the output plane.

matrices of Fig. 7a for all outputs into a single pattern with minimal spatial requirement as shown in Fig. 7b. It is interesting to note that the minterm pair {14, 17} can be shared by all outputs. But from Fig. 7b, we observe that this particular subpixel is shared among S, D, and B. If C is allowed to share this subpixel, then additional subpixels of C must be shared with other outputs, which is not possible in this design. Another attractive temptation would be to share subpixel I_{32} among the outputs B, C, and D, thereby reducing the size of I matrix from 5×4 to 4×4 (since the dimension D will change from 3×2 to 2×3, thereby releasing the bottom row of the I matrix). From the I matrix of Fig. 7b, however, we observe that only minterm 15 can be shared among these outputs. Therefore if I_{32} is asigned to minterm 15, then additional space will be required for minterm 16 of output D (refer to the pixel matrix of D). Also with this arrangement additional LEDs will be required at the source LED pattern, since overlap of both 2×3 and 3×2 pixel submatrices will be required in the output plane to generate the outputs. Thus to design the proposed multiprocessor, a 5×4 spatial matrix given by

$$
I = \begin{bmatrix} I_{11} & I_{12} & I_{13} & I_{14} \\ I_{21} & I_{22} & I_{23} & I_{24} \\ I_{31} & I_{32} & I_{33} & I_{34} \\ I_{41} & I_{42} & I_{43} & I_{44} \\ I_{51} & I_{52} & I_{53} & I_{54} \end{bmatrix}
$$

will be required for encoding the input overlap pixel for each input. To produce necessary overlap of the inputs in the output overlap plane for simultaneous generation of the outputs, a 3×2 (or 2×3) pixel matrix will be required for the source LED pattern:

$$
L = \begin{bmatrix} L_{11} & L_{12} \\ L_{21} & L_{22} \\ L_{31} & L_{32} \end{bmatrix}
$$

Thus the resulting output overlap pattern will consist of a 7×5 spatial pattern:

$$
D = \begin{bmatrix} D_{11} & D_{12} & D_{13} & D_{14} & D_{15} \\ D_{21} & D_{22} & D_{23} & D_{24} & D_{25} \\ D_{31} & D_{32} & D_{33} & D_{34} & D_{35} \\ D_{41} & D_{42} & D_{43} & D_{44} & D_{45} \\ D_{51} & D_{52} & D_{53} & D_{54} & D_{55} \\ D_{61} & D_{62} & D_{63} & D_{64} & D_{65} \\ D_{71} & D_{72} & D_{73} & D_{74} & D_{75} \end{bmatrix}
$$

In general, each subpixel of the source pattern will produce a shadow of the whole input overlap pattern on the output overlap pattern. This means that the output of each subpixel of the D matrix is determined by one or more subpixels of the L matrix after having overlapped with the I matrix. The innerside subpixels of the D matrix are influenced by more source LEDs when compared to the subpixels located near the boundary. For example, in the present design, the outputs of the D_{11}, D_{15}, D_{71}, and D_{75} subpixels are determined by only one of the L matrix subpixels L_{32}, L_{31}, L_{12}, and L_{11}, respectively. However, the innerside subpixels of the D matrix such as D_{32} are influenced by four source pattern subpixels, namely, L_{22}, L_{22}, L_{31}, and L_{32}. Considering another interior subpixel D_{33}, we observe that all six subpixels of the source pattern will influence this subpixel, and the whole source pattern and its overlap with the I matrix subpixels will be mapped into subpixel D_{33}.

Assume that the outputs C, B, S, and D are detected, for example, at the output overlap pattern (detector pattern) subpixels D_{32}, D_{33}, D_{34}, and D_{54} respectively. Then the projection equations corresponding to these subpixels become:

$$L_{11} \wedge I_{11} + L_{12} \wedge I_{12} + L_{21} \wedge I_{21} + L_{22}$$
$$\wedge I_{22} + L_{31} \wedge I_{31} + L_{32} \wedge I_{32} = D_{32} \quad (22)$$

$$L_{11} \wedge I_{12} + L_{12} \wedge I_{13} + L_{21} \wedge I_{22} + L_{22}$$
$$\wedge I_{23} + L_{31} \wedge I_{32} + L_{32} \wedge I_{33} = D_{33} \quad (23)$$

$$L_{11} \wedge I_{13} + L_{12} \wedge I_{14} + L_{21} \wedge I_{23} + L_{22}$$
$$\wedge I_{24} + L_{31} \wedge I_{33} + L_{32} \wedge I_{34} = D_{34} \quad (24)$$

$$L_{11} \wedge I_{33} + L_{12} \wedge I_{34} + L_{21} \wedge I_{43} + L_{22}$$
$$\wedge I_{44} + L_{31} \wedge I_{53} + L_{32} \wedge I_{54} \wedge = D_{54} \quad (25)$$

Using the projection Eq. (22) through (25), the POSC logic equation corresponding to each minterm of the outputs can be developed. For illustration, consider the development of POSC equations for S. From Figs 7a and 7b, we observe that the minterm pair $\{2^1, 4^2\}$ of the S matrix is assigned to the I_{13} subpixel of the I matrix. Therefore, from Eq. (24), the POSC equations corresponding to the minterms 2^1 and 4^2 become

$$L_{11} \wedge I_{13}(2^1) = H \quad (26)$$
$$L_{11} \wedge I_{13}(4^2) = V \quad (27)$$

Using the input conditions of minterms 2 and 4 from Table 5,

$$L_{11} \wedge \{\overline{X}_{13} \wedge \overline{Y}_{13} \wedge \overline{Z}_{13}\} = H \quad (28)$$
$$L_{11} \wedge \{\overline{X}_{13} \wedge \overline{Y}_{13} \wedge \overline{Z}_{13}\} = V \quad (29)$$

Since this input overlap pixel is also shared by the minterms 2^1 and 4^1 of B, the solution of the above equations is also applicable to B. Thus the subpixel I_{13} is shared by two minterms of S and two minterms of B. Although the POSC scheme allows two minterms to share the same memory location using orthogonal codes, in the present case, four minterms are sharing the same memory location. Here, 2^1 and 4^2 of S (2^1 and 4^1 of B) are assigned orthogonal codes H and V, since they share the same spatial memory location. Upon solving these two equations simultaneously, the POSC codes for the different elements are found to be $L_{11} = T$, $\overline{X}_{13} = Z_{13} = T$, $Y_{13} = V$, $\overline{Y}_{13} = H$, and $X_{13} = \overline{\overline{X}}_{13} = \overline{\overline{Y}}_{13} = \overline{Z}_{13} = \overline{\overline{Z}}_{13} = F$, where a single (double) bar over a variable is used to indicate that the variable takes on trinary value of 0 (2). The POSC equations corresponding to the other minterm pairs can be obtained and solved simultaneously using a similar procedure. The final coding for the source LED pattern and input overlap pattern can be obtained as

$$L = \begin{bmatrix} T & T & T \\ T & T & T \end{bmatrix}$$

$$X = \begin{bmatrix} H & F & F & V \\ F & H & F & V \\ F & H & H & H \\ F & F & F & F \\ F & F & V & V \end{bmatrix} \quad \overline{X} = \begin{bmatrix} F & F & T & F \\ F & V & T & F \\ F & F & F & F \\ F & F & V & T \\ F & F & F & F \end{bmatrix} \quad \overline{\overline{X}} = \begin{bmatrix} V & F & F & H \\ T & F & F & H \\ T & V & V & V \\ F & F & F & F \\ F & F & H & H \end{bmatrix}$$

$$Y = \begin{bmatrix} F & F & V & V \\ T & H & H & H \\ F & F & F & F \\ F & F & H & V \\ F & F & V & H \end{bmatrix} \quad \overline{Y} = \begin{bmatrix} V & F & H & F \\ F & F & F & V \\ V & F & F & T \\ F & F & V & V \\ F & F & H & F \end{bmatrix} \quad \overline{\overline{Y}} = \begin{bmatrix} H & F & F & H \\ F & V & V & F \\ H & T & T & F \\ F & F & F & H \\ F & F & F & V \end{bmatrix}$$

$$Z = \begin{bmatrix} T & F & T & F \\ H & T & F & T \\ V & V & T & F \\ F & F & F & F \\ F & F & H & H \end{bmatrix} \quad \overline{Z} = \begin{bmatrix} F & F & F & T \\ V & F & H & F \\ H & H & F & T \\ F & F & F & T \\ F & F & F & F \end{bmatrix}$$

Note from Table 5 that Z can only take the values 0 and 1. Accordingly, $\overline{\overline{Z}}$ must be false. The POSC system for the proposed trinary logic based multiprocessor is shown in Fig. 8.

If both the trinary full adder and the subtracter are implemented using separate POSC logic units, then two 3×2 source patterns (i.e., 12 LEDs), sixteen 3×3 input overlap patterns (i.e., 144 subpixels), and two detector masks will be required. In comparison, the proposed multiprocessor requires one 3×2 source pattern, eight 4×5 input overlap patterns (i.e., 160 pixels) and one detector mask (refer to Fig. 8). Notice the 2×2 pixel submatrix at the lower left corner of each input overlap pattern and that subpixel positions I_{21} are always false or unused. This means that these five subpixels can be used to realize additional arithmetic/logical operations. Thus the total number of subpixels actually used for encoding the inputs of the multiprocessor is 120. If the multiprocessor (trinary adder/subtracter) is designed using the algorithm of Ref. 14, 33% more source LEDs, 20% more input overlap space, and 30% more detector space will

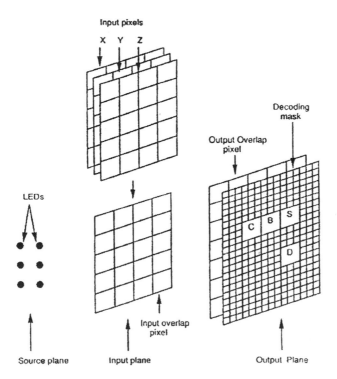

Figure 8 The POSC system for the trinary logic based multiprocessor.

be required when compared to our algorithm, thus making the system more bulky and inefficient.

5.2 Conclusion

In this section, we have presented the algorithm for the design of a multiple-valued logic based multiprocessor using POSC technique. By applying this algorithm, we have designed a trinary logic based multiprocessor that can perform trinary full addition and full subtraction in parallel. Notice that the logic circuit designed may be considered as the arithmetic unit of a multiple-valued logic based arithmetic logic unit. This algorithm is especially suitable for POSC logic circuits having large numbers of inputs and outputs, since it yields a minimized design, thereby demanding less space for encoding the inputs. Also, we have shown a technique by which four minterms can share the same memory location although traditional POSC algorithms permit only two minterms to share a particular memory location. However, this extra saving of memory space may vary from application to application. Other higher-order multiple-valued POSC logic circuits may also be designed using this algorithm.

6 POSC: AN EFFICIENT IMPLEMENTATION OF MODIFIED SIGNED-DIGIT ARITHMETIC

In an optical processor, which is inherently parallel, all the input digits simultaneously become available to the arithmetic unit that performs the arithmetic operations. When the input digits are represented using binary numbers, the computation speed decreases due to the formation and generation of carries, especially as the number of bits is increased [24,30]. The redundant signed-digit number representation [31] allows parallel arithmetic operations with a reduced number of carry propagation steps. In particular, the modified signed-digit number (MSD), which is a redundant signed-digit with radix = 2, limits the carry propagation to two positions to the left. In MSD addition, the to-be-added pair of numbers are mapped into an intermediate pair of numbers so that the latter numbers when added do not result in carry generation. The MSD arithmetic operations can be performed in parallel at all digit positions, and hence they fit elegantly with the optical computing architecture. Therefore the MSD carry-free property attracts a lot of interest to implement them optically [3,8,32–45]. In particular, optical symbolic substitution (SS) [3] and polarization-encoded optical shadow casting (POSC) techniques [2,8] have gained wide popularity recently. More recently, a new symmetrical recoding algorithm for the MSD number (SRMSD) was proposed to achieve efficient carry-free arithmetic operations [44] using optical SS. In this section, the SRMSD carry-free addition/subtraction operation is implemented using the POSC scheme with two design methods: direct implementation and truth-table partitioning.

6.1 MSD Arithmetic

The MSD number representation is a subset of the redundant signed-digit number representation with radix 2. A given decimal number D can be represented by an n-digit MSD number as $D = \sum_{j=0}^{n-1} x_i 2^j$ where the MSD variable X_i can take any value from the set $\{1, 0, \overline{1}\}$. Here $\overline{1}$ denotes -1. An MSD negative number is the MSD complement of the MSD positive number. For example, using primes to denote complementation, we have $\overline{1}' = 1$, $1' = \overline{1}$, $0' = 0$, and therefore $(-7)_{10} = [\overline{1}1\overline{1}\ \overline{1}]_{MSD}$ or, equivalently, $(-7)_{10} = [0\overline{1}\ \overline{1}\ \overline{1}]_{MSD}$. The addition (subtraction) of two MSD numbers involves three (four) steps: the first two (three) steps generate transfer and weight digits which are then summed; the third step gives the final result. The transfer and weight digits are generated as the operands flow through the processing architecture in parallel, making addition of any length operands occur in the same amount of time. In case of subtraction, however, the subtrahend is first complemented and then an addition is performed.

Exploring the redundancy of MSD numbers, a simple two-step addition (instead of three-step) can be obtained by eliminating consecutive 1 and $\overline{1}$ digits from the addend and the augend. The elimination process is achieved by using the special recoding algorithm shown in Table 6 where d denotes a don't care literal [24,44]. The recoding scheme (referred to as RMSD) may increase the number of MSD digits by one. Therefore when an n-digit MSD number $x = x_{n-1}x_{n-2} \ldots x_0$ is recoded using Table 6, an $(n + 1)$-digit number $z = z_n z_{n-1} z_{n-2} \ldots z_0$ is generated such that x and z are numerically equal. Also, the recoded number z will always have the property $z_i \times z_{j-1} \neq 1$. This recoding is essential to generate the simplified MSD addition truth table shown in Table 7. To illustrate the MSD recoding scheme (Table 6) and the addition operation (Table 7), let us consider an example:

Operand type	MSD representation	Recoded MSD representation	Decimal representation
Addend =	$\overline{1}1001001111\overline{1}\overline{1}$	$\overline{1}1001\overline{0}10\overline{1}01\overline{1}$	-3175_{10}
Augend =	$0111\ \overline{1}001\overline{1}001$	$01\overline{1}01\overline{1}001\overline{1}001$	1159_{10}
Sum =		$0\overline{1}00000100000$	-2016_{10}

Note that to apply Table 6, one leading zero (preceding the most significant digit) and three trailing zeros (after the least significant digit) need to be padded. Simi-

Table 6 Recoding Truth Table for MSD Numbers

x_i	x_{i-1}	x_{i-2}	x_{i-3}	z_i	x_i	x_{i-1}	x_{i-2}	x_{i-3}	z_i
$\bar{1}$	$\bar{1}$	$\bar{1}$	d	0	1	1	1	d	0
$\bar{1}$	$\bar{1}$	0	$\bar{1}$	0	1	1	0	1	0
$\bar{1}$	$\bar{1}$	0	1	1	1	1	0	$\bar{1}$	$\bar{1}$
$\bar{1}$	$\bar{1}$	1	d	1	1	1	$\bar{1}$	d	$\bar{1}$
$\bar{1}$	$\bar{1}$	0	0	1	1	1	0	0	0
$\bar{1}$	0	$\bar{1}$	d	1	1	0	1	d	$\bar{1}$
$\bar{1}$	0	0	d	$\bar{1}$	1	0	0	d	$\bar{1}$
$\bar{1}$	0	1	d	$\bar{1}$	1	0	$\bar{1}$	d	1
$\bar{1}$	1	$\bar{1}$	d	$\bar{1}$	1	$\bar{1}$	1	d	1
$\bar{1}$	1	0	$\bar{1}$	$\bar{1}$	1	$\bar{1}$	0	1	1
$\bar{1}$	1	0	0	0	1	$\bar{1}$	0	0	1
$\bar{1}$	1	0	1	0	1	$\bar{1}$	0	$\bar{1}$	0
$\bar{1}$	1	1	d	0	1	$\bar{1}$	$\bar{1}$	d	0
0	$\bar{1}$	$\bar{1}$	d	$\bar{1}$	0	1	1	d	1
0	$\bar{1}$	0	$\bar{1}$	$\bar{1}$	0	1	0	1	1
0	$\bar{1}$	0	0	0	0	1	0	0	1
0	$\bar{1}$	0	1	0	0	1	0	$\bar{1}$	0
0	$\bar{1}$	1	d	0	0	1	$\bar{1}$	d	0
0	0	d	d	0					

larly, to apply Table 7, trailing zeros are padded to the recoded numbers before the addition operation.

6.2 New Symmetric Recoding Algorithm

A close study of Table 6 entries reveals that the recoding truth table does not have a symmetrical complemented relationship between the input minterms (entries that produce an output of 1 or $\bar{1}$) and their corresponding output bits. For instance, for the input entry ($\bar{1}\,\bar{1}$00), the output bit $z_i = 1$. However, when this entry is complemented, i.e., for 1100, the output bit $z_i = 0$ (obviously, the output bit z_i is not complemented). Similar cases exist for the entries (100d), ($\bar{1}$00d), (0100), (0$\bar{1}$00), ($\bar{1}$100), and (1$\bar{1}$00). Based on these observations, a novel symmetrical recoding truth table for the MSD numbers (referred to as SRMSD) is developed as shown in Table 8. Note that the input numbers as well as their corresponding output bits are exactly bit-by-bit complementary to each other. This new algorithm still recodes an MSD number to yield a numerically equivalent MSD number such that $z_i \times z_{i-1} \neq 1$. Consequently, Table 7, the addition truth table, can still be applicable. Also, note that the new recoding truth table has

Table 7 Addition Truth Table for RMSD Numbers

x_i	y_i	x_{i-1}	y_{i-1}	z_i	x_i	y_i	x_{i-1}	y_{i-1}	z_i
$\bar{1}$	$\bar{1}$	0	0	0	1	1	0	0	0
$\bar{1}$	$\bar{1}$	0	1	0	1	1	0	$\bar{1}$	0
$\bar{1}$	$\bar{1}$	1	0	0	1	1	$\bar{1}$	0	0
$\bar{1}$	$\bar{1}$	1	1	$\bar{1}$	1	1	$\bar{1}$	$\bar{1}$	1
$\bar{1}$	0	0	d	$\bar{1}$	1	0	0	d	1
$\bar{1}$	0	1	$\bar{1}$	$\bar{1}$	1	0	$\bar{1}$	1	1
$\bar{1}$	0	1	0	$\bar{1}$	1	0	$\bar{1}$	0	1
$\bar{1}$	0	1	1	0	1	0	$\bar{1}$	$\bar{1}$	0
$\bar{1}$	1	0	$\bar{1}$	0	1	$\bar{1}$	0	1	0
$\bar{1}$	1	0	0	0	1	$\bar{1}$	0	0	0
$\bar{1}$	1	1	$\bar{1}$	0	1	$\bar{1}$	$\bar{1}$	1	0
$\bar{1}$	1	1	0	0	1	$\bar{1}$	$\bar{1}$	0	0
0	$\bar{1}$	d	0	$\bar{1}$	0	1	d	0	1
0	$\bar{1}$	$\bar{1}$	1	$\bar{1}$	0	1	1	$\bar{1}$	1
0	$\bar{1}$	0	1	$\bar{1}$	0	1	0	$\bar{1}$	1
0	$\bar{1}$	1	1	0	0	1	$\bar{1}$	$\bar{1}$	0
0	0	$\bar{1}$	$\bar{1}$	$\bar{1}$	0	0	1	1	1
0	0	1	$\bar{1}$	0	0	0	$\bar{1}$	1	0
0	0	d	0	0	0	0	0	d	0

fewer minterms than the old RMSD truth table. Further, the old RMSD algorithm recodes a negative and a positive number of the same magnitude differently, whereas the SRMSD generates exact recoded complement numbers. For example, $\overline{111}\,\bar{1}_{\text{MSD}} = (-7_{10})$ is recoded as $\bar{1}001_{\text{MSD}}$, and $11\overline{1}1_{\text{MSD}} = (7_{10})$ is recoded as $100\bar{1}_{\text{MSD}}$ using the SRMSD, while the RMSD algorithm recodes them as $\overline{1}01\bar{1}_{\text{MSD}}$ and $100\,\bar{1}_{\text{MSD}}$, respectively. To illustrate the symmetric recoding scheme, let us consider the recoding and the addition of the two MSD numbers of the previous example:

Operand type	MSD representation	Symmetrically recoded MSD representation	Decimal representation
Addend =	$\bar{1}\,100\bar{1}001111\,\bar{1}$	$\bar{1}010\bar{1}1010\bar{1}001$	-3175_{10}
Augend =	$011\bar{1}\,\overline{1}001\overline{1}00\bar{1}$	$01\bar{1}0010001011$	1159_{10}
Sum =		$\bar{1}100000100000$	-2016_{10}

Table 8 Symmetric Recoding Truth Table for MSD Numbers

x_i	x_{i-1}	x_{i-2}	x_{i-3}	z_i	x_i	x_{i-1}	x_{i-2}	x_{i-3}	z_i
$\bar{1}$	$\bar{1}$	$\bar{1}$	d	0	1	1	1	d	0
$\bar{1}$	$\bar{1}$	0	$\bar{1}$	0	1	1	0	1	0
$\bar{1}$	$\bar{1}$	0	1	1	1	1	0	$\bar{1}$	$\bar{1}$
$\bar{1}$	$\bar{1}$	1	d	1	1	1	$\bar{1}$	d	$\bar{1}$
$\bar{1}$	$\bar{1}$	0	0	0	1	1	0	0	0
$\bar{1}$	0	$\bar{1}$	d	1	1	0	1	d	$\bar{1}$
$\bar{1}$	0	0	d	1	1	0	0	d	$\bar{1}$
$\bar{1}$	0	1	d	$\bar{1}$	1	0	$\bar{1}$	d	1
$\bar{1}$	1	$\bar{1}$	d	$\bar{1}$	1	$\bar{1}$	1	d	1
$\bar{1}$	1	0	$\bar{1}$	$\bar{1}$	1	$\bar{1}$	0	1	1
$\bar{1}$	1	0	0	0	1	$\bar{1}$	0	0	0
$\bar{1}$	1	0	1	0	1	$\bar{1}$	0	$\bar{1}$	0
$\bar{1}$	1	1	d	0	1	$\bar{1}$	$\bar{1}$	d	0
0	$\bar{1}$	$\bar{1}$	d	$\bar{1}$	0	1	1	d	1
0	$\bar{1}$	0	$\bar{1}$	$\bar{1}$	0	1	0	1	1
0	$\bar{1}$	0	0	$\bar{1}$	0	1	0	0	1
0	$\bar{1}$	0	1	0	0	1	0	$\bar{1}$	0
0	$\bar{1}$	1	d	0	0	1	$\bar{1}$	d	0
0	0	d	d	0					

6.3 Carry-Free Adder Design I: Direct Implementation

For the direct implementation of the POSC adder, one has to determine the input encoding and the LED source patterns for both the recoding (Table 8) and the addition (Table 7) truth tables. However, in this section, only some details of the addition truth table design will be demonstrated, since a very similar procedure can be followed to generate the results of the recoding truth table. Table 7 is rewritten as shown in Table 9, where F, V, and H codes are assigned respectively for the output bits 0, $\bar{1}$, and 1. From Table 9, there are only 24 input conditions with nonzero outputs. The design will exclude the zero output conditions, since they tend to be satisfied by default [2]. With one pixel subcell assigned to each pair of rows, twelve pixel subcells will be necessary for satisfying the whole truth table. These pixel subcells can be arranged in either a 3 × 4 or a 4 × 3 spatial matrix form.

From Table 9, it is obvious that the pairs of input conditions are complementary to each other. This makes the generation of codes for X_i, Y_i, X_{i-1}, and Y_{i-1} very simple. The pairs of input conditions result in twelve pairs of POSC overlap equations. To illustrate, let us, respectively choose pixel subcells {1, 1}

Table 9 Truth Table and Output Coding for SRMSD Addition//Subtraction

Input condition	MSD				Output codes	
	x_i	y_i	x_{i-1}	y_{i-1}	S_i	D_i
1	1	1	$\bar{1}$	$\bar{1}$	V	F
2	$\bar{1}$	$\bar{1}$	1	1	H	F
3	$\bar{1}$	0	0	0	V	V
4	1	0	0	0	H	H
5	$\bar{1}$	0	0	1	V	V
6	1	0	0	$\bar{1}$	H	H
7	$\bar{1}$	0	0	$\bar{1}$	V	V
8	1	0	0	1	H	H
9	$\bar{1}$	0	1	$\bar{1}$	V	F
10	1	0	$\bar{1}$	1	H	F
11	$\bar{1}$	0	1	0	V	V
12	1	0	$\bar{1}$	0	H	H
13	0	$\bar{1}$	0	0	V	H
14	0	1	0	0	H	V
15	0	$\bar{1}$	1	0	V	H
16	0	1	$\bar{1}$	0	H	V
17	0	$\bar{1}$	$\bar{1}$	0	V	H
18	0	1	1	0	H	V
19	0	$\bar{1}$	$\bar{1}$	1	V	F
20	0	1	1	$\bar{1}$	H	F
21	0	$\bar{1}$	0	1	V	H
22	0	1	0	$\bar{1}$	H	V
23	0	0	$\bar{1}$	$\bar{1}$	V	F
24	0	0	1	1	H	F
25	$\bar{1}$	0	1	1	F	V
26	1	0	$\bar{1}$	$\bar{1}$	F	H
27	$\bar{1}$	1	1	$\bar{1}$	F	H
28	1	$\bar{1}$	$\bar{1}$	1	F	V
29	0	$\bar{1}$	1	1	F	H
30	0	1	$\bar{1}$	$\bar{1}$	F	V
31	0	0	$\bar{1}$	1	F	V
32	0	0	1	$\bar{1}$	F	H
For the remaining combinations					F	F

and $\{1, 2\}$, for satisfying the pairs of input conditions, $(1, 2)$ and $(3, 4)$. For simplicity, let us rename X_i, Y_i, X_{i-1}, and Y_{i-1}, as A, B, C, and D, respectively.

$$\overline{\overline{A}}_{11} \wedge \overline{\overline{B}}_{11} \wedge C_{11} \wedge D_{11} = H \tag{30a}$$

$$A_{11} \wedge B_{11} \wedge \overline{\overline{C}}_{11} \wedge \overline{\overline{D}}_{11} = V \tag{30b}$$

and

$$A_{12} \wedge \overline{B}_{12} \wedge \overline{C}_{12} \wedge \overline{D}_{12} = H \tag{31a}$$

$$\overline{\overline{A}}_{12} \wedge \overline{B}_{12} \wedge \overline{C}_{12} \wedge \overline{D}_{12} = V \tag{31b}$$

where $\overline{X}_{ij}$, X_{ij}, and $\overline{\overline{X}}_{ij}$, respectively represent MSD 0, 1, and $\overline{1}$ for the ijth pixel subcell of input X. The solution of Eqs. (30a) and (30b) is $\overline{A}_{11} = \overline{B}_{11} = \overline{C}_{11} = \overline{D}_{11} = F$, $A_{11} = B_{11} = V$, $C_{11} = D_{11} - II$, $\overline{\overline{A}}_{11} = \overline{\overline{B}}_{11} = H$, and $\overline{\overline{C}}_{11} = \overline{\overline{D}}_{11} = V$. However, in Eqs. (31a), (31b), B_{12}, C_{12} and D_{12} are common. Therefore $\overline{B}_{12} - \overline{C}_{12} = \overline{D}_{12} = T$, $A_{12} = H$, $\overline{\overline{A}}_{12} = V$, and $\overline{A}_{12} = B_{12} = C_{12} = D_{12} = \overline{\overline{B}}_{12} = \overline{\overline{C}}_{12} = \overline{\overline{D}}_{12} = F$. Note that the LED source at the corresponding pixel subcells is turned ON. Likewise the codes for the remaining pixel subcells can be obtained by solving the corresponding POSC overlap equations. The overall encoding pattern for the inputs is shown in Table 10.

Table 11 illustrates the input encoding and the LED source pattern needed to implement the symmetrical recoding truth table (Table 8). Note that Table 8 has 44 nonzero output entries. Therefore, 22 pixel subcells will be necessary for satisfying the whole truth table. Since 22 pixel subcells amount to an odd geometrical enclosure, a 6×4-pixel or 4×8 geometry is chosen for the input encodings.

Table 10 Input Encoding and Source Pattern for the Direct Implementation of the SRMSD Addition Truth Table

MSD logic	Inputs				Source pattern
	x_i	y_i	x_{i-1}	y_{i-1}	
0	F F F F	F T T T	F T T T	F T F F	
	F F T T	T T F F	F F T F	F T T T	
	T T T T	F F F T	F F T F	T F F F	
1	V H H H	V F F F	H F F F	H F V V	T T T T
	H H F F	F F H H	V V F V	H F F F	T T T T
	F F F F	H H H F	H H F H	F V V H	T T T T
$\overline{1}$	H V V V	H F F F	V F F F	V F H V	
	V V F F	F F V V	H H F H	V F F F	
	F F F F	V V V F	V V F V	F H H V	

Table 11 Input Encoding and Source Pattern for the Direct Implementation of the SRMSD Recoding Truth Table

MSD logic	Inputs				Source pattern
	x_i	y_i	x_{i-1}	y_{i-1}	
	F F F F	F F F F	T F F F	F T F F	
	F F F F	T T T T	F F F F	T F F T	
	F F F F	F T T T	F T T T	F T F F	
0	F F F F	T F F F	F F F F	F T F F	
	F T T T	F F F F	T F F F	F T V H	
	T T	F F	T T	F T	
	V V V V	V V V V	F H H H	H F H V	T T T T
	V V V V	F F F F	V V V F	F H V F	T T T T
	V V V V	F F F F	F F V V	H V F V	T T T T
1	V H H H	F V V V	V H H H	H F V H	T T T T
	H F F F	F F F F	T F F F	F T V H	T T T T
	T T	F F	T T	F T	T T
	H H H H	H H H H	F V V V	V F V H	
	H H H H	F F F F	H H H F	F V H F	
	H H H H	F F F F	F F H H	V H F H	
$\bar{1}$	H V V V	F H H H	H V V V	V F H V	
	V F F F	H V V V	F V V V	V F F F	
	F F	V V	F F	V F	

On the other hand, in MSD arithmetic, subtraction can be performed by addition if the subtrahend is complemented first. Therefore the same input codes of the above designed adder can be used, provided that the inputs to the adder are complemented first. Or, one can use the designed input codes for subtraction shown in Table 12. This POSC design for the SRMSD subtracter was derived by a similar procedure to the above SRMSD adder. However, entries of Table 9 corresponding to the difference bits (D_i) were used.

6.4 Carry-Free Adder Design II: Truth Table Partitioning

In order to reduce the number of pixel subcells in the input encoding that result from a large truth table, one could always partition the truth table columnwise into, say, two or more sections [11,45]. As a consequence of such partitioning the outputs could often be described in terms of the inputs of any one section (i.e., primary inputs, each combination of which would be referred to as a primary group), while the remaining inputs could be referred to as secondary inputs. The overlap of the secondary inputs is assigned a POSC code so that its overlap with

Table 12 Input Encoding and Source Pattern for the Direct Implementation of the SRMSD Subtraction Truth Table

MSD logic	Inputs				Source pattern
	x_i	y_i	x_{i-1}	y_{i-1}	
0	F F F F	T T T T	T T T F	T F F T	
	T T T T	F F F F	T F F T	T T T F	
	F F T T	T F F T	F F F F	F F F F	
1	H H H H	F F F F	F F F V	F V H F	T T T T
	F F F F	H H H H	F V H F	F F F V	T T T T
	H V F F	F H V F	V H H H	V V H V	T T T T
$\overline{1}$	V V V V	F F F F	F F F H	F H V F	
	F F F F	V V V V	F H V F	F F F H	
	V H F F	F V H F	H V V V	H H V H	

the code of the overlap of the primary inputs produces the code for the desired output. The overall input encoding algorithm of truth table partitioning can be summarized as follows.

Step 1. The truth table is partitioned.

Step 2. The nonzero output bits ($\overline{1}$ and 1) of an output of interests are assigned V and H codes, respectively.

Step 3. The primary input groups are assigned codes so that their overlap with the code of the secondary input group code may produce the desired output.

Step 4. From the POSC overlap equation of each group (which relates the primary input variables with the code of that primary code), the code for each of the primary input variables is determined. For the sake of saving memory (or the number of pixel subcells), groups having complementary codes are solved simultaneously whenever possible.

Step 5. A step similar to step 4 is repeated to determine the codes of the secondary input variables.

To apply this POSC partitioning algorithm to our SRMSD adder, Table 7 is partitioned into two sections: the first two input (X_i, Y_i) forming the primary section and the remaining two inputs (X_{i-1}, Y_{i-1}) forming the secondary section (see Table 13). Again, the output bits $\overline{1}$ and 1 are assigned V and H codes, respectively. A careful study of this table shows that there are only seven combinations of the inputs X_i and Y_{i-1}, namely, $\overline{1}0$, $0\overline{1}$, 01, 10, $\overline{1}\,\overline{1}$, 00, and 11, that produce output bits. A group code is assigned to the other two inputs (X_{i-1}, Y_{i-1}) so that their overlap produces the same output as the primary overlap group. Conse-

Table 13   Group Assignment Table for the Adder

Group number	x_i	y_i	Group code	x_{i-1}	y_{i-1}	S_i
1	$\bar{1}$	0	V	0	1	V
				1	0	
				0	$\bar{1}$	
				1	$\bar{1}$	
2	$\bar{1}$	0	V	0	0	V
3	0	$\bar{1}$	V	1	0	V
				0	1	
				$\bar{1}$	0	
				$\bar{1}$	1	
4	0	$\bar{1}$	V	0	0	V
5	0	1	H	$\bar{1}$	0	H
				$\bar{1}$	1	
				0	$\bar{1}$	
				1	0	
6	0	1	H	0	0	H
7	1	0	H	0	$\bar{1}$	H
				$\bar{1}$	1	
				$\bar{1}$	0	
				0	1	
8	$\bar{1}$	0	H	0	0	H
9	$\bar{1}$	$\bar{1}$	H	1	1	H
10	0	0	H	1	1	H
11	0	0	V	$\bar{1}$	$\bar{1}$	V
12	1	1	V	$\bar{1}$	$\bar{1}$	V

quently, for the 12 entries shown in Table 13, the design is reduced to solving only six POSC overlapping equations. These are input conditions (1, 7), (2, 8), (3, 5), (4, 6), (9, 12), and (10, 11). The resulting input codes are shown in Fig. 9a along with the code that must be provided by the secondary input overlap X_{i-1}, and Y_{i-1}. To illustrate, consider the pixel subcell $\{1, 1\}$, which satisfies the input condition (1, 7).

$$\overline{\overline{X}}_i \wedge \overline{Y}_i \wedge \overline{(X_{i-1} \wedge Y_{i-1})} = V \tag{32a}$$

$$X_i \wedge \overline{Y}_i \wedge (X_{i-1} \wedge Y_{i-1}) = H \tag{32b}$$

The solution of Eq. 32 is $\overline{\overline{X}}_i = \overline{(X_{i-1} \wedge Y_{i-1})} = V$, $X_i = (X_{i-1} \wedge Y_{i-1}) = H$, $\overline{Y}_{i-1} = T$, $\overline{X}_i = \overline{(X_{i-1} \wedge Y_{i-1})} = F$, and $Y_i = \overline{Y}_i = F$. To complete the adder design, the separate codes for X_{i-1} and Y_{i-1} must be found. These are shown in Fig. 9b.

	1	0	$\overline{1}$
x_i	H H F	F F T	V V F
	F V F	T F T	F H F
y_i	F F H	T T F	F F V
	H V Γ	ΓF T	V H F
$x_{i\text{-}1} \wedge y_{i\text{-}1}$	H F H	F T F	V F V
	F H H	T F F	F V V

(a)

	1	0	$\overline{1}$
$x_{i\text{-}1}$	H T H	T V T	V T V
	T H H	T F F	H V V
$y_{i\text{-}1}$	T H T	T T T	T V T
	H H H	T F F	V V V

(b)

Figure 9 Input encoding for the POSC adder: (a) primary group code along with the overlap code of the secondary and (b) the secondary input encodes separated.

Next, the symmetrical recoding table, Table 8, is also partitioned for an overall reduction in the number of pixel subcells of the input encodings. Table 14 shows the partitioned table along with the group assignment codes. Note that here we have 16 entries in the table for which eight possible POSC overlapping equations are obtained. The solutions of these equations are shown in Fig. 10. Finally, Table 15 is the partitioned table and the group assignment codes used in the design of SRMSD subtracter; and Fig. 11 shows the resulting input encodings for this design.

6.5 Conclusion

From the previous discussion, it is obvious that the second POSC adder design is better than the first one (it has fewer pixel subcells in the input encodings). The first design requires 24 and 16 pixel subcells in the input encodings for both the recoding and the addition/subtraction operations; while only 16 and 8 pixel subcells are needed in the second design, respectively. A previous two-step POSC adder design [8] requires a total of 18 pixel subcells for the input encodings. However, our second POSC adder/subtracter requires only a total of 14 pixel

Table 14 Group Assignment Table for the SRMSD Recoding Algorithm

Group number	$x_i x_{i-1}$	Group code	$x_{i-2} x_{i-3}$	Z_i
1	0 $\bar{1}$	H	$\bar{1}$ 0	H
			$\bar{1}$ 1	
			$\bar{1}$ $\bar{1}$	
			0 $\bar{1}$	
2	0 1	V	$\bar{1}$ 1	V
			1 0	
			1 1	
			0 1	
3	0 $\bar{1}$	H	0 0	H
4	0 1	V	0 0	V
5	1 1	H	$\bar{1}$ 0	H
			$\bar{1}$ 1	
			$\bar{1}$ $\bar{1}$	
6	$\bar{1}$ $\bar{1}$	V	$\bar{1}$ 1	V
			1 0	
			1 1	
7	1 1	H	0 $\bar{1}$	H
8	$\bar{1}$ $\bar{1}$	V	0 1	V
9	$\bar{1}$ 1	H	$\bar{1}$ 0	H
			$\bar{1}$ 1	
			$\bar{1}$ $\bar{1}$	
10	1 $\bar{1}$	V	$\bar{1}$ 1	V
			1 0	
			1 1	
11	$\bar{1}$ 1	H	0 $\bar{1}$	H
12	1 $\bar{1}$	V	0 1	V
13	$\bar{1}$ 0	H	$\bar{1}$ 1	H
			1 0	
			1 1	
			0 0	
			0 1	
			0 $\bar{1}$	
14	1 0	V	$\bar{1}$ 0	V
			$\bar{1}$ 1	
			$\bar{1}$ $\bar{1}$	
			0 0	
			0 1	
			0 $\bar{1}$	
15	1 0	H	$\bar{1}$ 1	H
			1 0	
			1 1	
16	$\bar{1}$ 0	V	$\bar{1}$ 0	V
			$\bar{1}$ 1	
			$\bar{1}$ $\bar{1}$	

	1	**0**	**$\bar{1}$**
x_i	F H V V	T F F T	F V H H
	F H V H	T F F F	F V H V
x_{i-1}	V H H F	F F F T	H V V F
	V H H F	F F F T	H V V F
x_{i-2}	V V V H	F F F F	H H H V
	F F F H	T T T T	F F F V
x_{i-3}	T T T T	T T T T	T T T T
	V V V T	T F F T	H H H T

Figure 10 Input encoding for the POSC symmetrical recoding truth table.

Table 15 Group Assignment Table for the Subtracter

Group number	$x_i y_i$	Group code	$x_{i-1} y_{i-1}$	D_i
1	$\bar{1}$ 0	V	0 $\bar{1}$	V
			1 0	
			0 1	
			1 1	
2	$\bar{1}$ 0	V	0 0	V
3	0 $\bar{1}$	H	$\bar{1}$ 0	H
			0 1	
			1 0	
			1 1	
4	0 $\bar{1}$	H	0 0	H
5	0 1	V	$\bar{1}$ 0	V
			1 $\bar{1}$	
			0 $\bar{1}$	
			1 0	
6	0 1	V	0 0	V
7	1 0	H	0 $\bar{1}$	H
			1 $\bar{1}$	
			$\bar{1}$ 0	
			0 1	
8	1 0	H	0 0	H
9	$\bar{1}$ 1	H	$\bar{1}$ 1	H
10	0 0	H	$\bar{1}$ 1	H
11	0 0	V	$\bar{1}$ 1	V
12	1 $\bar{1}$	V	$\bar{1}$ 1	V

	1	**0**	**$\overline{1}$**
x_i	H H V	F F F	V V H
	F F F	T T T	F F F
y_i	F F H	T T F	F F V
	V V F	F F T	H H F
$x_{i-1} \wedge y_{i-1}$	H F H	F T F	V F V
	F H H	T F F	F V V

(a)

	1	**0**	**$\overline{1}$**
x_{i-1}	V T H	T H F	H T V
	H V H	T T F	V H V
y_{i-1}	T V V	T V F	T H H
	H V V	T T F	V H H

(b)

Figure 11 Input encoding for the POSC subtracter: (a) primary group code along with the overlap code of the secondary and (b) the secondary input encoding separated.

subcells. On the other hand, a digital circuit can be easily designed to perform the recoding scheme [42] in parallel. Therefore the addition/subtraction process can be implemented in one step using POSC scheme with only six pixel subcells required in the input encodings. This represents a significant saving in the memory requirement of the system.

7 POSC: EDGE DETECTION USING THE ROBERTS OPERATOR

For image processing applications, it is often necessary to extract the outline feature of an image. An OSC based edge detector was proposed by Tanida and Ichioka [46] that involves a multistep sequential process and requires 36 light emitting diodes as well as a huge memory. Recently, POSC schemes have been used for edge detection by using a modified difference operator referred to as an L shaped operator [47]. However, a difference operator is not necessarily suitable for detecting edges with equal edge strength since it may produce a weak response for such cases. On the other hand, due to the symmetric nature of the Roberts mask [48], all the edges of an image will have equal strength. However, the Roberts operator suffers from directional dependency [48,49]. To overcome the aforementioned limitations, we implement in this paper a modified Roberts-operator based edge detector using the POSC technique. The performance of the proposed technique is verified by computer simulation.

7.1 Roberts Operator

The Roberts gradient is a simple nonlinear edge detector. It is employed by convolving an image with two 2×2 kernels that approximate the horizontal and vertical strength of the edge at each pixel location and are given by

$$R_+ = \begin{vmatrix} 0 & 1 \\ 1 & 0 \end{vmatrix} \quad \text{and} \quad R_- = \begin{vmatrix} 1 & 0 \\ 0 & -1 \end{vmatrix}$$

Thus the Roberts operator is especially useful in detecting edges of an image along the diagonal. Each pixel in the input image is replaced with the larger of the absolute values of these two operators, given by

$$f_{\text{edge}}(x, y) = \max\{|R_+|, |R_-|\} \tag{33}$$

The Roberts edge kernels are convolved with the input image to create the edge enhanced or sharpened image. It is obvious that the Roberts operator suffers from directional dependency. For example, $R1$ detects only $45°$ edges and $R2$ detects only $135°$ edges [48,49]. To avoid the directional dependency of the Roberts operator, a modified Roberts operator can be used. The truth table of the POSC edge detector using a modified Roberts mask is shown in Table 16, which shows the sixteen combinations corresponding to the 2×2 window of the Roberts mask. The input variables A, B, C, and D represent the binary image values at (i, j), $(i + 1, j)$, $(i, j + 1)$, and $(i + 1, j + 1)$ pixel locations, respectively. Notice that these input combinations include all possible edge orientations: vertical, horizontal, diagonal, no-edge, and corners. The absolute values of the responses $R1$ and $R2$ as well as their maxima are listed as shown in columns 6 through 8 of Table 16, where

$$|R1| = A - D \tag{34}$$

and

$$|R2| = B - C \tag{35}$$

From Table 16 it is evident that the edge detector can be implemented by using the input combinations that yield either the 1 outputs or the 0 outputs. Since the 0 output involves only four input conditions, it is preferable to design the edge detector using the minterms corresponding to the 0 outputs.

Table 16 Truth Table for the POSC Edge Detector Using Roberts Operator

Condition number	Inputs			Adder outputs		Subtracter outputs	
	X	Y	Z	S	C	D	B
1	0	0	0	0	0	0	0
2	0	0	1	1	0	2	1
3	0	1	0	1	0	2	1
4	0	1	1	2	0	1	1
5	1	0	0	1	0	1	0
6	1	0	1	2	0	0	0
7	1	1	0	2	0	0	0
8	1	1	1	0	1	2	1
9	0	2	0	2	0	1	1
10	0	2	1	0	1	0	1
11	2	0	0	2	0	2	0
12	2	0	1	0	1	1	0
13	2	2	0	1	1	0	0
14	2	2	1	2	1	2	1
15	1	2	0	0	1	2	1
16	2	1	0	0	1	1	0
17	1	2	1	1	1	1	1
18	2	1	1	1	1	0	0

7.2 Design

To determine the input encoding patterns of the POSC edge detector, we observe from Table 16 that only four minterms (i.e., 1, 7, 10, and 16) need to be generated corresponding to the 1 outputs. These four minterms may be paired as (1, 7) and (10, 16). Accordingly, a 1×2 spatial matrix will be required for each input encoding pattern and source LED pattern to generate the necessary overlap of shadows in the output plane. Assigning minterm pair (1, 7) to pixel subcell {1, 1} and minterm pair (10, 16) to pixel subcell {2, 1}, the four POSC overlap equations can be written as

$$\overline{A}_{11} \wedge \overline{B}_{11} \wedge \overline{C}_{11} \wedge \overline{D}_{11} = H \tag{36}$$

$$\overline{A}_{11} \wedge B_{11} \wedge C_{11} \wedge \overline{D}_{11} = V \tag{37}$$

$$A_{21} \wedge \overline{B}_{21} \wedge \overline{C}_{21} \wedge D_{21} = V \tag{38}$$

$$A_{21} \wedge B_{21} \wedge C_{21} \wedge D_{21} = H \tag{39}$$

where X_{ij} represents the ijth element of input X. Solving Eqs. (36) through (39), the POSC codes for the inputs can be written as

$$A = D = \begin{bmatrix} F \\ T \end{bmatrix} \tag{40}$$

$$B = C = \begin{bmatrix} V \\ H \end{bmatrix} \tag{41}$$

The POSC edge detector is established by overlapping four relatively shifted versions of the same input image, provided the four shifted versions are encoded, respectively, using the codes for A, B, C, and D. In the present case, two unpolarized LEDs are required in the source plane for the edge detector. If, for practical reasons, only unpolarized LEDs (T and F) are used for encoding, the input overlap pattern must be increased from a 1 × 2 pattern to a 2 × 2 pattern. Accordingly, the input encoding patterns for the four inputs can be obtained as

$$A = D = \begin{bmatrix} F & F \\ T & T \end{bmatrix} \tag{42}$$

$$B = C = \begin{bmatrix} F & T \\ F & T \end{bmatrix} \tag{43}$$

Notice that although the encoding for A and D (B and C) are same, every pixel of the images encoded for B,C, and D must be shifted relative to the image encoded for A (explained in Sec. 7.3) in order to generate the necessary overlap in the output plane, thus requiring four patterns in the input plane.

7.3 Simulation

The simulation experiment was conducted on a 24 × 36 pixel image as shown in Fig. 12. For the simplicity of encoding, the unpolarized codes used in Eqs. (42) and (43) were used. The input image of Fig. 12 was encoded four times using the codes for A, B, C, and D, respectively, as shown in Figs. 13a through 13d. The image encoded using the code for A is overlapped with the image encoded for B but shifted one pixel to the left and with the image encoded using the code for C but shifted one pixel up. The image encoded using the code for D is shifted one pixel to the left and one pixel up with respect to the corresponding encoded image for A. This process ensures the overlap of the neighboring pixels and establishes a 2 × 2 window. The overlapped input patterns are illuminated by four LEDs from the source plane, and the central pixels of the resulting output overlap pixels are masked out to yield the output of the edge detector. For illustration, the output of the Roberts-operator based POSC edge detector using Fig. 12 as the input is shown in Fig. 14. Notice that due to the symmetric nature of the Roberts mask, all edges have equal strength, which is obvious from Fig. 14. If a

Figure 12   Input image.

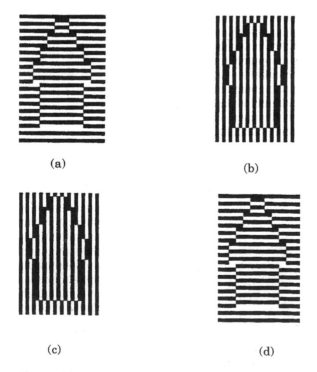

(a)

(b)

(c)

(d)

Figure 13   (a) Encoded input image using the code for A; (b) the code for B; (c) the code for C; and (d) the code for D.

Figure 14 Output image using the modified Roberts operator based POSC edge detector.

difference operator is used instead of the Roberts operator, then the output of the POSC edge detector is obtained as shown in Fig. 15. From Fig. 15, it is evident that a difference-operator based POSC edge detector cannot detect the edges with equal edge strength. Notice the weak response at both upper left corners and at 135° edges. Comparing Fig. 14 with Fig. 15, we observe that the Roberts-operator

Figure 15 Output image using the difference operator based POSC edge detector.

based POSC edge detector yields much better output than the difference-operator based edge detector.

7.4　Conclusion

In this section, we presented a POSC edge detector for image processing applications. The edge detector is designed using a modified version of the Roberts operator, which avoids the directional dependency of the Roberts operator and yields the entire edge-enhanced output of the input image in parallel. The proposed edge detector is found to yield better results than the previously reported shadow-casting based edge detectors. For example, the edge detector proposed by Tanida and Ichioka [46] involves multistep sequential processing and a large memory, which are entirely avoided in the present scheme. It also avoids the problems associated with the difference-operator based edge detectors. However, a difference-operator based POSC edge detector may be implemented with only two minterms. On the other hand, a Roberts-operator based edge detector requires four minterms.

8　POSC: MORPHOLOGICAL TRANSFORMATION

The optical symbolic substitution (SS) is a pattern transformation technique that realizes digital logic optically for information processing in two-dimensional images [3]. In this scheme, data are encoded as spatial patterns, and operators are seen as pattern transformation rules [50,51]. In its operation, SS functions by mapping a to-be-recognized pattern (recognition phase) into a new pattern (substitution phase). Because of its extensive parallelism as well as spatial orientation, SS appears particularly attractive for implementing image processing techniques. Optical symbolic substitution based edge detection [52,53], histogram equalization [54], morphological operations [55], image skeletonization [56], and arithmetic and/or logical operations [35–40] have already been realized. The SS technique consists of a series of image duplication, image shifting, and image overlapping in a particular and sequential order.

On the other hand, it has been shown that POSC is capable of performing any logical operation of two or more variables by changing the LED pattern or encoding [2,11]. In other words, a particular combination of input is selected by means of encoding and/or LED pattern and then a single [2] or multiple [11] outputs are substituted. Consequently, POSC can also be classified as a symbolic substitution technique. In its operation, we have shown [57] that, in a POSC scheme, the recognition and substitution phases take place by means of LEDs projecting shadows of the overlap pattern to the output plane. The location and the types of vertical [9,13] LEDs are chosen so that the transparent openings can project light in specific locations in the output plane. In other words, LEDs are chosen to cause predetermined shifts and overlap of the shadows of the input overlap plane. Therefore the POSC scheme can be seen as operating in a fashion similar to that of the SS scheme. More interestingly, one can perform any symbolic substitution logic using the optical shadow-casting scheme [58].

In this section, by combining the two powerful optical computing techniques, namely, optical symbolic substitution (SS) and polarization-encoded optical shadow-casting (POSC), morphological or shape transformation operations are demonstrated. Accordingly, erosion, dilation, opening, and closing operations are realized using both SS and POSC schemes. These morphological operations are used for noise removal in binary images [22]. Further, an optical morphological hit-or-miss transformation [23] is demonstrated that can be used in the recognition of perfect and imperfect shapes.

8.1 Morphological Operations

The image processing technique of shrinking and expanding a binary image to reduce or remove unwanted artifacts is very old but is still used. A single pixel shrink-and-expand operation, referred to as mathematical morphology [59], will be enough for cleaning an image when artifacts are small. When artifacts are large, however, several iterations may be necessary for sufficient image smoothing.

Algebraically, morphological analysis involving images consists of two basic set operations. Minkowski addition and subtraction. The image X can be conceived as an ensemble of black pixels constituting the object and white pixels constituting the background. The operations involve a specific binary operating image called a structuring element. The structuring element E determines the pixels of X that will be affected by the transformations.

The Minkowski sum or dilation is found by placing the reference pixel (origin) of the structuring element E over each of the activated pixels (those having a value of 0) of X and then taking the union of all the resulting copies. On the other hand, Minkowski subtraction or erosion can be described in terms of the structure element translation as follows. The origin of the structuring element E is placed successively over each activated pixel of X. If, for a given pixel, say (i, j), the activated pixel of the translated copy of E is a subimage of X, then the (i, j) pixel value is set to 0; otherwise, the (i, j) pixel is set to 1 in the eroded image. Two other operations that follow directly from dilation and erosion are opening and closing. A pixel is activated in the opened image if and only if it is a part of a fitted copy of the structuring element. Opening is achieved by having an erosion operation followed by a dilation. On the other hand, closing consists of a dilation followed by an erosion [59].

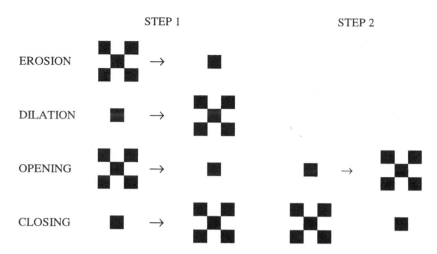

Figure 16   Symbolic substitution rules for morphological transformations.

The four morphological operations can be efficiently implemented using SS based on POSC. Figure 16 describes the necessary SS rules for implementing these operations for a specific structuring element. To perform an erosion, one needs to replace every occurrence of the structuring element (the recognition pattern) in the input image with a single opaque pixel (the substitution pattern). The dilation can also be achieved using SS rules by replacing every opaque pixel of the input image by the structuring element. Again, by combining erosion and dilation, the opening and closing of an image are realized.

8.2   POSC Based Morphological Transformations

An optical SS scheme for realizing morphological transformation can be realized using phase-only holograms and classical optical elements (e.g., mirrors, prisms) [50,51]. The basic system performs a series of operations in the following order: (1) splitting an input image, (2) shifting split images, (3) superimposing the shifted and unshifted images, (4) regenerating the superimposed image, (5) splitting the regenerated image, (6) shifting the resulting images, and (7) superimposing the shifted images.

However, an optical shadow-casting scheme provides faster image replication and spatial shifting. By choosing the spacing between the LEDs and the distance from the source plane to the input plane and from the input plane to the output plane [58], replicated and shifted copies of the input plane are obtained superimposed on the screen. The source LEDs and their ON–OFF states (representing the structure element E) determine the number of copies and their relative shifts.

Figure 17 shows a recognition unit using intensity coded binary logic where 0 and 1 are respectively represented as opaque and transparent. To meet the requirement of having a cross-shaped structuring element (Fig. 16), for instance, five LEDs of the source plane (denoting the five opaque pixels of the structural element) are turned ON accordingly. This configuration of LEDs yields five shifted and superimposed copies of the input image. The first, second, third, fourth, and fifth copies are respectively shifted down, shifted up, shifted down, shifted left, and not shifted by one pixel. The overlapping of these copies produces the output image. Obviously, this SS based OSC system actually realizes

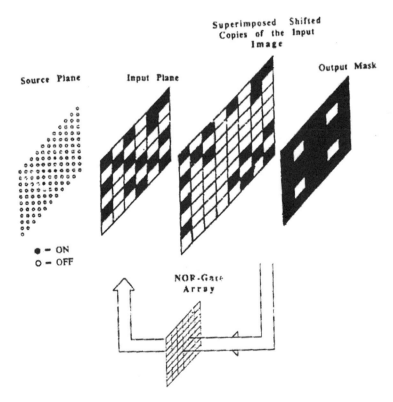

Figure 17 Optical symbolic substitution based optical shadow-casting system for morphological operations.

an erosion operation of the input binary image with a cross-shaped structuring element. Note that by turning ON or OFF different LEDs at the source plane, an erosion operation by many other structural elements is possible.

On the other hand, Fig. 17 can be also used to achieve a dilated image with the cross-shaped structuring element. The input images, in this case, is first complemented and then directed to the input plane of the POSC unit, where again five copies of the inverted input image are obtained. Shift operations similar to those encountered earlier are realized, and the overlapped copies consist of the partial output. Finally, the partial output image is incident on a NOR-gate array whose output is the inversion of the partial output image. Therefore the SS based POSC unit shown can be used for both erosion and dilation operations, provided that in the latter case, the input image and the resulting output image are inverted using a NOR-gate array. Further, by adjusting the LEDs at the source plane, many structure elements of any desired shape can be implemented. Furthermore, this system can be used to execute iteratively any morphological operation by simply feeding back the resulting output image and reintroducing it to the input of the system either inverted through a NOR-gate array or directly without inversion. Consequently, several morphological operations can be carried out sequentially one after the other, i.e., erosion followed by dilation followed by erosion, etc.

Since opening is equivalent to an erosion followed by a dilation, the output of the system of Fig. 17 is reintroduced (after inversion) as an input to the system. Similarly, one can obtain a closing operation from the system by executing a dilation followed by an erosion operation.

For illustration purposes, computer simulation results will be presented. A noisy binary image as shown in Fig. 18a is subjected to morphological transformations by the structuring element shown in Fig. 18b. Figure 19a shows the eroded image: the objects in the image have been reduced in size. Also, note that much of the "pepper" noise has been eliminated, while at the same time the

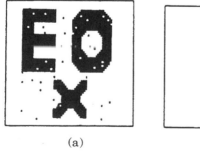
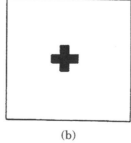

(a)   (b)

Figure 18 (a) Original image and (b) the structuring element.

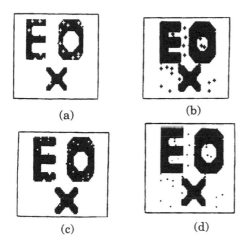

(a)

(b)

(c)

(d)

Figure 19 Morphological transformations: (a) eroded image, (b) dilated image, (c) opened image, and (d) closed image.

"salt" noise has been increased. Actually, every isolated white pixel takes the shape of the structuring element. Figure 19b shows the dilated image. Here, the sizes of the objects in the image have increased, and in contrast to the eroded image, most of the salt pixels have been removed while the pepper pixels have increased in size and have taken the shape of the structuring element. Figure 19c, on the other hand, shows the opened image. Note that in this figure the isolated pepper pixels and some of the salt pixels have been eliminated. Finally, Fig. 19d shows the closed image. Note that the holes inside the image have been filled. It is obvious that an erosion of this output would eliminate all of its isolated black pixels. It is worth mentioning that the results of morphological operations are extensively dependent on the shape of the structuring element.

8.3 POSC: Morphological Hit-or-Miss Transformation

Mathematical morphology has proved its usefulness in many image processing and computer vision problems. There are many important applications in machine vision and automated inspection where the shape of an object must be recognized [60–65]. In particular, the morphological hit-or-miss transformation is a special tool used to locate known objects. In this situation, the problem of shape recognition is equated to that of detecting occurrences of an object shape within an image. In this section, we propose a system that uses a noncoherent light to perform morphological hit-or-miss transformation.

The SS based POSC system can obtain replicated and shifted copies of the input image superimposed on the output screen simply by choosing the switching states of the source pattern (the LEDs) and the distances between the source plane, the input image plane, and the output plane. Further, the structure element of the morphological operation is represented by the ON–OFF states of the source pattern. This means that a data base presenting different structuring elements can be easily incorporated in our system.

The hit-or-miss transform is defined upon image X as $[X \ominus A] \cap [X^c \ominus B]$ where A and B are structural elements. Here $\ominus$ is the notation for erosion and X^c denotes the complement image of X. This operation is the intersection ($\cap$) or AND of the two erosions noted. Given that A represents a shape to be recognized within X, and B is a windowed complement of A, the resulting point set of hit-or-miss transformation consists of points each of which indicates one occurrence of shape A within X [65]. The function of the window is to guarantee that the range of the hit-or-miss operation on shape A does not overlap any portion of other objects that may be within the image.

The hit-or miss transform can be implemented in our system by first an erosion operation of the input image with the structure element (represented by the corresponding LEDs being turned ON) and saving the eroded image, and second by eroding the inverted input image by an appropriate windowed complement of the same structure element. Now the resulting two eroded outputs can be pipelined into an electronic device to get a bit-by-bit AND operation between them; or the two eroded images can be recoded and overlapped and used as a single image input to the POSC system, which finds optically the logical AND of the two eroded images [5].

It is known that the hit-or-miss transform contains a degree of redundancy. The redundancy is in the form of unnecessary computation during erosion operations. Recently it was shown that an essential subset of the structuring element A processes the same effect as if A itself were used [65]. In effect, the subset found to replace the shape or the windowed complement of the shape is the boundary of the object. Therefore erosion is reduced to using only those points on the boundary of the object shape. Consequently, a simpler LED source pattern is used. Since the shape to be recognized is known a priori, a boundary information can be easily recalled from a data base and used in our POSC system. Note that for a group of known shapes, it is possible to find an optimal subset that may be used for such recognition problems.

On the other hand, a major source of shape distortion is the nonlinear thresholding operation, which converts a gray-scale image into a binary one. In many situations, an object is not appropriately represented in the image. After all, the shape cannot be made exactly the same as in the sample space. Hence shape imperfections must be treated. One solution to this problem is to define a relevant sample space as a collection of indeterminate shapes that are variations of the shapes in the sample space [65].

8.4 Conclusion

In this section, we have extended and combined the capabilities of the two powerful techniques of SS and POSC to realize morphological operations. A simple POSC system combined with a NOR-gate array can be organized to generate the four basic morphological operations and to implement many other morphological operations such as hit-or-miss transform, thinning, and thickening.

REFERENCES

1. A. Louri. Parallel implementation of optical symbolic substitution logic using shadow-casting and polarization. Appl. Opt. 30, 540–548, 1991.
2. M. A. Karim, A. A. S. Awwal, and A. K. Cherri. Polarization encoded optical shadow-casting logic units: design. Appl. Opt. 28, 2720–2725, 1987.
3. A. Huang. Parallel algorithms for optical digital computers. Proceedings of the 10th International Optical Computing Conference, IEEE Computer Society, Los Angeles, 1983, pp. 13–17.
4. A. A. Sawchuk and T. C. Strand. Digital optical computing. Proc. IEEE 72, 758–779, 1984.
5. J. Tanida and Y. Ichioka. Optical logic array processor using shadowgrams. J. Opt. Soc. Am. 73, 800–809, 1983.
6. J. Tanida and Y. Ichioka. Optical parallel logic gates using a shadow-casting system for optical digital computing. Proc. IEEE 72, 787–801, 1984.
7. Y. Li, G. Eichman, and R. R. Alfano. Optical computing using hybrid encoded shadow-casting. Appl. Opt. 25, 2636–2638, 1986.
8. A. A. S. Awwal and M. A. Karim. Polarization encoded optical shadow-casting: design of a carry-free adder. Appl. Opt. 28, 785–790, 1989.
9. A. A. S. Awwal, M. A. Karim, and A. K. Cherri. Polarization encoded optical shadow-casting scheme: design of multioutput trinary combinational logic units. Appl. Opt. 26, 4814–4818, 1987.
10. D. A. Lu, R. M. Kent, J. M. Terhove M. A. Karim, and A. A. S. Awwal. Polarization encoded optical shadow-casting: Design of trinary multipliers. Appl. Opt. 29, 5242–5252, 1990.
11. A. A. S. Awwal and M. A. Karim. Polarization encoded optical shadow-casting programmable logic array: simultaneous generation of multiple outputs. Appl. Opt. 27, 932–936, 1988.
12. A. A. S. Awwal and M. A. Karim. Polarization encoded optical shadow-casting: design of a J-K flip-flop. Appl. Opt. 27, 3719–3722, 1988.
13. A. A. S. Awwal and M. A. Karim. Polarization encoded optical shadow-casting:

grey-level image encoding for serial and parallel operations. Appl. Opt. 28, 284–290, 1989.

14. A. A. S. Awwal and M. A. Karim. Multiprocessor design using polarization encoded optical shadow-casting. Appl. Opt. 29, 2107–2112, 1990.

15. R. A. Rizvi, K. Zaheer, and M. S. Zubairy. Separate and simultaneous generation of multioutputs in a polarization encoded optical shadow-casting scheme: design of half- and full adders and subtractors. Appl. Opt. 27, 5176–5180, 1988.

16. R. A. Rizvi, K. Zaheer, and M. S. Zubairy. Implementation of trinary logic in a polarization encoded optical shadow-casting scheme. Appl. Opt. 30, 936–942,1991.

17. M. S. Alam and M. A. Karim. Efficient combinational logic circuit design using polarization encoded optical shadow-casting. Opt. Comm. 93, 252–257, 1992.

18. M. S. Alam, M. A. Karim, and A. A. S. Awwal. Two-dimensional register design using polarization-encoded optical shadow-casting. Opt. Comm. 106, 11–18, 1994.

19. M. S. Alam and M. A. Karim. Multiple-valued logic based multiprocessor using polarization-encoded optical shadow-casting. Opt. Comm. 96, 164–170, 1993.

20. A. K. Cherri, M. A. Karim, and A. A. S. Awwal. Arithmetic logic units using optical shadow-casting systems. Proc. of the 30th Midwest Symp. on Circuits and Systems. New York, 1987, p. 816.

21. M. S. Alam, K. M. Iftekharuddin, and M. A. Karim. Polarization-encoded optical shadow-casting: edge detection using Roberts operator. Microwave and Optical Technology Letters 6, 190–193, 1993.

22. A. K. Cherri, A. A. S. Awwal, and M. A. Karim. Morphological transformation based on optical symbolic substitution and polarization-encoded optical shadow-casting systems. SPIE Proceedings 1215, 498–505, 1990.

23. A. K. Cherri and D. Zhao. Morphological hit-or-miss transform using polarization encoded optical shadow casting. In Machine Vision Applications, Architectures, and System Integration, SPIE Conference Proc. 1823, 14–24, Boston, November 15–20, 1992.

24. E. L. Johnson and M. A. Karim. Digital Design: A Pragmatic Approach. PWS-KENT, Boston, 1987.

25. M. A. Karim. Electro-Optical Devices and Systems. PWS-Kent, Boston, 1990.

26. C. Yeh. Handbook of Fiber-Optics Theory and Applications. Academic Press, San Diego, 1990.

27. J. L. Tocnaye and J. R. Brocklehurst. Parallel access read/write memory using an optically addressed ferroelectric spatial light modulator. Appl. Opt. 30, 179–180, 1991.

28. S. L. Hurst. Multiple-valued threshold logic: its status and its realization. Opt. Eng. 25, 44–55, 1986.

29. D. A. Lu, R. M. Kent, J. M. Terhove, M. A. Karim, and A. A. S. Awwal. Polarization encoded optical shadow-casting: design of trinary multipliers. Appl. Opt. 29, 5242–5252, 1990.

30. J. R. Jump and S. R. Ahuja. Effective pipelining of digital systems. IEEE Trans. Comput. C-27, 855–862, 1978.

31. A. Avizienis. Signed-digit number representations for fast parallel arithmetic. IRE Trans. Electronic Computers EC-10, 389–400, 1961.

32. B. L. Drake, R. P. Bocker, M. E. Lasher, R. H. Patterson, and W. J. Miceli. Photonic computing using the modified signed-digit number representation. Opt. Eng. 25, 38–43, 1986.

33. R. P. Bocker, B. L. Drake, M. E. Lasher, and T. B. Henderson. Modified signed-digit addition and subtraction using optical symbolic substitution. Appl. Opt. 25, 2456–2457, 1986.

34. Y. Li and G. Eichmann. Conditional symbolic modified-signed-digit arithmetic using optical content-addressable memory logic elements. Appl. Opt. 26, 2328–2333, 1987.

35. A. K. Cherri and M. A. Karim. Modified signed digit arithmetic using an efficient symbolic substitution. Appl. Opt. 27, 3824–3827, 1988.

36. A. K. Cherri and M. A. Karim. Simplified arithmetic for optical symbolic substitution. Microwave and Optical Technology Letters 1, 310–312, 1988.

37. A. K. Cherri and M. A. Karim. Symbolic substitution based flagged arithmetic using polarization-encoded optical shadow-casting system. Optics Communications 70, 455–461, 1989.

38. Y. Li, D. H. Kim, A. Kostrzewski, and G. Eichmann. Content-addressable-memory-based single stage optical modified-signed-digit arithmetic. Opt. Let. 14, 1254–1256, 1989.

39. S. Barua. Single-stage optical adder/subtracter. Opt. Eng. 30, 265–270, 1991.

40. M. S. Alam, M. A. Karim, A. A. S. Awwal, and J. J. Westerkamp. Optical processing based on conditional higher-order trinary modified signed-digit symbolic substitution. Appl. Opt. 31, 5614–5621, 1992.

41. T. K. Gaylord and M. M. Mirsalehi. Truth-table look-up processing: number representation, multilevel coding, and logical minimization. Opt. Eng. 25, 22–33, 1986.

42. B. Parhami. Carry-free addition of recoded binary signed-digit numbers. IEEE Trans. Comput. 7, 1470–1476, 1988.

43. A. A. S. Awwal. Recoded signed-digit binary addition-subtraction using opto-electronic symbolic substitution. Appl. Opt. 31, 3205–3208, 1992.

44. A. K. Cherri. Symmetrically recoded modified signed-digit optical addition and subtraction. Appl. Opt. 33, 4378–4382, 1994.

45. A. A. S. Awwal and M. A. Karim. Design of polarization-encoded optical shadow-casting logic units using truth-table partitioning. Can. J. Phys. 66, 841–843, 1988.

46. J. Tanida and Y. Ichioka. Programming of optical array logic I: image data processing. Appl. Opt. 27, 2926, 1988.

47. A. A. S. Awwal and M. A. Karim. Edge detection using polarization-encoded optical shadow-casting. Appl. Opt. 28, 3179, 1989.

48. R. C. Gonzalez and P. Wintz. Digital Image Processing, 2d ed. Addison-Wesley, 1987.

49. A. K. Cherri and M. A. Karim. Image enhancement using optical symbolic substitution. Opt. Eng. 30, 259–264, 1991.

50. K. H. Brenner, A. Huang, and N. Streibel. Digital optical computing with symbolic substitution. Appl. Opt. 25, 3054–3060, 1986.

51. K. H. Brenner. New implementation of symbolic substitution. Appl. Opt. 25, 3061, 1986.

52. A. K. Cherri and M. A. Karim. Optical symbolic substitution: edge detection using Prewitt, Sobel, and Roberts operators. Appl. Opt. 28, 4644–4648, 1989.
53. A. K. Cherri and M. A. Karim. Edge detection using optical symbolic substitution. Opt. Comm. 74, 10–14, 1989.
54. A. K. Cherri and M. A. Karim. Symbolic substitution based operations using holograms: multiplication and histogram equalization. Opt. Eng. 28, 638–642, 1989.
55. A. K. Cherri and M. A. Karim. Morphological transformation using optical symbolic substitution. Microwave and Optical Technology Letters 2, 282–285, 1989.
56. G. Eichmann, J. Zhu, and Y. Li. Optical parallel image skeletonization using content addressable memory-based symbolic substitution. Appl. Opt. 27, 2905, 1988.
57. A. A. S. Awwal, A. K. Cherri, and M. A. Karim. Architectural relationships of symbolic substitution and optical shadow-casting. SPIE Proceeding 1215, 506–516, 1990.
58. A. Louri. Efficient optical implementation method for symbolic substitution based on shadow-casting. Appl. Opt. 28, 3264, 1989.
59. J. Serra. Image Analysis and Mathematical Morphology. Academic Press, London, 1982.
60. D. G. Daut and D. Zhao. Mathematical morphology and its applications in machine vision. In Visual Communications and Image Processing 4, 181–191. Philadelphia, 1989.
61. T. R. Crimmins and W. R. Brown. Image algebra and automatic shape recognition. IEEE Trans. Aerospace Electron. Systems AES-21, 60–69, 1985.
62. C. R. Giardina and E. R. Dougherty. Morphological Methods in Image and Signal Processing. Prentice-Hall, Englewood Cliffs, NJ, 1987.
63. D. Casasent and R. Sturgill. Optical hit-or-miss morphological transforms for ATR. Proc. of SPIE on the Application of Digital Image Processing XII, 1153, 500, 1989.
64. E. Rotha, J. Richards, and D. Casasent. Optical laboratory morphological inspection processor. Appl. Opt. 28, 5342, 1989.
65. D. Zhao and D. Daut. Morphological hit-or-miss transformation for shape recognition. Journal of Visual Communication and Image Representation 2, 230–243, 1991.

Index